Radiophysical and Geomagnetic Effects of Rocket Burn and Launch in the Near-the-Earth Environment

Radiophysical and Geomagnetic Effects of Rocket Burn and Launch in the Near-the-Earth Environment

Leonid F. Chernogor
Nathan Blaunstein

CRC Press
Taylor & Francis Group
Boca Raton London New York

CRC Press is an imprint of the
Taylor & Francis Group, an **informa** business

CRC Press
Taylor & Francis Group
6000 Broken Sound Parkway NW, Suite 300
Boca Raton, FL 33487-2742

First issued in paperback 2017

ISBN-13: 978-1-4665-5113-8 (hbk)
ISBN-13: 978-1-138-03384-9 (pbk)

Library of Congress Cataloging-in-Publication Data

Chernogor, L. F. (Leonid Feoktystovych)
    Radiophysical and geomagnetic effects of rocket burn and launch in the near-the-earth environment / by Leonid F. Chernogor and Nathan Blaunstein.
        pages cm
    Includes bibliographical references and index.
    ISBN 978-1-4665-5113-8
    1. Rockets (Aeronautics)--Launching--Environmental aspects. 2. Artificial satellites--Launching--Environmental aspects. 3. Radio waves--Scattering. 4. Geomagnetism. I. Blaunstein, Nathan. II. Title.

TL784.L3C44 2014
538'.766--dc23   2013017527

Visit the Taylor & Francis Web site at
http://www.taylorandfrancis.com

and the CRC Press Web site at
http://www.crcpress.com

Contents

PART II DIAGNOSTICS OF PLASMA WAVE DISTURBANCES BY INCOHERENT SCATTER AND DOPPLER RADARS

PART IV   ROCKET BURN AND LAUNCH AND RADIO COMMUNICATION

PART V ECOLOGICAL PROBLEMS IN NEAR-THE-EARTH SPACE ACTIVITY

Preface

This book provides description of experimental and theoretical studies of the effects that rocket burn and launch have on the near-the-Earth environment, and on the effects of the geomagnetic field to differentiate and investigate them separately. The authors of the book illuminate the main geophysical and radiophysical effects that occur in the upper ionosphere and the magnetosphere surrounding the Earth, which accompany or are associated with rocket or cosmic apparatus burn and launch in space, from 1,000 to 10,000 km. What is very important to emphasize is that this book concentrates on the ecological consequences of such space exploration also.

The disturbances of plasma and the ambient magnetic and electric fields in the near-the-Earth environment, caused by rocket burns and launches from Plesetsk (Russia), Baikonur (Kazakhstan), Cape Canaveral (USA), as well as from China, France, and other cosmodromes worldwide, were analyzed based on numerous radiophysical methods and techniques. Thus, the measurements were carried out with high-frequency (HF) Doppler radar (DR), incoherent and coherent scatter radar systems, microwave radar, and magnetometer, as well as by using optical instrumentation and spectroscopy. These effects have been selectively analyzed from rocket burns and launches for the time period from 1975 to 2010 by the authors from rocket burn campaigns worldwide.

This book is a lasting reference for scientists in geophysics and radiophysics, specialists in rocket launch, and ecologists. It also serves as the fundamental handbook for graduate and postgraduate students taking courses in physics and cosmic sciences at the university level.

This monograph consists of five parts. Part I provides the diagnostics of plasma disturbances that are caused by rocket burn and launch in the middle and northern ionosphere by using radiophysical methods. Thereafter, the plasma perturbations are analyzed using distant radio sounding methods with DR operating at the decameter wavelength band. The statistical analysis of plasma perturbations transfer is also presented here.

Part II describes the results of radiophysical observations of plasma disturbances in the middle and outer ionosphere caused by rocket burn and launch using incoherent scattering (IS) and Doppler sounding (DS) methods.

Part III shows the geomagnetic effects of rocket burn and launch, calling the reader's attention to rocket burns and launches in Russia, the United States, China, France, and other cosmodromes worldwide.

Part IV describes the influence on the modified ionosphere caused by rocket burn and launch and by similar artificial and natural phenomena occurring in the ionosphere on ionospheric radio communication for various types of radio waves—from extremely low-frequency (ELF) bandwidth to very-high-frequency (VHF) bandwidth.

Part V describes the ecological problems of man-made cosmic activity and the effects of space exploration observed on the Earth's surface, in the near-the-Earth atmosphere, and in the near-the-Earth space. A perspective on cosmic activity and ways to minimize ecological disasters caused by current and future cosmic activity are briefly discussed here.

Nathan Blaunstein
Ben-Gurion University of the Negev

Acknowledgments

I write the acknowledgments for this book with kind permission of Leonid Chernogor, the first coauthor of this book.

This book is based on numerous observations of burns and launches of rockets and spacecrafts worldwide carried out by Professor Leonid Chernogor using a wide spectrum of ground-based facilities from Kharkov Radiophysical Observatory (KRO) and Radiophysical Faculty of Kharkov State University, Kharkov, Ukraine. On behalf of my coauthor, I thank all his colleagues since without their unwavering support and regular discussions he would have not obtained a high-level PhD, DSc, and Professor rank in physics of the ionosphere and the near-the-Earth cosmic space and geophysics. Based on numerous observations of the launch of space vehicles from worldwide cosmodromes (USSR, USA, India, France, Japan, China, etc.), a huge amount of statistical material was selected, analyzed, and then published in several corresponding books in Russian during the end of nineties to the beginning of the twenty-first century.

This book is based on the whole spectra of observed burns and launches worldwide, as well as on those that were observed in the recent decade.

Nathan Blaunstein, the second coauthor, for more than 30 years studied the ionosphere and conditions of radio wave propagation through the regular and perturbed ionosphere by different natural and man-made sources, using vertical and oblique radars and ionosondes, mostly for performance of long-path ionospheric and land-satellite channels. All these investigations were carried out at the Radiophysical and Geophysical Observatory of Moldavian State University, Moldova, initiated by the Radiophysical Research Institute (NIRFI), Gorky, former USSR, with help of State University, Yoshkar-Ola, former USSR, and by IZMIRAN, Moscow, Russia, where he wrote his PhD and DSc dissertations. He wishes to thank those with whom he published three books in the former USSR and Romania (1986, 1991, and 2006, respectively), and also his 2008 coauthor, Professor Eugeny Plohotniuc.

The idea to compare the same or similar features and processes occurring in the near-the-Earth coupling of the ionosphere–magnetosphere system observed in artificial active cosmic experiments (injection of ion clouds and ion beams in the ionosphere, heating of the ionosphere by powerful ground-based facilities, etc.), during observations of natural phenomena (meteor trails in the middle ionosphere, bubbles at the equator, magnetic storm effects at the northern ionosphere, etc.), and burns and launches of rockets and spacecraft initiated both the coauthors to write this book. Here, all geophysical (Chapters 1 through 10), radiophysical (Chapters 11 through 13), and ecological (Chapter 14) phenomena accompanied or associated with rocket burns and launches were illuminated in the consistent and physically vivid form.

In our opinion, such a book will broaden the existing theoretical frameworks, increase physical meaning (by introducing more precise explanation of the phenomena observed experimentally), and cover the books published earlier, which described only geophysical aspects of such rocket launches without entering deeper into radiophysical problems and, first of all, into the influence of such events on radio communication for various radio waves passing or propagating in the ionosphere.

Of course, we extend our thanks to the reviewers and technical editors of this book who made the more than usual number of grammatical and other changes in the text and did all the best to present the final text with clarity and precision.

Finally, this book would never have been finished and seen the light without the warm support of our families who surrounded us with a very kind atmosphere and plenty of understanding, which allowed us the time to write this book instead of enjoying with them the time we needed for this.

Nathan Blaunstein

Abbreviations

ACF	autocorrelation function
AE	active experiment
AES	artificial Earth satellite
AFT	adaptive Fourier transform
AGW	acoustic gravity wave
AW	Alfven wave
AWT	adaptive wavelet transform
CAGW	captured AGW (in the waveguide)
CIS	Commonwealth of Independent States
DFS	Doppler frequency shift
DMSP	Dense Meteorological Satellite Program
DMT	decreed Moscow time
DR	Doppler radar
DS	Doppler sounding
DSs	Doppler spectra
DSR	Doppler sounding radar
EAIM	Earth–atmosphere–ionosphere–magnetosphere (system)
ELF	extremely low frequency (extra-low frequency)
EO	extraordinary (wave)
eo.f	extraordinary fast (wave)
eo.s	extraordinary slow (wave)
EW	electronic whistler
FBI	Farley–Buneman instability
FFT	fast Fourier transform
GDI	gradient-drift instability
Glon	geographic longitude
GW	gravity wave
HF	high frequency
IEEE	Institute of Electrical and Electronics Engineers

IGW	inner gravity wave
IS	incoherent scattering
ISR	incoherent scatter radar
IW	ionic whistler
LET	local east time
LF	low frequency
LHC	left-hand circular
LHF	low-hybrid frequency
LHR	low-hybrid resonance
MF	median frequency
MGD	magnetic-gradient wave
MHD	magnetohydrodynamic (waves)
MLat	magnetic latitude
MLT	magnetic local time
MUF	maximum useful (usage) frequency
OME	orbit maneuver engine
OMS	orbit maneuver system
OR	ordinary (wave)
or.f	ordinary fast (wave)
or.s	ordinary slow (wave)
OS	orbit satellite
PLF	polarization loss factor
RHC	right-hand circular (polarization meaning)
RL	rocket launch
RPO	radiophysical observatory
RTI	Rayleigh–Taylor instability
SAW	shock-acoustic wave
SDF	spectral density function
SMHD	slow magnetohydromagnetic (wave)
SNR	signal-to-noise ratio
SS	Space Shuttle
ST	solar terminator
SV	space vehicle
TEC	total electronic content
UHF	ultra-high frequency
UHR	upper hybrid resonance
ULF	ultra-low frequency
UT	universal time
VGW	vertical gravity wave
VHF	very-high frequency
VLF	very-low frequency

VS	vertical sounding
VTEC	vertical total electronic content
WD	wave disturbance
WFT	windowed Fourier transform
WT	wavelet transform

Introduction

There are plenty of excellent books and articles in which the effects caused by rocket burns and launches have been described. Thus, the result of investigation of such phenomena has been summarized in numerous special textbooks and articles (see, e.g., Refs. [1–11] and the references therein). In the existing works, the effects observed along the active zone of the rocket trajectory have been discussed briefly. Unfortunately, very little work exists that is related to studies of plasma and geomagnetic field perturbations at ranges of 1,000–10,000 from the rocket trajectory. The authors of this book call the corresponding perturbations "large-scale" or "global" disturbances. As shown in the book, the lifetime of such perturbations is 80–90 min, covering ranges from 5,000 to 20,000 km of the rocket trajectory; that is, such disturbances are global both in time and space domains.

Up till now, there are no consistent physical–mathematical models that fully describe phenomena accompanied or associated with rocket burns and launches. Further, it is impossible to differentiate disturbances caused by the effects of rocket burns and launches from those observed during geomagnetic natural events, such as geomagnetic storms, meteor trails, ionospheric bubbles, and so on. Finally, it is not clear what mechanisms of global perturbation generation and transfer to significant distances from the rocket are taken to give a satisfactory explanation of the observed phenomena and its specific features.

Therefore, the main requirement to the research and the corresponding studies was to increase the volume of observations and search and differentiate the repeated phenomena precisely—the active man-made and the natural geophysical. This is the subject of the first group of problems put up by existing researches.

The second group of problems introduces the reader to a systematic investigation of the physical processes inside the global system: Earth–atmosphere–ionosphere–magnetosphere–planet–environment–Sun. The main goal of such investigations is the study of the reaction of Earth–atmosphere–ionosphere–magnetosphere (EAIM) on sources with huge energy. This yields the process of perturbation transfers to significant distances (~1,000–10,000 km), definition of the types of waves excited during rocket burns and launches, and so on. Moreover, different distances from cosmodromes, different conditions of launches, various types of rockets, and their engine power lead to variation of the "initial conditions" and to the adaptations of each physical model in an intensive study of the reactions and "responses" of the atmosphere and geocosmos environment (mostly the ionospheric plasma) on rocket burn and launch from the Earth's surface.

The third group of problems relates to the following aspects. At the boundary of the twentieth and twenty-first centuries, the increasing power of the rocket engine and the frequency of their launches reached such a situation, in which all observed perturbations caused in the atmosphere and the near-the-Earth cosmic environment could not be ignored. The study of these perturbations is an important issue for specialists in fields of geophysics, radiophysics, ecology of the atmosphere, and geocosmic space, as well as specialists investigating space weather.

The above-mentioned aspects define the actuality of experimental and theoretical investigations carried out by several separate groups all over the world, as well as by two groups of researchers led separately and jointly by the authors of this book, starting from 1970 till now.

References

1. Pinson, G. T., Apollo/Saturn 5 post flight trajectory—SA-513—Skylab 1 mission, *Tech. Rep. D5-15560-13*, Huntsville, AL: Boeing Co., 1973.
2. Mendillo, M., Hawkins, G. S., and Klobuchar, J. A., An ionospheric total electron content disturbances associated with the launch of NASA's Skylab, *Tech. Rep. 0342*, Bedford, MA: Air Force Cambridge Res. Lab., 1974.
3. Mendillo, M., The effect of rocket launches on the ionosphere, *Adv. Space Res.*, 1, 275–290, 1981.
4. Garret, H. B. and Pike, C. P., Eds., *Space Systems and Their Interaction with the Earth's Space Environment*, New York: Program of Astronautics and Aeronautics, AIAA, 1981.
5. Mendillo, M., Ionospheric holes: A review of theory and recent experiments, *Adv. Space Res.*, 8, 51–62, 1988.
6. Nagorski', P. M. and Tarashuk, Yu. E., Artificial modification of the ionosphere during rocket burns putted out at the orbits the cosmic apparatus, *Izv. Vuzov. Radiofizika*, 36, 98–106, 1993 (in Russian).
7. Bernhardt, P. A., Huba, J. D., Swartz, W. E., and Kelley, M. C., Incoherent scatter from space shuttle and rocket engine plumes in the ionosphere, *J. Geophys. Res.*, 103, 2239–2251, 1998.
8. Bernhardt, P. A., Huba, J. D., Kudeki, E., et al., The lifetime of a depression in the plasma density over Jicamarca produced by space shuttle exhaust in the ionosphere, *Radio Sci.*, 36, 1209–1220, 2001.

9. Hester, B. D., Chiu, Y.-H., Winick, J. R., et al., Analysis of space shuttle primary reaction-control engine-exhaust transient, *J. Spacecr. Rockets*, 46, 679–688, 2009.
10. Bernhardt, P. A., Ballenthin, J. O., Baumgardner, J. L., et al., Ground and space-based measurement of rocket engine burns in the ionosphere, *IEEE Trans. Plasma Sci.*, 40, 1267–1286, 2012.
11. Adushkin, V. V., Koslov, S. I., and Petrov, A. V., Eds. *Ecological Problems and Risks Caused by Rocket-Cosmic Technique on the Natural Environment*, Moscow: Ankil, 2000 (in Russian).

DIAGNOSTICS OF PLASMA PERTURBATIONS BY USING DOPPLER RADIO SOUNDING

I

Chapter 1

Perturbations in the Ionosphere Caused by Rocket Launches

1.1 Overview

Around the end of the twentieth century and up till the beginning of the twenty-first century, the rate of rocket launches (RLs) and the power of the rocket plume attained such a level that a question arose: Would the exhaust emissions from large rocket engines cause atmospheric perturbations such as those observed experimentally and, if so, should these perturbations be taken into account by the scientific community? Research and analysis of such possible perturbations is of great interest to experts in geophysics, radiophysics, and near-the-Earth space ecology and space weather.

Starting from the late fifties, numerous works investigated the effects of atmospheric and ionospheric perturbations caused by the burn and launch of rockets (see, e.g., Refs. [1–9]). Thus, in 1964, a comprehensive review of this subject was published in Ref. [10], which described the many possible ways rocket "pollutants" could have an environmental impact. The main assumption from this work is that the terrestrial atmosphere is sufficiently dense to absorb any conceivable shock that emanates from the resulting rocket plume. In the three decades of this publication, because of a large number of spacecraft and RLs in the United States, the former USSR, and other countries, which was called the "Shuttle and Soyuz era," increasing evidence was found to suggest that such launches affect the near-the-Earth environment.

Similarly, through the so-called active-space experiments for artificial modification of the ionosphere and the atmosphere, scientists were able to discover a wide range of physical and chemical processes for future aeronomic interest and for finding the physically vivid rocket exhaust effects on the atmosphere and the ionosphere (see bibliography presented in Ref. [11]). The effects observed experimentally in the active zone of the rocket's trajectory were discussed and explained as given in Refs. [1–9].

In this chapter we will discuss the processes that range from 1,000 to 10,000 km from the initial place of the RLs and their trajectories. The corresponding atmospheric and ionospheric perturbations were traditionally called large scale or global [12–14]. Further, as will be shown later, the time period of these perturbations is approximately 80–90 min and their speed about 1–3 km s^{-1}, which corresponds to their azimuthal elongations of 5,000–20,000 km. Therefore, these perturbations can be termed as *global*, as is given in Refs. [12–14].

It should be noted that until now there are no detailed physical–mathematical models of the processes that correspond to rocket exhaust effects. Moreover, it is often impossible to differentiate perturbations caused by the effects of RLs and the so-called natural effects caused by solar activity, magnetic storm, meteor trails, and so on. Until now the mechanisms of generation and propagation of such perturbations from the rocket trajectory at the global level cannot be understood and explained. This is because these disturbances are generated artificially by rocket plumes and do not differ much from natural disturbances. This can be done only by increasing the data and frequency of observation of the RLs, finding repeated features, and differentiating the difference between the two kinds of perturbations.

The peculiarity of this chapter is that in this, based on a unified framework and approach, we shall try to find the effects occurring in the middle of the ionosphere (at altitudes of 100–300 km) during the burn, launch, and flight of spacecraft and rockets (in our book we call these as *space vehicles* or *SVs*). It is important to note that all observations described in this chapter were carried out at different distances from the place of rocket burn and for different types of SVs, at various diurnal time periods, and in different geophysical conditions, following the results obtained in Ref. [15].

1.2 Properties and Parameters of the Lower, Middle, and Outer Ionospheres

The ionosphere has been defined by Plendl [16], and then by Kaiser [17], as that part of the upper atmosphere of the Earth which is reached from altitudes of 50 km to those on an order of 400–500 km, and can be divided formally as the lower (<50–100 km) and middle (~100–400 km) ionosphere.

This atmospheric altitudinal region is filled with partially ionized gas called *plasma*. The outer region of the ionosphere (~400–1000 km) is determined as the altitudinal region at which the concentration of charged particles of plasma, the electrons N_e and ions N_i (plasma is quasineutral, $N_e \approx N_i$), exceeds that of neutral molecules and atoms N_m. Thus, the concentration of the neutral particles, molecules, and atoms sharply decreases with altitude: from $\sim 10^{21}\,\text{m}^{-3}$ (at 60 km) and $\sim 10^{19}\,\text{m}^{-3}$ (at 100 km) to $\sim 10^{14}\,\text{m}^{-3}$ (at 400 km) and $\sim 10^{9}\,\text{m}^{-3}$ (at 1000 km) [18], respectively. At the same time, the concentration of the electrons or ions (because plasma is quasineutral) changes very slowly: from $\sim 10^{7} - 10^{8}\,\text{m}^{-3}$ (at 60 km) and $\sim 10^{11}\,\text{m}^{-3}$ (at 100 km) to $\sim 10^{12}\,\text{m}^{-3}$ (at 400 km) and $\sim 5 \times 10^{10}\,\text{m}^{-3}$ (at 1000 km) [18], respectively. Consequently, a degree of plasma ionization in the ionosphere sharply increases with altitude whereas in the lower ionosphere it is weakly ionized ($N_e / N_m \sim 10^{-14} - 10^{-10}$), and in the middle ionosphere it is partly ionized ($N_e / N_m \sim 10^{-8} - 10^{-3}$) whereas in the outer ionosphere it is fully ionized ($N_e / N_m \sim 10^{1} - 10^{2}$) [18].

The average temperature of plasma in the ionosphere increases with altitude: ~210–240 K in the lower ionospheric altitude, ~1500–2400 K in the middle ionospheric altitude, and ~3000–3500 K in the outer ionospheric altitude.

The ionosphere rises as a result of the influence of solar-ionized emission and high-energy particles of different gases in the Earth's atmosphere [16,19]. The structure and properties of the ionosphere depend essentially on processes occurring in the Sun (which are called *solar activity* [20]), variations in the Earth's magnetic field (called the *geomagnetic field effect* [21]), movements of neutral "wind" in the upper atmosphere due to the Earth's rotation, effects of electrical current and ambient electrical fields [22–25], density, and the content of the atmosphere at different altitudes and geographical latitudes [26,27], and so on.

Usually the ionosphere is separated into five independent regions sometimes called *layers* [11,18,26–28]. The bottom one, from 50 to 85–90 km, is called the *D*-layer, from 90 to 130 km is the *E*-layer, and above 150 km is the *F*-layer. The latter region is usually separated at the F_1-layer, which is reached from 130 to 200–250 km, and the F_2-layer is above 250 km. Apart from these layers, at altitudes of 90–120 km, the sporadic layer E_s is observed as having a small thickness in the vertical plane along the height. Its occurrence is usually explained by the influence of a neutral wind of atmospheric gases on the charged particles of plasma that result in exchange of plasma accompanied by a stratified wind structure along the height (for more detailed information, see Refs. [19,28]).

The physical processes in the ionosphere have several characteristics as functions of the main parameters—plasma density (concentration), temperature, and degree of ionization of each component of plasma, neutrals, electrons, and ions. These characteristics are lengths of the free path between interactions, average thermal velocity, and frequencies of collision for each component of plasma. As shown in Refs. [18,29–33], the length of the free path of neutral particles, λ_m, in the middle ionosphere is sufficiently long and ranges from ~80 m (at 200 km), increasing sharply from 300 km ($\lambda_m \approx 10^3$ m), to 1000 km ($\lambda_m \approx 8 \times 10^6$ m). The free path for electrons and ions is also sufficiently long. Thus, at 200 km $\lambda_e \approx 90$ m increased to 200 m and at 400 km $\lambda_m \approx 8 \times 10^3$ m increased to 1000 km, that is, more slowly with respect to that for neutral molecules and atoms. This is because at altitudes exceeding 200 km, the free path of electrons and ions is determined by the interaction of charged particles that describe the electrostatic far-range integral that exceeds the integral of interactions between neutral particles determined by a close-range gravity field.

The thermal velocity of neutral particles and ions is at the same level and not so high in the lower and middle ionosphere—from 400 (at 200 km) to 900 m s^{-1} (at 400 km), increasing in the outer ionosphere ($h = 1000$ km) to 2 km s^{-1}. As for the thermal velocity of electrons, it is sufficiently high, attaining 100 (in the middle ionosphere) to 500 km s^{-1} (in the outer ionosphere).

There is another group of plasma characteristics that is related only to charged particles, electrons, and ions. These are plasma frequency, gyrofrequencies of electrons and ions, and rate (or radius) of rotation of charged particles around ambient magnetic field lines. Larmor's radius of rotation of the electron or ion around a geomagnetic field characterizes the influence of the ambient magnetic field on various transport processes in plasma. The average Larmor's radius of electrons in all the areas of the ionosphere is sufficiently small—on the order of a few centimeters, whereas this characteristic for plasma ions varies from 2 m (in the lower ionosphere) to 20 m (in the outer ionosphere).

1.3 Peculiarities of the Interaction of Moving SVs with the Ionosphere

Why are the characteristics mentioned earlier very important in the problem of SV launch and flight interaction with the ionospheric plasma? This is because the peculiarities of this interaction in the ionosphere depend on the relations among the length, dimensions, and velocity of the moving vehicle and the kinetic and electromagnetic parameters and characteristics of the ionospheric plasma mentioned earlier. As will be shown later, the length and dimensions of the rockets and spacecraft do not exceed several tens of meters to several hundred meters, but a free path of plasma particles is much larger (see the discussions earlier). Therefore, plasma cannot be considered as a continuous medium in the proximity of the moving vehicle—it can be considered as plasma that is only an ionized gas, consisting of separate particles such as neutrals, electrons, and ions. Therefore, for the description of transport of plasma processes in the proximity of the moving vehicle, the elements of kinetic theory should be used instead of those of hydrodynamic and aerodynamic theories.

Moreover, collisions between plasma particles do not affect plasma disturbances around the moving vehicle, because its dimensions are smaller than the free path lengths of the plasma components. The effects become important at long ranges from the moving rocket, that is, at ranges on the order of free path lengths of the plasma particles. Therefore, as mentioned earlier, we will analyze plasma perturbations that range from the proximity of the rocket to distances of hundreds and thousands of kilometers.

As is well known, neutral particles reflected from the vehicle's body in the lower and middle ionospheres can ionize and heat gas in front of the moving vehicle because of collisions. As shown in Ref. [29], the energy of the reflected neutral molecules (atoms) is only on the order of 10–20 eV ($1\,eV = 1.6 \times 10^4\,K$). Their velocity is much lesser than the velocity of the electrons within the atoms. In such conditions the probability of ionization is sufficiently small with a cross-section of ionization that does not exceed $10^{-44}\,m^2$ [29]. Moreover, if the constraint $\lambda_m v_m \gg V_0 L_0$ occurs, heating of the gas is not so important. Thus, the total energy of the particles, E_{refl}, reflected during time Δt from the moving body approximately can be written as follows [29]:

$$E_{refl} \sim M V_0^2 N_m L_0^2 V_0 \Delta t, \qquad (1.1)$$

where:
$v_m = \left(2 k_B T / M\right)^{1/2}$ is the thermal velocity of gas molecules
V_0 and L_0 are the velocity and the scale of the moving vehicle, respectively, (in meters)
M and T are the mass and temperature (in Kelvin, K) for neutral particles, respectively
$k_B = 1.38 \times 10^{-23}\,J/K$ is the Boltzmann constant

Those molecules (atoms) fly without collision to distances on the order of free path length λ_m. If we suppose that $V_0 \Delta t \gg \lambda_m$, these neutral particles will be concentrated within the volume of $\sim V_0 \Delta t \lambda_m^2$ with the total number of particles inside it on the order of $\sim N_m \lambda_m^2 (V_0 \Delta t)$. Therefore, the average energy of the neutral particles of the nonperturbed gas can be increased due to interactions with the particles reflected from vehicle's surface to the value $\Delta E \sim (MV_0^2) L_0^2 / \lambda_m^2$, and, finally, the constraint $\Delta E \ll k_B T$ leads to $\lambda_m v_m \gg V_0 L_0$. As shown in Ref. [29], for vehicles with dimensions exceeding 1 m, the corresponding constraint is correct in the middle ionosphere, that is, at altitudes above 200 km.

Next, comparing the thermal velocity of plasma components, neutrals, and ions, in the middle ionosphere and the velocity of the moving vehicle, V_0, which is on the order of 8×10^3 m/s, one can find that $V_0 / v_{m,i}$ is on the order of 5–10, whereas for electrons $V_0 / v_e \approx 0.03 - 0.06$. Thus, we can state that because the velocity of the vehicle exceeds the velocity of the sound waves, that is, $V_0 \approx (5-10) v_{m,i}$, it can generate shock sound waves in the ionospheric plasma (this subject will be discussed later). If the wavelength of such waves exceeds the free path length of plasma particles, they attenuate in the ionosphere weakly, and their amplitude is also weak because the vehicle's dimensions are smaller than the free path of plasma particles. In the case of ion plasma waves, excited at wavelengths lesser than the free path of ions, their attenuation in ionospheric plasma is sufficiently strong [28]. As for electron plasma waves, *a priori* they cannot be excited in ionospheric plasma [30].

A few general assumptions can also be made regarding the stability of the disturbed area in the proximity of the moving vehicle. This question is related to the analysis of instabilities in plasma—their increment of excitation or decrement of decay. There are many instabilities occurring in the inhomogeneous ionospheric plasma, where the spatial distributions of plasma particles and their temperatures are inhomogeneous and lead to creation of a wide spectra of unstable plasma waves in ambient electric and geomagnetic fields (description of these types of instabilities occurring in the ionosphere, lower and middle, is fully presented in [18]). We do not broach this subject because it is out of the scope of this book, mentioned only because in the areas of instability behind the moving vehicle, a turbulent wake is usually observed that can be recorded as an electromagnetic noise at frequencies corresponding to the frequencies of unstable waves generated in plasma by the moving vehicle. This global turbulence leads to an anomalous increase of plasma diffusion across the geomagnetic field, and as a result the turbulent wake created by the moving vehicle in the perturbed ionospheric region spreads (due to diffusion, drift, and thermodiffusion) much faster, finally decreasing the length of the perturbed region behind the vehicle.

All these peculiarities will be taken into account further in our description of the interaction of the moving rockets and spacecraft in the lower, middle, and outer ionosphere.

Let us now introduce the reader to the various types of rockets and spacecraft, and give some brief information on the chief cosmodromes of their launch.

1.4 Cosmodromes and Types of Rockets

Observations were carried out on more than 10 types of rockets, the majority of them being presented in Appendix 1. Among these, the more powerful rockets have carriers of Energia type (former USSR) and Space Shuttle type (USA). The force of the engine of the first stage exceeds 3.5×10^7 N and 2.5×10^7 N, respectively, the power is $\sim 10^{10}$ W, engine energy $\sim 10^{12}$ J, and working time ~ 2 min. The force of the engine for further stages was several times less, but its working time exceeded ~ 8 min. Closer to the above-mentioned energy parameters of the rockets were Proton (former USSR) and Ariane (France). A rocket vehicle much smaller in power was Pegasus (USA) launched from an aircraft. The force of the engine of the first stage (step) of this rocket was $\sim 6 \times 10^5$ N; the power and energy of engines were $\sim 10^8$ W and $\sim 10^8$ J, respectively. Generally, we will differentiate rockets by accounting for their initial mass. Thus, we define rockets with a mass of 1000 t as ultra heavy, 300–1000 t as heavy, 100–300 t as medium, 30–100 t as light, and 10–30 t as ultra light.

Observations have been carried out for different distances R from cosmodromes (see Appendix 2). As can be seen from the data presented there, minimum R is about 700 km and maximum is about 10,000 km. The farthest cosmodromes are at Edwards and Vandenberg (USA), whereas the closest is at Kapustin Yar (Russia). The Plesetsk cosmodrome (Russia) is more "northern." For description of the variations of the reflected signal observed experimentally it is necessary to have information on the kinematical characteristics of SV launches and their orbits. In Table 1.1 some of the approximate characteristics are presented for the Space Shuttle, for which many observations were carried out.

Table 1.2 presents some of the approximate characteristics of Molnia and Soyuz rocket types (former USSR) for comparison. Their initial mass is ~ 305 t and trajectory sharper than that of the Space Shuttle.

We use the data given in Refs. [34,35] for construction of these two tables.

Table 1.1 Cinematic Characteristics of the Space Shuttle

Stages of Lift Off	Time (s)	Altitude (km)	Distance from Cosmodrome (km)	Velocity (km s⁻¹)
Command to setting fire	0	0.06	0	0
Beginning of lift off	0.2	0.06	0	0
Maximum dynamic pressure (2 × 10⁴ Pa)	54	8	~10	0.30
Entering the stratosphere	73	14	~20	0.45
Entering the ionosphere (day)	160	60	~40	1.3
Entering the ionosphere (night)	300	90	~100	2.5
Passing the *E*-region of the ionosphere. Switching off of the main engine	512	115	~750	3.8
Separation of outer fuel tank	530	119	~850	7.8
First switching on of the maneuver engine or ME	632	126	~1650	7.8
Switching off of the ME	721	132	~2350	7.8
Second switching on of the ME	2640	242	~17,400	7.8
Switching off of the ME	2717	243	~18,000	7.8

Table 1.2 Cinematic Characteristics of Soyuz

Stages of Lift Off	Time (s)	Altitude (km)	Distance from Cosmodrome (km)	Velocity (km s⁻¹)
Vertical lift off	8	~1	0	~0.1
Switching off of the engines of the packet, separation of the zero stages	119	48	44	1.8
Throwing off of the head of the rocket body's contour	151	78	96	2.1
Switching off of the engine of the first stage	286	171	451	3.9
Switching off of the engine of the second stage	522	200	1680	7.7
Separation of the SV	527	200	1715	7.8

1.5 Large-Scale Perturbations in RL and Flight—Doppler Spectra Analysis

Vertical sounding (VS) Doppler radar (DR) was used for investigating the non-stationary processes occurring in the middle ionosphere. Its main characteristics are presented in Appendix 3. Let us consider the variations of the Doppler spectra (DSs) taking place in the RL and flight to distances up to 2500 km. An example of the results of diurnal observations is shown in Figure 1.1.

The Soyuz launch happened at 13:22 h on October 18, 1999 (here, the time is given in universal time or UT). Before the launch, the ionosphere was normal and not disturbed. The DSs were of one-mode type with sharply indicated maxima. Approximately between 12:55 and 13:03 h, the main and additional modes of the reflected signal were observed and a quasiperiodical process with a period of $T \approx 10$ min started. After 13:45 h, the amplitude of oscillations increased essentially, achieving a Doppler shift $f_{Da} = 0.5\,\text{Hz}$, and oscillations with a period of $T \approx 20$ min were observed. This process was observed for 30 min. From 14:12 to 14:40 h, the ionosphere was practically unperturbed. In the short-term period of 14:40–14:50 h, additional mode to main mode was registered that was shifted

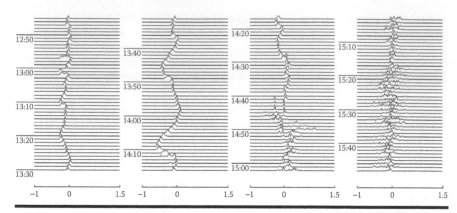

Figure 1.1   Variations in the DS with frequency 3.5 MHz at the start of Soyuz rocket on October 18, 1999. As in all other figures along the horizontal axis, DFS (Hz) is indicated here, and UT is given along the vertical axis.

at the Doppler frequency of $f_D \approx -0.3\,\text{Hz}$. After 14:50 h, the stable maximum in Doppler frequency domain disappeared. This fact states the existence (or amplification of intensity) of sporadic plasma inhomogeneities in the ionosphere lower than the region of regular reflection of radio waves of about 150 km. Such a type of disturbance was observed for more than an hour. Additionally, we should note that the sunset at the ground level of the observation terminal was at about 15:20 h, but at the region of reflection the sunset was observed an hour later.

An example of DSs variation during the heavy Proton RL is shown in Figure 1.2. The launch started on July 5, 1999 at 13:32 UT. This was interesting because at 277 s from the start of this rocket an explosion took place in the reservoir of the engine at the second stage at an altitude of 110 km, and then at 410 s of the rocket's flight an explosion of the falling rocket was seen at a height of 25 km. The sunset at the Earth level took place at about 19:00 h, and before launch the ionosphere was not disturbed. Short-term effects of the spread of the spectra were observed only between the periods of 12:54–12:58 h and 13:08–13:12 h. From 13:15 h and up to 14:50 h any perturbations of the ionosphere were absent. Approximately from 14:54 to 14:58 h additional mode shifting at −0.12 Hz was observed. A weak mode was also registered in the short-time periods of 15:03–15:05 h and 15:08–15:09 h. After that, the ionosphere returned to quite a steady-state condition during the long period of observation.

An example of the result of nocturnal observation of processes occurring in the ionosphere that accompany the launch of medium-heavy rockets is shown in Figure 1.3, which relates to Zenit-2 RL (Ukraine) at 20:29 UT, on September 9, 1998.

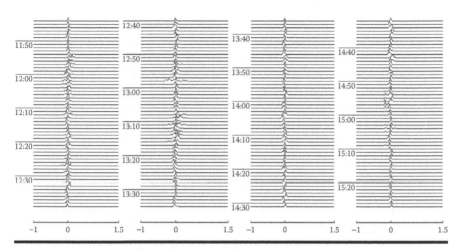

Figure 1.2   Variations of the DS with frequency 3.0 MHz at the start of Proton rocket on July 5, 1999.

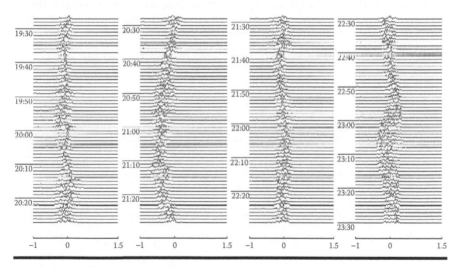

Figure 1.3   Variations of the DS with frequency 3.0 MHz at the start of Zenit-2 rocket on September 9, 1998.

It should be noted that the ionosphere was perturbed during this night and sporadic small-scale plasma inhomogeneities were observed both before and after the start of the rocket. In this case, the maximum in DSs disappeared, but spectra have a tendency to spread (called in Ref. [36] as "spilling" process). This burn of the rocket is interesting because at 272 s into the flight, switching off of the rocket's computers was registered and, finally, the rocket started to fall down from an altitude of 160 km. Due to strong natural perturbations in the ionosphere, search for the effects related to the RL became very difficult. Despite this fact, we can state that before 20:10 UT a median frequency (MF) of DS, $f_{Da} \approx 0$, of the width of the spectrum of (−0.1 to 0.5 Hz) was seen. From 20:10 to 20:30 h the DS spilled, and from 20:30 to 20:45 h it became diffused, despite the fact that its width decreased from 0.7 to 0.3 Hz. A weak quasiperiodical process was indicated for a period of $T \approx 10$ min and then, more strongly, for $T \approx 20$ min. A significant broadness of the DS (up to 0.8 Hz) was observed in the time interval of 20:29–21:37 h. After 21:40 UT, the width of DSs achieved 0.3–0.5 Hz.

An example of the results of nocturnal observations of effects that accompanied the heavy-type Proton rocket burn and launch (Russia) at 00:09 UT on March 21, 1999, is shown in Figure 1.4. Before the start, the ionosphere was in quite a steady-state regime, that is, it was not perturbed. After 23:13 h the quasiperiodical process was observed with $f_{Da} \approx 0.2\,\text{Hz}$ and period $T \approx 18$ min. The duration of this process was observed as $\Delta T \approx 40$ min. From 01:12 to 01:18 h an additional mode of the reflected signal was observed that shifted at 0.2 Hz, and from 01:28 to 02:15 h the effect of spectrum spilling was also observed. It should be noted that sunrise at the ground level was observed approximately at 02:30 h,

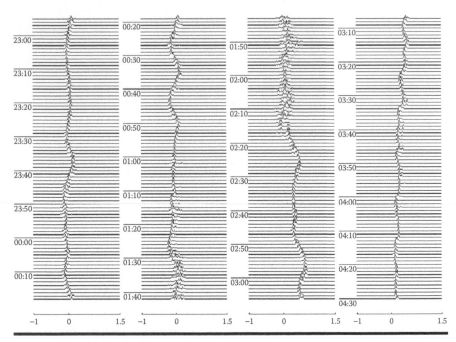

Figure 1.4 Variations of the DS with frequency 3.0 MHz at the start of Proton on March 20 and 21, 1999.

that is, an hour before the mentioned effects occurred in the lower part of the *F*-region of the ionosphere. After 02:15 h, the reflected signal from the sporadic irregularities of the ionosphere had single-mode form and quasiperiodical processes were observed: from the beginning, with $T \approx 18$ min and $f_{Da} \approx 0.25$ Hz, and, then, with $T \approx 7–8$ min and $f_{Da} \approx 0.15$ Hz.

During Energia ultra-heavy type RL (former USSR) at 17:30 UT on May 15, 1987, the quasiperiodical oscillations of the reflected signal with periods of $T \approx 10–15$ min were observed after 66 min and these continued for about an hour (observations were done at the carrier frequency $f_c = 4.6$ MHz [14]). About the same scenario was observed during the burn and launch of the same rocket on November 15, 1987.

Now let us briefly describe the effects in the ionosphere that accompanied the burns and launches from Plesetsk cosmodrome (Russia). Thus, on September 9, 1999 evening, the ionosphere was in quite a stable state. Only during the time interval of 17:00–17:50 UT weak perturbations were observed that were caused by the passing of the evening terminator. The Soyuz rocket was launched at 18:00 UT. Approximately from 18:08 to 18:45 h the essential broadness of DS (up to 0.5 Hz) took place, but from 18:55 to 19:17 h decrease of DS at $f_{Da} \approx 0.3$ Hz was observed. During the time period 19:00–20:50 h a broadness of DS at 0.2 Hz was registered. After that the ionosphere was quiet.

The above results, which will be discussed later, show the efficiency of DR sounding of the perturbed ionospheric large-scale sporadic plasma structures to analyze the processes that accompany the burn, launch, and flight of rockets of various weight types.

Now we introduce the reader to the DR technique for investigation of global perturbations induced by rockets in the middle ionosphere.

1.6 Global Perturbations in Rocket Flights and SV Maneuvers

For the analysis of possible global effects in the ionosphere, it was convenient to launch rockets from cosmodromes located in the United States and France (under the patronage of the European Space Agency). As for cosmodromes in China, India, and Japan, their launches were rare with small-power rockets compared to those mentioned earlier.

An example of the temporal variations of DSs in diurnal time is shown in Figure 1.5. The start of Space Shuttle Endeavour took place at 08:35 UT on December 4, 1998, in the United States. On this day the ionosphere was strongly perturbed and during the entire time of observation the effect of spectrum spilling occurred. Before the rocket started and soon after its start, a quasiperiodical

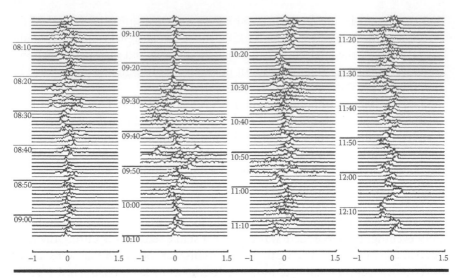

Figure 1.5 **Variations of the DS with frequency 2.8 MHz at the start of Space Shuttle Endeavour on December 4, 1998.**

process in the Doppler frequency domain was registered by the DR with time period $T \approx 10$ min and Doppler spread $f_{Da} \approx 0.1$ Hz. From 08:50 to 09:25 UT the ionosphere was in quite steady state. However, after 09:25 h and until 10:00 h, the essential broadness of the DS as well as a significant quasiperiodical process with period $T \approx 15$–20 min was fixed. In the short-time period of 10:05–10:10 h, an additional mode with maximum shifting in frequency at 0.5 Hz was registered. After 10:22 h for a long time the effect of spectrum spilling took place with the quasiperiodical process and amplitude achieving 0.2 Hz in the frequency domain and period $T \approx 10$ min.

An example of the variations of DSs accompanying the Space Shuttle Endeavour launch and flight (start at 19:19 UT on October 29, 1998) is shown in Figure 1.6. Sunset at ground level was at about 15:00 UT and at the lower part of the F-layer of the ionosphere at about 16:00 UT.

It is assumed that all effects relating to the movements of the terminator ended before 18:50 UT. In the period of 30 min from the RL in the ionosphere a weak quasiperiodical process with $T \approx 7$ min accompanied a DS, $f_{Da} \approx 0.10$–0.15 Hz. Moreover, from 19:13 to 19:45 h, a two-mode reflected signal was registered, despite the fact that the ionosphere remained weakly perturbed until 20:30 UT. After this time, the reflected signal started to attain the two-mode form again with $\Delta T \approx 15$ min during the observation time of 50 min. From 21:10 to 21:55 h the DS remained practically constant. At the time interval of 21:55–22:00 h a short-period spectrum spilling was registered, as can be clearly seen from Figure 1.6.

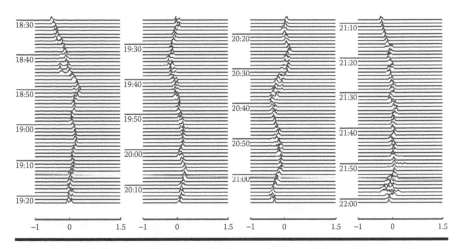

Figure 1.6   Variations of the DS with frequency 3.0 MHz at the start of Space Shuttle Discovery on October 29, 1998.

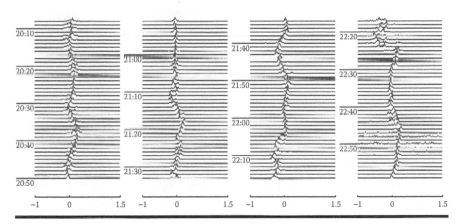

Figure 1.7   Variations of the DS with frequency 3.0 MHz at the start of Delta on February 7, 1998.

The start of Delta medium-weight-type rocket took place at 21:04 UT on February 7, 1999. The corresponding spectral characteristic in two-dimensional (2D) time–frequency domain is shown in Figure 1.7. Approximately at 21:10 UT, an additional mode arose with a frequency shift of $\Delta f_D = -0.2$ Hz, a quasiperiodical process with $T \approx 8$–10 min, $\Delta T \approx 40$ min, and DS $f_{Da} \approx 0.1$ Hz as well. After 21:50 h, the ionosphere remained quite calm, but at the time interval of 22:00–22:25 h a sharp decrease of f_{Da} was registered at 0.4–0.6 Hz, accompanied by spectrum spilling. After 22:25 h the ionosphere remained sufficiently calm; only a weak quasiperiodical process was found with $T \approx 25$ min and $f_{Da} \approx 0.10$–0.15 Hz.

The next example relates to observations at nighttime (see Figure 1.8 where DS in 2D time–frequency domain is shown). The start of Atlas-type rocket took place at 01:45 h on February 16, 1999. Until 00:30 h the ionosphere remained weakly perturbed. During the time interval of 00:35–01:05 h, a quasiperiodical process with $T \approx 20$–25 min and $f_{Da} \approx 0.2$ Hz was registered. Exactly before the start of the rocket, the period of DS variations decreased up to 10 min (with $\Delta T \approx 20$ min). From 01:55 to 02:05 h a two-mode signal was observed, and after 02:32 h a strong quasiperiodical process with $f_{Da} \approx 0.3$–0.4 Hz was fixed, with changing period from 12 to 24 min and duration more than 2.5 h. Note that sunrise at the ground level took place at about 4 o'clock (i.e., in early morning), and at the lower part of F-layer of the ionosphere at about 3 o'clock.

Another example relates to observation of the rocket's effects basically at the night–morning time. Thus, the launch of Titan-II-type rocket took place at 02:15 UT on June 20, 1999. Sunrise at the ground level was at about 23:50 UT,

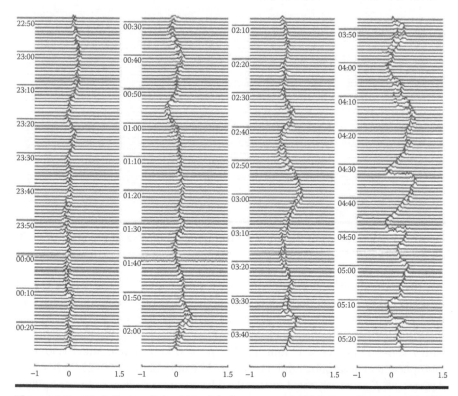

Figure 1.8 Variations of the DS with frequency 3.0 MHz at the start of Atlas on February 16, 1999.

but at the lower F-layer of the ionosphere it was an hour earlier. Suppose that all transfer processes were finished by 2 o'clock. Before the rocket's launch the ionosphere remained weakly perturbed, but DSs broadness of 0.15–0.20 Hz was observed. After 02:20 h an increase of f_{Da} occurred at 0.2–0.3 Hz that continued for 30 min. At 02:52 h the quasiperiodical process appeared with $T \approx 6$–7 min, $\Delta T \approx 20$ min, and $f_{Da} \approx 0.1$–0.2 Hz. During the time period 03:45–04:05 h, a weak effect of signal spectrum spilling was observed.

The next example also corresponds to the morning period of observations. The launch of Space Shuttle Columbia took place at 04:28 UT on July 23, 1999. Sunrise at the ground level was observed at about 00:20 h, but at the lower part of the F-layer of the ionosphere about an hour earlier. By the time interval 01:00–02:00 h, all transfer processes relating to the terminator movements were completed. That morning the ionosphere was quite calm. Only from 03:30 to 03:48 h did the decrease of the Doppler frequency up to -0.8 Hz register, which was quickly finished (after 1–3 min). Before launch, a weak quasiperiodical process was observed for 30 min with $T \approx 8$ min and $f_{Da} \approx 0.1$–0.2 Hz. At 04:40 h the short-term effect of signal spectrum spilling was fixed, which was repeated at time intervals of 05:29–05:35 h and 05:55–06:05 h. The ionosphere remained quite calm at other times.

The presented example relates to the launch of the ultra-light rocket Pegasus during nighttime at 00:02 UT on October 23, 1998. Approximately an hour before launch the ionosphere was weakly perturbed: a weak quasiperiodical process with DS broadness not exceeding 0.2–0.3 Hz being observed. Around the time interval of 00:14–00:24 h this spread had achieved 0.5 Hz; the same thing was repeated from 00:40 to 01:40 h. A more essential spread of DS (from 0.6 to 0.7 Hz) occurred from 01:10 to 01:25 h, whereas the decrease of f_D achieved 0.5 Hz. Later on, the DS remained a one-mode type with $f_D \approx 0.3$ Hz. Sunrise at the lower part of the F-layer of the ionosphere began at approximately 02:25 h and later the essential variations of DSs, which continued for not less than 1.5 h, were registered.

All the above-mentioned rockets were launched from cosmodromes in the United States. Finally, we will briefly describe the variations happening in the ionosphere during nighttime from March 21 and 22, 2000. Ariane RL happened from Kourou cosmodrome (French Guiana) and took place at 23:28 h, as shown in Figure 1.9.

From 22:00 to 23:00 h the signal spectrum was mostly of multimode type, but from 23:10 to 00:20 h it was basically unimode type. The broadness of DS was observed only from 23:32 to 23:40 h. After 00:25 h and for approximately 40 min the well-expressed quasiperiodical process with varied period and frequency amplitude of 0.4 Hz was fixed. For a further 20 min the ionosphere remained quite calm, but after 01:40 h it started showing the effects of the morning terminator.

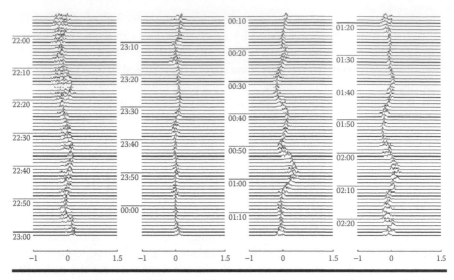

Figure 1.9 **Variations of the DS with frequency 3.5 MHz at the start of Ariane on March 21 and 22, 2000.**

1.7 Global Perturbations during Landing of SVs

SVs should apply brake pulse in time during landing. Thus, for a Space Shuttle this time did not exceed 10^1–10^2 s and the brake pressure was 10^7–10^8 N s^{-1}. The power of the reactive jet was 10^8–10^9 W and its energy 10^9–10^{10} J. Considering that brake pressure is transmitted to the engines while hidden directly within the plasma, its energy is sufficient to create large-scale and even global disturbances in the ionosphere.

For example, Figure 1.10 shows variations of DSs that accompany switching off of the engines during the braking of Space Shuttle Discovery, which took place at 18:00 UT on June 12, 1998. At the time interval of 15:50–16:45 h the ionosphere was disturbed and, therefore, multiray phenomena and quasiperiodical process in joint time–frequency domain showed up with $T \approx$ 8–10 min and $f_{Ds} \approx$ 0.1–0.2 Hz, respectively. From 16:47 to 17:00 h the signal spectrum spilling that was repeated was fixed (but with weaker effects) at intervals of 17:07–17:18 h and 17:37–17:56 h. Approximately at 18:10 h and for 25 min the ionosphere remained weakly disturbed. From 18:15 to 19:30 h the effects of signal spectrum spilling were observed again, and after 19:30 h these effects were indicated as weakening. Additionally, we noticed that sunset at the ground level took place at 19:11 h and at the level of reflection of radio wave an hour later. Figure 1.11 shows variations of DSs corresponding to the landing of the same type of SV, taking place at 17:04 h on November 7, 1998.

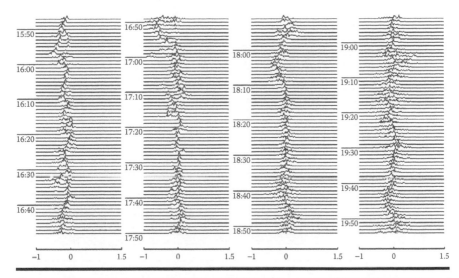

Figure 1.10 Variations of the DS with frequency 3.5 MHz at the landing of Space Shuttle Discovery on June 12, 1998.

At the time interval of 14:35–15:24 h, a quasiperiodical process with $T \approx 10$ min and $f_{Da} \approx 0.2 - 0.4$ Hz registered in the ionosphere. The signal remained either of one-mode or two-mode type with a small frequency shift of the second mode. From 15:24 to 16:05 h a sharp fall in f_D at 0.5–0.6 Hz, and from 15:46 to 15:50 h, even at 1 Hz, was observed. After 17:00 h the ionosphere was weakly disturbed for an hour. At the time interval of 18:00–18:50 h, essential variations of f_D that do not exceed 0.5 Hz were again registered, as well as the existence of two-mode type of the signal. Note that sunset at the ground level took place approximately at 14:30 h, but at the lower part of the F-layer of the ionosphere about an hour later.

We will finally consider nighttime variations of DS that accompany the landing of Space Shuttle Discovery at 00:01 UT on December 28, 1999. These variations are presented in 2D joint time–frequency domain in Figure 1.12. Sunset at the ground level was observed at about 14:00 h. Therefore, from 18:00 to 19:00 h all transfer processes caused by the night terminator were completed successfully. During the whole night the DS had a width of 0.2–0.3 Hz. From 21:40 to 22:30 h, the Doppler frequency shift (DFS) was sufficiently small (≤ 0.1–0.2 Hz). The quasiperiodical process took place with $T \approx 12$–15 min and $f_{Da} \approx 0.2$ Hz at 22:36 h for about an hour.

It is interesting to note that the constant component in this long-periodic spectrum decreased approximately at 0.3 Hz, and new harmonics with periods of 2, 4, and 8 min were observed in this spectrum that were never observed in the regular F-layer of the ionosphere. At time intervals of 23:40–00:10 h

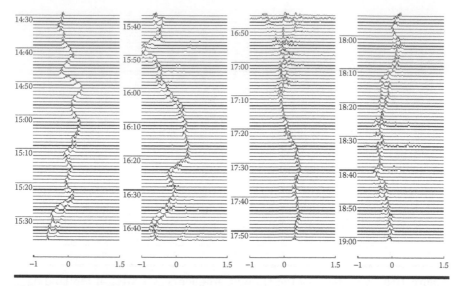

Figure 1.11 Variations of the DS with frequency 3.5 MHz at the landing of Space Shuttle Discovery on November 7, 1998.

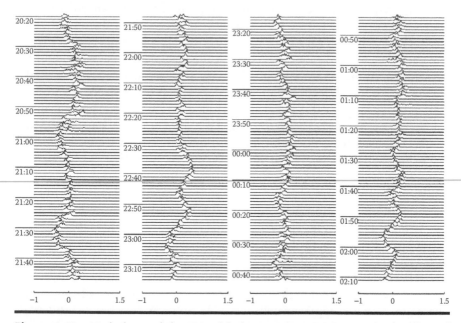

Figure 1.12 Variations of the DS with frequency 3.5 MHz at the landing of Space Shuttle Discovery on December 27 and 28, 1998.

and 00:45–01:30 h, the value of the Doppler frequency was constant, that is, $f_D \approx$ constant. From 00:10 to 00:45 h and after 01:30 h quasiperiodical processes with amplitude in frequency $f_{Da} \approx 0.2$ Hz and periods $T \approx 20$ min and $T \approx 12$ min, respectively, were observed again.

1.8 Discussion of Results

1.8.1 Large-Scale Disturbances

Now we return to the results shown in Figures 1.1 through 1.4. The same general peculiarity shown in all these figures is that an additional mode is seen at the 2D time–frequency spectra with a time delay of 60 min (during night) to 80 min (during day) with shifting at 0.2–0.3 Hz to the lower band of Doppler frequencies, that is, to signal spectra spilling. The time duration of these effects is about 5–10 min. With the range between the cosmodrome and the observation point being $R \approx 2055$ km and the time difference being $\Delta t \approx 60-80$ min, it is possible to estimate the velocity of propagation of the plasma disturbance caused by the rocket. Thus, such a velocity can be achieved, according to the corresponding estimations, to about 410–570 m s^{-1}. These velocities are on the same order of magnitude as those related to acoustic gravity waves (AGWs). We notice that the acoustic power the rocket engines can attain is $(1-3) \times 10^7$ W. The wave energy propagating from a place of rocket burn and launch, initially as a shock wave and then, as the AG wave, attains an altitude of $z \approx 150$ km for a time of $\delta t \approx 6-7$ min, whereas the rocket attains these altitudes earlier at $\delta t_1 \approx 4$ min. It is important to note that the velocity of the rocket significantly exceeds sound wave velocity in the ionosphere. Therefore, SVs can be considered as a source of shock waves. In the atmosphere usually acoustic-type waves propagate, from sound to AGWs. Sound wave has a velocity of

$$v = \left(\frac{\gamma k_B T}{M} \right)^{1/2},$$

(1.2)

where:

$\gamma = 1.4$ is the adiabatic coefficient
$k_B = 1.38 \times 10^{-23}$ J K^{-1} is the Boltzmann constant
T is the temperature (in Kelvin, K)
M is the average mass of molecules

The time of sound wave propagation in the ionosphere along the horizontal line (i.e., parallel to the Earth's surface) during the diurnal period is equal to

$\Delta t' = \Delta t - \delta t_1 \approx 76\text{min}$, where, as mentioned above, $\delta t_1 \approx 4\text{min}$. At the same time, when moving toward the east the rocket will be $R = 2300$ km from the observation point. Then, the corrected value of the velocity of the rocket will be, $v' = R / \Delta t' \approx 500\,\text{ms}^{-1}$. For $M \approx 4.6 \times 10^{-26}$ kg, the temperature is $T \approx 575$ K, which corresponds to gas temperature at an altitude of $z \approx 140$ km. This altitude is close to that of sounding wave reflection during diurnal time periods.

During the Energia rocket burn and launch (Russia), the following parameters $\Delta t' \approx 62$ min, $v' \approx 700\,\text{ms}^{-1}$, and $T \approx 880$ K were estimated. Such a temperature takes place at an altitude of $z \approx 175$ km that is approximately the same as the height of the radio wave reflection operating at $f = 4.6$ MHz. It is important to note that with the increase of frequency f, the altitude z of the wave reflection also increases and thereby, values of temperature T and velocity v' increase too.

During the launch at nighttime periods, $\Delta t \approx 57$–58 min, $v' \approx 700\,\text{ms}^{-1}$, and $T \approx 1000$ K. Such a temperature is observed at an altitude of $z \approx 250$ km, where the reflection of sounding radio wave takes place.

The same scenario is observed for the start of the rocket from Plesetsk cosmodrome (Russia). Disturbance of the acoustic type at the diurnal period has a delay of about $\Delta t \approx 55$ min; the corrected value is $\Delta t' \approx 51$ min. For $R = 1500$ km, we get $v' \approx 500\,\text{ms}^{-1}$, that is, practically the same value obtained during RLs from Baikonur cosmodrome (Russia) at nighttime with the following parameters: $\Delta t \approx 36$ min, $v' \approx 700\,\text{ms}^{-1}$, and $T \approx 1000$ K. The acoustic wave is the initial wave, after which the dispersive AGW arrives, having a velocity of several hundred meters per second. As a result inhomogeneities of various scale and dimension occur in the ionosphere, which lead to the signal spectra spilling observed. This process has periods from 10 min up to 1–2 h (Figures 1.1 and 1.4).

Faster perturbations that accompany rocket burn and launch cannot be observed regularly using the proposed methodology of observation. Thus, during experimental observations carried out on March 21 (Figure 1.1) and October 18 (Figure 1.4), 1999, after the RL with a delay of $\Delta t \approx 2$–4 min, quasiperiodical processes in the ionosphere were seen clearly. The abovementioned delay corresponds to a velocity of $v \approx 10$–20 km s^{-1}. Actually, such velocities correspond to gyrotropic waves that propagate in the magnetoactive plasma in the E- and F-layers of the ionosphere [37].

Between March 21 and July 5, 1999, a rocket of the same type as Proton was launched but on July 5, immediately after the start of the SV, quasiperiodical processes were not observed. This was because the rocket launched on July 5 exploded at an altitude of 110 km, and its pieces exploded at an altitude of 25 km. The energy, E, yielded from these two explosions was about 550 and 730 GJ (Giga Joule), respectively. We can estimate the characteristic dimension of the area of perturbation, R_{sw}, and the pulse duration of the shock wave, τ_{sw}, using the following relations:

$$R_{sw} = \left(\frac{E}{p_0}\right)^{1/2}, \ \tau_{sw} = \frac{R_{sw}}{v_s}, \tag{1.3}$$

where:

p_0 is the atmospheric pressure at the height of explosion

v_s is the sound velocity

Assuming the value of pressure from the first and the second explosions to be equal to $p_0 = 3 \times 10^{-2}$ Pa and $p_0 = 3.5 \times 10^3$ Pa, respectively, we get R_{sw} to be 26 and 0.6 km, respectively, and τ_{sw} to be 79 and 1.8 s, respectively, for the first and the second explosions. Because for the first explosion the standard atmospheric height scale, $H \approx 8$ km, was less than $R_{sw} \approx 26$ km, the products of the explosions passed through the atmosphere.

Reaction from the first explosion was observed at the time intervals 14:54–14:58 h and 15:03–15:05 h (local time). The magnitude of the quasiperiods was 7–9 min. The reaction of the atmosphere on the second explosion that occurred at a lower altitude, was of short duration, and was observed at 15:09 h (local time). The characteristic velocities of propagation of explosion-excited perturbations from both explosions were between 470 and 500 m s^{-1}.

1.8.2 Global Disturbances

Twenty-three launches of SVs of various types were analyzed, including 19 launches from cosmodromes in the United States (see Table 1.3). Despite the fact that Pegasus is a small-power rocket that is not likely to generate much global perturbation, observations of its effects on the ionosphere were done only for comparison with the effects caused by more powerful rockets having bigger mass and higher power.

Sufficiently stable reaction of the ionosphere for SVs starting from cosmodromes in the United States and France was observed generally after 60–80 min of their launch (see Table 1.3). The trajectories of the rockets differed and this fact can explain the variations of time delay from 60 to 100 min.

Suppose that on average $R \approx 10,000$ km, then we get a virtual velocity $v \approx 2$–3 km s^{-1}. Such velocities in the ionosphere have slow magnetohydrodynamic (SMHD) waves [37]. They are weakly attenuated waves and therefore can be responsible for global perturbations. The real mechanism of transfer of such perturbations is as follows. Such perturbations are not caused by the work of engines of zero and first stages (steps) in the near-the-Earth atmosphere, but by the functioning of other stages such as maneuvering engines. Despite the fact

Table 1.3   Parameters of Global Perturbations

Types of Rockets (SV)	Number of Observations	Time Delay (min)	Time Duration (min)
Space Shuttle	5	60–80	5–15
Ariane	4	60–65	40–60
Delta	8	70–80	5–10
Atlas	4	70–100	~10
Titan	1	70	5
Pegasus	1	40–90	10–20

that the power of the later engines was significantly lesser, they inject products of burn directly into plasma, causing effective excitation of electromagnetic and magnetohydrodynamic (MHD) waves which should be accounted for. Nonrare perturbations were observed with a typical time delay of about 6–7 min with $R \approx 9,000-10,000$ km and duration of time 10 min. Such perturbations correspond to velocity on the order of 25 km s^{-1}. The gyrotropic waves mentioned earlier have such a velocity.

1.8.3 Global Effects during Landing of SVs

Six landings of Space Shuttle–type rockets were analyzed. Sufficiently stable reaction appeared 70–90 min before landing, and its duration was 10 min. Thus, during the landing of the SV on December 28, 1999, the delay on account of perturbations relative to the switching on of the brake pressure that took place at 22:48 h on December 27 was about 12 min. This delay was defined via the decrease in the constant component inside the long-periodic spectrum of the DS frequency and the appearance of new harmonics with $T = 2, 4$, and 8 min that do not characterize the F-region of the ionosphere. The expected velocity of the propagation of perturbations is about several tens of kilometers per second. This means that during launches gyrotropic waves are responsible for the transfer of disturbances. The second group of quasiperiodical perturbations has delay $\Delta t \approx 80$ min, $\Delta T \approx 40$ min, and $T \approx 10$ and 20 min. If these processes related to the switching on of the burn engines, then their expected velocity of propagation comes to be on the order of a few kilometers per second. Such a velocity, as was mentioned earlier, has slow MHD waves.

1.9 Main Results

1. Acoustic perturbations (sound and AGW) from rocket engines propagate in the ionosphere to distances not less than 2300 km, which correspond to heavy and medium rockets of average weight. Often, after arrival of such perturbations, the ionosphere becomes turbulent at the *F*-layer altitudes, resulting in the spreading of DSs. Faster perturbations and thereafter acoustic waves are observed sometimes at distances of up to 2300 km from the rocket. Generally, their velocity is 10–20 km s^{-1}. In this case the transfer is done by gyrotropic waves. This assumption needs additional investigation.

2. At the start of launch of an SV in the ionosphere, perturbations are observed at a distance of about 10,000 km. The mechanism of their generation can be related either to the work of more powerful maneuvering engines in the near-the-Earth atmosphere, or to the functioning of less powerfully maneuvering engines, or to the last stage engines. The latter statement can be easily understood because it is clear how electromagnetic waves and MHD waves are generated in plasma by a reactive stream. It can be suggested that the waves corresponding to the first mechanism are gyrotropic waves. According to the observations made, their velocities lead to the range of values from 10 to 25 km s^{-1}.

3. Approximately 60–80 min before the landing of the SVs, perturbations with duration of several tens of minutes registered in the ionosphere. These perturbations can be explained by the switching on of the braking (maneuvering) engines. The virtual velocity of these perturbations is about 10–20 km s^{-1}. The second group of perturbations has velocity $v \approx 2 - 3$ km s^{-1}.

4. During the experiments mentioned in this chapter three groups of velocities: from 0.5 to 0.7 km s^{-1} and less, from 2 to 3 km s^{-1}, and from 10 to 25 km s^{-1} were observed. It is possible that these correspond to AG, SMHD, and gyrotropic waves, respectively.

References

1. Mendillo, M., Ionospheric holes: A review of theory and recent experiments, *Adv. Space Res.*, 8, 51–62, 1988.
2. Booker, H. G., A local reduction of *F* region ionization due to missile transit, *J. Geophys. Res.*, 66, 1073–1081, 1961.
3. Mendillo, M., Hawkins, G. S., and Klobuchar, J. A., A sudden vanishing of the ionospheric *F* region due to the launch of "Skylab," *J. Geophys. Res.*, 80, 2217–2228, 1975.
4. Mendillo, M., The effects of rocket launches of the ionosphere, *Adv. Space Res.*, 1, 275–290, 1981.

5. Bernhardt, P. A., A critical comparison of ionospheric depletion chemicals, *J. Geophys. Res.*, 92, 4617–4628, 1987.

6. Karlov, V. D., Kozlov, S. I., and Tkachev, G. N., Large-scale perturbations in the ionosphere caused by flight of rocket with working engine, *Cosmic Res.*, 18, 266–277, 1980 (in Russian).

7. Cotton, D. E., Donn, W. L., and Oppenheim, A., On the generation and propagation of shock waves from "Apollo" rockets at orbital altitudes, *Geophys. J. Roy. Astron. Soc.*, 26, 1496–1503, 1971.

8. Nagorskii, P. M. and Taraschuk, Yu. U., Artificial modification of the ionosphere by rocket launching transported on the orbit the space vehicles, *Izv. Vuzov. Phys.*, 36, 98–107, 1993 (in Russian).

9. Nagorskii, P. M., Analysis of the short-wave signal response on ionospheric plasma perturbation caused by shock-acoustic waves, *Izv. Vuzov. Radiofizika*, 42, 36–44, 1999 (in Russian).

10. Francis, S. H., Global propagation of AGWs: A review, *J. Atmos. Terr. Phys.*, 37, 1011–1054, 1975.

11. Belikovich, V. V., Benediktov, E. A., Tolmacheva, A. V., and Bakhmet'eva, N. V., *Ionospheric Research by Means of Artificial Periodic Irregularities*, Katlenburg-Lindau: Copernicus GmbH, 2002.

12. Garmash, K. P., Rozumenko, V. T., Tyrnov, O. F., Tsymbal, A. M., and Chernogor, L. F., Radio physical investigations of processes in the near-the-earth plasma disturbed by the high-energy sources, *Foreign Radioelectronics: Success in Modern Radioelectronics*, 7, 3–15, 1999 (in Russian).

13. Chernogor, L. F., Garmash, K. P., Kostrov, L. S., et al., Perturbations in the ionosphere following U.S. powerful space vehicle launching, *Radiophys. Radioastron.*, 3, 181–190, 1998.

14. Garmash, K. P., Rozumenko, V. T., Tyrnov, O. F., Tsymbal, A. M., and Chernogor, L. F., Radio physical investigations of processes in the near-the-earth plasma disturbed by the high-energy sources, *Foreign Radioelectronics: Success in Modern Radioelectronics*, 8, 3–19, 1999 (in Russian).

15. Kostrov, L. S., Rozumenko, V. T., and Chernogor, L. F., Doppler radio sounding of perturbations in the middle ionosphere accompanied burns and flights of cosmic vehicles, *Radiophys. Radioastron.*, 4, 227–246, 1999 (in Russian).

16. Plendl, H., Concerning the influence of eleven-year solar activity period upon the propagation of waves in wireless technology, *Proc. Inst. Radio Eng.*, 20, 520–539, 1932.

17. Kaiser, T. R., The first suggestion of an ionosphere, *J. Atmos. Terr. Phys.*, 24, 865–872, 1962.

18. Blaunstein, N. and Plohotniuc, E., *Ionosphere and Applied Aspects of Radio Communication and Radar*, Boca Raton, FL: Taylor and Francis/CRC Press, 2008.

19. Ratcliffe, J. A., *Physics of the Upper Atmosphere*, New York: Academic Press, 1960.

20. Hulburt, E. O., Ionization in the upper atmosphere of the Earth, *Phys. Rev.*, 31, 1018–1037, 1928.

21. Chapman, S. and Ferraro, V. R., A new theory of magnetic storms, *Terr. Magn. Atmos. Electr.*, 38, 79–96, 1933.

22. Martyn, D. F., Electric currents in the ionosphere, III. Ionization drift due to winds and electric fields, *Phil. Trans. Roy. Soc. Lond. A*, 246, 306–320, 1953.

23. Maeda, K., Dynamo-theoretical conductivity and current in the ionosphere, *J. Geomagn. Geoelect.*, 4, 63–82, 1952.
24. Baker, S. G. and Martyn, D. F., Electric currents in the ionosphere, I. Conductivity, *Phil. Trans. Roy. Soc. Lond. A*, 246, 281–294, 1953.
25. King, J. W. and Kohl, H., Upper atmosphere winds and ionospheric drifts caused by neutral air pressure gradients, *Nature*, 206, 899–701, 1965.
25. Ratclife, J. A., The formation of the ionospheric layers F-1 and F-2, *J. Atmos. Terr. Phys.*, 8, 260–269, 1956.
27. Yonezawa, T., A new theory of formation of the $F2$ layer, *J. Radio Res. Labs*, 3, 1–16, 1956.
28. Gershman, B. N., *Dynamics of the Ionospheric Plasma*, Moscow: Nauka, 1974 (in Russian).
29. Alpert, Ya. L., Gurevich, A. V., and Pitaevsky, L. P., *Artificial Satellites in the Rare Dense Plasma*, Moscow: Nauka, 1964 (in Russian).
30. Gurevich, A. V., *Nonlinear Phenomena in the Ionosphere*, Berlin: Springer-Verlag, 1978.
31. Whitten, R. C. and Poppoff, I. G., *Physics of the Lower Ionosphere*, New York: Prentice Hall, 1965.
32. Rees, H., *Physics and Chemistry of the Upper Atmosphere*, Cambridge: Cambridge University Press, 1989.
33. Filip, N. D., Blaunshtein, N. Sh., Erukhimov, L. M., Ivanov, V. A., and Uryadov, V. P., *Modern Methods of Investigation of Dynamic Processes in the Ionosphere*, Kishinev: Shtiintza, 1991 (in Russian).
34. Encyclopedia. *Cosmonautics*, Moscow: Soviet Encyclopedia, 1985, 528 (in Russian).
35. Gethand, K., *Cosmic Technique*, Moscow: Mir, 1986 (in Russian).
36. Gorely, K. I., Lampey, V. K., and Nikol'sky, A. V., Ionospheric effects of cosmic vehicles burn, *Geomagn. Aeronom.*, 4, 158–161, 1994.
37. Sorokin, V. M. and Fedorovich, G. V., *Physics of Slow MGD-Waves in the Ionospheric Plasma*, Moscow: Energoatomizdat, 1982, 134 (in Russian).

Chapter 2

Perturbations in
the Background
of a Magnetic Storm

2.1 Overview

It was stated in Chapter 1 that in the recent past two decades increasing interest has been observed regarding investigations into disturbances caused in the ionosphere by the launch and flight of space vehicles (SVs). The References section of Chapter 1 as well as Refs. [1–10] given in this chapter shows the willingness of scientific society to understand the processes that accompany such events occurring in the ionosphere. The main results of these investigations were: (1) decrease in plasma concentration N (electrons or ions; plasma is quasineutral) up to tens of percentage in the F-region of the ionosphere (called the "ionospheric holes" in the literature); (2) generation of plasma instabilities and, finally, plasma inhomogeneities with horizontal dimensions of L ~ 100–1000 km; (3) generation of dense waves with velocities of 0.1–1 km s^{-1}. Moreover, it was established that disturbances in the ionosphere can be not only of local type (of L ~ 10–100 km) but also of large-scale type (of L ~ 100–1000 km) [3]. The possibilities of the appearance of global-scale perturbations (with L ~ 1,000–10,000 km) have been analyzed in Refs. [8–10]. It is still not clear until now as to what are the mechanisms by which transport of plasma disturbances happens at the global scale, the velocities of their propagation, and the magnitude and character of such perturbations. We will describe these aspects in later chapters.

The main goal of this chapter is the description of the results of experimental investigations into possible perturbations in the lower and middle ionosphere that are caused by the start, launch, and flight of the Space Shuttle on May 15, 1997. The main peculiarity of this experiment is that it took place at the beginning of the main phase of the magnetic storm. Thus, in Ref. [11], it was suggested that natural perturbations in the ionosphere–magnetosphere system are the cause of its destability, yielding of its artificial source of energy, and even its weak intensity; this can cause significant perturbations in the ionospheric plasma. In other words, a triggering mechanism can occur from this supply of energy. Before delving into the description of the results of observations and its corresponding technique, let us introduce the reader to the two main aspects that should be understood: (1) What is meant by "magnetic storm"? (2) How can a moving SV disturb the ambient geomagnetic field?

2.1.1 Magnetic Storm Phenomenon

It is well known from the literature that the outer geomagnetic field surrounding the Earth can be modeled in such a manner that it can be described by the magnetic dipole located at the center of the Earth. Then, distribution of the corresponding geomagnetic field can be schematically presented as distribution of magnetic lines created by this dipole, as shown in Figure 2.1.

The magnitude Φ of such magnetic dipole is related to the angle I of geomagnetic field inclination, $\Phi = \tan^{-1}\left[(1/2)\tan I\right]$. Based on such geomagnetic line structure, and accounting for the magnetosphere–ionosphere coupling and interactions, the ionosphere is categorized as high-latitude (with $|\Phi| > 55°–60°$), middle-latitude (with $30° \le |\Phi| \le 55°$), and lower-latitude (with $|\Phi| < 30°$) ionosphere, with a special zone of geomagnetic equator (with $|\Phi| \le 5°–10°$) [12–16].

The magnetosphere can be characterized as a fully ionized plasma structure. Due to magnetosphere–ionosphere coupling, the outer geomagnetic field continuously spreads over 100,000 km, covering both the magnetosphere and the ionosphere, as is schematically seen in Figure 2.2.

High-energy charged particles and radiation, as can be seen from Figure 2.1, mainly penetrate at high latitudes, also called the *polar cap* and the corresponding regions of the high-latitude ionosphere that are called subauroral, auroral, and polar [17–21]. As clearly seen from Figure 2.2, magnetic lines passing through the high-latitude ionosphere cover the magnetosphere by creating a global zone of magnetopause. Along these lines, solar wind energy dissipates in the

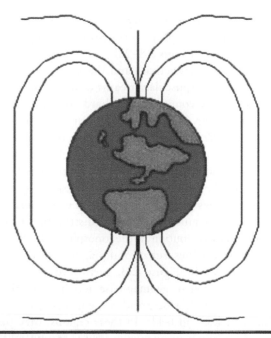

Figure 2.1 Schematically presented structure of geomagnetic field lines surrounding the Earth created by a magnetic dipole located at its center.

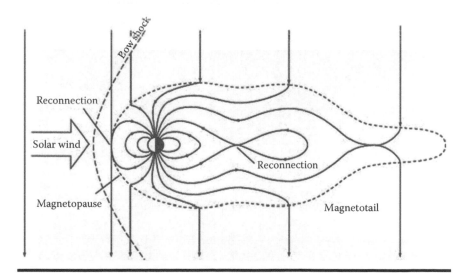

Figure 2.2 Schematically presented distribution of the outer magnetic field lines covering the magnetosphere and the ionosphere of the Earth.

magnetosphere as a stream of high-energy protons and other charged particles, and then is transferred at ionospheric altitudes by heating flows, precipitating particles, and longitudinal currents [18–22].

Unfortunately, due to solar activity and the activity of the galactic objects close to the solar system, corpuscular streams of heavy protons or other high-energy charged particles accompanied by hard solar and galactic radiation, called *solar wind* and *cosmic rain*, respectively, can create irregular distribution of regular variations of the geomagnetic field lines covering the magnetosphere and the ionosphere, as is seen clearly from Figure 2.3. Variations of the parameters of the solar wind and galactic rain are closely correlated with the dynamic processes occurring in the magnetosphere, which lead to spatial and temporal irregularity of energy flows penetrating from the magnetosphere to the ionosphere, and finally, the irregularity of spatial distribution of the geomagnetic field and plasma density distribution in the ionosphere [19,20].

As was observed by satellites, subauroral magnetic field lines map the inner magnetosphere, as shown in Figure 2.3, earthward of the electron plasma sheet. During intensification of geomagnetic activity, electric fields permeate the inner magnetosphere with widespread consequences for the global distributions of plasmas, particles, and fields. In these regions, they perturb the ionospheric plasma and the outer geomagnetic field surrounding the Earth. Such phenomena are called magnetic storms in literature [23–25]. Strong magnetic storms with large conventional electric fields associated with magnetic storms can be considered as the source of a large number of ring current ions that

Figure 2.3 The influence of solar wind and galactic rain on the outer geomagnetic field surrounding the magnetospheric structure and the Earth's ionosphere.

enter the plasmasphere. The corresponding perturbed regions of the ionosphere, caused by magnetic storm and the corresponding auroral particle precipitation, expand toward the equator through the middle latitudes [25–27]. The effects of ionospheric plasma perturbations caused by magnetic storms will be discussed further in our description of the matter. Here, we only note that depending on solar activity and other galactic events, the energy efficiency of precipitated charged particles can achieve 0.1–10 MeV, and therefore, can perturb dramatically (up to thousands of nanoTesla, where 1 nT = 10^{-9} T) the outer background geomagnetic field. Often, in literature the unit used usually is Tesla and the present magnetic density distribution is given in nanoTesla, as shown in Figure 2.4 extracted from Ref. [26], in which the upper panel presents the distribution of the magnetic field amplitude (in nanoTesla) and the bottom panel presents the magnetic field variations defined by K_p, index of magnetic activity.

The results presented in Figure 2.4 were observed during satellite observations of a magnetic storm whose peak was registered in the morning of September 23, 1999, and in which the relative magnetic field amplitude attained a value of −164 and the corresponding K_p index of magnetic field activity attained a value of 8.0.

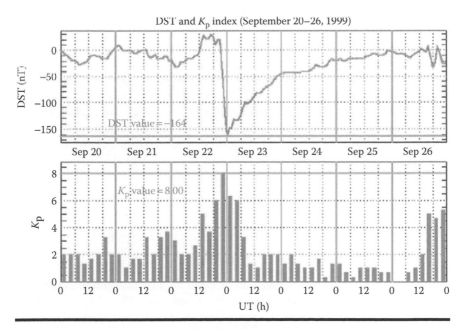

Figure 2.4 The index of magnetic disturbances (DST) and K_p-index variations during a magnetic storm that occurred between September 20 and 26, 1999.

2.1.2 Perturbation of the Ambient Geomagnetic Field by a Moving SV

As shown previously in pioneering researches (see Ref. [28]), SVs generate a moving spatial charge during their movements. Movements of charges, electrons, and ions create an electrical current in the ionospheric plasma, which finally leads to the generation of additional magnetic field (intrinsic) that is fully determined by this current. This intrinsic magnetic field is considered in the literature as a perturbation of the ambient geomagnetic field. This additional field can significantly change the dynamics of the movement of electrons and ions in the background of ionospheric plasma. To investigate all the peculiarities of the additional magnetic field generation surrounding the moving vehicle in plasma, the elements of kinetic theory and derivation from mathematically complicated formulas can be used. These aspects are out of the scope of this book. At the same time, we will show the reader numbers that allow us to suggest: whether we need or not to account for these local magnetic field disturbances with respect to those occurring during global magnetic storms. It was shown by Alpert et al. [28] that the perturbations $\delta \mathbf{B}$ of the ambient magnetic field, $\mathbf{B}_0$, are proportional to the vector product of the vehicle's velocity, $\mathbf{V}_0$, and the total electrical field created surrounding the moving SV, $\mathbf{E} = \mathbf{E}_0 + \delta \mathbf{E}$, that is, $\delta \mathbf{B} \sim \mathbf{V}_0 \times \mathbf{E}$. Because the total electric field far from the SV decreases with distance R to reach $\sim R^{-3}$, $\delta \mathbf{B}$ also decreases with distance under the same law far from the vehicle. Moreover, for average concentration of plasma in the middle ionosphere of $N \sim 10^{12}\,\mathrm{m}^{-3}$, the perturbation of the ambient magnetic field caused $\delta \mathbf{B}$ by the moving SV to differ for $E = 1\text{--}10$ mv m^{-1} dimensions of the vehicle for 1–100 m in the range of $\sim 0.4\text{--}40$ nT, that is, sufficiently weak. These estimations, as well as the estimations of $\delta \mathbf{B}$ obtained during the magnetic storm described earlier, allow us to note that it is problematic to record the local effects of the moving vehicle on the outer geomagnetic field with respect to the effects of the not even strong magnetic storm.

2.2 Results of Observations Using Different Radiophysical Techniques

The Space Shuttle was taken as a source of ionospheric perturbation (see Appendix 1), starting at 08:07 h on May 15, 1997, from Cape Canaveral, Florida. It moved approximately in the southwest direction from the cosmodrome. After 44 min from launch, when the SV was at the eastern part of the Indian Ocean at an altitude of 250 km, the corrected

Table 2.1 Geophysical Conditions

		Roentgen Blasts			Estimations of Planetary Indexes								
Date	F *(10.7 cm)*	W	C	M	X	A_p	K_p						
May 12, 1997	72	12	1	0	0	3	1	0	1	1	1	2	1	1
May 13, 1997	74	15	0	0	0	3	0	0	0	0	1	2	2	1
May 14, 1997	74	17	0	0	0	6	1	0	0	1	2	3	3	2
May 15, 1997	73	15	0	0	0	53	3	3	6	7	7	5	3	2

Note: A_p, cumulative index of planetary activity.

engines (at about 80 s of working regime) were switched on. The inclination of the orbit was 57°.

The condition of the ionosphere was characterized as quiet or weakly perturbed (after 09:00 h). The critical frequency of the F_2-layer, f_0F_2, was changed from 4.2 to 5.1 MHz, and the minimum observed frequency recorded by the ionospheric station was $f_{min} = 1.2 - 1.4$ MHz. According to the data from the Center of Cosmic Environment, perturbations of the Sun were not observed. The density of the radio radiation stream, F, at the wavelength of 10.7 cm was recorded as 73, and Wolf's number as $W \approx 15$. Geomagnetic activity was classified as high-level type (see Table 2.1, columns 2 and 3). The start of the SV was done at the beginning of the main phase of magnetic storm.

Index K_p increased from 3 (before the start of the rocket) to 6–7 (after the start of the rocket). As a control, measurements (for calibration) were used from the other days of May 1997.

2.2.1 Partial Reflection Method

Measurements of amplitudes of the partially reflected (PR) signal of ordinary A_o (o-wave) and extraordinary A_x (x-wave) polarizations and the corresponding noises, A_{no} and A_{nx}, were done at a frequency of 2.3 MHz. An example of temporal dependence of $\langle A_{o,x}^2 \rangle$, $\langle A_{no,nx}^2 \rangle$, and the spectral component of $\langle A_o^2 \rangle$ with a period of 5.3 min is shown in Figure 2.5a–d. As can be seen from Figure 2.5a and b, the significant (up to seven times) increase of $\langle A_{o,x}^2 \rangle$ takes place around 08:00 h and from 09:00 to 10:00 h. The same happens regarding the parameter $\langle A_{no,nx}^2 \rangle$. Spectral analysis of the series of $\langle A_o^2 \rangle$ at intervals of $\Delta t_2 = 32$ min showed that the harmonic with a period on the order of 5 min was increased much more (up to 2–3 times). In Figure 2.5c, the temporal dependence of this harmonics is shown, normalized at the mean square deviation $\langle A_o^2 \rangle$, derived at the same time intervals.

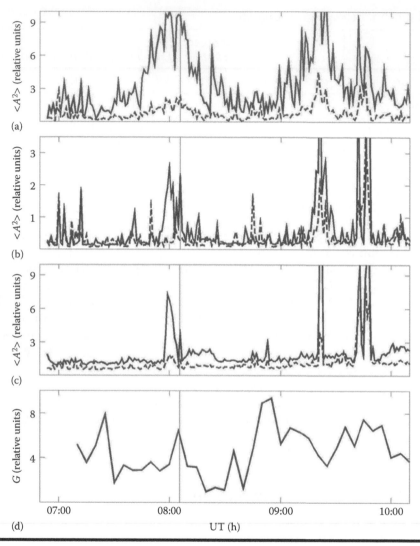

Figure 2.5 Time dependences of normalized square-averaging amplitudes of the signal from different altitudes: (a) 84 km, (b) 81 km, and (c) noise, as well as normalized spectral component $G = <A^2>/<A_0^2>$ of the signal (with period $T \approx 5.3$ min), and (d) received from an altitude of 84 km. The solid curve is the ordinary component and the dashed curve is the extraordinary component. The vertical line is for the time moment 08:07 h here and at other panels indicates the moment of the rocket launch. The amplitude of spectral component, estimated at the time interval of 32 min, related to the middle of this interval.

For derivation of ionospheric plasma profiles $N(z)$, where z is the altitude, the method of differential absorption was used along with the following parameter:

$$R(z) = \frac{\langle A_x^2 \rangle - \langle A_{nx}^2 \rangle}{\langle A_o^2 \rangle - \langle A_{no}^2 \rangle}. \tag{2.1}$$

Averaging of amplitudes was done at intervals of 5 and 10 min.

On the other hand, the theory of the PR method gives the relation between the "profile" $R(z)$ estimated from observation and the real profile $N(z)$:

$$\ln\left(\frac{R_0(z)}{R(z)}\right) = \int_{z_0}^{z} K(z')N(z')dz', \tag{2.2}$$

where:

z_0 is the altitude of the lower boundary of the ionosphere

$K(z')$ is the nuclear of the integral equation

$R_0(z)$ is a well-known function obtained theoretically [11]

As a profile of the collision frequency $\nu(z)$, the model profile from Ref. [29] was used. Because the inverse problem in the PR method is not correct mathematically, it was used in Ref. [11] as the algorithm of regulations described in Ref. [30]. We briefly describe this algorithm below. Numerical differentiation of Equation 2.2 on variable z gives the solution to this equation in the following form:

$$N(z) = K^{-1}(z)\frac{d\tilde{R}}{dz}, \quad \tilde{R} = \ln\frac{R_0}{R}. \tag{2.3}$$

The quality of the obtained profiles depends on the accuracy of $\langle A_{o,x}^2 \rangle$ and $\langle A_{no,nx}^2 \rangle$ estimation, as well as on the stability of the algorithm of differentiation to the mistakes of $R(z)$. The algorithm of regulation gives, instead of the solution to the problem (Equation 2.2), differentiation of $N(z)$ via minimization of the functional

$$\Phi[N, R, \alpha] = \left| \int_{z_0}^{z} K(z')N(z')dz' - \tilde{R}(z) \right|^2 + \tilde{\alpha}\Omega[N], \tag{2.4}$$

where:

$\Omega[N]$ is the stabilization factor

$\tilde{\alpha}$ is the parameter of regulation

In Ref. [11], the stabilization factor was chosen in the form of

$$\Omega[N] = \left| \left[(N / \tilde{N}) - 1 \right] \left(q_{max} / q \right)^{1/4} \right|^2, \tag{2.5}$$

with

$$q^{-1} = q_o^{-1} + q_x^{-1}, \, q_{max} = \max\left[q(z) \right], \tag{2.6}$$

where:

$q_{o,x} = \left\langle A_{o,x}^2 \right\rangle / \left\langle A_{no,nx}^2 \right\rangle$ is the signal-to-noise ratio (SNR) for o- and x-waves, respectively

$\tilde{N}(z)$ is the initial approximation for $N(z)$

We note that the stability factor $\Omega[N]$ does not depend on N strongly and decreases the influence of data obtained for altitudes where $q_{o,x}$ is sufficiently small. It was found that using the regulation algorithm allows decreasing the statistical error of reconstruction of profile $N(z)$ using $R(z)$ 3–5 times, and increasing the range of altitudes up to 10 km, where reconstruction is possible. Further, the optimal time interval of averaging, ΔT, for estimation of $\left\langle A_{o,x}^2 \right\rangle$ was 5–10 min. For a smaller interval ΔT, statistical errors can increase significantly, whereas for larger interval ΔT, it is impossible to check the time variations of N because of smoothing effect.

Error of reconstruction for $N(z)$ attained 50% and 30% at the averaging intervals of 5 min and 10 min, respectively. The profile $N(z)$ was obtained for the altitudinal range of $z \approx 75$–85 km. Below and above this range, the SNR, $q_{o,x}$, was much lower than thought of (i.e., <1).

Time dependence of N for different altitudes is shown in Figure 2.6 (continuous curve). The start of the SV was at 08:07 h. It is seen that at about 1 h before the start and 45 min after the start, $N(z)$ practically remained constant. At about 08:50 h, quasiperiodical changes of profile N with period $T \sim 30$ min appeared, having an amplitude of 30%–70% (the error of N estimation does not exceed 30%) and a time duration of $\Delta T \approx 2$ h at altitudes of $z \sim 75$–85 km.

It is important to note that the time delay of the occurrence of perturbations relating to rocket start at different altitudes was about 45–55 min. Additionally, we should note that during control days such perturbations were absent (dashed curves in Figure 2.6).

2.2.2 Doppler Sounding Method

Corresponding measurements were carried out at a frequency of 4.1 MHz. The Doppler spectra (DSs) were estimated at time interval $\Delta t_1 \approx 50$ s (resolution at the frequency was about 0.02 Hz).

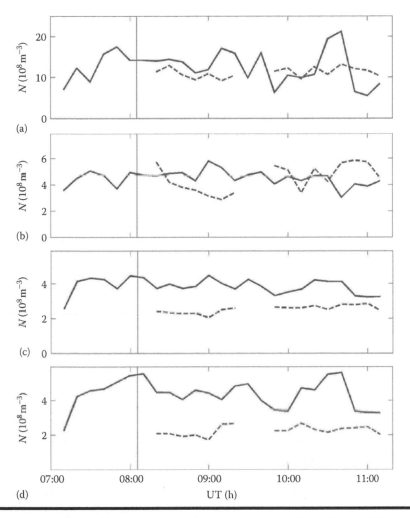

Figure 2.6 Time dependence of electron concentration (the solid curve corresponds to the date May 15, 1997; dashed curve corresponds to the date May 24, 1997) at different altitudes: (a) 84, (b) 81, (c) 78, and (d) 75 km.

Spectral analysis has shown that generally speaking the recording signal was of multimode type (Figure 2.7). Figure 2.7a shows that before the start of the Space Shuttle (at ~08:00 h) for the main mode, $f_d(z) \approx$ constant, the spectral width (at the level of 0.5) was about 0.1–0.2 Hz. At 08:54 h the signal turned to two-mode type (the difference of Doppler frequencies attained 0.4 Hz, as seen from Figure 2.7b). Repetition of the second mode was observed from

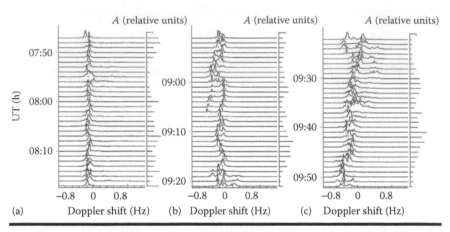

Figure 2.7 The examples of DSs are: (a) before rocket launch and the first few minutes after that, (b) occurrence of the additional mode and quasiperiodical variations of $f_d(t)$, and (c) quasiperiodical variations of $f_d(t)$ and destruction of spectra. Horizontal lines on the right indicate the normalized amplitude of the signal.

09:18 to 09:27 h. During other time intervals, the DSs broadened up to 0.4 Hz (Figure 2.7b and c).

The dependence of the Doppler frequency, $f_d(z)$, for the main mode is shown in Figure 2.8. It is seen that from 06:00 to 08:00 h the average value was $\langle f_d(z) \rangle \approx -0.1\,\text{Hz}$, and the value of its deviations did not exceed $\Delta f_d \approx \pm 0.2\,\text{Hz}$. During the first 40 min after the start, the average values were $\langle f_d(z) \rangle \approx -0.1\,\text{Hz}$ and $\Delta f_d \approx \pm 0.1\,\text{Hz}$. After 08:50 h, the average value $\langle f_d(z) \rangle$ continuously decreased up to -0.7 Hz for 70 min. Quasiperiodical perturbations additionally occurred with such a behavior of $\langle f_d(z) \rangle$ in periods of tens of minutes.

Spectral analysis at the interval $\Delta t_2 \approx 64\,\text{min}$ with shifting of 15 min showed that 45–50 min after the start of the Space Shuttle (i.e., at ~08:50 h) in the F-region of the ionosphere ($z \sim 250$–300 km) the amplification of wave process amplitude was about 1.5 times higher with period $T \sim 13$ min (Figure 2.8b). If before the start of the SV the amplitude of harmonics was about -0.1 Hz, 40–100 min after the start it attained -0.2 Hz. The amplitude of harmonics with $T = 32$ min and $T = 64$ min increased to ~30% and ~100%, respectively (Figure 2.8c and d). The time duration of quasiperiodical perturbations was more than an hour. It is interesting to note that the time delay of perturbations was about 45–60 min, that is, it was on the same order as that at the lower ionosphere described earlier.

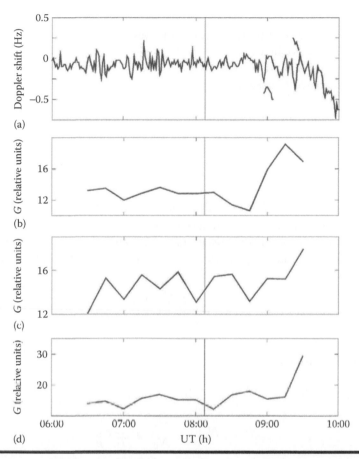

Figure 2.8 (a) Time dependences of f_d of the main mode and its normalized spectral component, (b) for $T \approx 12.8$ min, (c) for $T = 32$ min, and (d) for $T = 64$ min. Amplitudes of the spectral components, estimated at a definite interval, relate to the middle of the interval.

2.3 Discussions on the Result of Observations

2.3.1 Effects at the F-Region of the Ionosphere

As already known (see, e.g., Ref. [31]), during vertical lift Doppler sounding can be calculated according to the formula:

$$\frac{f_d}{f} = -\frac{2}{c} \int_0^{z_r} \frac{dn}{dt} dz, \qquad (2.7)$$

where:

z_r is the altitude of reflection of the radio wave
n is the refractive index
c is the speed of light
f is the radiated frequency

This expression can be presented in another form according to Ref. [31], that is,

$$\frac{f_d}{f} = -2\frac{v}{c} - \frac{2}{c}\int_0^{z_r}\frac{dn}{dt}dz. \tag{2.8}$$

In this equation, the first term describes the contribution of movements of the probing signal reflection region with velocity $v = dz/dt$. The second term describes the contribution of nonstationary media below the region of reflection. Here, without accounting for the ambient magnetic field and collisions of particles, we get

$$\frac{dn}{dt} \approx -\frac{1}{2}\left(\frac{f_p}{f}\right)^2\left(\frac{1}{N}\frac{\partial N}{\partial t}\right), \tag{2.9}$$

where:

f_p is the plasma frequency

For periodical changes of $N(t)$, according to the law

$$N(t) = N_0\left(1 + \frac{\Delta N}{N_0}\sin\frac{2\pi}{T}t\right). \tag{2.10}$$

From Equations 2.8 and 2.10 for the amplitude of the velocity movements, we get

$$v_a = 2\pi\frac{l}{T}\frac{\Delta N}{N_0}, \tag{2.11}$$

where:

l is the scale of changes of $N(z)$

This was defined from ionograms and was ~50 km.

If we suggest that the Doppler shift of the main mode is caused by periodical changes of radio wave reflection altitude, it is very easy to obtain the following expression for the amplitude, f_{da}:

$$\frac{f_{da}}{f} = 2\frac{v_a}{c}. \tag{2.12}$$

Now, from Equations 2.11 and 2.12, we get

$$\frac{\Delta N}{N_0} = \frac{cT}{4\pi l}\frac{f_{da}}{f}.$$ (2.13)

Thus, for $T \approx 13\,\text{min} \approx 8 \times 10^2\,\text{s}$, $f = 4.1 \times 10^6\,\text{Hz}$, $f_{d0} = 0.25\,\text{Hz}$, we get: $\Delta N / N_0 \approx 3 \times 10^{-2} \to 3\%$.

If the observed Doppler effect is caused by temporal changes of N, according to law (Equation 2.10) below the region of reflection, then for the amplitude of the harmonic with period T we get

$$\frac{f_{da}}{f} \approx 2\pi\left(\frac{\overline{f_p}}{f}\right)^2 \frac{\Delta z}{cT}\frac{\Delta N}{N_0},$$ (2.14)

where:

Δz is the thickness of the disturbed region

$\overline{f_p}$ is the average value of plasma frequency f_p in this region, which equals 4.1 MHz.

From Equation 2.13 it follows that

$$\frac{\Delta N}{N_0} = \frac{cT}{2\pi\Delta z}\left(\frac{f}{\overline{f_p}}\right)^2 \frac{f_{da}}{f}.$$ (2.15)

For $\Delta z = 50\,\text{km}$, we get $\Delta N / N_0 \approx 3\%$. We see that the Doppler effect is caused by oscillations of the reflection altitude, as well as by changes of N below this region. In any case this effect is caused by quasiperiodical changes of N of about 3% as a result of the occurrence of plasma density waves with a period of $T \approx 13$ min passing through this disturbed region of the ionosphere.

Let us note that such waves are generated in natural conditions during a magnetic storm and propagate from high latitudes to lower latitudes with velocities ~1 km s^{-1} [32] (see Section 2.1 and the results described therein). If we assume that the wave passed the distance of $R \sim 3000$–4000 km, then it would have been created at high latitudes at around 50–70 min before the recording of quasiperiodical disturbances (i.e., at about 08:50–08:55 h). Exactly at this time period the beginning of the main phase of the magnetic storm is observed.

At the same time it should be kept in mind that such wave perturbations with $T \sim 10$–100 min are generated (more strictly, amplified) in the ionosphere by seismic events, explosions, and so on [32]. It is reasonable to suggest that such types of waves can be generated during the start of powerful rockets, such as the Space Shuttle. It is hard to say that these observed perturbations were caused by the main engines of the rocket because they were switched on in the near-the-Earth atmosphere, and the amount of energy transferred to ionospheric perturbations is insignificant. For the time delay of ~50 min, their "expected" (e.g., "virtual") velocity should be about $v \sim 3$ km s^{-1}.

At the same time, these perturbations can be caused by the switching off of the corrected engines 44 min after the start of the SV. In this case, the time delay of such perturbations, defined by $f_d(t)$, can be 7 ± 2 min. Then, for $R \approx 12,000$ km, we get $v \approx 29 \pm 9$ km s^{-1}. Despite the fact that the corrected engines have power hundred times lesser than the main engines, and their energy yielding, related to the operation of these engines, is thousand times lesser with respect to the main engines, perturbations in the ionosphere from the corrected engines can be sufficient. This happens because the energy of these engines enters directly into the ionospheric plasma at altitudes of 250–300 km.

Additionally, let us note that the observed perturbations in the middle ionosphere can hardly be explained by their transfer from the magnetically conjugated regions through which the SV passes during the time period of about 08:40–08:45 h by switching off its engines.

2.3.2 Effects in the D-Region of the Ionosphere

As was mentioned earlier, in the lower ionosphere at altitudes of $z \sim 75–85$ km, $\Delta N / N_0 \approx 30\%–70\%$, respectively. It is hard to assume that such an amplitude has plasma density waves propagating from the region of the rocket start (in this case, their velocity should be on the order of about 3 km s^{-1}) or from the magnetosphere. More realistically, these wave perturbations are caused by quasiperiodical precipitations of charged particles (e.g., high-energy electrons) arriving from the magnetosphere. If so, let us estimate their parameters by following this hypothesis.

The density of flow of energy from the precipitating particles equals [33]:

$$\Pi = \varepsilon p = 2\varepsilon_i \Delta z \Delta q_i, \tag{2.16}$$

where:
  p is the density of the stream of precipitating particles
  ε is their energy
  $\varepsilon_i \approx 10^{-18}$ J is the potential of ionization of atmospheric molecules
  $\Delta z \approx 10$ km is the thickness of the region where these particles are braked
  Δq_i is the change of ionization speed of electrons

In the quasistationary stage,

$$q_{io} = \alpha_0 N_0^2, \; q_i = \alpha N^2. \tag{2.17}$$

Then, for $\alpha \approx \alpha_0$

$$\Delta q_i \approx \alpha \left(N^2 - N_0^2 \right) = \alpha N_0^2 \left(N^2 / N_0^2 - 1 \right). \tag{2.18}$$

Let us account for the predominance of electron recombination with ion clusters at an altitude of $z \leq 75$ km, for which $\alpha \approx 2 \times 10^{-11} \mathrm{m^3 s^{-1}}$. Above $z \sim 100$ km, recombination of electrons with molecular ions usually occurs, for which $\alpha \approx 2 \times 10^{-13} \mathrm{m^3 \, s^{-1}}$. In the intermediate region from 75 to 100 km the parameter of recombination α decreases smoothly from $\alpha \approx 2 \times 10^{-11} \mathrm{m^3 s^{-1}}$ to $\alpha \approx 2 \times 10^{-13} \mathrm{m^3 s^{-1}}$. Let us consider that at $z = 84$ km, $\alpha \approx 2 \times 10^{-12} \mathrm{m^3 s^{-1}}$. Then from Equations 2.17 and 2.18 it follows that

$$\Pi = 2\varepsilon_i \Lambda z \alpha_0 N_0^2 \left(N^2 / N_0^2 - 1 \right) = 2\varepsilon_i \Lambda z q_{io} \left(N^2 / N_0^2 - 1 \right). \tag{2.19}$$

The results of the estimation of Π and other parameters are presented in Table 2.2.

It is seen from Table 2.2 that to guaranty the necessary variations of N in the D-region of the ionosphere, the flow density of charged particles, $p \sim 10^6 - (4 \times 10^7) \mathrm{m^{-2} s^{-1}}$, is required with energies of $\varepsilon \sim 10^2 - 10 \mathrm{keV}$, respectively. Such flow densities of participating particles are not so large even for the middle latitude ionosphere (at the high latitudes $p \sim 10^{10} - (4 \times 10^{13}) \mathrm{m^{-2} s^{-1}}$ [33]).

Finally, we need to answer the question: What causes such flows of participating particles—the magnetic storm or the SV launch? The question allows us to state the hypothesis more realistically that the powerful magnetic storm is accompanied by the effects related to the rocket start, which acts as a triggering mechanism for the purpose of supplying energy.

Table 2.2 Parameters of the Ionosphere and Participating Particles

z (km)	α $(m^3 s^{-1})$	N_0 (m^{-3})	ΔN (m^{-3})	q_{io} $(m^{-3} s^{-1})$	Δq_i $(m^{-3} s^{-1})$	Π $(Jm^{-2} s^{-1})$	ε (keV)	p $(m^{-2} s^{-1})$
75	10^{-11}	3×10^8	10^8	9×10^5	7×10^5	1.4×10^{-8}	10^2	1.4×10^6
84	10^{-12}	10^9	7×10^8	10^6	2×10^6	4×10^{-8}	10	4×10^7

2.4 Main Results

1. Analysis of radio sounding of plasma disturbances in the middle and lower ionosphere accompanied the start of Space Shuttle 10,000 km from the place of observation. The start was done at the beginning of the main phase of the magnetic storm.
2. Quasiperiodic perturbations of plasma density were created with periods of about 30 min in the lower ionosphere (e.g., at the D-layer) and at about 13, 32, and 64 min in the middle ionosphere (e.g., at the F-layer) 45–50 min after the rocket started in the D- and F-regions of the ionosphere. The time duration of such plasma density N variations was 1–2 h.
3. The amplitude of N variations in the lower ionosphere was about 30%–70% [or $10^8 - (7 \times 10^8) \text{m}^{-3}$] and in the middle ionosphere was about 3% ($\sim 3 \times 10^9 \text{ m}^{-3}$).
4. It is impossible to unilaterally state that the observed quasiperiodical disturbances related directly to the start of the rockets and not to magnetic storms accounting for the estimations presented in Section 2.1. At the same time it can be suggested that a magnetic storm can amplify perturbations caused by the rocket start and its flight, namely, by switching on the corrected engines of the SV at an altitude of about 250 km.

References

1. Medillo, M., Hawkins, G. S., and Klobuchar, J. A., A sudden vanishing of the ionospheric F-region due to the launch of Skylab, *J. Geophys. Res.*, 80, 2217–2228, 1975.
2. Zasov, G. F., Karlov, V. D., Romamnchuk, T. E., et al., Observations of disturbances at the lower ionosphere during experiments on the program Soyuz–Apollo, *Geomagn. Aeronom.*, 17, 346–348, 1977.
3. Karlov, V. D., Kozlov, S. I., and Tkachev, G. N., Large-scale perturbations in the ionosphere occurring during flight of the rocket with working engine (issue), *Cosmic Res.*, 18, 266–277, 1980 (in Russian).
4. Karlov, V. D., Kozlov, S. I., Kudryavtsev, V. P., et al., On one type of large-scale perturbations in the ionosphere, *Geomagn. Aeronom.*, 24, 319–322, 1984.
5. Misyura, V. A., Pahomova, O. V., Piven, L. A., and Chernogor, L. F., On possibility to investigate epizodic short-term disturbances in the lower ionosphere by help of method of vertical sounding, *Geomagn. Aeronom.*, 27, 677–679, 1987.
6. Misyura, V. A., Pahomova, O. V., and Chernogor, L. F., Investigations of global and large-scale disturbances in the ionosphere by a net of ionosondes, *Cosmic Sci. Tech.*, 4, 72–75, 1989 (in Russian).
7. Pahomova, O. V. and Chernogor, L. F., Virtual velocities of propagation of perturbations in the near-the-Earth cosmic space, *Cosmic Sci. Tech.*, 5, 71–74, 1990 (in Russian).

8. Gritchin, A. I., Denisov, V. I., Kapanin, I. I., et al., Complex radiophysical investigations of ionospheric disturbances caused by launches and flights of space-crafts, *Proc. Int. Seminar—Physics of Cosmic Plasma*, Kiev, Ukraine, 1994, 161–170.

9. Kapanin, I. I., Kostrov, L. S., Martynenko, S. I., et al., Complex radiophysical investigations of large scale and glob disturbances of ionosphere plasma and variations of radiowave characteristics, *Proc. Contributed Papers of Int. Conf.—Physics in Ukraine*, Kiev, Ukraine, June 22–27, 1993, 126–129.

10. Rozumenko, V. T., Kostrov, L. S., Martynenko, S. I., et al., Studies of global and large-scale ionospheric phenomena due to sources of energy of different nature, *Turkish J. Phys.*, 18, 1193–1198, 1994.

11. Garmash, K. P., Kostrov, L. S., Rozumenko, V. T., et al., Global disturbances of the ionosphere caused by launching of spacecrafts at the background of magnetic storm, *Geomagn. Aeronom.*, 39, 72–78, 1999.

12. Ratcliffe, J. A., *Physics of the Upper Atmosphere*, New York: Academic Press, 1960.

13. Gershman, B. N., *Dynamics of the Ionospheric Plasma*, Moscow: Nauka, 1974 (in Russian).

14. Volland, H., *Atmospheric Electrodynamics*, Heidelberg: Springer-Verlag, 1984.

15. Gershman, B. N., Erukhimov, L. M., and Yashin, Yu. Ya., *Wave Phenomena in the Ionosphere and Cosmic Plasma*, Moscow: Nauka, 1984 (in Russian).

16. Richmond, A. D., Ionospheric electrodynamics, in *Handbook of Atmospheric Electrodynamics*, Volland, H., Ed., vol. II, Boca Raton, FL: CRC Press, 1995, 249–290.

17. Mizun, Yu. G., *Polar Ionosphere*, Leningrad: Nauka, 1980 (in Russian).

18. Mizun, Yu. G., *Lower Ionosphere of High Latitudes*, Leningrad: Nauka, 1983 (in Russian).

19. Titheridge, J. E., Plasmapause effects in the topside ionosphere, *J. Geophys. Res.*, 81, 3227–3233, 1976.

20. Sverdlov, Yu. L., *Morphology of Radio Aurora*, Leningrad: Nauka, 1982 (in Russian).

21. Marubashi, K., Structure of topside ionosphere in high latitudes, *J. Radio Res. Lab.*, 75, 7175–7181, 1970.

22. Gel'berg, M. G., *Inhomogeneities of High Latitude Ionosphere*, Novosibirsk: Nauka, 1986 (in Russian).

23. Fel'dshtein, Ya. I., Levitin, A. E., Afonina, R. G., and Belov, B. A., Magnetosphere-ionosphere relations, in *Interplanetary Medium and Magnetosphere*, Moscow: Nauka, 1982, 64–116 (in Russian).

24. Foster, J. and Rich, F., Prompt midlatitude electric field effects during severe magnetic storms, *J. Geophys. Res.*, 103, 26367–26373, 1998.

25. Mishin, E. V., Burke, W. J., Basu, Su., et al., Stormtime ionospheric irregularities in SAPS-related troughs: Causes of GPS scintillations at mid latitudes, *AGU Fall Meeting 2003*, Abstract SH52A-07, Colorado, January, 11–14, 2003.

26. Mishin, E. V. and Burke, W. J., Stormtime coupling of the ring current, plasmosphere and topside ionosphere: Electromagnetic and plasma disturbances, *J. Geophys. Res.*, 110, 7209–7216, 2005.

27. Mishin, E. and Blaunstein, N., Irregularities within SAPS-related troughs and GPS radio interference at mid latitudes, in *AGU Geophysical Monograph*, Kintner, P., Ed., doi:10.1029/2008BK000687, 2008.

28. Alpert, Ya. L., Gurevich, A. V., and Pytaevsky, L. P., *Artificial Satellites in the Rarefied Plasma*, Moscow: Nauka, 1964 (in Russian).
29. Gurevich, A. V., *Nonlinear Phenomena in the Ionosphere*, Berlin: Springer-Verlag, 1978.
30. Garmash, K. P. and Chernogor, L. F., Profiles of electron concentration of D-layer of the ionosphere at calm and perturbed conditions using data of partial reflections, *Geomagn. Aeronom.*, 36, 75–81, 1996.
31. Davies, K., *Ionospheric Radio*, London: Peter Pelegrimus Ltd., 1990.
32. Sorokin, V. M. and Fedorovich, G. V., *Physics of Low MHD-Waves in Ionospheric Plasma*, Moscow: Energoizdat, 1982 (in Russian).
33. Lyazky, V. B. and Mal'zev, Yu. P., *Magnetosphere–Ionosphere Interaction*, Moscow: Nauka, 1988 (in Russian).

Chapter 3

Wave Disturbance Transport Caused by Rocket Launches Accompanied by Solar Terminator

3.1 Overview

In the previous chapter, the essential characteristics of investigation into the physical processes that occur in geospace environment caused by rocket launches (RLs) were observed [1–7]. The main objective of such investigations concerns the analysis of the reaction of the ionosphere on ambient sources with essential energy transmission. The proposed theoretical and experimental frameworks allow us to examine the transport of ionospheric plasma disturbances globally to large distances from the source, defining the mechanisms of their transport as well as the physical aspects of such sources of transportation, and finally find peculiarities of interaction of subsystems in the global system: Earth–atmosphere–ionosphere–magnetosphere [1].

In this chapter, the solar terminator (ST) is shown to be used as a natural source of energy radiation. Usually, the ST is defined as the boundary between "lightness" and "darkness" occurring during the Earth's rotation around the Sun. Thus, for each longitudinal region on Earth, the dynamics of this boundary (called *sunrise* and *sunset*) leads to changes in the ionospheric plasma content and in redistribution of ionospheric layers, from D to F, which, finally, leads to plasma density disturbances of varying degrees in the ionosphere. The ST differs from other natural sources by its regular appearance, precise accuracy of prediction of the time of its occurrence, and essential energy "hidden" in the ionosphere [2–4].

As for the artificial sources of ionospheric plasma perturbations that are caused by the starting of powerful space vehicles (SVs), in this chapter we will discuss only about those that were launched in the period of 1999–2002 from various points on Earth that are located far from the SV start, up to 1,500–10,000 km. For any observer, this source of energy allows investigation into the reaction of ionospheric plasma on RL, without waiting for the moment when the desired natural phenomena occurs. Different ranges from cosmodromes, different power of rockets, fuel content, and various trajectories of SV—all of these features guarantee various "initial conditions" and flexibility in investigation of the "reaction" of geospace environment on acts of such artificial sources [5–7].

The aim of this chapter is to describe the results of statistical analysis of the characteristics of the Doppler radar (DR) signals (see Appendix 3), reflected from the middle ionosphere caused by SV launches accompanied by the dynamics of the ST passing through the observed regions of the ionosphere. In this chapter, we also discuss the types of waves generated by artificially and naturally induced plasma disturbances, which are transported long distances, far and globally, following the results described in Ref. [8].

3.2 Results of Observations

3.2.1 Rocket Launches

Statistical processing and corresponding analysis of Doppler spectra (DSs) variations were carried out for RLs during the period 1999–2002 (see Table 3.1).

In total, 72 RLs were analyzed, starting from cosmodromes in Russia, the United States, France, and Japan. Further, we will discuss DS variations that were observed prior to the launches as well as those that followed them after the launches.

An example of DS variations during Soyuz-V launch from Plesetsk cosmodrome is shown in Figure 3.1. This event happened at 21:00 h on

Table 3.1 Information on RLs and the Geomagnetic Conditions on Rocket Start Days

Date	Time of Launch (h)	Title of SV	Title of Cosmodrome	Country	Amplitude of Magnetic Activity, A_p	Summary Index of Magnetic Activity, ΣK_p
January 3, 1999	22:21	Delta 2	Cape Canaveral	USA	4	9
February 16, 1999	03:45	Atlas 2	Cape Canaveral	USA	4	8
February 20, 1999	06:18	Soyuz	Baikonur	Russia	3	4
March 15, 1999	05:06	Soyuz	Baikonur	Russia	14	22
March 21, 1999	02:09	Proton	Baikonur	Russia	5	9
May 27, 1999	03:49	Discovery	Cape Canaveral	USA	7	14
June 20, 1999	05:15	Titan 2	Vandenberg	USA	4	10
June 24, 1999	18:44	Delta 2	Cape Canaveral	USA	4	10
July 5, 1999	16:32	Proton	Baikonur	Russia	2	5
July 16, 99	19:36	Soyuz	Baikonur	Russia	4	8
July 23, 1999	07:28	Columbia	Cape Canaveral	USA	11	19
September 9, 1999	21:00	Soyuz	Plesetsk	Russia	5	18
October 7, 1999	15:51	Delta 2	Cape Canaveral	USA	5	12

(Continued)

Table 3.1 (Continued)   Information on RLs and the Geomagnetic Conditions on Rocket Start Days

Date	Time of Launch (h)	Title of SV	Title of Cosmodrome	Country	Amplitude of Magnetic Activity, A_p	Summary Index of Magnetic Activity, ΣK_p
October 18, 1999	16:32	Soyuz	Baikonur	Russia	6	12
October 19, 1999	09:22	Ariane	Kourou	France	6	9
November 15, 1999	09:29	H2	Tanegashima	Japan	5	9
November 23, 1999	06:06	Atlas 2	Cape Canaveral	USA	19	26
December 20, 1999	02:50	Discovery	Cape Canaveral	USA	3	6
January 21, 2000	03:03	Atlas 2	Cape Canaveral	USA	2	2
January 25, 2000	03:04	Ariane	Kourou	France	7	14
February 1, 2000	08:47	Soyuz	Baikonur	Russia	8	16
February 4, 2000	01:30	Atlas	Cape Canaveral	USA	5	11
March 12, 2000	06:07	Proton	Baikonur	Russia	19	22
April 19, 2000	03:29	Ariane	Kourou	France	12	21
November 21, 2000	20:24	Delta 2	Vandenberg	USA	9	18
November 30, 2000	21:59	Proton	Baikonur	Russia	6	13
December 20, 2000	14:26	Ariane	Kourou	France	3	5

January 11, 2001	00:09	Ariane 4	Kourou	France	4	10
January 24, 2001	06:28	Soyuz	Baikonur	Russia	12	16
January 30, 2001	09:55	Delta 2	Cape Canaveral	USA	3	6
February 8, 2001	01:05	Ariane 44	Kourou	France	5	12
February 8, 2001	01:13	Atlantis	Cape Canaveral	USA	5	12
February 26, 2001	10:09	Soyuz	Baikonur	Russia	5	11
February 27, 2001	23:20	Titan 4	Cape Canaveral	USA	8	14
March 19, 2001	00:33	Zenit	Sea Platform	Ukraine	19	24
April 7, 2001	18:02	Delta 2	Cape Canaveral	USA	17	23
April 18, 2001	13:13	GSLV-D1	Sriharikota	India	22	25
April 28, 2001	10:37	Soyuz	Baikonur	Russia	34	32
May 15, 2001	04:11	Proton	Baikonur	Russia	10	17
May 18, 2001	20:45	Delta 2	Cape Canaveral	USA	11	19
June 09, 2001	08:45	Ariane 44	Kourou	France	19	25
June 16, 2001	04:49	Proton	Baikonur	Russia	4	9
June 19, 2001	07:41	Atlas 2	Cape Canaveral	USA	12	18

(Continued)

Table 3.1 (Continued)   **Information on RLs and the Geomagnetic Conditions on Rocket Start Days**

Date	Time of Launch (h)	Title of SV	Title of Cosmodrome	Country	Amplitude of Magnetic Activity, A_p	Summary Index of Magnetic Activity, ΣK_p
July 23, 2001	10:23	Atlas 2	Cape Canaveral	USA	6	16
August 6, 2001	10:28	Titan 4	Cape Canaveral	USA	14	23
August 8, 2001	19:13	Delta 2	Cape Canaveral	USA	7	14
August 11, 2001	00:15	Discovery	Cape Canaveral	USA	2	5
September 8, 2001	18:25	Atlas 2	Vandenberg	USA	5	10
September 26, 2001	02:21	Ariane 44	Kourou	France	13	20
October 6, 2001	05:38	Titan 4	Cape Canaveral	USA	6	13
October 11, 2001	03:32	Atlas 2	Cape Canaveral	USA	13	19
October 18, 2001	21:51	Delta 2	Cape Canaveral	USA	2	6
January 15, 2002	23:28	Titan 4	Cape Canaveral	USA	9	19
February 11, 2002	19:44	Delta 2	Vandenberg	USA	6	12

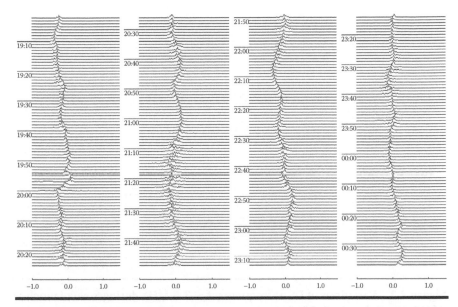

Figure 3.1 Variations of DSs at frequency *f* = 2.8 MHz during Soyuz launch at 24:00 h on September 9, 1999. Here, as in other figures, the Doppler frequency shift is indicated along the horizontal axis in Hz, and the local time in Kiev, Ukraine along the vertical axis.

September 9, 1999 (in this chapter the time is indicated according to Ukrainian local time). The total mass of the rocket was about 300 t and the initial pulling force was 4 MN (meganewton).

The working time of the zero, first, and second stages were equal to 118, 236, and 250 s, respectively. Switching off of these stages was done at altitudes of 48, 171, and 200 km, which corresponded from the cosmodrome to the range 44, 451, and 1680 km, respectively. The velocities of SV for all three locations were 1.8, 3.5, and 7.7 km s^{-1} [6]. The Soyuz-type rocket corresponds mainly to heavy rockets with an engine power of $\sim 10^{10}$ W and energy transfer not less than $\sim 3 \times 10^{12}$ J.

From Figure 3.1, it can be seen that prior to and during the launch of the rocket, weak variations of DS were observed, probably related to the passage of the evening ST (see below). The corresponding spectrum was unimodal. In the period from 21:10 to 22:00 h, the spectrum was sharply broadened, but its amplitude was practically constant. Further changes in the character of DS variations were observed from 22:00 to 22:35 h appeared and amplified the oscillations of Doppler shifting frequency, f_d. At the same time the spectrum shape remained unimodal.

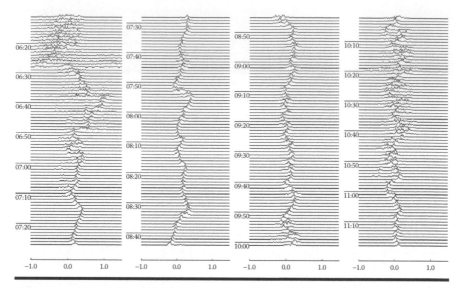

Figure 3.2 **Variations of DS at frequency** f = 3.5 MHz during Soyuz launch at 08:47 h on February 1, 2000.

An example of DS variations during another Soyuz-V launch is shown in Figure 3.2. The event happened at Baikonur cosmodrome at 08:47 h on February 1, 2000. As can be seen from Figure 3.2, approximately prior to 07:00 h strong variations of DS that became lesser during the time interval of 07:00–09:00 h were observed. The Doppler shift frequency was positive, that is, $f_d > 0$. DS practically were not changed from 09:08 to 09:40 h. After that variations of the average frequency and essential spread of DS were observed; these phenomena continued until 11:00 h.

We now consider the launch of Space Shuttle Discovery from Cape Canaveral cosmodrome that happened at 00:15 h on August 2001. The total mass of the rocket was about 2030 t, initial pulling force was 26 MN, time duration of the accelerated block and engines was 124, 480, and 480 s. These rockets related to the class of super-heavy rockets; the power of their engines is about 10^{11} W with an energy output not less than 10^{13} J [6]. Approximately 1 h prior to the event of launch the ionosphere became quiet and in steady state. Only a quasiperiodical process with $T \approx 20$ min was observed with the amplitude of Doppler shifting $f_{da} \approx 0.1$ Hz (Figure 3.3). The constant component of the long-periodical spectrum $f_d(t)$ was negative and did not exceed 0.1 Hz. About the same scenario was observed for a further 85 min.

About 01:40 h, the character of the signal changed: constant components became positive with $f_{d0} \approx 0.1$ Hz and for about 20 min the value of

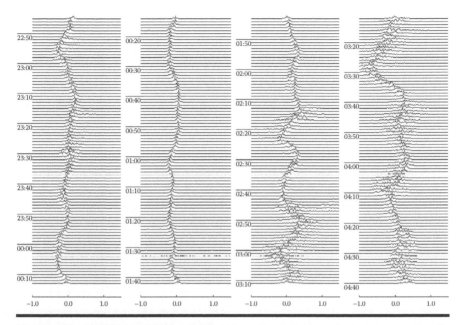

Figure 3.3   Variations of DS at frequency $f = 3.5$ MHz during Space Shuttle Discovery launch at 00:15 h on August 11, 2001.

the quasiperiod remained the same. After 02:15 h, a quasiperiodical process occurred in the ionosphere with varying period (15–20 min), $f_{da} \approx 0.3$ Hz, and the essential spreading of DS (from 0.1 to 0.5 Hz). Such a tendency was observed up to 04:40 h (i.e., until the end of observations).

For comparing the heavy and super-heavy RLs mentioned earlier, let us describe the possible effects that accompany rockets of medium weight. The Atlas 2AS-type rocket was launched from Cape Canaveral cosmodrome at 18:27 h on September 8, 2001. Its full mass was 234 t. The working times of the zero, first, second, and third stages were 56, 172, 283, and 392 s, respectively, and their initial pulling forces were 4×0.48, 2.09, 0.39, and 0.18 MN, respectively. For about 165 min before launch the ionosphere was quiet and in steady state (Figure 3.4). The first notable changes in the character of the signal—a jump of $f_d(t)$ at -0.2 Hz, sporadic occurrences of double-ray effects, were observed at 18:45 h (i.e., when $\Delta t_0 = 20$ min). Such a "reaction" of the ionosphere was observed for not more than 40 min. At 19:35 h, essential changes in the character of the signal were observed—the quasiperiodical disturbances in DS appeared. The time delay of $\Delta t_1 \approx 70$ min corresponded to these effects. After 20:00 h, additional effects caused by the passage of the evening ST came on.

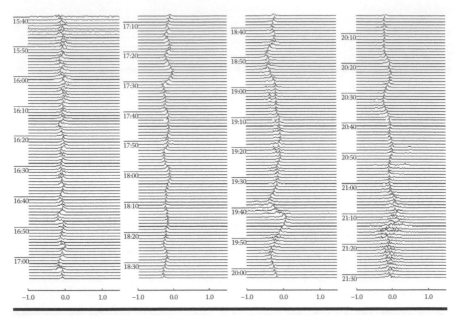

Figure 3.4 Variations of DS at frequency $f = 3.5$ MHz during Atlas 2AS launch at 16:25 h on September 8, 2001.

It is interesting to find the effects of GSLV-D1 launch (with a weight of 400 t) that was started at 13:13 h on April 18, 2001 from the Sriharikota cosmodrome (India). The range from the cosmodrome to the observation point was about 5,600 km. Prior to 13:00 h, the ionosphere was relatively perturbed, and DS was "spreading." After that, the signal was practically unimodal. Changes in its character occurred at 13:46 h. The Doppler frequency shift sharply increased from 0 to 0.13 Hz, and then for 8 min it decreased up to -0.25 Hz. Further, a sharp transfer from this value to $f_d = 0$ was observed. During the rest period, some essential changes in the signal characteristics were not observed.

From the analysis presented above, it follows that the described variations of DSs principally do not differ prior to and after the launches—the signal characteristics were changed many times during each time of observations. Therefore, to understand the "reaction" of the ionosphere on SV launches, it is obvious to statistically analyze the large amount of data of the observations carried out.

3.2.2 Passage of the Solar Terminator

In total, 35 observations were carried out from which only 31 gave good quality distribution of DSs in the time domain. In registration of DS during January 25, 2000, strong natural perturbations were presented. The processes during the passage of the 21 morning and the 10 evening STs were analyzed precisely. Further, we describe DS variations that can be related to plasma disturbances caused by passage of the STs.

First, we consider an example of ionospheric reaction on the action of the morning ST shown in Figure 3.3. About 02:15 h a strong quasiperiodical process was observed with a varying period in the ionosphere. The sunrise at that moment took place at an altitude of about 450 km. At about 02:45 h, the character of signal was changed and the sunrise was observed at an altitude of 350 km. The time period of the process was $T \approx$ 15–20 min. After sunrise at an altitude of about 250 km (time of observation was 03:13 h), the absolute magnitude of the constant component of DS, f_{d0}, was increased from 0.0–0.1 to 0.3 Hz. This process continued for 20 min. Then, f_{d0} was changed from −0.3 to +0.1 Hz, and this process was observed during about 30 min. Further, $f_{d0} \approx 0$. At the same time, a quasiperiodical process with $f_{da} \approx$ 0.1–0.2 Hz was registered with period $T \approx$ 20 min. Hence, the reaction of the ionosphere relating to the morning ST was observed for not less than 2 h.

Another example of ionospheric reaction initiated by passing of the evening ST is also shown in Figure 3.4. Sunset at the ground level took place at about 19:06 h. For about 30 min the ionosphere remained quiet and in steady state. After 19:35 h, quasiperiodical process registered with period $T \approx$ 17 min. The constant component f_{d0} decreased from −0.1 to −0.2 Hz. The same scenario remained until 20:10 h, prior to sunset at an altitude of 120 km. In the time interval from 20:15 to 20:40 h, a weak oscillation with $f_{da} \approx$ 0.1 Hz and period $T \approx$ 15 min occurred. Further, from 20:40 to 21:00 h (i.e., from the moment of sunset at an altitude of 300 km) the ionosphere remained quiet and in steady state. After this duration, $f_{d0} \approx 0$, the DS stretched essentially, from 0.1 to 0.5 Hz. This process was observed for not less than 40 min.

3.3 Results of Statistical Analysis

3.3.1 Rocket Launches

From 72 observations selected, there were only 51 records whose quality of registered signal was satisfactory for statistical analysis. During eight observations—November 15, 1999; April 19, 2000; August 6, 2000; October 20, 2000; November 21, 2000; November 30, 2000; May 16, 2001; and September 26, 2001—the ionosphere was strongly perturbed by the natural sources (see Table 3.1). Therefore, the rest of the 43 observations were analyzed statistically to obtain pure effects from RLs. Using changes in signal characteristics, we registered delay of reaction of the environment, Δt, and computed

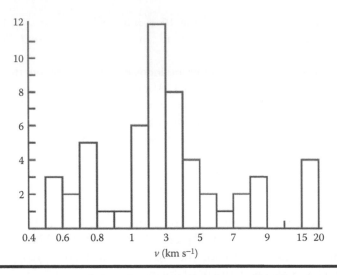

Figure 3.5 **Histogram of virtual velocities for plasma disturbance propagation that accompanies SV launch (the scale along the horizontal axis is not equally regular).**

noncorrelated values of velocities, $v = R_0/\Delta t$, where R_0 is the distance from the cosmodrome and the observation setup location. Such velocities will be defined as "virtual" (e.g., "seeming"). The histogram of these velocities is shown in Figure 3.5. It is seen that four groups of velocities can be differentiated: 0.5–0.8, 1–6, 7–9, and 15–20 km s⁻¹.

The first group of velocities was observed only during launch from the Plesetsk and Baikonur cosmodromes—10 cases in total. The second group of velocities, from 1 to 6 km s⁻¹, was registered in 32 cases mostly during launches from cosmodromes in the United States and France. For this group the average velocity was $v = 3.1 \pm 0.2$ km s⁻¹. The third group with velocities from 7 to 9 km s⁻¹ was registered in only five cases. Only in four cases was the fourth group with velocities from 15 to 20 km s⁻¹. We should note that the velocities of the last two groups were rarely observed only during launch from cosmodromes that ranged far from the observation points at about 10,000 km.

3.3.2 Passing of the Solar Terminator

As was mentioned earlier, 31 events were analyzed statistically that related to the ionosphere strongly perturbed by natural sources. Essential changes in signal spectral characteristics accompanied by passing of ST were recorded in all cases. These changes were related to the increase in magnitude of the constant component, f_{d0}, Doppler frequency shift during morning time and

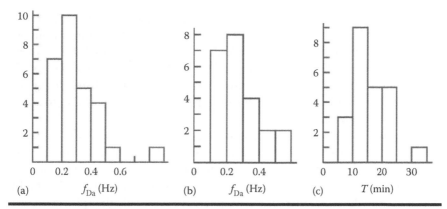

Figure 3.6 Histograms of changes in the magnitude of constant component of Doppler frequency shift (a) for amplitudes, (b) for periods of quasiperiodical variations of the central frequency in DS, and (c) when accompanied by the passing of the solar terminator.

its decrease during evening time accompanied by generation of the quasiperiodical disturbances in the ionospheric plasma, with the occurrence in some cases of the two-mode and the multimode shaped signals, and finally, with distortion of DS.

Histograms of the absolute values $|f_{d0}|$, DS amplitudes, f_{da}, and the periods T of quasiperiodical oscillations are shown in Figure 3.6. It can be seen that the $|f_{d0}|$ is changed from 0.1 to 0.9 Hz with the average value $|f_{d0}| \approx 0.27 \pm 0.02$ Hz. The average duration of f_{d0} was about 1–2 h. The histogram of oscillation period T has a sharp maximum at the range from 10 to 15 min with the average value $\bar{T} \approx 15.2 \pm 1.1$ min. The amplitude of the quasiperiodical variations, f_{da}, of the DS is changed from 0.1 to 0.6 Hz, with the average value $\bar{f}_{da} \approx 0.28 \pm 0.03$ Hz.

3.4 Transportation of Perturbations—Types of Waves

As was shown earlier, RLs are accompanied by large-scale global perturbations in the ionosphere. The questions that remain are the following: How these disturbances are generated? In what manner are they transferred to distances up to 1,000–10,000 km? What is the magnitude of such disturbances?

As is well known [9–12], the atmosphere at altitudes up to 100 km can be considered a comparably dense environment. Here, the work of rocket engines and movement of SVs are accompanied with generation of shock-acoustic wave (SAW). In its power, P_a may be converted to about 10^{-3}–10^{-2} parts from the total power of the reactive jet of fuel. The power of the latter for heavy and super-heavy rockets is $10^{10}-10^{11}\,\text{W}$, respectively. Therefore, we can estimate that $P_{\text{a}} \approx 10^{8}-10^{9}\,\text{W}$. The optimal altitudes for generation of SAW range from 100 to 130 km. At the same time, this region of altitudes lies above the mesospheric waveguide, which effectively captures and channels acoustic-type waves [13–17], and here the atmosphere is sufficiently dense. On the other hand, heavy and super-heavy rockets move here with super speed of sound that relates to conditions that generate shock waves.

Moreover, within the above height range, the rocket trajectory is close to horizontal. The truth of this statement was proved by observations [9] and computations [10]. It is important to note that at the region of its generation SAW yields the relative changes of electron concentration by about 10%–30% [9]. The shock wave in the atmosphere and ionosphere is a source of atmospheric-acoustic waves (AAWs) [18–21] and the acoustic gravity waves (AGWs) [11,12].

Further, both super-sound broadening of plasma by a reactive stream of rocket fuel and super-sound movements of plasma caused by SVs in gyrotropic ionosphere generate electromagnetic and magnetohydrodynamic (MHD) waves.

We will consider all types of waves further in more detail.

3.4.1 Acoustic Gravity Waves

For propagation of AGW in isothermally weak inhomogeneous atmosphere, the dispersion equation is as follows:

$$\omega^4 - \omega^2 v_{\text{s}}^2 \left(k_x^2 + k_z^2 + \frac{1}{4H^2} \right) + (\gamma - 1) g^2 k_x^2 = 0, \tag{3.1}$$

where:

ω is the wave frequency

k_x and k_z are the projections of the wave vector, where the z-axis is directed vertically and the x-axis is directed horizontally

$H = kT_{\text{a}} / M_{\text{g}}$ is the scale height of the atmosphere

k is Boltzmann's constant

M is the mass of molecule

g is the acceleration of the free fall

$\gamma = 1.4$ is the adiabatic parameter

T_{a} is the temperature of the atmosphere (in Kelvin)

Here, the speed of sound can be presented as [12]

$$v_{\text{s}} = \left(\gamma \frac{kT_{\text{a}}}{M} \right)^{1/2}. \tag{3.2}$$

From Equation 3.1 the expression for the refractive index of AGW is as follows:

$$n^2 = \frac{\omega_A^2 - \omega^2}{\omega_B^2 \cos^2 \alpha - \omega^2} = \frac{\beta_0^2 Y - 1}{Y \cos^2 \alpha - 1}, \tag{3.3}$$

where:

$\beta_0^2 = \omega_A^2/\omega_B^2 = \gamma^2/4(\gamma - 1) \approx 1.225,\ Y = (\omega_B/\omega)^2$

α is the angle between the x-axis and the wave vector

$$\omega_A = \left(\frac{\gamma g}{4H} \right)^{1/2} \tag{3.4}$$

is the limit (e.g., minimal) frequency of AAW

$$\omega_B = \left(\frac{(\gamma - 1)g}{\gamma H} \right)^{1/2} \tag{3.5}$$

is the Brunt–Väisälä frequency.

As defined in Ref. [21], this is the natural frequency of unforced oscillation of a parcel of air. For $\omega > \omega_A$ the waves relate to acoustic waves, and for $\omega < \omega_B \cos \alpha$ to inner gravity waves (IGWs) [19,20].

For large-scale AGWs with horizontal wavelengths of 1000 km or more, horizontal velocities of 500–1000 m s^{-1}, typical periods of 60 min and more, the wave frequency is much smaller than the Brunt–Väisälä frequency, that is, $\omega \ll \omega_B$. For such waves the wave fronts are almost horizontal. Therefore, the study of large-scale AGWs is relatively simple and many critical questions are answered easily by simple theory [13–15].

The problem with medium-scale AGWs is more complicated. These waves have horizontal wavelengths of several hundred kilometers, horizontal velocities between 100 and 250 m s^{-1}, and periods less than 1 h. The closer the period of an AGW is to the oscillation period of the atmosphere, the closer is its horizontal phase velocity to the mean wind velocity; it is complicated to interpret the behavior of an AGW in terms of simple linear theory [17–21]. Thus, if the Brunt–Väisälä period increases with height, then an upward propagating wave with intrinsic frequency in the region of generation slightly less than ω_B can reach a height where $\omega = \omega_B$. Conversely, if there is a large wind shear in the background atmosphere, the effective period of a wave, propagating upward in a joint system with a traveling local mean wind, Doppler shifting to the Brunt–Väisälä period can occur, and propagation becomes impossible. As a result, medium-scale waves from the troposphere and stratosphere, which propagate upward, often cannot propagate in the mesosphere [21]. There are other alternative scenarios related to AGW propagation in the atmosphere that are out of the scope of this book and instead are fully described in Refs. [20,21].

Let us now describe large-scale disturbances occurring in the ionosphere that, in our opinion, are more important for the subject matter of this chapter. The most interesting for us is on the situation when the angle $\alpha \approx 0$. In this case,

$$n^2 = \frac{\beta_0^2 Y - 1}{Y - 1}, \qquad (3.6a)$$

$$v_{ph} = \frac{v_s}{n}, \qquad (3.6b)$$

$$v_g \equiv v_{gx} = v_s \frac{n(Y - 1)}{n^2 + \beta_0^2 Y - 2}, \qquad (3.6c)$$

where:

v_{ph} and v_g are the phase and group velocities of AGW, respectively

For sufficiently long IGW (for which $Y \gg 1$), we get

$$n = \beta_0, v_{ph} \approx v_g \approx \frac{v_s}{\beta_0} \approx 0.9 v_s. \qquad (3.7)$$

The results of computation of the main parameters of IGW are shown in Tables 3.2 and 3.3.

As is known, at $z \leq 100$ km altitude in the ionosphere, the ionized component of plasma is small with respect to the neutral component, and dynamic processes in this field are under the control of neutral gases, not the plasma. It is easy to show that variations of electron concentration N are connected to relative changes of neutral gas pressure p and concentration of neutral molecules (atoms) N_n via the following expression:

$$\frac{\Delta N}{N} \approx \frac{H}{L_N} \frac{\Delta p}{p} \approx \frac{H}{L_N} \frac{\Delta N_n}{N_n}, \qquad (3.8)$$

where:

$L_N = N |dz/dN|$ is the characteristic scale of changes of N

Temporal variations of ΔN can be registered by DR.

Table 3.2 Parameters of the Atmospheric Model and the IGWs[a]

z (km)	H (km)	v_s (m s⁻¹)	$\omega_B \times 10^2$ (s⁻¹)	T_B (min)
100	6.5	300	2.08	5.0
150	25	590	1.06	9.9
200	40	740	0.84	12.5
250	50	830	0.75	14.0
300	60	910	0.68	15.4
350	63	930	0.67	15.7
400	65	940	0.66	16.0

[a] $\alpha = 0$.

Table 3.3 Dependence of v_{ph} and v_g on T_B/T[a]

T_B/T	v_{ph}/v_s	v_g/v_s
0.9	0.68	0.20
0.8	0.78	0.47
0.7	0.83	0.64
0.6	0.86	0.75
0.5	0.88	0.81
0.4	0.89	0.85
0.3	0.90	0.88
0.2	0.90	0.89
0.1	0.90	0.90
0.0	0.90	0.90

[a] $\alpha = 0$.

3.4.2 Slow MHD Waves

It is well known that fully ionized plasma are excited and propagate MHD waves with a characteristic velocity which equals (see, e.g., Ref. [22])

$$v_a = \frac{B_0}{\sqrt{\mu_0 \rho_p}} = \frac{B_0}{\sqrt{\mu_0 N M_i}}, \tag{3.9}$$

where:

B_0 is the induction of the ambient magnetic field
μ_0 is the magnetic permeability
M_i is the mass of the ion
ρ_p is the density of plasma

In the weakly ionized plasma the analog of MHD waves is slow MHD (SMHD) waves, introduced in Ref. [23]. The characteristic velocity of these waves is

$$v_{as} = \frac{B_0}{\sqrt{\mu_0 \rho}} = \frac{B_0}{\sqrt{\mu_0 N_n M}} = v_a s^{1/2}, \qquad (3.10)$$

where:

ρ is the gas density
N_n is the concentration of neutral molecules (atoms)
$s = N/N_n$ is the degree of plasma ionization

At all altitudes of the ionosphere, $s \ll 1$ and $v_{as} \ll v_a$. The results of computation of v_{as} are shown in Table 3.4.

Table 3.4 Parameters of the Atmospheric and Ionospheric Model as Also Characteristic Velocity of SMHD Waves for a Day and a Night

z (km)	ρ (kg m^{-3})	s	v_{as} (km s^{-1})
50	10^{-3}	10^{-15} (10^{-17})	1.6×10^{-3}
100	10^{-6}	10^{-8} (10^{-9})	5×10^{-2}
120	10^{-7}	10^{-7} (10^{-8})	0.16
150	10^{-8}	10^{-6} (10^{-7})	0.5
200	10^{-9}	10^{-5} (10^{-6})	1.6
250	3×10^{-10}	10^{-4} (10^{-5})	3
300	10^{-10}	10^{-3} (10^{-4})	5
350	3×10^{-11}	3×10^{-3} (10^{-3})	10
400	10^{-11}	10^{-2} (5×10^{-3})	16

Note: The nighttime values are given in parenthesis.

   Expression for the refractive index of SMHD waves can be presented in the following form [23]:

$$\tilde{n}_{\pm} = \frac{c}{v_{as}}\left(1 \pm \alpha - i\beta\right), \tag{3.11}$$

where:
   c is the speed of light in vacuum
   $\alpha = \omega/s\Omega_B, \beta = \omega/sv_1,$
   ω is the wave frequency
   $v_1 = v_{ni}/(1 + v_{in}v_e/\omega_B\Omega_B)$
   ω_B and Ω_B are the gyrofrequencies of electrons and ions, respectively
   v_e is the collision frequency of electrons with ions (v_{ei}) and with molecules
       (v_{en}), v_{in} is the collision frequency of ions with molecules
   $v_{ni} = v_{in}s$
   The $\pm$ sign corresponds to two types of wave polarization.

From Equation 3.11, it is possible to obtain expressions for the refractive index

$$n_{\pm} - \frac{c}{\sqrt{2}v_{as}}\frac{a_{1\pm}}{a_{2\pm}} \tag{3.12a}$$

and for the coefficient of absorption

$$\kappa_{\pm} = \frac{c}{\sqrt{2}v_{as}}\frac{\beta}{a_{1\pm}a_{2\pm}}, \tag{3.12b}$$

where

$$a_{1\pm} = \left(1 \pm \alpha + a_{2\pm}\right)^{1/2}, \; a_{2\pm} = \left[\left(1 \pm \alpha\right)^2 + \beta^2\right]^{1/2}.$$

For computation of the main characteristics of SMHD waves, phase velocity, $v_{ph\pm}$, depth of attenuation, $L_{\pm} = c/\omega\kappa_{\pm}$, and parameters of the ionospheric model were used; these are presented in Table 3.5.
   The results of computation of SMHD wave characteristics for two altitudes, 150 and 250 km, are presented in Tables 3.6 and 3.7, respectively. These altitudes were selected because most contributions in Doppler effects for vertical sounding give an ionospheric layer in the proximity of the region of reflection [5]. For frequencies of radio waves ranged from 2.8 to 3.5 MHz, this region lies approximately at heights of 150 and 250 km in diurnal and nocturnal time, respectively.
   From Tables 3.6 and 3.7, it can be seen that plasma disturbances can be transported to distances ~1000 km for $\omega \leq 10^{-2}$ Hz (periods of $T \geq 10$ min)

Table 3.5 Parameters of the Ionospheric Plasma for a Day and a Night

z (km)	$\nu_{in}(s^{-1})$	$\nu_e(s^{-1})$	$\nu_i(s^{-1})$	$\Omega_B(s^{-1})$	$\omega_B(s^{-1})$
50	3×10^7	7×10^7 (7×10^7)	2.1×10^{-14} (2.1×10^{-16})	170	8.8×10^6
100	1.2×10^4	5×10^4 (4×10^4)	8.5×10^{-5} (0.9×10^{-5})	170	8.8×10^6
120	10^3	10^4 (6×10^3)	10^{-4} (10^{-5})	170	8.8×10^6
150	10^2	1.5×10^3 (8×10^2)	10^{-4} (10^{-5})	180	8.8×10^6
200	12	6×10^2 (10^2)	1.2×10^{-4} (1.5×10^{-5})	250	8.8×10^6
250	4	7×10^2 (40)	4×10^{-4} (4×10^{-5})	320	8.8×10^6
300	1.5	8×10^2 (1.2×10^2)	1.5×10^{-3} (1.5×10^{-4})	330	8.8×10^6
350	0.6	7×10^2 (2×10^2)	2×10^{-3} (6×10^{-4})	335	8.8×10^6
400	0.3	6×10^2 (2.7×10^2)	3×10^{-3} (1.5×10^{-3})	340	8.8×10^6

Note: The nighttime values are given in parenthesis.

Table 3.6 Characteristics of SMHD Waves at an Altitude of 150 km for Diurnal Time Periods

ω (Hz)	v_{ph+} (km s^{-1})	v_{ph-} (km s^{-1})	L_+ (km)	L_- (km)
10^{-1}	19	33	330	200
10^{-2}	6	10	1000	630
10^{-3}	2	3	3600	1900
10^{-4}	0.7	0.6	25,000	94,000

and to global distances for $\omega \le 10^{-3}$ Hz (periods of $T \ge 100$ min) with the help of SMHD waves. In this case, the phase velocity equals 6–10 km s^{-1} and less.

3.4.3 Gyrotropic Waves

These waves were introduced in Ref. [23]. Naturally, they are crossing electromagnetic waves in the gyrotropic plasma with periods of $T \sim 10^1 - 10^4$ s. The magnetic component of such a wave yields variations of the geomagnetic field on the Earth's surface with magnitudes of $\Delta B \approx 1-10$ nT [23]. Gyrotropic waves

Table 3.7   Characteristics of SMHD Waves at an Altitude 250 km for Diurnal and Nocturnal Time Periods

ω (Hz)	v_{ph} (km s^{-1})	L (km)
10^{-1}	66	660
10^{-2}	21	2100
10^{-3}	6	8600
10^{-4}	3	24,000

propagate horizontally in the natural waveguide at the *E*-layer altitudes of the ionosphere ($z \approx 100$–130 km) practically without attenuation. The refractive index of gyrotropic waves can be presented as [24]:

$$n = \frac{c}{v_{gt}\cos^2\alpha}\left[x(1 + \coth x)\right]^{-1/2}, \tag{3.13}$$

where:

α is the angle between the wave vectors $\mathbf{k}_{gt}$ and $\mathbf{B}_0$

$x = \left|\mathbf{k}_{gt}\right| z_0$, z_0 is the height of the waveguide with depth *l* above the Earth's level

$$v_{gt} = \frac{1}{\mu_0 \sigma_H \sqrt{z_0 l}} \tag{3.14}$$

is the characteristic velocity of gyrotropic waves

$$\sigma_H = \tilde{\sigma}_H \varepsilon_0, \tilde{\sigma}_H = \frac{\omega_p^2 \omega_B}{\omega_B^2 + v_{en}^2} + \frac{\Omega_p^2 \Omega_B}{\Omega_B^2 + v_{in}^2}, \tag{3.15}$$

where:

σ_H is Hall's conductivity

ε_0 is the electrical permittivity

At *E*-layer altitude, the effects of ions in σ_H can be neglected. Moreover, here $\omega_B^2 \gg v_{en}^2$. Then,

$$\tilde{\sigma}_H \approx \frac{\omega_p^2}{\omega_B}, \sigma_H = \varepsilon_0 \frac{\omega_p^2}{\omega_B}. \tag{3.16}$$

In this case,

$$v_{gt} = \frac{1}{\mu_0 \varepsilon_0 \tilde{\sigma}_H \sqrt{z_0 l}} = \frac{c^2}{\tilde{\sigma}_H \sqrt{z_0 l}} = \frac{c^2 \omega_B}{\omega_p^2 \sqrt{z_0 l}}. \tag{3.17}$$

Assuming $z_0 \approx 120$ km, $l = 30$ km, $\omega_B = 8.8 \times 10^6 \, s^{-1}$, $\omega_p^2 = (3.2 \times 10^{13}) - (3.2 \times 10^{14}) s^{-2}$, ($N = 10^{10} - 10^{11} m^{-3}$ for nocturnal and diurnal ionosphere, respectively), we get $v_{gt} \approx 410 \, km \, s^{-1}$ (nighttime) and $v_{gt} \approx 41 \, km \, s^{-1}$ (daytime).

Let us consider Equation 3.13 for two limited cases. Thus, for $x \ll 1$, that is, for large period T, $\coth x \approx 1/x$, and we get

$$n \approx \frac{c}{v_{gt} \cos^2 \alpha}, v_g \equiv v_{ph} = \frac{c}{n} = v_{gt} \cos^2 \alpha. \tag{3.18}$$

For too small T we get $x \gg 1$, $\coth x \approx 1$, and

$$n \approx \frac{c}{\sqrt{2x} v_{gt} \cos^2 \alpha}. \tag{3.19}$$

Because x does not depend directly on n, we finally get:

$$n \approx \frac{c}{\left(2 z_0 v_{gt}^2 \omega \cos^4 \alpha\right)^{1/3}}, \tag{3.20a}$$

$$v_{ph} = \left(2 z_0 v_{gt}^2 \omega \cos^4 \alpha\right)^{1/3}, \tag{3.20b}$$

$$v_g = \frac{3}{2} v_{ph}. \tag{3.20c}$$

Using well-known relation, $\omega = 2\pi/T$, we can rewrite Equation 3.20b as follows:

$$v_{ph} = \left(\frac{4\pi z_0}{T} v_{gt}^2 \cos^4 \alpha\right)^{1/3}. \tag{3.21}$$

For some critical value $T = T_{cr}$ estimations of v_{ph} using Equations 3.20b and 3.21 should be on the same order. This takes place for

$$T_{cr} = \frac{4\pi z_0}{v_{gt} \cos^2 \alpha}. \tag{3.22}$$

Thus, for propagation along meridian waves above Kharkov (Ukraine), where the inclination angle $\alpha \approx 66.4°$, we get from Equation 3.22: $T_{cr} = 10 \, s$ (night)

Table 3.8 Phase Velocity of Gyrotropic Waves for $T \gg T_{cr}$ and Hantadze Waves in the E-Layer of the Ionosphere

$N \times 10^{-10}$ (m^{-3})	v_{ph} $(km\ s^{-1})$			v_H $(km\ s^{-1})$
	$\alpha = 0$	$\alpha = 30$	$\alpha = 60$	$\alpha = 90$
0.6	680	510	170	4.7
0.8	513	385	128	3.5
1	410	308	103	2.8
2	205	154	51	1.4
4	103	77	26	0.7
6	64	48	16	0.5
8	51	38	13	0.4
10	41	31	10	0.3
20	21	16	5	0.1

and $T_{cr} = 100$ s (day). For Doppler sounding, Equation 3.20a is useful because such a method allows detection of wave processes with $T > 1–2$ min. The results of the computation of phase velocity of gyrotropic waves for various values of N in the ionospheric E-region are presented in Table 3.8. It is important to note that for these types of waves the dispersion and attenuation processes are practically absent.

3.4.4 Magnetic-Gradient Waves

A new branch of the excited oscillations of the magnetically active ionosphere was studied in Refs. [25,26]. The existence of these types of waves relates to their occurrence in the dynamo region of the ionosphere ($z \approx 100–130$ km), the latitudinal gradient of the geomagnetic field vertical projection, $\partial B_{0z}/dy$, where the y-axis is directed along the corresponding meridian. Due to the passing of these waves in the E-layer of the ionosphere, only electrons oscillate with magnetic strength lines freezing into the electronic gas. Expression for the phase velocity of the magneto-gradient waves (Hantadze waves) propagating along the parallel, can be presented in the following form:

$$v_H = \left(\frac{c}{\omega_p}\right)^2 \left|\frac{\partial \omega_{Bz}}{\partial y}\right| \approx \left(\frac{c}{\omega_p}\right)^2 \frac{\omega_B}{L_B}, \qquad (3.23)$$

where:

$L_B = |B_0 \partial y / \partial B_{0z}|$ is the characteristic scale of changes of the vertical component B_{0z}

Equation 3.23 can be rewritten as follows:

$$v_H = (\mu_0 e N)^{-1} \left| \frac{\partial B_{0z}}{\partial y} \right| \approx (\mu_0 e N L_B)^{-1} B_0. \qquad (3.24)$$

Let us estimate L_B. Accounting for the fact that at the equator $B_{0z} \approx 0$ and near the polar cap $B_{0z} \approx 6 \times 10^{-5}$ T, we get $|\partial \omega_{Bz}/\partial y| \approx 1 \text{ m}^{-1}\text{s}^{-1}$ and $L_B \approx 10^7$ m. The results of estimation of the Hantadze waves are shown in Table 3.8, which shows that v_H can achieve magnitudes of 0.1–0.5 and 0.5–5 km s^{-1} for diurnal and nocturnal times, respectively.

The amplitude of the magnetic component of Hantadze waves equals:

$$\Delta B \approx \frac{e N \mu_0}{k_x} v_e = \frac{v_e T}{2\pi} \frac{\partial B_{0z}}{\partial y} = \frac{B_0}{2\pi} \frac{v_e T}{L_B}, \qquad (3.25)$$

where:

$v_e = E_i / B_0$ is the drift velocity of plasma electrons
$k_x = 2\pi/\lambda$, $\lambda = v_H T$ is the wavelength of Hantadze's wave
T is its period

For $E_i = 1 \text{ mV m}^{-1}$, we get $v_e = 20 \text{ m s}^{-1}$, $\Delta B = 2$–200 nT for $T = 10^2$–10^4 s.

3.5 Discussion of Results

3.5.1 Passing of Solar Terminator

The following effects are observed for the action of the ST in the ionosphere:

- Shifting of the constant component of Doppler frequency, f_{d0}; as a rule, $f_{d0} < 0$ (during the evening) and $f_{d0} < 0$ (during the morning).
- Generation and amplification of quasiperiodical disturbances in the ionosphere with varying periods from 10 to 30 min
- Multiray phenomenon, spreading and distortion of DSs

During sunrise, the reaction of the ionosphere is frequently started 40–50 min prior to the passing of ST at the height of reflection of the sounding radio wave. Perhaps, this phenomenon takes place due to the incoming wave disturbances (WDs), which arrive faster than the beginning of ST at *F*-layer altitudes in the ionosphere. In fact, the velocity of the latter at the latitude of the area around the Radio Physical Observatory (RFO) (Kharkov, Ukraine) was about 300 m s⁻¹, whereas the phase and group velocities of IGWs, which travel in before ST, are smaller than v_s for $T > T_B$ (see Table 3.3). As is well known, for $z \leq 200–250$ km the IGWs are weakly attenuated and their amplitude decreases twice at distances of several thousand kilometers [12].

At the same time, the beginning of the ionospheric reaction on the passing of the ST had already been observed ~50 min after sunset at the ground level, that is, much earlier than this event occurs at the region of wave reflection. Variations of DS were registered also for 50–60 min after passing of the ST through the sounding wave region of reflection. We suppose that this phenomenon is related to incoming IGWs generated before ST.

The average period of observed oscillations of central frequency inside the DS was about 15 min. This value slightly exceeded the Brunt–Väisälä period (see Equation 3.5):

$$T_B = \frac{2\pi}{\sqrt{\gamma - 1}} \frac{v_s}{g}, \tag{3.26}$$

whose computation results are given in Table 3.2.

3.5.2 Rocket Launches

The effects accompanying SV flights differ due to their variety. They depend on the distance to the SV trajectory, its height, type of fuel, and, of course, on the power of its engines, as well as on the conditions in the ionosphere [2,3,6,7].

Let us return again to Soyuz-V flight (see Figure 3.1). Changes in DS variations occurred at 21:00 and 22:00 h. If we suppose that they were affected by SV, then the corresponding time delays of $\Delta t_0 \approx 10$ min and $\Delta t_1 \approx 60$ min should be related to them. Radio waves with frequency 2.8 MHz at time interval 21:00–22:00 h were reflected at 160–180 km altitudes. If we assume that the disturbance of the source propagates nearly horizontally with respect to the ground surface at the same altitude, then the time of the upswing of the rocket, δt, at the height of 170 km is about 5 min. The corrected values were $\Delta t_0' \approx 5$ min and $\Delta t_0' \approx 55$ min. For time $\delta t \approx 5$ min the SV moves away in the east direction from the cosmodrome at about 500 km (the distance to the place of registration was $R \approx 1700$ km). Thus, $v_0' = R/\Delta t_0' \approx 5.7$ km s⁻¹ and $v_1' = R/\Delta t_1' \approx 515$ m s⁻¹. The first value of the velocity is close to the velocity of SMHD waves with frequency $10^{-3}–10^{-2}$ Hz (time period 10–100 min). For these types of waves, the depth of attenuation is about several thousand kilometers (see Tables 3.6 and 3.7).

The second value of the velocity corresponds to the speed of sound in the atmosphere for $T_a \approx 610\,\text{K}$ (see Equation 3.2). This is a typical value of the temperature in the ionosphere at altitudes of 160–180 km during evening time.

Let us return to the Soyuz-V launch that happened on February 1, 2000. As mentioned earlier, significant changes of DSs took place at time intervals from 09:00 to 09:43 h and from 09:43 to 11:10 h. If they were caused by the SV, the delays of environmental reaction were $\Delta t_0 \approx 23\,\text{min}$ and $\Delta t_1 \approx 56\,\text{min}$. The radio wave with $f = 3.5$ MHz in the morning was reflected at an altitude of about 170 km. Then, the corrected values of environmental reaction delays were $\Delta t_0' \approx 18\,\text{min}$ and $\Delta t_1' = \Delta t_1 - \delta t \approx 51\,\text{min}$. During time δt the rocket moves away from the cosmodrome to the east at a distance $\Delta R \approx 450$ km. Then the distance between the places of wave generation to the places of observation was $R = R_0 + \Delta R$, where $R_0 \approx 2050$ km is the distance from Baikonur cosmodrome to RFO in Kharkov. Now, for $R \approx 2500$ km the corrected velocities of WD propagation equal $v_0' = R/\Delta t_0' \approx 2.3\,\text{km s}^{-1}$ and $v_1' = R/\Delta t_1' \approx 820\,\text{m s}^{-1}$. The first value of the velocity also corresponds to the velocity of SMHD waves having $\omega \approx 10^{-4}\,\text{Hz}$. For these waves, $L_\pm \approx 1000–3600$ km in diurnal periods for altitude $z \approx 150$ km (see Table 3.6).

The second value of the velocity is close to speed of sound $i\ v_s\ n$ the ionosphere. This value of v_s in the year, close to the solar activity year, really can be observed at altitudes of 180–190 km in daytime.

A reasonable approach to v estimation shows that between September 9, 1999 and February 1, 2000, ionospheric reactions to SV launches were observed. The average duration of disturbances was close to 120 min. Oscillations accompanying SV launches have a quasiperiod of $T \approx 15$ min. This value, as it should be in reality, slightly exceeds the Brunt–Väisälä period. Thus, for $v_s = 515–683\,\text{m s}^{-1}$, we get $T_B \approx 8.7–11.5$ min.

During Space Shuttle Discovery launch the character of the signal changed (it became $f_d > 0$) with delays of $\Delta t_1 \approx 85\,\text{min}$ and $\Delta t_2 \approx 120\,\text{min}$. The SV achieved altitudes of 100 and 150 km during times $\delta t_1 \approx 7\,\text{min}$ and $\delta t_2 \approx 15\,\text{min}$, and ranged from the cosmodrome at distances of about 300 and 4000 km and the place of observations—at 10,000 and 6,000 km, respectively. Then, the velocities of the WD movement, $\Delta t_1' = \Delta t_1 - \delta t$ in the E-layer and the F-layer, were estimated as ~2.1 and ~1.4 km s^{-1}, respectively.

Perturbations caused by the SV flight continued up to 02:15 h. Then, another wave process was "hidden" into the first one as it was related to the arrival of the second wave process. As the result, the DSs were sharply broadened—up to 0.15 Hz. Such an ionospheric reaction was observed for not less than 40 min. SMHD waves appear in the F-layer of the ionosphere at a velocity of 1–3 km s^{-1}, if $\omega \le 10^{-4}$ Hz. In such conditions the depth of attenuation of waves reaches more than 10,000 km.

Let us discuss separately about oscillations with large amplitude Doppler frequency shift that was observed after 02:15 h. If $\Delta t_2 \approx 120\,\text{min}$, then $\Delta t_2' \approx 105$ min and $v' \approx 2.2\,\text{km}\,\text{s}^{-1}$. Such a velocity also has SMHD waves in the nocturnal F-layer of the ionosphere $\omega \leq 10^{-4}\,\text{Hz}$.

We will now check the possibility of transporting the disturbances by Hantadze waves, which propagate along latitudes [25,26]. Suppose that the rocket was launched to the south-east from Cape Canaveral cosmodrome, it can attain an altitude of 130 km approximately at the end of 12th minute, for which the SV was away from the cosmodrome at a distance of about 2,350 km. The distance to the place of observation was $R = 7,650$ km. The rocket speed v was about 1.2 km s^{-1}; approximately, the same velocity that Hantadze waves have at evening time periods when the plasma concentration in the E-layer of the ionosphere is $N \approx 3 \times 10^{10}\,\text{m}^{-3}$. Note that the SV launch happened at about 16:00 h (local time of the cosmodrome).

We suppose that in these measurements, the WDs could be transported by SMHD waves and by Hantadze waves. However, a disadvantage with the latter waves is that they can propagate only along latitudes. The large amplitude of oscillations could be related to the fact that after 02:15 h, the reaction on SV launch and flight was amplified by the effects of the passing morning terminator.

Atlas 2AS launch (see Table 3.1) was accompanied by disturbances with time delays of $\Delta t_0 \approx 20\,\text{min}$ and $\Delta t_1 \approx 70\,\text{min}$ (see Figure 3.4). If these related to the working of the rocket engines, then we need to subtract the time of SV movement to the corresponding region of the ionosphere from these delays. Thus, for $z \approx 100$ km, $\delta t \approx 4$ min and $\Delta R \approx 200$ km. In this case $v_0' = R/\Delta t_0' \approx 10\,\text{km}\,\text{s}^{-1}$, if disturbances propagate in the E-layer of the ionosphere (the wave was reflected at an altitude of about 150 km). For velocity $v_1' = R/\Delta t_1'$, we get ~2.5 km s^{-1} for propagation in the same region of the ionosphere.

The velocity, $v_0' \approx 10\,\text{km}\,\text{s}^{-1}$, differs only slightly from that of gyrotropic waves in the diurnal E-region of the ionosphere, where it equals about 10 km s^{-1} for concentration $N \approx 10^{11}\,\text{m}^{-3}$ and angle $\alpha = 60°$. The value $v_1' \approx 2.5\,\text{km}\,\text{s}^{-1}$ is close to the velocity of SMHD waves (propagating with $v \approx 2–3$ km s^{-1} at an altitude of about 200 km (see Tables 3.6 and 3.7), if $\omega \approx 10^{-3}\,\text{Hz}$ ($T \approx 100$ min). Just for frequencies such as ω, the wave can reach the place of observation, that is, the constraint $L_\pm \geq 10,000$ km should be satisfied.

It is not possible to state confidently that we note a disturbance with $\Delta t_1 \approx 70\,\text{min}$ because the reaction caused by the SV flight was masked by a reaction caused by the passing of the evening terminator. Meanwhile, there is a reason for the transport of the observed disturbances by gyrotropic and SMHD waves.

During the SV launch from Sriharikota cosmodrome (India) on April 18, 2001, $\Delta t_1 \approx 33\,\text{min}$. Thus, for $R_0 \approx 5,600$ km we get $v_1 \approx 2.8\,\text{km}\,\text{s}^{-1}$. For $\delta t \approx 5$ min we get $v_1' \approx 3.3\,\text{km}\,\text{s}^{-1}$. This velocity is close to the velocity of SMHD waves.

All data mentioned earlier were in agreement with those described in Ref. [27]. Here, we add that in 43 cases of SV launches, confidence reaction was observed in all cases.

Let us now discuss histograms of "virtual" velocities. As mentioned earlier, four groups of velocities were observed: (1) 0.5–0.8; (2) 1–6; (3) 7–9; and (4) 15–20 km s^{-1}. It is without doubt that the first group is related to SAWs and AGWs.

The average velocity was $\bar{v} \approx 3.1 \pm 0.2$ km s^{-1} in the second group. Such a velocity will have slow MHD waves with frequencies $\omega \approx 10^{-3}$ Hz (period $T \approx 100$ min; see Tables 3.6 and 3.7). Hantadze waves also have a close value at nocturnal time periods, but they propagate only along latitudes.

Velocities of the third group correspond to gyrotropic waves propagating in diurnal time periods when $N \approx (1-1.5) \times 10^{11}$ m^{-3} in the E-layer of the ionosphere. Real velocities $v' \approx 30$ km s^{-1} for $R \approx 10,000$ km and $\delta t \approx 5$ min correspond to virtual velocities $v = 15-20$ km s^{-1} (with time delays of 10 min). The magnitudes of v' are also close to gyrotropic waves when N was decreased up to $N \approx 4 \times 10^{10}$ m^{-3} (at morning and evening time periods) in the E-layer of the ionosphere. Everything mentioned previously is correct for $\alpha = 60°$. We note that the inclination of the geomagnetic field is 66.4° at the RFO in Kharkov, Ukraine. Therefore, velocities of 7–9 and 15–20 km s^{-1} should be related to the same group that corresponds to the velocities of gyrotropic waves for different time periods round the clock.

3.5.3 Magnitude of Disturbance

Passing of the ST. First of all, we consider passing of the ST. Usually, the duration of ST was not less than 1–2 h. The time of passing of the terminator through the beam diagram of the radar that was 200 km wide with about 300 m s^{-1} velocity was approximately 700 s. To this time period, we need to add the lifetime of electrons caused by the photochemical processes occurring in the ionosphere. Thus, at 150–200 km altitudes it equals 300–500 s [28–30]. Then, the characteristic time of plasma concentration N formation in the proximity of the radio wave reflection can be estimated as $t_r \approx 1000-1200$ s. If so, we will use the formula to estimate f_d [5]:

$$f_d = \frac{f}{c} \frac{L_g}{t_r}, \qquad (3.27a)$$

where:

L_g is the group path in the region where the refractive index n significantly differs from the unit (its depth approximately equals L_N defined earlier)

r is the index, which indicates that the considered parameter relates to the region of reflection

Assuming $f = 3$ MHz and $L_g \approx 30$ km, we get $f_d \approx 0.25$–0.3 Hz. The observations gave close results with the average value of $\bar{f}_d \approx 0.27 \pm 0.02$ Hz.

Doppler frequency shift caused by the passing of the ST can be explained also as a shift caused by movements of the reflection region with velocity v_r. In this case,

$$f_d = \frac{2 f v_r}{c}. \tag{3.27b}$$

From Equation 3.27b, it follows that for $f_d = 0.3$ Hz and $f = 3$ MHz, we get $v_r \approx 15$ ms^{-1}. This means that for time t_r the region of reflection shifted at 15 km.

Effects of SV launches. Let us consider the effects of SV launches. The sensitivity of Doppler approach attains a value of about 0.1 Hz that corresponds to the minimum values of plasma density variations $(\Delta N/N)_{min} \approx (5 \times 10^{-3}) - (5 \times 10^{-2})$ $(T \approx 10^2 - 10^3 \text{s})$ recorded by the DR. The plasma density disturbance ΔN can be derived from the equation of continuity:

$$\frac{\partial N}{\partial t} + \text{div}\left(N\mathbf{v}_d\right) = 0, \tag{3.28}$$

where:
$\mathbf{v}_d$ is the speed of plasma movement

For propagation of electromagnetic waves in the ionosphere, we get for the absolute value of vector $\mathbf{v}_d$:

$$\left|\mathbf{v}_d\right| \equiv v_d \approx \frac{\Delta E}{B_0}, \tag{3.29}$$

where:
ΔE is the strength of the electrical component of the total field yielding the drift of plasma electrons in the crossing $\Delta \mathbf{E} \times \mathbf{B}_0$ fields

Replacing derivatives in Equation 3.28 by their estimations, we get

$$\frac{\Delta N}{N} \approx \frac{v_{dz} T}{2 \pi L_N}, \tag{3.30}$$

where:
v_{dz} is the vertical projection of the drift velocity $\mathbf{v}_d$

From Equation 3.30, we get:

$$v_{dz} = 2\pi \frac{\Delta N}{N} \frac{L_N}{T}. \tag{3.31}$$

For typical $L_N \approx 10\text{--}100$ km we get $v_{dz\,min} \approx 3\text{--}30\ \mathrm{m\,s^{-1}}$. Such drift velocities, as follows from Equation 3.29, can be realized only for $\Delta E_{min} \approx (1.5 \times 10^{-4}) - (1.5 \times 10^{-3})\mathrm{V\,m^{-1}}$. Because for SMHD, gyrotropic, and Hantadze waves $n \approx 10^4\text{--}10^5$, the induction of the magnetic component of radio waves $\Delta B_{min} \approx \Delta E_{min} n/c = \Delta E_{min}/v_{ph} \approx (1.5 \times 10^{-10}) - (1.5 \times 10^{-8})\mathrm{T}$ happens. Magnetic variations with amplitude $\Delta B \approx 0.1\text{--}10\ \mathrm{nT}$ do not seem to be exotic. Let us explain this.

The velocity of the reactive flow from the engine fuel is about 3–4 km s^{-1}. In the direction of the fuel exit, the jet spreads with a thermal velocity on the order of ~1 km s^{-1} (the temperature of gases in the fuel flow is about 3000–4000 K). It is important to note that these velocities exceed the velocities of acoustic waves, v_s and v_{as}. As a result of such balanced relation of velocities, shooting up of gas pressure, Δp, electronic concentration, ΔN, and magnetic field, ΔB, are generated. The fuel jet causes accelerated movement of the ionospheric plasma. Electrons at E-layer altitudes are magnetized whereas ions are not [29]. This yields the current in plasma, which results in generation of disturbances of the geomagnetic field disturbances, ΔB.

From the second Maxwell equation for the curl vector of magnetic field **B**, it follows that the amplitude of its pulsations is

$$\Delta B = \mu_0 j l_p, \tag{3.32}$$

where:

j is the current density caused basically by the movement of electrons
l_p is the characteristic scale of the perturbed region

Because

$$j = \sigma_H E_\perp = \sigma_H v_\perp B_0 \approx \varepsilon_0 \frac{\omega_p^2}{\omega_B} v_\perp B_0, \tag{3.33}$$

we get for ΔB

$$\Delta B = \varepsilon_0 \mu_0 \frac{\omega_p^2}{\omega_B} v_\perp l_p B_0 = \frac{\omega_p^2 v_\perp l_p}{c^2 \omega_B} B_0, \tag{3.34}$$

where:

$v_\perp$ is the magnitude of the velocity component of carriers across the magnetic field $\mathbf{B}_0$ affected by the movement of fuel jet

$E_\perp$ is the magnitude of the vector of the electric field in the plasma perpendicular to both $\mathbf{v}_\perp$ and $\mathbf{B}_0$ (i.e., $\mathbf{E}_\perp = \mathbf{v}_\perp \times \mathbf{B}_0$)

Other parameters were introduced earlier in Section 3.4. Assuming $v_\perp = 1\ \mathrm{km\,s^{-1}}$, $l_p = 3$ km, and $\omega_p^2 = (3 \times 10^{13}) - (3 \times 10^{14})\mathrm{s^{-2}}$ for nocturnal and diurnal time periods, we get respectively, $\Delta B = 5$–50 nT. Fields on such an order should be generated in the ionospheric plasma by engine fuel flows. Due to spreading of waves, their amplitude at the place of registration can be several times lesser. But as it occurs, it is sufficient for essential perturbation of plasma density N that can actually be recorded by the Doppler method.

As for generation of AGWs, the magnitude of Doppler frequency is on the order $f_d \approx 0.1$–0.2Hz, which corresponds to the magnitude of the particle velocity in the wave

$$v_z = \frac{1}{2} \frac{f_d}{f} c \qquad (3.35)$$

on the order of 5–10 m s^{-1}. Such values of v_z relate to real observed data.

Because $\Delta N_n/N_n \approx v_z/v_s$, from Equation 3.8 we get

$$\frac{\Delta N}{N} \approx \frac{H}{L_N} \frac{v_z}{v_s}, \qquad (3.36)$$

at an altitude of 150 km in diurnal time periods $L_N \approx 10$ km, $H \approx 25$ km, $v_s \approx 600\ \mathrm{m\,s^{-1}}$. Then, $v_z \approx 5$–10 m s^{-1}, we get $\Delta N/N \approx 2 \times 10^{-2} - 4 \times 10^{-2}$ (or 2%–4%). As mentioned earlier, such disturbances can be recorded and registered by Doppler method.

Let us summarize what we mentioned earlier.

■ Disturbances in the ionospheric plasma occur as the result of generation of SAWs, as well as waves of electromagnetic nature (such as MHD waves).

■ The super-sound movements of the SV and the reactive fuel flow cause the appearance of SAWs, as well as broadening of the ionized substance of fuel flow in the presence of the geomagnetic field.

■ All these effects cause quasiperiodical disturbances of plasma density ΔN, electric ΔE and magnetic ΔB fields with amplitudes of ~100%: $\Delta E \approx 5\ \mu\mathrm{V\,m^{-1}}$–5 mV m^{-1} and $\Delta B \approx 5$–50 nT in the proximity of the SV.

■ These disturbances propagate at the *E*- and *F*-layer altitudes in directions close to horizontal, decaying mostly due to spatial spreading of waves not more than on 1–2 orders.

■ Plasma perturbations with amplitudes of $\Delta N/N \approx 10^{-2}$ by using the Doppler method occur at the place of registration which is sufficient for their existence.

3.6 Main Results

1. As a result of the statistical analysis of variations of DSs that accompanied 43 SV launches, four groups of "virtual" velocities of plasma disturbance propagation in the ionosphere were found: 0.5–0.8, 1–6, 7–9, and 15–20 km s^{-1}. The reaction on SV launch was registered confidently and practically in all cases observed.

   The first group of velocities was observed during SV launches from Plesetsk and Baikonur cosmodromes. Weakly attenuated IGWs propagate with such velocities whose period was 10–15 min. These are strongly dispersive waves.

   The second group of velocities was registered during SV launches from cosmodromes in Russia, the United States, and France (where $R_0 \approx 2,000$–$10,000$ km). Approximately, the same velocities have SMHD and Hantadze waves. The phase velocity and the depth of attenuation (i.e., the distance of propagation) of the first waves depend significantly on the wave frequency. The velocity of propagation of Hantadze waves is determined basically by the magnitude of electron concentration in the *E*-layer of the ionosphere. These waves practically do not attenuate and are not dispersive.

   The third and the fourth groups (they in fact have the same group of velocities) appeared during SV launches at far distances from the cosmodromes—around ~10,000 km. Evidently, it is supposed that the gyrotropic waves transport plasma disturbances in this case. Their velocity significantly depends on daytime periods and on the orientation of the wave vector with respect to the geomagnetic field.

2. The amplitude of disturbances of the environmental parameters caused by the broadened reactive fuel flow of SVs was estimated. It is shown that these amplitudes can achieve the essential values. Finally, these features guarantee the disturbances in the ionosphere observed at distances on the order of 1,500–10,000 km.

3. Statistical analysis of 31 cases of ST passage shows that the essential reaction on its dynamics was registered in all cases of observation.

It registered increase (or decrease) of the magnitude of the constant component of Doppler frequency shift at morning time periods (or at evening time periods) for a value of 0.27 ± 0.02 Hz with a duration of 1–2 h, as well as generation of quasiperiodical disturbances in the ionosphere with the average period $T \approx 15.2 \pm 1.1$ min and a mean amplitude of the central frequency in DS oscillations of 0.28 ± 0.03 Hz.

References

1. Chernogor, L. F., Physics of the Earth, atmosphere and geospace in the lightness of system's paradigm, *Radiophys. Radioastron.*, 8, 59–106, 2003 (in Russian).
2. Garmash, K. P., Gokov, A. M., Kostrov, L. S., et al., Radiophysical investigations and modeling of processes in the ionosphere perturbed by sources of different nature, *Radiophys. Electron.*, 405, 157–177, 1998; 417, 3–22, 1999 (in Russian).
3. Garmash, K. P., Rozumenko, V. T., Tyrnov, O. F., Tsymbal, A. M., and Chernogor, L. F., Radiophysical investigations of processes in the near-Earth plasma perturbed by high-energy sources, *Foreign Radioelectronics: Success in Modern Radioelectronics*, 7, 3–15, 1999; 8, 3–19, 1999 (in Russian).
4. Chernogor, L. F., Energetic of the processes in the atmosphere and near-Earth space in lightness of the project "Popeledgennia," *Cosmic Tech. Technol.*, 5, 38–47, 1999 (in Russian).
5. Kostrov, L. S., Rozumenko, V. T., and Chernogor, L. F., Doppler radio sounding of the naturally-perturbed middle ionosphere, *Radiophys. Radioastron.*, 4, 209–226, 1999 (in Russian).
6. Kostrov, L. S., Rozumenko, V. T., and Chernogor, L. F., Doppler radio sounding of the disturbances in the middle ionosphere accompanied launchings and flights of cosmic vehicles, *Radiophys. Radioastron.*, 4, 227–246, 1999 (in Russian).
7. Chernogor, L. F. and Rozumenko, V. T., Wave processes, global- and large-scale disturbances in the near-Earth plasma, in *Proc. Int. Conf. "Astronomy in Ukraine-2000 and Perspective," Kinematics and Physics of Sky Bodies*, Annex K, 514–516, 2000.
8. Burmaka, V. P., Kostrov, L. S., and Chernogor, L. F., Statistical characteristics of signals of the Doppler radar during sounding of the middle ionosphere perturbed by launchings of rockets and the solar terminator, *Radiophys. Radioastron.*, 8, 143–162, 2003 (in Russian).
9. Afraimovich, E. A. and Perevalova, N. P., *GPS-Monitoring of the Upper Earth's Atmosphere*, Irkutsk: GU NZ PBX VSNZ SO RAMN, 2006 (in Russian).
10. Li, Y. Q., Jacobson, A. R., Carlos, R. C., et al., The blast wave of the Shuttle plume at ionospheric heights, *Geophys. Res. Lett.*, 21, 2737–2740, 1994.
11. Ponomarev, E. A. and Erushenkov, A. I., Infrasound waves at the Earth's atmosphere (special issue), *Izv. Vuzov. Radiofizika*, 20, 1773–1789, 1977 (in Russian).
12. Gossard, E. E. and Hooke, W. H., *Waves in the Atmosphere: Atmospheric Infrasound and Gravity Waves: Their Generation and Propagation*, Amsterdam: Elsevier, 1975, 476.
13. Hines, C. O., Internal AGWs at ionospheric heights, *Can. J. Phys.*, 38, 1441–1481, 1960.

14. Klostermeyer, J., Gravity waves in the *F* region, *J. Atmos. Terr. Phys.*, 31, 25–45, 1969.
15. Yeh, K. C. and Liu, C. H., AGWs in the upper atmosphere, *Rev. Geophys. Space Sci.*, 12, 193–216, 1974.
16. Francis, S. H., A theory of medium-scale traveling ionospheric disturbances, *J. Geophys. Res.*, 79, 5245–5260, 1974.
17. Francis, S. H., Global propagation of AGWs: A review, *J. Atmos. Terr. Phys.*, 37, 1011–1054, 1975.
18. Millward, G. H., Quegan, S., Moffett, R. J., and Fuller-Rowell, T. J., Effects of an atmospheric gravity wave on the mid-latitude ionospheric F layer, *J. Geophys. Res.*, 98, 10173–10179, 1993.
19. Mercier, C., Genova, F., and Aubier, M. G., Radio observations of AGWs, *J. Ann. Geophys.*, 7, 195–202, 1989.
20. Williams, P. J. S., van Eyken, A. P., and Bertin, F., A test of the Hunes dispersion equation for AGWs, *J. Atmos. Terr. Phys.*, 44, 573–576, 1982.
21. Williams, P. J. S., Tides, Atmospheric gravity waves and traveling disturbances in the ionosphere, in *Modern Ionospheric Science*, Kohl, H., Ruster, R., and Schlegel, K., Eds. Katlenburg-Lindau, Germany: Max-Planck-Institut für Aeronomie 1996, 136–180.
22. Ginzburg, V. L., *Propagation of Electromagnetic Waves in Plasmas*, New York: Pergamon Press, 1964.
23. Sorokin, V. M. and Fedorovich, G. V., *Physics of Slow MHD-Waves in the Ionospheric Plasma*, Moscow: Energoizdat, 1982 (in Russian).
24. Sorokin, V. M., Middle-latitude long-periodical oscillations of the geomagnetic field and their relation to wave disturbances in the ionosphere, *Geomagn. Aeromn.*, 27, 104–108, 1987.
25. Hantadze, A. G., On a new branch of own oscillations of the electro-conductive atmosphere, *Rep. Acad. Sci.*, 376, 250–252, 2001 (in Russian).
26. Hantadze, A. G., Electromagnetic planetary waves in the Earth's ionosphere, *Geomagn. Aeronom.*, 42, 333–335, 2002.
27. Chernogor, L. F., Kostrov, L. S., and Rozumenko, V. T., Radio probing of the perturbations originating in the near-Earth plasma from natural and anthropogenic energy sources, in *Proc. Int. Conf. "Astronomy in Ukraine-2000 and Perspective," Kinematics and Physics of Sky Bodies*, Annex K, 497–499, 2000.
28. Rees, H., *Physics and Chemistry of the Upper Atmosphere*, Cambridge: Cambridge University Press, 1989.
29. Uryadov, V., Ivanov, V., Plohotniuc, E., Eruhimov, E., Blaunstein, N., and Filip, N., *Dynamic Processes in the Ionosphere—Methods of Investigations*, Iasi: Tehnopress, 2006 (in Romanian).
30. Blaunstein, N. and Plohotniuc, E., *Ionosphere and Applied Aspects of Radio Communication and Radar*, New York: CRC Press, 2008, 577.

DIAGNOSTICS OF PLASMA WAVE DISTURBANCES BY INCOHERENT SCATTER AND DOPPLER RADARS

II

Chapter 4

Perturbations in the Middle and Outer Ionosphere

4.1 Overview

Investigation of wave disturbances (WDs) is one of the priorities in the field of physics of the atmosphere and the ionosphere [1–28]. There are several reasons for this.

First of all, the near-the-Earth environment can rarely be quiet, as there are many perturbations including those having the nature of waves. The WDs accompany the perturbations of the Sun, geospace storms, powerful atmospheric processes, volcanic activity, earthquakes, and so on. As mentioned in Chapters 2 and 3, WDs are generated also by the launches and flights of space vehicles (SVs) and rockets, powerful explosives, and other high-energy sources affecting the near-the-Earth environment. We should note that the frequency of the occurrence of the WDs is very difficult to estimate. It depends on the methods of detection of the WDs, the sensitivity of the detectors that record the relative amplitude $\delta = \Delta N / N$ of the electron density of N variations, the frequency of occurrence of the sources of these WDs, as also other factors. We can only declare that the frequency of appearance of the WDs does not exceed several tens of percent from the entire time of observations of the near-the-Earth environment (see, e.g., Refs. [5,19]).

Second, WDs play an essential role in the transport of energy from the lower atmosphere to the middle and upper atmosphere [18]. This means that the WDs are responsible for the interactions of subsystems in the whole system of Earth–atmosphere–ionosphere–magnetosphere (EAIM) coupling [20]. It is impossible to prepare the corresponding physical and mathematical models of the EAIM system without knowledge of the peculiarities and details of such interactions.

Third, WDs significantly influence the precise characteristics of various radio systems, quality of operational characteristics of telecommunication networks, and so on.

Fourth, study of the features and parameters of the WDs in the ionosphere, as well as their sources is far from complete. As a rule, observation of all wave phenomena occurring in the near-the-Earth environment was carried out episodically (even sporadically) with the required sensitivity of the proposed techniques and methods of the selection, detection, and identification of various WDs. Moreover, existing radiophysical methods allow detection of only the parameters of the WDs having plasma concentration (electrons or ions—plasma is quasineutral [21]), ignoring its parameters such as the temperature of particles, gas pressure, and other parameters of ambient plasma.

Therefore, for a systematic and complete study of the WDs in the near-the-Earth environment, it is necessary to employ the whole arsenal of existing techniques and methods, including the method of incoherent scattering (IS). This technique, as is well known [22–27], relates to the complex methods of diagnostics of the parameters of ionospheric plasma.

The main goal of this chapter is to show the reader the results of observation of the WDs in the ionosphere, which accompany the natural (such as solar terminal) and the artificial (such as rocket launches, RLs) sources based mostly on the results obtained and described in Ref. [28]. For the detection and identification of WDs with relatively small amplitudes ($\delta_N = \Delta N/N \sim 0.1\%-1\%$), a technique of measurements and observed statistical data processing was performed as given in Ref. [28]. In this, the potential possibilities of the IS methodology were analyzed to achieve the main aim of the investigation. As follows from the statistical theory of signal parameters estimation, "potential possibilities" can be defined as the minimum magnitude of the relative amplitudes of WDs, δ_N, which can be measured with satisfactory accuracy and minimum signal-to-noise ratio (SNR) that can be recorded by the corresponding radar system as well as other parameters of the signal [29].

The corresponding accuracy of such an approach and the corresponding results obtained experimentally were proved by the special simultaneous observations by using the Doppler sounding radar (DSR) of radio waves in the decameter waveband (see Chapter 3).

As earlier shown in Chapter 3, a study of the reaction of the ionosphere on solar terminal dynamics and RLs was described in numerous special articles which are impossible to refer in this chapter. What is important to note is that, as given in Chapter 3 references, the effects occurring along the trajectories of rocket flights were basically described (see also, e.g., Refs. [30–42]). All mentioned results were summarized in a special handbook [43].

Only a small amount of research concentrated on large-scale and global perturbations in the near-the-Earth environment (see, e.g., Refs. [4,19,44–48]). This is because, after dealing with the large-scale (~1000 km) and the global (~10,000 km) effects, the researchers should answer the following complicated questions:

- How are such disturbances generated?
- What is the way to transport such disturbances to distances 1,000–10,000 km long?
- What is the magnitude of such disturbances?

The practical response to these questions was done systematically in Refs. [19, 48–51]. In particular, it was shown that in the quiet ionosphere large-scale and global perturbations, as a rule, are stable as observed. Thus, as was shown in Ref [48], from 43 RLs the expected phenomenon was observed in all 43 cases. But in the case of transferred daytime duration and those of strongly perturbed environment we found that the WDs effects are a very complicated problem, which we will demonstrate to the reader in this chapter as a natural continuation of Refs. [19,28,44–51] mentioned earlier.

4.2 Method of Identification of Wave Disturbances

4.2.1 Method of IS

This well-known method is related to the most informative methods of near-the-Earth plasma parameter investigation [22–27]. It is used basically for detecting the regular parameters of the desired environment. To use the method of IS for studying the short-term ionospheric plasma disturbances, its inhomogeneous structure, and wave processes, the researchers encountered some difficulties. These related to the relatively small magnitude of disturbances of electron concentration N (usually $\delta_N \sim 1\%$ and rarely $\delta_N \sim 10\%$) as well as peculiarities of the proposed IS technique. For the latter we can relate the following:

- The stochastic character of the recorded signal
- Low SNR ($q = P_s / P_n$ and P_s, P_n are the power of the signal and the noise inside the measured system, respectively)

■ Statistical relations between the characteristics of the signal and parameters of the observed environment and its multiparametric character (i.e., the character of the signal is determined by many parameters of the atmosphere and the ionosphere)
■ The object under investigation (called the *radiolocation target*) is spatially distributed.

The points mentioned above yield the necessity of statistical averaging of P_s and P_n at the time interval ΔT, which for low values of $q = P_s / P_n$ are achieved in 10 min. At the same time, ΔT does not exceed the lifetime of sporadic irregularities, $t_{\Delta N}$, in the ionosphere and the period T of WDs of N. Limits from 1 to 10^4 s for irregularities on a scale of ~1–100 km change in the F-region of the ionosphere. Usually, the second time $T \approx 10$–100 min. Therefore, study of irregularities N with $t_{\Delta N} > 15$ min or $T > 15$ min is not very difficult. The results of such studies were carried out, for example, in Ref. [49]. The difficulties occur for $t_{\Delta N}$ and T lower than the above-mentioned time limits and for small values of $\delta_N \sim 0.1\%$–1%. In this case, special optimization is required for parameters of the IS radar, methods of measurements, and signal processing, as well as a special estimation of the accuracy and sensitivity of the technique of recording $\Delta N/N$.

Further, we will consider the identification method of wave processes. We assume that registration of the combined IR signal and noise, P_{sn}, takes place with frequent repetition of sounding pulses, F, and with discrete steps along the height, Δz_0. During the time period ΔT, the $n = F\Delta T$ discrete outputs P_{sn} are registered. Let us consider that the number of outputs of the noise power for each radiolocation scan equals m. Then, during the time interval ΔT, the mn outputs of P_n are registered.

At the preliminary stage (during the time interval, $\Delta T_0 = 1 - 1.5\,\mathrm{min}$), the basic estimations of P_{n0} and $P_{s0} = P_{sn0} - P_{n0}$ can be obtained for each height. The variances of these estimations equal:

$$\sigma_{s0}^2 = P_{s0}^2 \frac{(q+1)^2 + m^{-1}}{n_0 q^2}, \sigma_{n0}^2 = \frac{P_{n0}^2}{m n_0}, \tag{4.1}$$

where:

$n_0 = F\Delta T_0$

q is the SNR that will be defined later

At the first stage, the average values of powers, P_{s1} and P_{n1}, are calculated for n_1 records, that is, $P_{s1} = \langle P_{s0} \rangle$ and $P_{n1} = \langle P_{n0} \rangle$ (the angular parentheses indicate results of averaging), as well as the signal difference, $\delta P_{s1} = P_{s1} - P_{s0}$.

At the second stage, the average values of δP_{s1}, P_{s1}, and P_{n1} are calculated for n_2 records, that is, $\delta P_{s2} = \langle \delta P_{s1} \rangle$, $P_{s2} = \langle P_{s1} \rangle$, and $P_{n2} = \langle P_{n1} \rangle$. The variance of estimation of δP_{s2} equals:

$$\sigma_s^2 = \sigma_{s0}^2 \frac{1 + n_1}{n_1 n_2}. \tag{4.2}$$

For the study of wave processes in the ionosphere, it is interesting to estimate the relative amplitude of quasiperiodical variations of electron concentration, $\delta_{Nm} = \Delta N_m / N_0$, where ΔN_m is the recorded amplitude of ΔN during observation. It can be estimated as follows:

$$\delta_{Nm} \approx \frac{\delta P_{s2m}}{P_{s2}} = \frac{\langle \delta P_{s1} \rangle_m}{\langle \delta P_{s1} \rangle}, \tag{4.3}$$

where:

$\delta P_{s2m} = \langle \delta P_{s1} \rangle_m$ is the amplitude of oscillations in time dependence $\delta P_{s2}(t)$

Then, the relative error in estimation of δ_N equals:

$$\tilde{\sigma}_N = \frac{\sigma_{s0}}{P_{s1}} \frac{1}{\sqrt{n_1}} \sqrt{1 + \frac{1}{\delta_{Nm}^2} \frac{1 + n_1}{n_2}} = \sqrt{\frac{(1 + q)^2 + m^{-1}}{q^2 n_0 n_1} \left(1 + \frac{1}{\delta_{Nm}^2} \frac{1 + n_1}{n_2}\right)}, \tag{4.4}$$

where:

$q = P_{s2}/P_{n2}$

Usually $m \gg 1$, $n_1 \gg 1$. Then we get

$$\tilde{\sigma}_N = \frac{1 + q}{q} \sqrt{\frac{1}{n_0 n_1} \left(1 + \frac{n_1}{\delta_{Nm}^2 n_2}\right)}. \tag{4.5a}$$

As a rule, $\delta_{Nm}^2 n_2 \ll n_1$. Then

$$\tilde{\sigma}_N = \frac{1 + q}{q \delta_{Nm}} \frac{1}{\sqrt{n_0 n_2}}. \tag{4.5b}$$

Definition of the sensitivity of WD identification method yields the computation:

$$\delta_{N \min} = \frac{1 + q}{q \tilde{\sigma}_{N \max} \sqrt{n_0 n_2}}. \tag{4.6}$$

For $q \leq 1$ detection of WDs in the ionosphere is possible only for $\delta_{N \min} \geq 1.4\%$. To further increase the sensitivity of the method of identification of such disturbances it is necessary to use the third stage of signal processing that is based on the averaging procedure of δ_N over the whole range of heights, where the wave processes were observed, or over only its part, where these wave processes are close to the cophase (sine-phase) processes (e.g., where the maximum phase shift does not exceed 10%–20%).

At the next stage, we estimate the potential possibilities of the method under consideration for detecting WDs N with the period T and the relative amplitude $\delta_{N\min}$. For $\delta_N^2 n_2 \ll n_1$, we get

$$\delta_{Np} = \frac{1+q}{q\tilde{\sigma}_{N\max}\sqrt{n_0 n_2 n_3}}, \qquad (4.7)$$

where:

$(n_0 n_2)_{\max} = FT_{\min}$, $n_3 = \Delta z/\Delta z_s$

Δz is the range of altitudes where quasiperiodical cophase disturbances of N occur

$\Delta z_s = c\tau$ is the height range of the scattering area

For example, for $q = 0.1$, $F = 25$ Hz, $T_{\min} = 10^3$ s, $n_3 = 10$, $\tilde{\sigma}_{N\max} = 0.5$, we get $\delta_{Np} \approx 4\%$. For $q = 3$ and similar other parameters, we get $\delta_{Np} \approx 0.1\%$. The parameters of the IS radar, arranged at Kharkov University, Ukraine, are shown in Annexure 3.

4.2.2 Method of Doppler Sounding

Usually, the temporal variations of the Doppler spectra (DSs) are studied by using this method. Each of the spectrum was obtained at the time interval of 51.2 s (see details in Refs. [19,52]. For definition of WD periods, the temporal dependences of the average frequency in DS were processed by fast Fourier transform (FFT) at time intervals of 64 and 128 min. The parameters of the DSR are presented in Appendix 3.

In the Doppler sounding method, the minimum Doppler frequency shift is $f_{d\min} \approx 0.01$–0.1 Hz. Such magnitudes are not limited. The potential accuracy of measurements of f_d is limited by error, for which, for some simplifications taken from the general rules of the statistical theory of signal parameter estimation [29], one can obtain the following relation:

$$\Delta f_{dp} = \left(\frac{12}{qT_d^2}\right)^{1/2}, \qquad (4.8)$$

where:

T_d is the time interval at which the spectral estimation is made

Following the results of Ref. [19], it can be shown that the potential sensitivity of the DS method for identifying WDs will equal:

$$\delta_{Np} \approx \frac{T}{4\pi}\frac{c}{L_N}\frac{\Delta f_{dp}}{f} = \frac{1}{\pi}\frac{\lambda}{L_N}\frac{T}{T_d}\sqrt{\frac{3}{q}}, \tag{4.9}$$

where:

L_N is the range of the height region giving the main effect in Doppler shift,
$\lambda = c/f$

For example, for $f = 3$ MHz, $q \approx 10^6$, $T_d = 51.2$ s, $T = 10^3$ s, and $L_N = 30$ km, we get $\Delta f_{dp} \approx 3.4 \times 10^{-5}$ Hz and $\delta_{Np} \approx 1.8 \times 10^{-3}\%$. It is necessary to understand that for this case short-term (at limits of ~1 min) relative frequency instability should not be less than 10^{-11}. It is reasonable that the error Δf_{dp} should not be very less than that permitted for the frequency δf_d, which corresponds to the Doppler estimation method. Without the method giving such an express permission, we have $\delta f_d \approx 1/T_d$, for which, the error of estimation equals $\delta f_d / 2$. Thus, for $T_d = 51.2$ s, we get $\delta f_d \approx 2 \times 10^{-2}$ Hz. To achieve $\delta f_d \approx 10^{-4} - 10^{-3}$ Hz, it is necessary to increase T_d up to $10^3 - 10^4$ s, respectively. But for this case, the vivid constraint $T \geq T_d$ or $T \geq 2T_d$ should be satisfied (due to Kotelnikov's theorem). Therefore, for $T_{min} \approx 600$ s, we get $T_{dmin} \approx 300$ s, $\delta f_d \approx 3 \times 10^{-3}$ Hz, $\delta f_{dmin} \approx 1.5 \times 10^{-3}$ Hz, and $\delta_{Nmin} \approx 0.05\%$. In our observation $\Delta f_{dmin} \approx 0.01$ Hz, to which $T = 10$ min and $L_N = 30$ km correspond $\delta_{Nmin} \approx 0.3\%$.

The permitted time capability of the Doppler sounding method is about 1 min, that is, the same as in the vertical sounding (VS) method using ionosondes. Due to the random nature of the IS signal, this method is weaker than the VS and Doppler sounding methods. For the method of incoherent sounding, it equals 10–15 min. But for potential sensitivity, the IS method surpasses the VS and DS methods (for equality of SNR). Of course, the IS method exceeds both the other methods due to its high capacity of information. However, for this purpose, the cost of the IR radar is more expensive and has a more complicated structure.

4.3 Processes in the Ionosphere That Accompany Soyuz Launch

The aim of the observations was to find WDs during the launch of comparably large and powerful SVs ranged at a relatively close distance from the point of registration. The parameters of incoherent scatter radar (ISR) were chosen in such a manner to give the magnitude of the SNR $q = 0.3$–3. For $\tilde{\sigma}_{Nmax} = 0.5$ detection of WDs with $\delta_{Nmin} \approx 0.3\% = 0.8\%$–2%, respectively, was guaranteed.

The RL from Baikonur cosmodrome happened at 03:11 UT (here and in further instances time is given in Universal Time or UT) on October 30, 2002. The distance between the cosmodrome and the place of observation was $R = 2050$ km. The rocket comes in the class of heavy rockets with a total mass of 305 t (see Appendices 1 and 2).

4.3.1 Observations on Use of ISR

Measurements for the altitudinal range of 108–610 km were carried out in the time interval of 02:00–06:00 h. The parameters of ISR are presented in Table 4.1.

The start of the SV was related to the night-to-day transfer time. The level of power, P_{s0}, of the ISR signal, averaged over $\Delta T_0 = 1$–1.5 min time interval, increased from about 1 to 40 (here and in further explanation the power level is shown in relative units). This effect is clearly seen from Figure 4.1 (upper panel). It is also seen from Figure 4.1 that the power P_{s0} monotonically increases with some insufficient fluctuations of P_{s0}.

Here, the power of the noise, P_{n0}, is also presented in the upper panel by dotted lines, which averaged over time interval. As can be seen, after starting the measurements, the fall in the noise power P_{n0} from 6.7 to 5.1 (here the noise level is also in relative units) occurs, and then slowly grows up to 6.4. After about 05:30 h a slight decrease in noise power P_{n0} was observed.

The SNR (see Figure 4.1, middle panel) changed during the experiment in limits of 0.02–13 for heights of 108–610 km, respectively, whereas at the moment of the start of the SV this parameter changed from 0.05 to 1.2 at the same range of altitude.

We should note additionally that the SV launch was "hidden" in the environmental effects, such as the morning solar terminator (ST). Sunrise was observed near 02:30 h (at an altitude of 300 km), 02:40 h (at 250 km), 02:50 h (at 200 km), 03:00 h (at 150 km), 03:15 h (at 100 km), 03:35 h (at 50 km), and 04:20 h (at the Earth's level).

Table 4.1 Parameters of the ISR, Interval of Preliminary Processing (ΔT_0), and Ranging along Altitude (Δz_0)

Date	P_i (MW)	$\bar{P}$ (kW)	τ (μs)	F (Hz)	ΔT_0 (min)	Δz_0 (km)
October 30, 2002	2.2	7.2	135	24.4	1	4.5
October 7, 2002	1.8	2.8	65	24.4	1.5	10
October 7, 1999	2.2	42.9	800	24.4	1	18

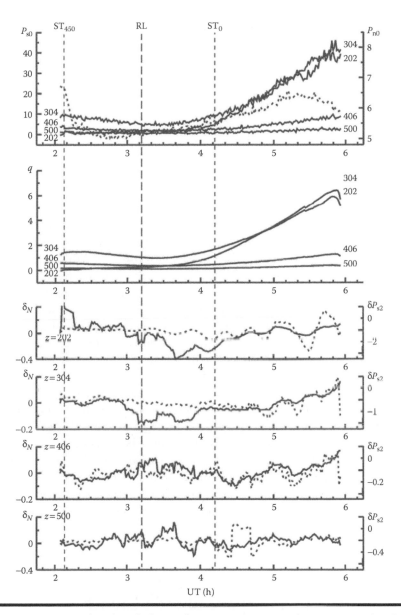

Figure 4.1 Example of variations of P_{s0}, P_{n0}, q_1, δ_N, and δP_{s2} versus the time; P_{n0} and δP_{s2} are presented by a dotted line; the numbers are shown at the right-hand side scale. Here, and in other figures, CP is the moment of RL, ST_0, ST_{100}, and ST_{450} are the moments of passing of the solar terminator at the ground level, and at altitudes of 100 and 450 km, respectively.

In the lower panel of Figure 4.1, the time dependences of $\delta_N(t)$ are presented for a net of altitudes as prior to and after the SV launches. It is important to note that the parameters of oscillations for each altitude were different. At the time interval of observations at 02:00–06:00 h, the parameters of the ISR supported the NSR by not less than 0.4–7 (at an altitude of 200 km), 1.4–7 (at 300 km), and 0.5–1.2 (at 400 km). This means that at the indicated time interval without averaging of δ_N over altitudes, WDs were observed with the relative amplitude not less than 3% (200 km) to 1% (400 km), respectively.

4.3.2 Observations by the DR

Measurements were carried out at frequency of 3.5 MHz from 23:20 (October 29, 2002) to 10:00 h (October 30, 2002). Their results are presented in Figure 4.2. Prior to 02:20 h slow changes of DSs appeared; the amplitude of oscillations f_{da} of the average frequency in DS did not exceed 0.1–0.2 Hz. The constant component of the Doppler frequency shift f_{d0} basically was positive and achieved about 0.1–0.2 Hz. Such a situation corresponds to a quiet ionosphere. In this case, the index of the magnetic activity K_p achieved the magnitude of 2–3. Approximately, after 02:40–03:30 h $f_{d0} \approx -0.2$ Hz and $f_{da} \approx -0.1$ Hz. At the time interval from 03:30 to 06:30 h, a vivid quasiperiodical process was observed with a period of $T \approx 10$ min and $f_{da} \approx 0.13$ Hz.

At the time interval from 6:20 to 10:00 h, strong variations of the DS with $f_{d0} \approx 0$ Hz occurred. Sometimes, nonharmonic oscillations of $f_d(t)$ registered with periods of 5–10 min. At the time interval from 06:00 to 12:00 h, magnetic activity was sufficiently high with $K_p = 5$ and $K_p = 4$, respectively. This means that at high latitudes of about ~70°N WDs could be generated which travel in the equatorial direction. For the average velocity of about 500 m s^{-1} (see, e.g., Ref. [3]), they can achieve a place of observation located at ~50°N with a time delay close to 70 min. As can be seen in Figure 4.2, the character of the WDs changed significantly with respect to that observed in the previous time interval of 03:00–06:00 h approximately after 07:20 h.

4.4 Processes in the Ionosphere Accompanying Space Shuttle Launch

The distinguishing peculiarity of the observations of measurements carried out for large distances (~10,000 km) from the place of rocket start with high resolution along altitudes and with low magnitudes of SNR—from 0.1 to 1. The SV was launched at 19:48 h, on October 7, 2002, from Cape Canaveral cosmodrome (the distance to the point of observations was $R \approx 9300$ km). The rocket

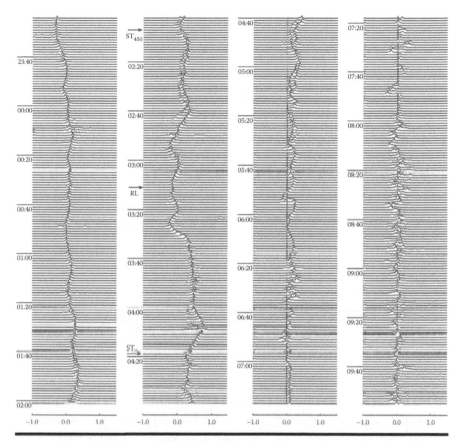

Figure 4.2 Variations of DS at frequency 3.5 MHz during Soyuz launch at 03:11 h on September 30, 2002 (virtual altitude was 315–390 km). In this and other figures along the horizontal axis, the Doppler frequency shift is indicated in Hertz along the vertical axis in universal time.

is the largest one of all till now. It can be also defined as a super-heavy rocket. Its total mass equals 2029.633 t (see Appendices 1 and 2).

4.4.1 Observations by the ISR

Special measurements at altitudes from 83 to 604 km were carried out at 18:00 h, on October 7, 2002, up to 01:00 h, on October 8, 2002. Sunset was observed at about 15:03 h (for the level of the ground), at 15:47 h (at 50 km), 16:05 h (at 100 km), 16:19 h (at 150 km), 16:31 h (at 200 km), 16:41 h (at 250 km), 16:50 h (at 300 km), 16:58 h (at 350 km), 17:06 h (at 400 km), and 17:14 h (at 450 km), respectively. To the moment of observation of SV launch, the processes caused by movements of ST were finished. The parameters of the ISR are presented in Table 4.1. The RL was during the changeover from day to night. At all altitudes the power level P_{s0} was decreased by 1 order of magnitude. This effect was registered basically during 1–1.5 h after the SV launch (Figure 4.3). This fact,

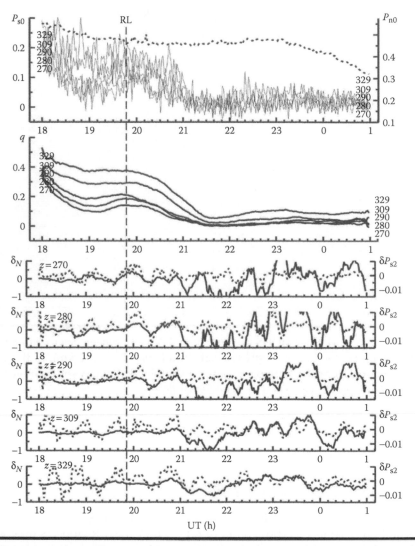

Figure 4.3 **Example of variations of P_{s0}, P_{n0}, q_1, δ_N, and δP_{s2} for altitudes 270, 280, 290, 309, and 329 km, according to ISR data during Space Shuttle launch at 19:46 h on October 7, 2002.**

together with the small SNR (of $q \sim 0.02$–0.1), sharply complied detection of the WDs that accompanied this launch. It is because of such small magnitudes of q, the value of $\delta_{N\,min}$ equals 8%–38%. It is impossible that such strong WDs can be observed at distances of 10,000 km from the place of SV launch. To increase the sensitivity of the proposed methodic, summation of P_{s0} along altitudes was made. After this procedure, the magnitude of the SNR was $q \equiv q_\Sigma \approx 0.5$ and $\delta_{N\,min} \approx 2\%$.

The results of such a signal processing showed that magnitudes of $\delta_{N\Sigma}$ did not exceed 2%–3% before the SV launch, and after the launch, increased by up to 10%–20%. It is to be noted that such growth of $\delta_{N\Sigma}$ is caused not only by the SV launch, but mostly by decrease of P_{s0}. This effect explains also strong disturbances of δ_N observed at all altitudes during the time interval from 21:30 to 22:30 h.

4.4.2 Observations by the DR

On the same day, as given earlier, measurements were carried out at frequency $f = 3.5$ MHz during the period from 15:50 to 21:10 h (Figure 4.4). Approximately, from 16:00 to 19:00 h strong (up to 1 Hz) variations of f_d registered, and sometimes quasiperiodical WDs with a period of 10–20 min were recorded. The shift of the constant component of DS was achieved ~0.4 Hz. At the time interval of 19:00–19:50 h, variations of f_d did not exceed 0.2–0.3 Hz. Near 19:56 h, the Doppler frequency shift became negative; the DS, as a rule, was diffuse.

More strong variations of DS were registered after 20:40 h. The absolute value of frequency shift exceeded 1 Hz and the DS was fully destroyed. Then, a strong quasiperiodical process was observed with a period of about 25–30 min.

4.5 Processes in the Ionosphere Accompanying Delta Launch

The aim of the observations was to study the possibilities of detecting WDs caused by medium weight SV launches, whose effects can be observed at distances far from the place of observation (~10,000 km). For achieving sufficient magnitude of the SNR in the method of IS, very long pulses (of about 800 μs) and large steps along the height (of 18 km) were used. In this case, depending on the altitude and time of observation, SNR, q, was 5–35, and the sensitivity of the method of detecting WDs (without summation along altitudes) was about 0.6%.

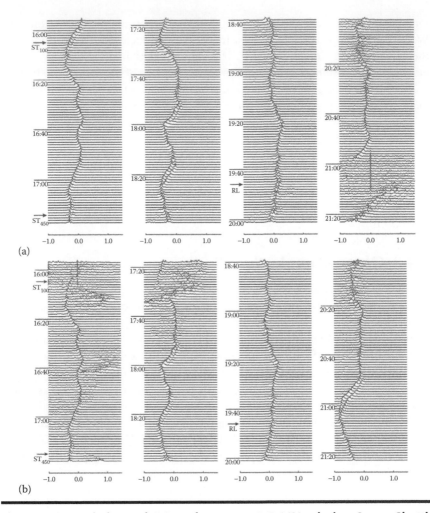

Figure 4.4 **Variations of DS at frequency 3.5 MHz during Space Shuttle launch at 19:46 h on October 7, 2002: (a) the virtual altitudes—315–390 km, (b) the virtual altitudes—390–465 km.**

An SV was launched at 12:61 h on October 7, 1999, from Cape Canaveral cosmodrome ($R \approx 9300$ km). The rocket was of medium weight with a total mass of 231.870 t (see Appendices 1 and 2).

4.5.1 Observations by the ISR

Measurements for the height range of 125–1760 km were carried out at the time interval of 13:00–21:00 h, that is, in the interregnum between day and night.

The sunset was near 15:03 h (for the ground level), at 15:47 h (at 50 km), 16:05 h (at 100 km), 16:19 h (at 150 km), 16:31 h (at 200 km), 16:41 h (at 250 km), 16:50 h (at 300 km), 16:58 h (at 350 km), 17:06 h (at 400 km), and 17:14 h (at 450 km), respectively. Unfortunately, measurements started only 9 min after the SV launch (Figure 4.5). Therefore, it was impossible to identify those WDs whose time delay was at the same level of magnitude.

Examples of temporal variations of P_{s0} for separate altitudes, as well as of P_{n0} and q are shown in Figure 4.5, where it is seen that at the time interval of 14:00–17:00 h a quick decrease of P_{s0} and q occurs. The dependence of $P_{n0}(t)$ was nonmonotonic.

Thus, from 15:00 to 17:20 h, a growth of P_{n0} from 1.5 to 2.1 (as shown earlier, in relative units) was observed, and then basically a decrease of P_{n0} from 2.1 to 1.6 was observed. Such behavior of $P_{n0}(t)$ is related to the noisy nonthermal radiation of the galaxy.

Temporal variations of δ_N are shown in Figure 4.5 (lower panel). From the presented illustrations WDs were observed at all altitudes during the entire period of observations, though the character of the WDs sometimes varied. The relative amplitude usually was 1%–2%, that is larger than $\delta_{N\min} \approx 0.6\%$. It is important to note that October 7, 1999, as well as the previous days, was magnetically quiet: during the day of the observation, the magnetic index $K_p \approx 1$–2, whereas earlier it had changed from 1 to 3. The ionosphere was also not perturbed during these days.

4.5.2 Observations by the DR

During the same day, as mentioned before, Doppler sounding was carried out at frequency $f = 3.5$ MHz at the time interval from 12:00 to 15:00 h. About an hour before launch, the Doppler frequency shift was $f_d \approx 0$. Sometimes a weak quasiperiodical process occurred with $f_d \leq 0.1$ Hz (Figure 4.6).

At 13:00 h, that is, after $\Delta t_1 \approx 9$ min, the quasiperiodical disturbance with the period of $T \approx 5$ min increased in the ionosphere. From 14:15 to 14:19 h and from 14:30 to 14:50 h essential broadening of the DS was observed with the weakly expressed quasiperiodical process.

4.6 Discussion of Results

4.6.1 Soyuz Launch

Let us consider the results of the observation by using the IS method. Nearly an hour before launch, at chaotic variations of δP_{s2} and δ_N quasiperiodical processes with varied period were added in which oscillations with $T \approx 12$–20 min were predominant (see Figure 4.1). The relative amplitude δ_N did not exceed 5%.

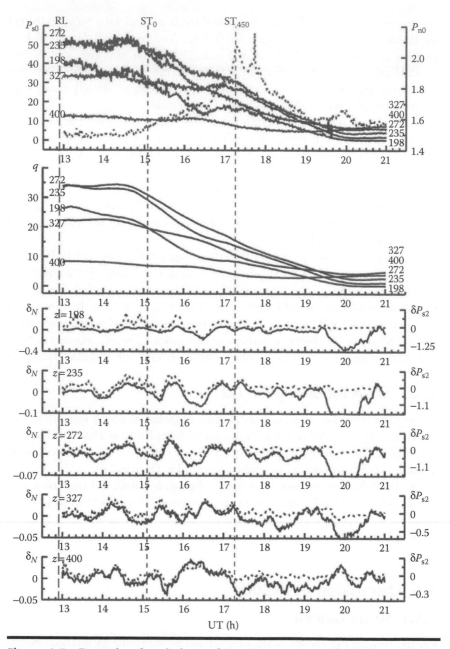

Figure 4.5 Example of variations of P_{s0}, P_{n0}, q_1, δ_N, and δP_{s2} for altitudes 198, 235, 272, 327, and 400 km, according to ISR data during Delta launch at 12:51 h on October 7, 1999.

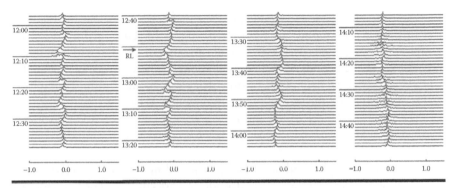

Figure 4.6   Variations of DS at frequency 3.5 MHz during Delta launch at 12:51 h on October 7, 1999; working altitudes—315–390 km.

It decreased with the increase in altitude. The degree of ordering of these processes grew at the moment of sunrise in the corresponding altitude.

The first change of the signal character at all heights was simultaneously registered near 03:15 h, or accounting for the interval of slide averaging 15 min, close to 03:21 h, that is, 10 min after the SV launch. Let us suppose that $\Delta t_1 \approx 10$ min. During this time, the quasiperiodical process increased significantly. Its relative amplitude δ_N achieved 16% at an altitude of 186 km, and decreased with increase in height: at an altitude of 250 km, it was already ~3%, and higher $\delta_N \approx 1\%$ (see Figure 4.1). Along with decrease of δ_N, the quasiperiod also decreased from 40 to 21 min with increase of altitude from 186 to 273 km.

The second change of the signal character came at the time interval from 03:25 to 03:45 h for altitudes of 186–273 km, respectively. Accounting for slide averaging, the time period delays of 52–42 min for the same altitudes corresponded. Thus at lower altitudes (176–210 km) the relative amplitude of the oscillations of plasma density N became lesser, but at upper altitudes (241–273 km) it increased by up to 6%. Between these altitudes, at 210 and 240 km, imposition of disturbances of the second type on those of the first type did not result in change in the amplitude of quasiperiodical disturbances. Here, and in our further explanations, the disturbances of the first of the second type are those which correspond to the first and the second changes of the signal character after the moment of the SV launch, respectively. The magnitude of the quasiperiod depended weakly on altitude and was markedly dependent on the time of observation. Thus, the long-periodical disturbances (with $T \approx 30$ min) registered first and then, the short-periodical disturbances (with $T \approx 20$ min). Hence, for the disturbances of the second type, it is characterized by a delay Δt_2 of about 40–50 min and the essential frequency dispersion.

The third change of the signal character practically coincides in time with the sunrise at the ground (at about 04:10–04:25 h). During this period the amplitude of oscillations decreased several times (up to 1%–3%) with simultaneous change in the periods of excited oscillations. If these effects were related to the SV launches, the magnitude of time delay was about 70 min. Examples of WDs of the first and the second types are shown in Table 4.2.

The results of the calculated and measured magnitudes of WD velocities of the second type are shown in Table 4.3. It is seen that these magnitudes coincide with each other.

Let us return to the method of Doppler sounding. Before the sunrise event the radio wave was reflected at an altitude of 250 km, for which sunrise was observed at about 02:40 h. Practically from the beginning of this event, $f_d(t)$ was changed according to quasiperiodical law. The moment of occurrence of these WDs is vivid proof that these variations were generated by movement of the ST. After 03:30 h, for about 3 h, the Doppler frequency shift was positive, and during the first hour f_d achieved 0.4–05 Hz. Such a shift is caused

Table 4.2 Magnitudes of the Relative Amplitude of WDs of the First (δ_{N1}), Second (δ_{N2}), and Third (δ_{N3}) Types for the Corresponding Periods, and of the Time Delay of WDs of the Second Type (Δt_2) during the Day of the Soyuz Launch

z (km)	δ_{N1} (%)	T (min)	δ_{N2} (%)	T (min)	δ_{N2} (%)	T (min)	Δt_2 (min)
186	16	36	4	30	2.5	30	58
194	14	40	6	27	3	30	55
202	13	40	6.5	29	2.5	20	52
210	12	38	6	29	3	20	49
218	10	40	5	30	2.5	20	48
225	10.5	40	8	30	2	20	47
233	8	35	6	30	1	20	46
241	5	25	6	29	2	20	46
249	3	20	6	30	1	20	45
257	2.5	20	6	26	1	20	45
265	2.5	20	5	30	2	20	44
273	2	20	5	28	2	20	43

Table 4.3 Calculated (v_s) and Observed (v'_2) Magnitudes of Velocities of WDs of the Second Type during the Day of Soyuz Launch, as Well as the Parameters of the Neutral Atmosphere

z (km)	M_m (×10²⁶ (kg))	T_m (K)	v_s (m s⁻¹)	v'_2 (m s⁻¹)
186	3.5	760	652	603
194	3.5	810	673	636
202	3.4	860	704	673
210	3.4	905	722	714
218	3.4	950	740	729
225	3.3	990	769	745
233	3.3	1025	780	761
241	3.3	1060	793	761
249	3.2	1080	813	778
257	3.2	1090	817	778
265	3.2	1095	819	795
273	3.2	1100	821	814

Note: M_m is the average mass of molecules, T_m is the temperature of molecules, and v_s is the sound velocity in the atmosphere. The magnitudes of T_m were used in computations as obtained from the MSIS model of the atmosphere.

by the downward movement in the morning of the region of reflection with a constant speed that can be estimated from the well-known relation: $v = cf_d/2f$. For $f_d = 0.3$ Hz and $f = 3.5$ MHz, we get $v \approx 17$ m s⁻¹.

The downward process of the region of reflection continued for about 100 min. During this time, the height of reflection changed from 100 km to about 150 km. It can be assumed that close to 05:00 h the effects caused by the morning rearrangement of the ionosphere were completed. The further sufficiently fast variations of DS may be raised by magnetic activity (that was mentioned earlier). These variations continued until 10:00 h, that is, up to the end of the observations.

Detection of the effects of SV launch in the backdrop of the WDs described earlier is very difficult if the Doppler sounding method is used. Despite this, it can be assumed that with the SV launch changes in signal character can be related close to 03:20 h when the quasiperiodical change of f_d with $T \approx 20$–30 min occur with a duration of not less than 30 min (which corresponds to the time delay of $\Delta t_1 \approx 10$ min), as well as close to 04:00 h (with the time delay of $\Delta t_2 \approx 50$ min). In the latter case, inside the DS a second mode was observed, which shifts at the smaller magnitudes f_d, shifted approximately at 0.1–0.5 Hz. The duration of such a DS is about 10 min. The time delays Δt_1 and Δt_2 correspond to the corrected magnitudes of the virtual velocity of the propagation of WDs, $v_1' \approx 5.8 \, \mathrm{km \, s^{-1}}$ and $v_2' \approx 760 \, \mathrm{m \, s^{-1}}$, respectively. As given in Refs. [48,52], here we used the corrected magnitudes of the distance of a rocket from the place of observation, $R' \approx 2{,}100$ km, and the time of the rocket's movement, 3 min, to the region of the ionosphere where an effective generation of waves (~100–150 km) was excited. It is clear that v_2' corresponds to the average velocity of traveling of the shock acoustic gravity wave (AGW), propagating inside the atmospheric waveguide at altitudes of 200–250 km (see, e.g., Refs. [53,54]). The sound wave was reflected exactly from these altitudes at the time interval of 03:00–04:00 h. The velocity of the acoustic wave, as a rule, does not exceed the speed of sound in the atmosphere, $v_s = \sqrt{\gamma k T_m / M_m}$, where T_m and M_m are the temperature and the mean mass of molecules of the atmosphere, and γ is the ratio of the relative gas thermocapacities.

The velocities close to v_s were earlier observed by the ISR at Kharkov University, Ukraine [49]. The reaction of the ionosphere on the SV launches and powerful chemical explosions was studied here. It is well known (see, e.g., Ref. [43]) that both the events were accompanied by generation of shock AGWs. During the chemical explosion the power of the generated acoustic wave was $P_a \approx 10^9$ W. For such magnitudes of P_a, the AGW propagates at distances not less than 3000–4000 km [49].

Velocities close to v_1' have slow magnetohydrodynamic (MHD) waves whose refractive and absorbed indexes have different polarizations (denoted by $\pm$, as in Chapter 2) can be presented, respectively, as follows [6,48]:

$$n_\pm = \frac{c}{\sqrt{2} \cdot v_{as}} \frac{a_{1\pm}}{a_{2\pm}}, \; \kappa_\pm = \frac{c}{\sqrt{2} \cdot v_{as}} \frac{\beta}{a_{1\pm} a_{2\pm}}, \qquad (4.10)$$

where v_{as} is the characteristic velocity of slow MHD waves equaling 1.6–2.5 km s^{-1} for altitudes of 200–250 km (i.e., at heights of reflection of sounding radio wave during diurnal and nocturnal periods, respectively),

$$a_{1\pm} = \left(1 \pm \alpha + a_{2\pm}\right)^{1/2}, a_{2\pm} = \left[(1 \pm \alpha)^2 + \beta^2\right]^{1/2} \qquad (4.11a)$$

and

$$\alpha = \frac{\omega}{s \Omega_H}, \; \beta = \frac{\omega}{v_1}, \qquad (4.11b)$$

where:

ω is the frequency of slow MHD wave

$\Omega_H \approx 300 \text{ s}^{-1}$ is the gyrofrequency of ions

s is the degree of ionization of plasma

$v_1 \approx 10^{-3} \text{s}^{-1}$ is for altitudes of 200–250 km [6]

At the indicated altitudes, often $\beta^2 \gg \alpha^2$, $\beta^2 \gg 1$, $a_{1\pm} \approx \sqrt{\beta}$, $a_{2\pm} \approx \beta$, and $n_\pm \approx \kappa_\pm \approx c/\sqrt{2\beta} \cdot v_{as}$. For this case, the phase velocity and the depth of the wave attenuation equal, respectively, $v_{ph} = v_{as}\sqrt{2\beta}$ and $L = v_{ph}/\omega$. For $v_1 \approx 10^{-3}\text{s}^{-1}$, $v_{as} \approx 2.5\,\text{km s}^{-1}$, $s = 3 \times 10^{-5}$, and $\omega \approx 3.5 \times 10^{-3}\text{s}^{-1}$, we can estimate the phase velocity $v_{ph} \approx 5.9$ km s^{-1} for an altitude of 250 km. The observed value of $v_1' \approx 5.8$ km s^{-1} is close to the estimated magnitude of $v_{ph} \approx 5.9$ km s^{-1}, for which we have $L \approx 1700$ km.

Hence, the slow MHD wave generated in the proximity of the rocket moving in the ionosphere can reach the place of observation. Due to absorption, the amplitude of WDs decreases by 3–3.5 times. The amplitude of disturbances decreases also due to the spatial spreading of the wave. Thus, spherical wave δ_N decreases in R/R_0 times, where R_0 is the characteristic scale of the region of the wave generation. For example, the magnitude of δ_N decreases 20 times for $R_0 = 100$ km and $R \approx 2000$ km. The generated wave, however, is close to the cylindrical wave type because the source of such waves is the reactive fuel flow from the moving rocket. If so, for $R_0 \approx 200$ km and $R \approx 2000$ km, we get $(R/R_0)^{1/2} \approx 3$.

4.6.2 Space Shuttle Launch

Let us return to the results of the observation obtained by IS method. As shown in Figure 4.3, the first notable increase of δP_{s2} and δ_N was observed after $\Delta t_1 \approx 10$–20 min, whereas the second one after $\Delta t_2 \approx 65$ min. Here, we consider that due to averaging at the time interval of 15 min, increase of δP_{s2} and δ_N outstripped the "reaction" of the ionosphere in 7.5 min time. But the increase of δP_{s2} and δ_N during this observation, as was mentioned earlier, is not a part of the reaction of the ionosphere on SV launch. Maybe this reaction indicates the changes in the period of the WDs. The time interval for this transfer is about 2 h. It can be assumed that the reaction on the SV launch continued for about 2 h, which consisted of two types of WDs with delays of about 10–20 and 65 min and with periods of ~35 and ~45 min for each type, respectively. To these delays, the corrected magnitudes of velocities of 20–10 and 2.4 km s^{-1} corresponded, respectively. Because WDs mostly propagate from the place of their generation (close to the cosmodrome) to the place of observation in the diurnal ionosphere, so that the magnitude of v_1' corresponds to the velocity of gyrotropic waves traveling in the waveguide at the E-region altitude of the ionosphere, where $N \approx 10^{11}\,\mathrm{m^{-3}}$ and $\alpha = 60°$. Let us prove this statement through the following estimations. Thus, for periods of waves which are more than 1 min in length [48]:

$$v_{\mathrm{ph}} \approx v_{\mathrm{gr}} \cos^2 \alpha, \tag{4.12a}$$

where:

$$v_{\mathrm{gr}} = \frac{c^2 f_{\mathrm{B}}}{2\pi f_{\mathrm{p}}^2 \sqrt{z_0 l}} \tag{4.12b}$$

is the characteristic velocity of gyrotropic waves, z_0 is the height of the waveguide with an effective width of l. For $f_{\mathrm{B}} \approx 1.5\,\mathrm{MHz}$, $f_{\mathrm{p}} \approx 2.7\,\mathrm{MHz}$, $z_0 \approx 100\,\mathrm{km}$, and $l \approx 30\,\mathrm{km}$, we get $v_{\mathrm{gr}} \approx 53\,\mathrm{km\,s^{-1}}$, then for $\alpha = 60°$, we get $v_{\mathrm{ph}} \approx 13\,\mathrm{km\,s^{-1}}$. We add here that gyrotropic waves are sufficiently well localized along the E-region altitude of the ionosphere and, therefore, their observation at F-region altitudes is complicated.

For a magnitude of $v_2' \approx 2.4\,\mathrm{km\,s^{-1}}$ slow MHD waves localized at altitudes of 150–200 km correspond to $\omega \sim 10^{-3}\,\mathrm{s^{-1}}$ [48]. For this type of wave the depth of attenuation is not less than 5000 km [48]. In earlier discussions, the WDs were observed at least at the height range of 270–330 km.

Hence, the possible reaction on the SV launch during this day was displayed weakly. This occurred because of low-registered SNR and the simultaneous observation of the time interval of WDs propagation caused by the launch and the time interval of the essential decrease P_{s0}. We add here that the second WD was registered and identified better than the first one. Experimentally proved estimations of the magnitudes of velocities in WD propagation are good witnesses of the artificial nature of these WDs.

Next, we will compare the results of the observations obtained earlier by using ISR with those obtained by using the DSR method. The day of the test, October 7, 2002, is related to the magnetically perturbed day. During this day, according to the corresponding classification [20], a strong magnetic storm was registered that, as a rule, was accompanied by generation of wave processes that were moving toward the Equator at high latitudes (65–75 N) with a velocity of 1 km s⁻¹. These waves reached the place of observation 1–2 h after their generation. We suppose that this phenomenon explains a strong perturbation of the ionosphere at the time interval 16:00–18:40 h. However, 1 hr before the SV launch, the ionosphere was relatively quiet. The first essential change of character of the recording signal was registered at about 19:56 h, that is, $\Delta t_1 \approx 10\,\mathrm{min}$ after the SV launch, and continued for about 30 min. The DS was mostly diffuse. The next essential change of character of the recording signal registered at about 20:50 h, that is, with a delay of $\Delta t_2 \approx 64\,\text{min}$ after the SV launch. To these two delays two magnitudes of the velocities, $v_1' \approx 21\,\text{km}\,\text{s}^{-1}$ and $v_2' \approx 2.5\,\text{km}\,\text{s}^{-1}$ relate. The first velocity is close to the velocity of the gyrotropic waves for the electronic concentration of $N \approx 5 \times 10^{10}\,\text{m}^{-3}$ in the E-region of the ionosphere and for the average angle between the wave vector and the vector of the magnetic field of $\alpha \approx 60°$ [46]. The second velocity is close to the velocity of slow MHD waves with a frequency of $\omega \sim 10^{-3}\,\text{s}^{-1}$ occur during nocturnal periods, for which $L \sim 10,000$ km.

Let us compare the results obtained by the IS and DS methods. For the second type of disturbances, the observed and estimated velocities of their propagation were very close (2.4 and 2.5 km s^{-1}), that is, the difference between them did not exceed the accuracy of the velocity estimations. As for the second type of disturbances, the velocities obtained by these two methods were significantly different. This, in our opinion, can be due to subjective definition of the moments of the beginning of the disturbances. For sufficiently small time delay, the error in 5–10 min leads to significant differences in the velocity v_1'. Perhaps the first type of disturbance generally does not relate to SV launches and has a natural origin.

4.6.3 Start of SV Delta

Let us return again to the results of observation by the IS method. Essential change of character in the signal was registered at about of 13:45 and 13:20 h at altitudes of ~200 and ~400 km, respectively (Figure 4.5). During this time period WDs with large amplitude, and generally speaking, with other periods were added to the existing wave process. If these WDs relate to the SV launch, the delay of 62 and 37 min (accounting for the interval of 15 min on average) corresponds to the WDs, and $v_2' \approx 2.5$–$4.4\,\text{km}\,\text{s}^{-1}$ (for $z \approx 200$–400 km). It is important to notice that with growth of altitude, the velocity also grows. Approximately, slow MHD waves with frequency of $\omega \sim 10^{-3}\,\text{s}^{-1}$ have velocities of such magnitude [48]. The duration of these types of waves was 1.5–2 h. Then a WD was registered with $\delta_N \approx 1.5\%$–2%, an essentially large period (up to 1 h), and duration of 2 h. Its relative amplitude and period slowly decreased with growth in altitude, from 2% to 1% and from 60 to 40 min with increase of z from 200 to 400 km. This WD was caused by movements of the evening ST. After 16:30 h (i.e., after passing of the terminator), the WD did not disappear though its amplitude and period continuously decreased (the period is up to 20–25 min). At altitudes above 350 km, the WDs were weak enough ($\delta_N \leq 0.6\%$) and their identification became impossible.

Let us address the method of DS again. As mentioned earlier, the day of October 7, 1999 was magnetically quiet. After the SV launch, the character of the signal changed twice: at 13:00 and at 14:15 h (Figure 4.6). To these periods correspond delays of $\Delta t_1 \approx 9$ min and $\Delta t_2 \approx 84$ min. If these changes were caused by the SV launch, the WDs propagated with the corrected magnitudes at velocities of 25 and 2 km s^{-1}, respectively. The first magnitude is close to the velocities of gyrotropic waves at the E-region of the ionosphere for $N \approx 4 \times 10^{10}$ m^{-3} and for $\alpha \approx 60°$. The second magnitude corresponds to the velocity of slow MHD waves with $\omega \leq 10^{-3}$ s^{-1} at diurnal time periods with the range of travel at $L \geq 4000$ km. At the same time, $v'_2 \approx 2.0$ km s^{-1} is close to the Hantadze wave velocity [55,56] propagating in the E-region of the ionosphere (where $N \approx 2 \times 10^{10}$ m^{-3}) in latitude-ward direction (i.e., for $\alpha \approx 90°$).

Delta, as mentioned earlier, is a rocket of median weight. Therefore, all effects in the ionosphere caused by this rocket are sufficiently small with respect to those caused by Space Shuttle launch. Therefore, we cannot accept that the disturbances may have the natural origin. However, because the magnitudes of the velocities described earlier are close to those identified earlier in Ref. [48] as having artificial origin, we can finally outline that the observed WDs after the Delta launch also have artificial origin and are caused by its launch. Because sunset at the ground level was occurred at 15:03 h, its effects were not observed by DSR.

Now we compare the results obtained by the ISR and the DSR. The WDs of the second type had the same-order delays, as well as the velocities of propagation. Because the main effect in Doppler frequency shift gives the height range of 120–150 km, the velocity v'_2 obtained by the method of DS is somewhat less than that measured by the IS method at higher attitudes (200–400 km).

Now we will discuss the role of the ST, because all three launches described earlier occurred during the passing daytime. Passing of the terminator, as follows from Chapters 2 to 3, is usually accompanied by generation of WDs with period of 5–100 min and $\Delta N/N \approx 1\%$–10%. Their duration is 2–2.5 h (see, e.g., Ref. [19]). WDs caused by RLs have closed characteristics. This circumstance seriously limits differentiation of ionospheric reactions on ST passage and SV launch, and sometimes results in their poor identification.

Moreover, the reaction caused by ST sometimes is interpreted as the reaction caused by the SV launch, namely, as was assumed by the authors of the classical work [44]. According to their opinion, using ionosondes it can be found that global disturbances accompanied SV launches at distances ~10,000 km. Unfortunately, the time of the desired reaction, described in Ref. [44], was exactly the time of passing of the evening terminator [at about 23:00–24:00 h of Moscow decreed time (MDT)] at altitudes of 200–300 km. The wave processes were observed at one of the stations in the time interval from 23:00 to 01:00 h MDT. The range of plasma density deviations was $\Delta N/N \approx 7\%$–11%. The time of reaction and its duration, the magnitude of $\Delta N/N$ and the period of oscillations—are vivid witnesses that the observed reaction of the ionosphere was caused by ST—and not by the SV launch. As for the effects accompanying Soyuz launches ($R \approx 2000$ km) and described by the authors of Ref. [44], their existence is under no doubt. The positive aspect by the authors of Ref. [44] is that their study stimulated search on the reaction of the ionosphere on launches of SV at the "far zones" by other researchers (see, e.g., Refs. [46,57] and references there).

Now, we will briefly discuss the accuracy of estimations of $\delta_N = \Delta N/N$. It, as a rule, did not exceed 1%–2%. Increase of δ_N in the nocturnal time period relates not only to increase of the amplitude of oscillations of ΔN_m, but also to decrease of P_{s1} and increase of its variance. This leads to increase of variance $\tilde{\sigma}_N^2$ describing deviations of the magnitude of δ_N. Together with this, $\Delta N_m/N$ increase is also registered at the nocturnal time period, despite the fact that magnitudes of ΔN_m are changed lesser than N during the whole day.

Summarizing the discussions, here we note that a perspective of the IS method for the study of small intensity WDs ($\delta_N \sim 1\%$), having periods of more than 10 min. The proof of the obtained results was confirmed by DS method. There is a high possibility of identifying WDs from the given source "hidden" in the background of natural disturbances. We also note that during the quiet time periods of the day, WDs in the ionosphere were confidently observed (as in Ref. [58], from 43 cases of the SV launch—43 observations of WDs). At transferred daytime periods during strong perturbations of the medium, it is difficult to decide on the existence of WDs caused by SV launches. However, the time delays, velocities of the observed waves (that are given in Ref. [58]), and their periods and decrements of attenuation, all correspond to the theoretical calculations made in Ref. [58], which allow the authors to state with a definite degree of confidence that in all the cases described in Ref. [48], the observed WDs related to the SV launches.

4.7 Main Results

1. The method, close to the optimal, was performed for identification of disturbances of electron concentration in the ionosphere based on measurements of the power of the IS signal. It was shown that the proposed method allows detecting and identifying WDs with the relative amplitude from several portions of percentages and more for changes of SNR, q, from 3 to 0.1. It is interesting to notice that further increase of q (of more than 3–5) does not lead to significant increase of sensitivity of the proposed method.

2. By using the IS and DS cluster methods, observations carried out during time periods close to the three RLs from space ports showed that the distances differed by about five times, and also that their mass and power differed substantially. The energy characteristics (the average power) of the IS radar in different observations differed by more than in one order. The geophysical environment and situation also differed in all the three launches.

3. The observations of the Soyuz launch were exactly at the same time of the morning ST occurrence, which essentially complicated identification of the effects caused by the launch. We suppose that the launch was accompanied by the generation of two types of disturbances, whose velocities were about 5.8 km s^{-1} and 700–800 m s^{-1}. Such magnitudes of the velocity are close to velocities of slow MHD waves with frequency of $\omega \approx 10^{-3}$ s^{-1} as also to velocities of acoustic type waves. This allows us to state with a definite degree of confidentiality that the WDs related to the SV launch were observed during observations.

4. Observations of Space Shuttle launch were carried out at the transferred daytime period, when the decrease of SNR seen in the IS method reached 1 order of magnitude. Disturbances of the second type having average velocity along the trace of propagation of about 2.4–2.5 km s^{-1} were observed more confidently. Such a velocity corresponds to the velocity of slow MHD waves. Due to this reason, these WDs were generated by the SV launch.

5. Observations of the Delta launch were carried out before the evening terminator, but that related exactly to its movements. Using the IS and DS methods, disturbances with definite amount of confidence can be related to the SV launch. The velocities of these WDs are close to the velocity of slow MHD waves. Moreover, disturbances propagating with an "apparent" velocity of 25 km s^{-1} were observed via the DS method. Perhaps these WDs are caused by the generation of gyrotropic waves during the launch.

   WDs caused by the passing of the evening terminator were observed vividly. Their relative amplitude was about 1.5%–2%, the period of oscillations about of 1 h, and the life-time more than 2 h.

6. Generally, we demonstrated the possibility to observe the sufficiently weak WDs (with $\delta_N \sim 1\%$) with periods of 10 min and more by ISR. The error in estimations of δ_N usually did not exceed 1%. The results obtained by the IS and DS methods were agreed with each other.

7. The obtained magnitudes of propagating WD velocities accompanying the SV launches are in a good agreement with the results by other authors (see, e.g., Refs. [6,30–37,54]), as well as from our earlier studies (see, e.g., Refs. [47–49,51,52]).

References

1. Dikiy, A. A., *Theory of Earth's Atmosphere Oscillations*, Moscow: Hydrometeoizdat, 1969 (in Russian).
2. Beer, T., *Atmospheric Waves*, New York: Prentice-Hall International, 1974.
3. Gossard, E. E. and Hooke, W. H., *Waves in the Atmosphere: Atmospheric Infrasound and Gravity Waves, Their Generation and Propagation*, Amsterdam: Elsevier Scientific Publishing Co., 1975, 476.
4. Kazimirovsky, E. S. and Kokourov, V. D., *Movements in the Ionosphere*, Novosibirsk: Nauka, 1979 (in Russian).
5. Avakyan, S. V., Drobgev, V. I., Krasnov, V. M., et al., *Waves and Radiation of the Upper Atmosphere*, Alma-Ata: Nauka, 1981 (in Russian).
6. Sorokin, V. M. and Fedorovich, G. V., *Physics of Slow MGD-Waves in the Ionospheric Plasma*, Moscow: Energoatomizdat, 1982 (in Russian).
7. Drobgev, V. I., Ed., *Wave Disturbances in the Ionosphere*, Alma-Ata: Nauka, 1987 (in Russian).
8. Booker, H. G., A local reduction of F region ionization due to missile transit, *J. Geophys. Res.*, 66, 1073–1079, 1961.
9. Cotton, D. E., Donn, W. L., and Oppenheim, A., On the generation and propagation of shock waves from "Apollo" rockets at orbital altitudes, *Geophys. J. Roy. Aston. Soc.*, 26, 1496–1503, 1971.
10. Chernogor, L. F. and Rozumenko, V. T., Wave processes, global- and large-scale disturbances in the near-Earth plasma, in *Proc. Int. Conf. Astronomy in Ukraine-2000 and Perspective, Kinematics and Physics of Sky Bodies*, Annex K, Kiev: Academy of Sciences, No. 3, 514–516, 2000.
11. Kellog, W. W., *Space Sci. Rev.*, 3, 275–283, 1964.
12. Nagorskii, P. M., Analysis of the short-wave signal response on ionospheric plasma perturbation caused by shock-acoustic waves, *Izv. Vuzov. Radiofizika*, 42, 36–44, 1999 (in Russian).
13. Chernogor, L. F., Garmash, K. P., Kostrov, L. S., et al., Perturbations in the ionosphere following U.S. powerful space vehicle launching, *Radiophys. Radioastron.*, 3, 181–190, 1998.
14. Gorely, K. I., Lampey, V. K., and Nikol'sky, A. V., Ionospheric effects of cosmic vehicles burn, *Geomagn. Aeronom.*, 4, 158–161, 1994.
15. Rozumenko, V. T., Kostrov, L. S., Martinenko, S. I., et al., Studies of global and large-scale ionospheric phenomena due to sources of energy of different nature, *Turkish J. Phys.*, 18, 1193–1198, 1994.
16. Garmash, K. P., Kostrov, L. S., Rozumenko, V. T., et al., Global disturbances of the ionosphere caused by launching of spacecrafts at the background of magnetic storm, *Geomagn. Aeronom.*, 39, 72–78, 1999.
17. Chernogor, L. F., Kostrov, L. S., and Rozumenko, V. T., Radio probing of the perturbations originating in the near-Earth plasma from natural and anthropogenic energy sources, in *Proc. Int. Conf. Astronomy in Ukraine-2000 and Perspective, Kinematics and Physics of Sky Bodies*, Annex K, Kiev: Academy of Sciences, No. 3, 497–499, 2000.

18. Danilov, A. D., Kazimirovsky, E. S., Vergasova, G. V., and Hachikyan, G. Ya., *Meteorological Effects in the Ionosphere*, Leningrad: Hydrometeoizdat, 1987.
19. Kostrov, L. S., Rozumenko, V. T., and Chernogor, L. F., Doppler radio sounding of the naturally perturbed middle ionosphere, *Radiophys. Radioastron.*, 4, 209–226, 1999 (in Russian).
20. Chernogor, L. F., Physics of the earth, atmosphere and geospace in the lightness of system's paradigm, *Radiophys. Radioastron.*, 8, 59–106, 2003 (in Russian).
21. Blaunstein, N. and Plohotniuc, E., *Ionosphere and Applied Aspects of Radio Communication and Radar*, New York: Taylor and Francis, 2008.
22. Evans, J., Theoretical and applied questions of study of the ionosphere by methods of incoherent scattering of radio waves, *Proc. IEEE*, 57, 139–175, 1969.
23. Evans, J., Study of the ionosphere by powerful radio locators, *Proc. IEEE*, 63, 520, 1975.
24. Brunelli, B. E., Kochkin, M. I., Presnyakov, I. N., Tereshenko, E. D., and Tereshenko, V. D., *Method of Incoherent Scattering of Radio Waves*, Leningrad: Nauka, 1979 (in Russian).
25. Thome, G. D., Incoherent scatter observations of TIDs, *J. Geophys. Res.*, 69, 4047–4049, 1964.
26. Wang, R. H., Semidiurnal tide in the *E* region from incoherent scatter measurements at Aresibo, *Radio Sci.*, 11, 641–652, 1976.
27. Farley, D. T., Incoherent scatter radar probing, in *Modern Ionospheric Science*, Kohle, H., Ruster, R., and Schlegel, K., Eds., Katlenburg-Lindau: Max-Plank Institute of Aeronomy Press, 1996, 415–439.
28. Burmaka, V. P., Taran, V. I., and Chernogor, L. F., Wave disturbances in the ionosphere accompanied rockets' launchings at the background of natural transferred processes, *Geomagn. Aeronom.*, 44, 518–534, 2004.
29. Amiantov, I. N., *Selected Questions of Statistical Theory of Communication*, Moscow: Sov. Radio, 1971 (in Russian).
30. Booker, H. G., A local reduction of *F* region ionization due to missile transmit, *J. Geophys. Res.*, 66, 1073–1079, 1961.
31. Arendt, P. R., Ionospheric undulations following "Apollo-15" launching, *Nature*, 231, 438–439, 1971.
32. Arendt, P. R., Ionospheric shock front from "Apollo-15" launching, *Nature*, 236, 8–9, 1972.
33. Mendillo, M., Hawkins, G. S., and Klobuchar, J. A., A sudden vanishing of the ionospheric *F* region due to the launch of "Skylab," *J. Geophys. Res.*, 80, 2217–2228, 1975.
34. Mendillo, M., The effects of rocket launches of the ionosphere, *Adv. Space Res.*, 1, 275–290, 1981.
35. Mendillo, M., Modification of the ionosphere by large space vehicle, *Adv. Space Res.*, 2, 150–156, 1982.
36. Mendillo, M., Ionospherie holes: A review of theory and resent experiments, *Adv. Space Res.*, 8, 51–62, 1988.
37. Nagorskii, P. M. and Taraschuk, Yu. U., Artificial modification of the ionosphere by rocket launching transported on the orbit the space vehicles, *Izv. Vuzov. Phys.*, 36, 98–107, 1993 (in Russian).

38. Garmash, K. P., Leus, S. G., Chernogor, L. F., and Shamota, M. A., Geomagnetic pulsations accompanied starts of rockets from various cosmodromes of the world, *Cosmic Sci. Technol.*, 15, 31–43, 2009.

39. Nagorskii, P. M., Inhomogeneous structure of *F*-region of the ionosphere generated by rockets, *Geomagn. Aeronom.*, 38, 100–106, 1998.

40. Deminov, M. G., Oraevsky, V. N., and Rugin, Yu. Ya., Ionospheric-magnetospheric effects of launchings of rockets in directions to the high latitudes, *Geomagn. Aeronom.*, 41, 772–781, 2001.

41. Afraymovich, E. L., Chernuhov, V. V., and Korushkin, V. V., Spatial-temporal characteristics of ionospheric disturbance caused by shock acoustic waves generated by launchings of rockets, *Radiotech. Electron.*, 46, 1299–1307, 2001 (in Russian).

42. Afraymovich, E. L., Perevalova, N. P., and Plotnikov, A. V., Generation of ionospheric responses on shock acoustic waves generated by launching of rocket-carriers, *Geomagn. Aeronom.*, 42, 790–797, 2002.

43. Adushkin, V. V., Kozlov, S. I., and Petrov, A. V., Eds., *Ecological Problems and Risks of Actions of Rocket-Cosmic Technique on Natural Environment*, Moscow: Ankil, 2000 (in Russian).

44. Zasov, G. F., Karlov, V. D., Romamnchuk, T. E., et al., Observations of disturbances at the lower ionosphere during experiments on the program Soyuz–Apollo, *Geomagn. Aeronom.*, 17, 346–348, 1977.

45. Noble, S. T., A large-amplitude traveling ionospheric disturbance excited by the Space Shuttle during launch, *J. Geophys. Res.*, 95, 19037–19044, 1990.

46. Karlov, V. D., Kozlov, S. I., and Tkachev, G. N., Large-scale perturbations in the ionosphere caused by flight of rocket with working engine, *Cosmic Res.*, 18, 266–277, 1980 (in Russian).

47. Garmash, K. P., Rozumenko, V. T., Tyrnov, O. F., Tsymbal, A. M., and Chernogor, L. F., Radio physical investigations of processes in the near-the-earth plasma disturbed by the high-energy sources, *Foreign Radioelectronics: Success in Modern Radioelectronics*, 7, 3–15, 1999 (in Russian).

48. Burmaka, V. P., Kostrov, L. S., and Chernogor, L. F., Statistical characteristics of signals of Doppler HF radar during sounding the middle ionosphere perturbed by rockets' launchings and solar terminator, *Radiophys. Radioastron.*, 8, 143–162, 2003 (in Russian).

49. Taran, V. I., Pod'yachii, Yu. I., Golovin, V. I., Vashenko, V. I., and Arkad'ev, I. D., Traveling ionospheric disturbances found by method of incoherent scattering, *Ionos. Res.*, 27, 102–110, 1979 (in Russian).

50. Taran, V. I., Pod'yachii, Yu. I., and Maksimov, A. A., Long-periodical disturbances of the ionosphere of technogeny nature, *Ionos. Resp. Join. Sci. Tech. J.*, 32–41, 1991 (in Russian).

51. Garmash, K. P., Rozumenko, V. T., Tyrnov, O. F., Tsymbal, A. M., and Chernogor, L. F., Radio physical investigations of processes in the near-the-earth plasma disturbed by the high-energy sources, *Foreign Radioelectronics: Success in Modern Radioelectronics*, 8, 3–19, 1999 (in Russian).

52. Kostrov, L. S., Rozumenko, V. T., and Chernogor, L. F., Doppler radio sounding of perturbations in the middle ionosphere accompanied burns and flights of cosmic vehicles, *Radiophys. Radioastron.*, 4, 227–246, 1999 (in Russian).

53. Nalesso, G. F. and Jacobson, A. R., On a mechanism for ducting of acoustic and short-period acoustic-gravity waves by the upper atmospheric thermocline, *J. Ann. Geophys.*, 11, 372–376, 1993.

54. Jacobson, A. R. and Carlos, R. C., Observation of acoustic-gravity waves in the thermosphere following Space Shuttle ascents, *J. Atmos. Terr. Phys.*, 56, 525–528, 1994.

55. Hantadze, A. G., On a new branch of own oscillations of the electro-conductive atmosphere, *Rep. Acad. Sci.*, 376, 250–252, 2001 (in Russian).

56. Hantadze, A. G., Electromagnetic planetary waves in the Earth's ionosphere, *Geomagn. Aeronom.*, 42, 333–335, 2002.

57. Taran, V. I., Study of the ionosphere in the natural and artificially excited states by method of incoherent scattering, *Geomagn. Aeronom.*, 41, 659–666, 2001.

58. Akimov, L. A., Grigorenko, E. I., Taran, V. I., Tyrnov, O. F., and Chernogor, L. F., Complex radiophysical and optical studies of dynamic processes in the atmosphere and geo-space caused by eclipse of the Sun on August 11, 1999, *Foreign Radioelectronics: Success in Modern Radioelectronics*, 25–63, 2002.

Chapter 5

Results of Similar Types of Experiments

5.1 Overview

As mentioned in Chapters 2–4, despite the fact that a study of ionospheric phenomena caused by the launch and flight of space vehicles (SVs) is fully illuminated in many articles and specific works, some of which were summarized in a special textbook and reports [1–3], there are many problems that were missed and were out of the scope of the discussion. As a rule, the effects observed only along the active part of the rocket trajectories were described in existing works (see, e.g., Refs. [4–6]). Much fewer works dealt with the study of the effects at distances of ~1–10,000 km from the rocket's trajectory [7–14]. Until now, detailed physical–mathematical models that described the corresponding processes accompanying rocket launches (RLs) and flights were not prepared. Moreover, it is impossible to differentiate the disturbances created by RL and flight; it is not clear what are the mechanisms of generation and transfer of disturbances, at significant and global distances from the rocket. Most problems are because of difficulties in differentiating natural and artificial perturbations excited in the upper atmosphere [10–14].

Therefore, the main requirement for the investigators was to increase the number of observations (i.e., enormous statistic of recording data) and to search repeated phenomena more precisely. This was the first aspect of the problems that existed in the current studies.

The second aspect relates to the following problems. In recent decades, great attention was paid to the systematic study of physical processes in the system that is the Earth–atmosphere–ionosphere–magnetosphere–interplanetary medium–Sun (see, e.g., Refs. [15–17]). The main point in such investigation was study of the reaction of the atmosphere and the geospace environment on artificial sources with a high yield of energy. This aspect allowed the researchers to search and track the transfer process of disturbances from short to long distances (~1–10,000 km), define the types of waves responsible for creating such disturbances, as well as define and understand the peculiarities of interaction of each subsystems (atmosphere, ionosphere, etc.) in the whole system mentioned earlier.

As an artificial source of energy radiation, it is convenient to observe powerful RLs at different points all over the world that are ranged far from the place of observation at about 1–10,000 km. For the observer this source is "free of charge." At multiple ranges away from the cosmodromes, various types of rocket power, various content and consistency of fuel, and various trajectories of SVs support variations of the "initial conditions" and flexibility in the intensive study of reaction of the atmosphere and geospace environment (mostly the ionospheric plasma) on such a source. Because the processes accompanying launches and their parameters have not been sufficiently studied, it is obvious from the start to organize a systematic analysis of the possible reaction of media on launch of similar types of rockets from the same cosmodrome.

The third aspect is as follows (we discuss it in the Introduction to this book). Close to the demarcation between the twentieth and twenty-first centuries, increase of rocket power and the frequency of their launches attained such magnitude that it was necessary to account for the disturbances excited in the atmosphere and the geospatial environment. The study of these perturbations is an interesting subject for specialists in the fields of geophysics, radiophysics, atmospheric ecology, geospace, and space weather.

All the above-mentioned aspects define the actuality of our investigations, which can be considered as continuation of the work carried out by other authors [9–14,18–26,31–33].

5.2 Method of Identification of Wave Disturbances

As given in our description below, we base our theory on the method of identification of wave disturbances (WDs) described in Chapter 4 and in Refs. [13,14]. This method considers statistical averaging of massive amounts of data as a combination of "incoherent scatter signal plus noise" and "pure noise," to obtain basic estimations of signal power P_s and noise power P_n at time interval $\Delta T_0 = 1.0-1.5$min. After eliminating the trend $\overline{P}_s$ calculated at time interval $n_1 \Delta T_0$ with a slicing step of ΔT_0 (usually $n_1 = 40-60$), the time series $\delta P_s = P_s - \overline{P}_s$ was found. For the possibility of comparing the amplitudes of WDs at different altitudes z, dependence $P_s \propto z^{-2}$ was taken into account (or the dependence of SNR, $q \propto z^{-2}$, see Chapter 4). Here also dependence $P_s \propto (1+T_e/T_i)$ was eliminated, where T_e and T_i are the temperatures of electrons and ions, respectively. The sets δP_s used the so-called smoothness procedure at a time interval $n_2 \Delta T_0$ with a slicing step of ΔT_0 (usually $n_2 = 10-15$). The smoothing magnitudes of $\langle \delta P_s \rangle$ obtained in such a manner are proportional to the absolute values of variations ΔN of electron concentration N at the desired altitude. Finally, we get the relative amplitudes $\delta_N = \Delta N/N \approx \langle \delta P_s \rangle / \langle P_s \rangle$.

The dependence of P_s consists of nonregular and regular (quasiperiodical) components. Using the smoothness procedure at time interval $n_2 \Delta T_0$ allows us to find the WD if its amplitude is essentially lower than the mean square deviation of the nonregular component. The benefit in the signal-to-noise ratio (SNR) for this case was guaranteed by a correlation interval τ_c of the nonregular component (i.e., "noise") equaling ΔT_0, which is smaller than period T of the regular component (i.e., "signal"). Usually $T/\tau_c = 30-100$ and $n_2 \Delta T_0/\tau_c = n_2 = 10-15$.

For the chance of comparing the amplitudes of WDs, dimensionless magnitudes of type $\langle \delta q \rangle = \langle \delta P_s \rangle / \langle P_n \rangle$ were formed in different measuring campaigns, where $\langle P_n \rangle$ is the average noise power during the entire time of observations (usually for a day). If different magnitudes of P_s, τ_c, and Δf were used in various measuring campaigns, the parameters $\langle \delta q \rangle$ were unified to obtain the same magnitudes in different campaigns.

The necessity of normalization for $\langle P_n \rangle$ is because, as is known, measurement of P_s and P_n is usually done in relative magnitudes by the incoherent scattering (IS) method. The profile of electronic concentration obtained by this method is normalized with the help of measurements carried out by ionosondes, which leads to the estimation error $\langle \delta P_s \rangle / P_{s0} \approx 1\%-3\%$. It should be noted that measurement of P_s and P_n in absolute values could be done by using a special calibrator. Because of lack of it, we studied diurnal and seasonal variations of the statistical characteristics of P_n. Later knowledge allowed us to substantiate the possibility of normalizing the characteristics P_s and δP_s on $\langle P_n \rangle$.

For identification of WDs with the prevailing amplitudes, a corresponding spectral analysis of the time series of magnitudes $\langle \delta q(t) \rangle$ and $\delta_N(t)$ was carried out.

If before launch quasiperiodical variations occurred, their average amplitude and period were estimated and the corresponding process $\delta q_0(t)$ was formed.

The amplitude and period were estimated through its temporal dependence, as well as by the help of spectral analysis at a time interval that equals two real periods, that is, $2T$, or even exceeding $2T$. As a degree of the influence of launch at the level of WDs in the ionosphere, process $s_q(t) = \langle \delta q(t) \rangle - \delta q_0(t)$ was chosen. It is evident that for larger deviations of signal magnitudes $s_q(t)$ from zero level, more intensive "reaction" of the environment corresponded to SV launch.

5.3 Results of Observation of RLs from Russian Territory

5.3.1 Soyuz Launch on October 18, 2003

The observation is described as carried out in mid autumn. Its aim was to find WDs in the morning. The condition of magnetic activity during this and further launches is presented in Table 5.1. The radar parameters guaranteed SNR, $q = 0.1$–5, for nocturnal and diurnal time periods, respectively. In this case, the minimum value of δ_N, determined with relative accuracy of $\pm 50\%$, was $\delta_{N\,min} \approx 8\%$ (night) and 1% (day). The main parameters of the rocket are presented in Appendix 1.

The RL happened at 05:37 h (time is in Universal Time or UT here and later in the text). Measurements for the height range of 125–1000 km were carried out from 21:00 h on October 17, 2003, to 15:00 h on October 18, 2003. We should note that the time interval shown in later figures can be less than the time of measurements. SV launch happened in the morning; the SNR, q, in the middle (up to 300 km) and in the outer ionosphere corresponded to 3–5 and 0.3–1, respectively. In this case, $\delta_{N\,min} \approx 3\%$–$1\%$.

The characteristics of the working regimes of the radar are presented in Table 5.2. The effects of the morning terminator were sufficiently weak during the day and they were not observed at the moment of the start.

For 3–4 h after sunrise electron concentration had grown significantly in the ionosphere. During this time 5–10 times growth was also observed in the absolute amplitude of WDs, ΔN (Figure 5.1). The oscillations of $\langle \delta q \rangle$ became more regular. At altitudes of $z \approx 140$–180 km, the period of the main oscillations was 40–45 min. At the same time, the additional period $T \approx 110$ min was vividly registered at altitudes of 200–500 km. The duration of the process with $T \approx 110$ min was about 5 h. We noticed that this process became clear 13–20 min after the SV launch. Changes in the signal character after 06:00 h are seen clearly from the temporal behavior of the process $s_q(t)$ at an altitude of 320 km (Figure 5.1). The amplitude of variations s_{qm} increased approximately

Table 5.1 Geomagnetic Environment during Measurements

Date	Time of Observation [UT (h)]	K_p								A_p
		0–3	3–6	6–9	9–12	12–15	15–18	18–21	21–24	
October 18, 2003	02:14	4	5	4	4	4	3	4	4	27
December 10, 2003	14:24	5	4	5	6	5	5	5	4	42
December 11, 2003	00:02	5	5	6	5	4	5	3	4	40
December 27, 2003	15:24	3	3	2	3	3	3	3	4	12
December 28, 2003	00:03	3	3	2	3	3	3	3	2	12
January 29, 2004	11:17	1	2	2	4	4	3	2	2	10
March 15, 2004	21:24	4	3	2	3	3	3	3	2	13
March 16, 2004	00:03	3	2	1	3	2	3	2	2	8
May 25, 2004	11:16	2	1	3	3	3	2	2	2	8
June 17, 2004	21:24	1	3	2	2	2	2	3	2	7
June 18, 2004	00:03	1	2	2	3	2	3	2	2	8
October 29, 2004	18:24	0	0	1	3	3	2	3	1	7
October 30, 2004	00:03	4	3	3	4	3	3	2	3	17
February 3, 2005	00:06	1	4	3	1	1	2	1	1	8
March 29, 2005	19:24	0	1	1	0	1	2	2	3	5

Table 5.2   Characteristics of the Working Regimes of the Radar

Dates	P_i (MW)	τ (μs)	ΔF (kHz)
October 18, 2003	2.1	135	6
December 10–11, 2003	2.2	135	6
December 27–28, 2003	2	780	9.5
January, 2004	2	780	9.5
March 15–16, 2004	1.8	780	9.5
May 25, 2004	1.8	640	9.5
June 17–18, 2004	1.8	640	9.5
October 29–30, 2004	2	640	9.5
February 3, 2005	2.2	640	9.5
March 29, 2005	2.2	640	9.5

Note: The interval of the previous processing, $\Delta T_0 = 1$ min, and the step along the height, $\Delta z_0 = 10$ km; the height range is 125–1005 km.

three times. If before launch the relative amplitude δ_N was 2%–4%, after launch it slightly increased up to 5%–10%. Thus, for oscillations with $T \approx 110$ min, $\delta_N \approx 10\%$ (see Figure 5.1).

5.3.2 Soyuz Launch on December 27, 2003

The Soyuz launch took place at 21:30 h. The aim of this observation was to find WDs excited by SVs in the nocturnal period after winter solstice. Pulses of 780-μs duration were used for achieving sufficient magnitudes of SNR, q (see Table 5.2). For that, depending on the altitude and time of observation, $q \approx 0.2$–10 and $\delta_{N\min} \approx 4\% - 0.8\%$. Measurements for the altitudinal range of 125–1000 km were carried out in the time period from 13:00 h on December 27, 2003, to 01:00 h on December 28, 2003. Temporal variations of $\langle \delta q \rangle$ and δ_N are shown in Figure 5.2. It is seen that significant increase of amplitude $\langle \delta q \rangle$ was registered at altitudes of 250–350 km soon after the SV launch. These changes continued for more than 5 h. This phenomenon has also been indicated by the behavior of $s_q(t)$ at an altitude of 320 km where the amplitude of the signal, s_{qm}, grew in the order of value.

As can also be seen from Figure 5.2, weak variations of δ_N, which are supposed to be related to the passage of the evening terminator, have been

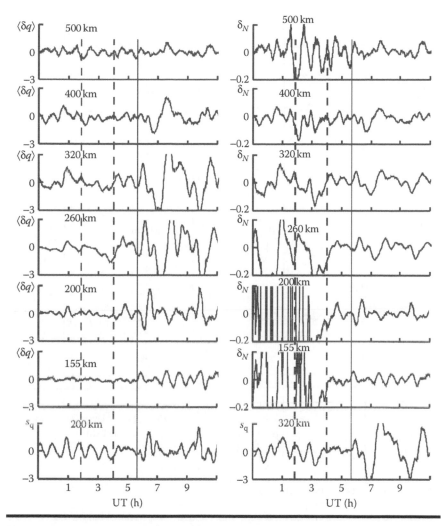

Figure 5.1 **Temporal variations of the amplitudes (left panel) and relative amplitudes (right panel) of WD electronic concentration accompanying Soyuz launch on October 18, 2003. The moment of start is indicated by vertical solid line. The moments of sunrise at an altitude of 450 km and the ground level are indicated by vertical dashed lines (here, and at the observatory near Kharkov, Ukraine).**

registered first (near 17:00 h). Then, for 4–5 h quasiperiodical variations of δ_N were observed with relative amplitude not exceeding 10%. After 21:00 h, its magnitude increased up to 30%. Oscillations with periods of about 65 and 110 min prevailed. The oscillations continued for not less than 4 h, and the corresponding WDs, observed after 21:30 h, probably were related to SV launches.

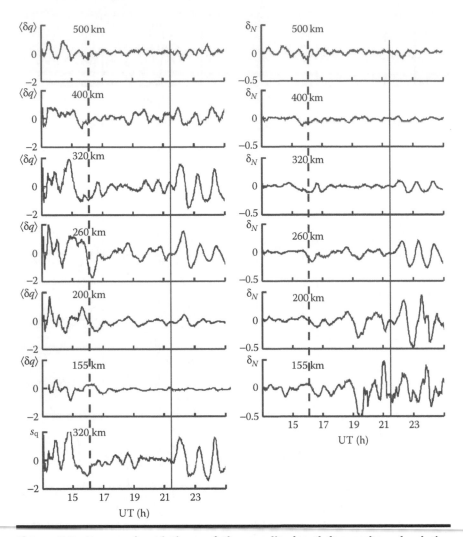

Figure 5.2 Temporal variations of the amplitudes (left panel) and relative amplitudes (right panel) of WD electronic concentration accompanying Soyuz launch on December 27, 2003. The moment of start is indicated by vertical solid line. The moments of sunset at an altitude of 450 km and at ground level are indicated by vertical dashed lines.

Their amplitude depended on the altitude of observation. Thus, for the first 50 km, starting from 155 km, the amplitude δ_N increased with increase of altitude, and then at the range of altitude from 200 to 400 km, it decreased by more than 1 order of magnitude. The maximum value of $\delta_N \approx 40\%-50\%$ was registered at an altitude of about 200 km. This altitude can be considered as the height of the atmospheric waveguide. The effective width of this waveguide (at the level of 0.5 $\delta_{N\,max}$) was about 120 km, where the top and bottom "walls" of such a thermospheric waveguide were located at altitudes of about 140 and 260 km, respectively.

5.3.3 Soyuz Launch on January 29, 2004

The aim of this observation was to find WDs in the diurnal time period and in the beginning of the second half of winter. SNR, q, achieved during this period of observation was about 4, 12, 4, 0.8, 0.3, and 0.05 at altitudes of 120, 200, 300, 400, 500, and 700 km, respectively. The relative error in estimation of δ_N was about 15% (at 120 km), 10% (at 200 km), 10% (at 300 km), 25% (at 400 km), 50% (at 500 km), and 80% (at 700 km).

The rocket started at 14:00 h. Measurements were carried out in the time interval from 11:00 to 16:00 h. Time dependence $N(t)$ has maximum overall height at the time interval from 13:00 to 13:30 h. WDs took place both before and after the SV launch (Figure 5.3). Wave processes of 55 and 95 min periods were observed before the launch. Relative amplitude δ_N attained 10%. The RL implemented the changes in the character of temporal variations of $\langle \delta q(t) \rangle$ and $\delta_N(t)$ and suppression of existing oscillations in the ionosphere. This phenomenon was observed at a height range of 125–500 km, and more clearly at 200–260 km. The relative amplitude δ_N changed from 5% to 10%. A delay of about 40–50 min corresponded to maximum changes in the behavior of $\langle \delta q(t) \rangle$ and $\delta_N(t)$. More fast disturbances were observed less definitely. These disturbances are indicated by an essential increase of $\langle \delta q \rangle$ in the first wave, which follows after launch at altitudes of 180–300 km (in Figure 5.4 at an altitude of 200 km). The delay of this disturbance in that case was about 10 min.

5.3.4 Soyuz Launch on May 25, 2004

The observation was carried out at the end of spring. Its aim was to study WDs that accompanied SV launches at diurnal time periods in conditions of weak magnetic activity. In the experiment, sounding pulses with duration 640 μs were used, which allow obtaining height resolution close to 100 km. Magnitudes of SNR, q, were changed from 0.5 to 6. The relative error in estimation of δ_N was 50%–10%, respectively.

Before launch WDs were observed at all altitudes (Figure 5.4). Their amplitude was maximal at altitudes from 200 to 250 km. The relative amplitude if the WD was insignificant was not more than 4%. Oscillations with a

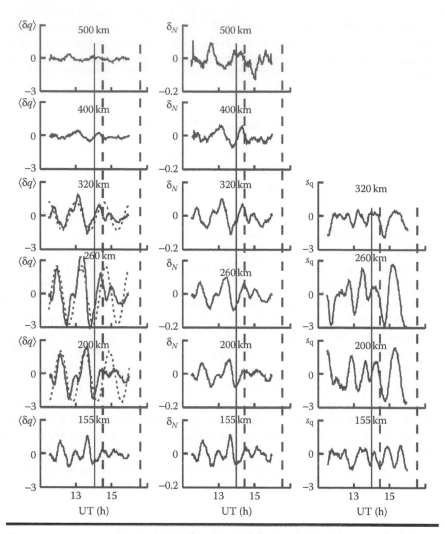

Figure 5.3 **Temporal variations of the amplitudes and relative amplitudes of WD electronic concentration accompanying Soyuz launch on January 29, 2004. The moment of start is indicated by solid vertical line. The moments of sunset at an altitude of 450 km and at ground level are indicated by vertical dashed lines. The oscillation that was predominant before the launch is indicated by a dotted line.**

period of 30 min prevailed. The SV launch started at 12:34 h. For 60–70 min after launch at altitudes of 150–200 km "distortion" of the wave process remained in the ionosphere. This process continued for about 40–50 min. After 14:00 h another wave process was observed with a larger quasiperiod (40–60 min).

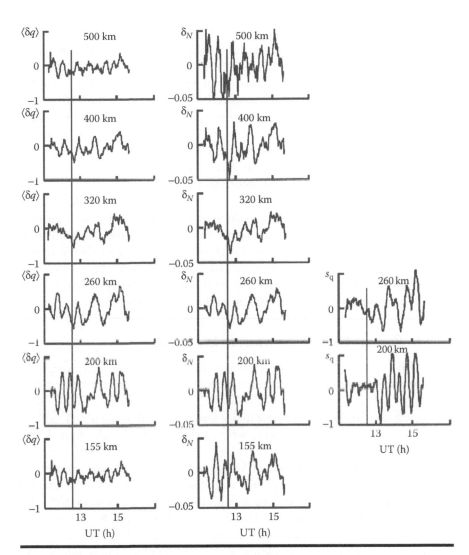

Figure 5.4 Temporal variations of the amplitudes and relative amplitudes of WD electronic concentration accompanying Soyuz launch on May 25, 2004. The moment of start is indicated by solid vertical line.

5.3.5 Proton Launch on December 10, 2003

The launch happened at 17:22 h on December 10, 2003. The main parameters of Proton SV are presented in Appendix 1. The aim of observation was to study the possibility of finding WDs excited by launch at the nocturnal time intervals during winter, when magnitudes of electron concentration were sufficiently low ($N \leq 10^{11}\,\text{m}^{-3}$) that led to low magnitudes of SNR, q. Measurements were carried out at the height range of 125–1000 km from 14:00 h on December 10, 2003, to 06:00 h on December 11, 2003. The characteristics of the working regimes of the radar are presented in Table 5.2. Because the SV launch was done at nocturnal time period, the SNRs, sufficient for detection ($q \geq 0.1$–0.2), were registered only at the height range of $z \approx 290$–455 km. It is important to note that in the winter night period, P_s is higher than during the transferred periods of the day. This happens more clearly at an altitude of $z > 350$ km. Such a behavior of $P_s(t)$ was caused by night flows of plasma from the plasmosphere.

The WDs in the ionosphere were observed both before and after the SV launch (Figure 5.5). The absolute values of their amplitudes were changed weakly in time. After launch, the magnitudes of the relative amplitude at an altitude of $z \leq 350$ km have increased slightly. Apparently, with the SV launch, changes in the signal character were related, having delays of $\Delta t_1 = 15$–20 min and $\Delta t_2 = 60$–70 min. These changes were observed better at altitudes of 308–418 km and at a height of 320 km, as shown in Figure 5.5.

5.3.6 Proton Launch on March 15, 2004

The main peculiarity of this observation was the study of WDs that accompanied SV launches at nocturnal time periods during the spring equinox and the SNR, q, had magnitude from 0.2 to 1, whereas during the diurnal time period q attained a magnitude of 10. The relative error in δ_N estimation at altitudes of 125–350 km did not exceed 20%. Measurements were carried out at altitudes from 125 to 1000 km.

WDs with a prevailing period of 55 min were observed before launch at all altitudes (Figure 5.6). Their amplitude was maximal for $z \approx 320$–350 km. The relative amplitude of WDs attained maximum magnitudes, close to 30%–40% at altitudes of 150–200 km. The launch occurred at 23:06 h. The structure of the existing WDs was "destroyed" about 65 min post launch. The effect of "distortion" was observed more clearly at $z \approx 250$–350 km. The amplitude $\langle \delta q \rangle$ decreased by 3–5 times. The temporal behavior of the process $s_q(t)$ testifies to this (see Figure 5.6).

5.3.7 Proton Launch on June 17, 2004

The peculiarity of this observation was that it happened near the summer solar solstice. Moreover, the launch at 22:27 h happened in the background of natural perturbations in the ionosphere caused by the evening terminator (Figure 5.7).

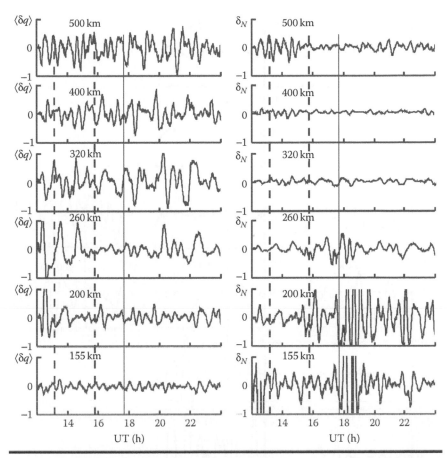

Figure 5.5 **Temporal variations of the amplitudes (left panel) and relative amplitudes (right panel) of WD electronic concentration accompanying Proton launch on December 10, 2003. The moments of sunset at an altitude of 450 km and at the Earth's level are indicated by vertical dashed lines.**

The geomagnetic environment was quiet during the launch. Magnitudes of q were changed in limits from 1 to 10. Usually, the relative error in δ_N estimation was about 15% but sometimes achieved 50%.

Before launch, WDs were observed only at altitudes of 150–400 km (see Figure 5.7). Their period was about 30–40 min. The amplitude of WDs was maximal at altitudes of 250–300 km. The relative amplitude of WDs were maximal (~30%) at an altitude of 155 km, but the higher it went the quickly it decreased. Probably these WDs were related to the evening solar terminator. The amplitude of the WDs changed significantly at the time interval of 22:30–23:30 h, that is, after the SV launch. Further variations of WDs were observed after about 30 min.

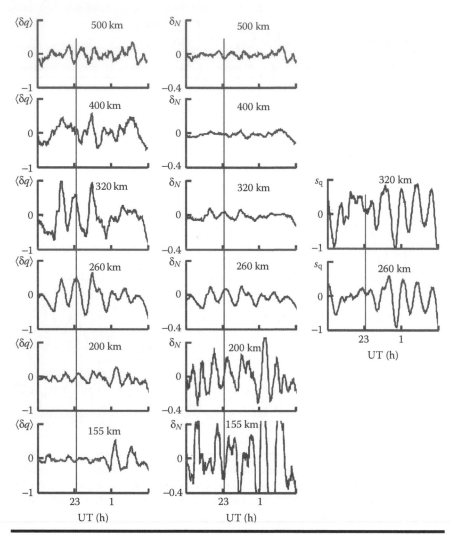

Figure 5.6 Temporal variations of the amplitudes and relative amplitudes of WD electronic concentration accompanying Proton launch on March 15, 2004.

5.3.8 *Proton Launch on October 29, 2004*

The peculiarity of this observation was that the launch happened after local half-night time (at 01:11 or 22:11 UT). Geomagnetic perturbations were absent during the launch. SNR was 0.2 (at 200 km), 2 (at 300 km), 1 (at 400 km), and 0.3 (at 500 km). The relative error in δ_N estimation did not exceed 10%

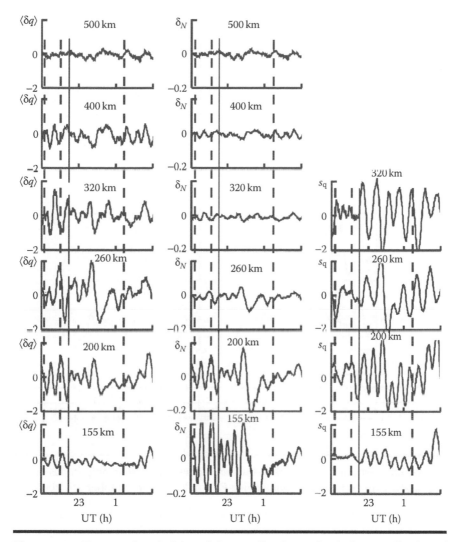

Figure 5.7 **Temporal variations of the amplitudes and relative amplitudes of WD electronic concentration accompanying Proton launch on June 17, 2004. The moments of sunset at an altitude of 450 km and sunrise at an altitude of 450 km and the ground level are indicated by vertical dashed lines.**

at altitudes of 200–300 km, continuously increasing with height and attaining 40%–50% at an altitude of 500 km. Before launch, oscillations with a main period of 60 min were observed at all altitudes (Figure 5.8). The relative amplitude of WDs at 150–200 km attained 30%–40%; above this height range δ_N decreased and at 500 km it did not exceed 6%.

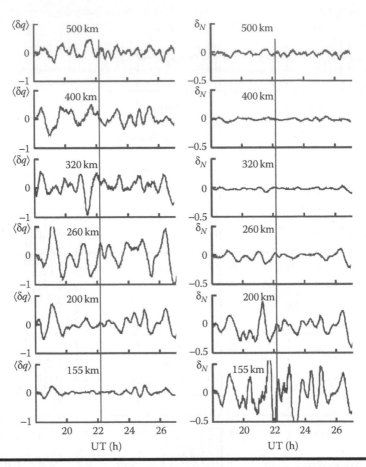

Figure 5.8 Temporal variations of the amplitudes and relative amplitudes of WD electronic concentration accompanying Proton launch on October 29, 2004.

The character of the WDs changed essentially after the launch: the period of the predominant oscillations became about 30 min and their amplitude decreased 1.5–2 times. It is important to note that this was observed mainly at the height range of 200–500 km. Magnitudes of δ_N also decreased (this can be seen at altitudes of 200–320 km more clearly, see Figure 5.8). "Suppression" of wave processes was also observed.

5.3.9 Proton Launch on February 3, 2005

The launch of Proton happened at 02:27 h, practically at the moment of sunrise in the outer ionosphere (altitudes of 400–1000 km). The geomagnetic perturbations were negligible. SNR was 0.5–9 (at 150 km), 1–9 (at 200 km), 0.5–2 (at 300 km),

0.2–0.5 (at 400 km), and 0.1–0.2 (at 500 km). The relative error in δ_N estimation did not exceed 10%–50% at altitudes of 200–500 km, respectively.

As shown in Figure 5.9, WDs have registered in the earlier launches (from 00:00 to 02:30 h). Close to 03:00–04:30 h the "suppression" of WDs was observed. After 04:30 h the amplitude of the WDs increased 5–10 times. The essential wave activity was registered during the entire "lightness" time period. Stronger amplitudes of WDs were recorded in the period from 05:00 to 09:00 h when the relative amplitude of WDs achieved 10%.

5.3.10 Proton Launch on March 29, 2005

The Proton launch happened at 22:31 h in the outer ionosphere, that is, 4 h after sunset. SNR, q, was about 0.2 (at 150 km), 1 (at 200 km), 2 (at 300 km), 0.6 (at 400 km), and 0.2 (at 500 km). The relative error in δ_N estimation did not exceed 5%–50% depending on the altitude and magnitude of q. The period of observation was the magnetically quiet one (see Table 5.1).

WDs were observed clearly at altitudes of 180–500 km and their maximal amplitude attained 1 at the height range of 200–300 km (Figure 5.10). In this case, δ_N changed in limits of 30%–5%. Wave processes were observed both before and after the launch. At first they were observed 10–20 min after the launch, then they decreased several times, but after 50–60 min of the launch, increased 5–6 times; the period of oscillation also changed essentially. The duration of the described variations attained several tens of minutes.

5.4 Results and Discussions

To find the reaction of the geospace environment on RLs, we analyzed the temporal dependences of P_s, electron concentration N, temperatures of electrons T_e, and ions T_i. The dependences of $P_s(t)$ or their derivatives, δP_s and $\langle \delta q \rangle$, have much less error than for other parameters of plasma. This aspect relates to the influence of additional errors during resolution of inverse problem in the IS method that is used for receiving these parameters of plasma, that is, N, T_e, and T_i. Errors in estimation of the latter parameters are several times larger than those in estimation of δP_s and $\langle \delta q \rangle$. Therefore, in this chapter, the results of analysis of the dependences $\delta P_s(t)$ and $\langle \delta q(t) \rangle$ are presented and discussed.

For the second phase of signal processing, we use and present spectral analysis at the time interval of 2–4 h with different slicing intervals (from 0.2 to 0.5 h). Using such an approach, it is impossible to determine the time delay Δt of WDs occurring with respect to the moment of the SV launch. Slightly better results give the wavelet analysis. For the estimation of Δt, it seems more preferable to

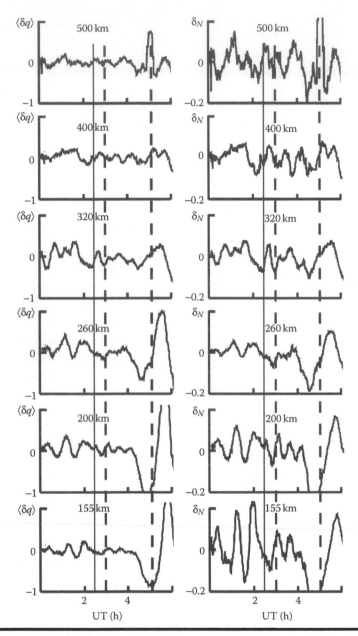

Figure 5.9 Temporal variations of the amplitudes and relative amplitudes of WD electronic concentration accompanying Proton launch on February 03, 2005. The moments of sunrise at an altitude of 450 km and at the Earth's level are indicated by vertical dashed lines.

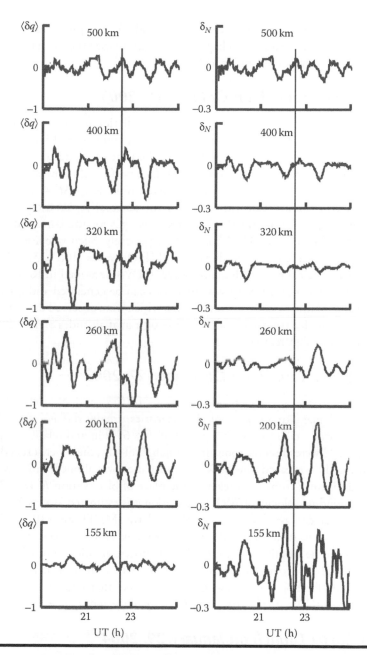

Figure 5.10 **Temporal variations of the amplitudes and relative amplitudes of WD electronic concentration accompanying Proton launch on March 29, 2005.**

analyze the temporal dependence of $\delta P_s(t)$ and $\langle \delta q(t) \rangle$. Just these results are presented and discussed in this chapter. Let us summarize the main features that accompanied launches of similar rockets.

5.4.1 Soyuz Launch on October 18, 2003

Actually, this rocket launch was related with oscillation of the recording signal having a period of 110 min. It appeared in the period $\Delta t = 13\text{--}20$ min after launch. As was done earlier, supposing that the time of the rocket's movement to the region of effective generation of WDs (at $z = 100\text{--}120$ km) is $\Delta t_0 = 3$min, following Refs. [10,11,13], we get the corrected time delay $\Delta t' = \Delta t - t_0$. Then, the corrected distance from the rocket to the place of registration will equal R'. For $\Delta t_1' = 10\text{--}17$min and $R' \approx 2100$ km, we get for the corrected velocity of WDs traveling of $v' \approx 3.5\text{--}2.1$ km s^{-1}, respectively. Such velocities are close to those of slow magnetohydrodynamic (SMHD) waves [10,11,13,14,27]. The latter can also be called the Hantadze waves (HWs) described in Refs. [28–30] (see details in Chapter 2). Very slow processes can be generated and travel with the velocities of inner gravity waves (IGWs), which were observed during the day of the launch with a time delay of $\Delta t_2' \approx 50\text{--}60$min at altitudes of 400–200 km, respectively (see Figure 5.1).

5.4.2 Soyuz Launch on December 27, 2003

Before launch, the already chaotic variations of $\langle \delta q \rangle$ and δ_N were observed; the maximal magnitudes of these characteristics usually did not exceed 10%. Quasiperiodical oscillations with a period of $T \approx 65$ min and relative amplitude achieving 43% were observed soon after launch. Probably, these were excited by the SV launch. The corrected time of WDs propagation was about 13 min. With the corrected distance from the rocket to the place of observation $R' \approx 2100$ km, the corresponding velocity of WDs propagation was estimated as $v' \approx 2.7$ km s^{-1}. Such velocities for wave frequency $\omega \sim 10^{-1}$s^{-1} have SMHD waves [10,11,27]. The IGWs were not clearly registered during observations. Probably, these waves caused an increase in the amplitude of the oscillation process at altitudes of 250–300 km (see Figure 5.2) due to arrival of faster waves, for which the time delay was about 50 min. Thus, for $\Delta t_2' \approx 47$min, we get $v_2' \approx 740$ m s^{-1} that actually corresponds to the velocity of IGWs at altitudes in the F-region of the ionosphere.

5.4.3 Soyuz Launch on January 29, 2004

Maximum variations are there in the absolute and relative amplitudes of WDs occurring 40–50 min after the SV launch. Velocities varying in the range of $v' \approx 740\text{-}950$ m s^{-1} correspond to such delays. Probably, this disturbance was excited by the generation and propagation of acoustic shock waves and IGWs. Faster disturbances have been rarely registered. It is probably because of these disturbances that the delay of about 10 min corresponded to $v' \approx 5$ km s^{-1}. Such a velocity in the ionosphere may have SMHD and MG waves.

5.4.4 Soyuz Launch on May 25, 2004

In this observation, the "distortion" caused by the existence of WDs in the ionosphere was registered 60–70 min after the RL. In this case velocities of WDs propagating at $v' \approx 520$–610 m s^{-1} were registered. A different wave process appeared after the "distortion" in the first one because of oscillations and time delay of about 90 min, to which the velocity of $v' \approx 400$ m s^{-1} corresponded. We can assume that these disturbances were caused by the flight of the SV.

5.4.5 Proton Launch on December 10, 2003

The effects of the SV launch in this case were fully masked by the WDs existing before the RL. Changes in the character of the signal—knock off of the phase of oscillations, changes in the signal amplitude, and the period, were observed 15–20 and 60–70 min after launch. This can be seen from variations of $\langle \delta q \rangle$ and δ_N at altitudes of 200–350 km (see Figure 5.5). If these variations were caused by the launch of the rocket, then we get velocities $v' \approx 2.9$–2.1 km s^{-1} and $v' \approx 610$–520 m s^{-1}, respectively. Such orders of magnitude have SMHD and MH waves with $\omega \sim 10^{-3}$ s^{-1} and IGWs in the F-region of the ionosphere [10,11,27].

5.4.6 Proton Launch on March 15, 2004

The structure of existing WDs was "destroyed" 65 min after the launch at altitudes of 150–350 km. The velocity of $v' \approx 600$ m s^{-1} corresponds to this time delay. Such a velocity has IGW. Therefore, there was evidence that the abovementioned disturbance was excited by the RL and flight of the rocket.

5.4.7 Proton Launch on June 17, 2004

The first essential change in the parameters of the wave process was started at about 10 min after launch, and the second after 60 min, to which the velocities $v' \approx 5$ km s^{-1} and $v' \approx 600$ m s^{-1} correspond. The common duration of disturbances in both types was about 120 min. Because the faster disturbance was observed after sunrise, the fact that it could be generated by the passage of the morning terminator cannot be ignored. The second disturbance, of course, was excited by the rocket's flight.

5.4.8 Proton Launch on October 10, 2004

Changes in the character of WDs were observed at 10–15 min after launch. If that disturbance related to the launch and flight of the SV, it corresponds to the corrected magnitude of velocity $v' \approx 3$–5 km s^{-1}.

5.4.9 Proton Launch on February 3, 2005

Comparison of the character of WDs taking place between February 2 and 3, 2005 (see Figure 5.9) allows us to conclude as follows. In the time interval 00:00–03:00 h the WDs were distinguished as insignificant. From about 03:00 to 04:30 h on February 3, 2005, the WD amplitude at 150–350 km altitudes decreased several times. Above 400 km, it changed insignificantly. The described "suppression" of WDs obviously was related to the RL. The character of WD variations after 04:30 h was caused by the passage of the morning terminator. The duration of strong WDs was observed as being not less than 4 h. The described effects were evidently seen at the height range of 125–350 km.

5.4.10 Proton Launch on March 29, 2005

In this instance, the RL was probably related with the increase in amplitude of the WDs that appeared 50–60 min after the launch (see Figure 5.10). The velocity of about 0.7–0.6 km s^{-1} and the period of 45 min correspond to the wave process. Another WD with a period of $T \approx 30$ min, delay of about 10–20 min, and velocity of $v' \approx 5$–2 km s^{-1} registered before this WD.

It is interesting to compare the parameters of WDs described earlier with those obtained by other authors. Thus, in Ref. [23], two types of WDs that sprang from the cosmodrome at 1275 km were observed by vertical sounding method. The first type was registered as an aperiodical disturbance of the electron concentration N in the E- and lower part of the F-region of the ionosphere. The duration of this type of disturbance slowly decreased from 20–25 min to 0 min with increase in altitude. The time of its delay Δt depended weakly on altitude and was about 11 min. For $\Delta t_1' \approx 9$ min and $R \approx 1275$ km, we get $v' \approx 2.3$–2.4 km s^{-1}.

The second type of WDs, observed both in the E- and F-region of the ionosphere, was periodical. With increase in altitude, the magnitude of the period continuously increased from 15 to 40 min. For $\Delta t \approx 37$ min, $\Delta t_1' \approx 35$ min, we get $v' \approx 610$ m s^{-1}.

Ref. [24] shows the IS method used for the diagnosis of WDs that accompanied the launches of four spacecraft. For $R \approx 1100$ km WDs with $T \approx 15$–75 min, having velocities of 703–735 m s^{-1} and $\delta_N \approx 5\%$ were observed. In Ref. [8], the authors show the velocities of propagation and the periods of WDs approximately.

References

1. Adushkin, V. V., Koslov, S. I., and Petrov, A. V., Ed., *Ecological Problems and Risks Caused by Rocket-Cosmic Technique on the Natural Environment,* Moscow: Ankil, 2000 (in Russian).
2. Garret, H. B. and Pike, C. P., Ed., *Space Systems and Their Interaction with the Earth's Space Environment,* New York: Program of Astronautics and Aeronautics, AIAA, 1981.
3. Pinson, G. T., Apollo/Saturn 5 post flight trajectory–SA-513-Skylab mission, *Tech. Rep. D5-15560-13,* Huntsville, AL: Boeing Co., 1973.
4. Nagorskii, P. M. and Tarashuk, Yu. E., Artificial modification of the ionosphere during rocket burns putted out at the orbits the cosmic apparatus, *Izv. Vuzov. Phys.,* 36, 98–106, 1993 (in Russian).
5. Nagorskii, P. M. and Taraschuk, Yu. U., Artificial modification of the ionosphere by rocket launching transported on the orbit the space vehicles, *Izv. Vuzov. Phys.,* 36, 98–107, 1993 (in Russian).
6. Nagorskii, P. M., Inhomogeneous structure of *F*-region of the ionosphere generated by rockets, *Geomagn. Aeronom.,* 38, 100–106, 1998.
7. Zasov, G. F., Karlov, V. D., Romamnchuk, T. E., et al., Observations of disturbances at the lower ionosphere during experiments on the program Soyuz–Apollo, *Geomagn. Aeronom.,* 17, 346–348, 1977.
8. Karlov, V. D., Kozlov, S. I., and Tkachev, G. N., Large-scale perturbations in the ionosphere caused by flight of rocket with working engine, *Cosmic Res.,* 18, 266–277, 1980 (in Russian).
9. Garmash, K. P., Kostrov, L. S., Rozumenko, V. T., et al., Global disturbances of the ionosphere caused by launching of spacecrafts at the background of magnetic storm, *Geomagn. Aeronom.,* 39, 72–78, 1999.
10. Kostrov, L. S., Rozumenko, V. T., and Chernogor, L. F., Doppler radio sounding of perturbations in the middle ionosphere accompanied burns and flights of cosmic vehicles, *Radiophys. Radioastron.,* 4, 227–246, 1999 (in Russian).
11. Burmaka, V. P., Kostrov, L. S., and Chernogor, L. F., Statistical characteristics of signals of Doppler HF radar during sounding the middle ionosphere perturbed by rockets' launchings and solar terminator, *Radiophys. Radioastron.,* 8, 143–162, 2003 (in Russian).
12. Burmaka, V. P., Taran, V. I., and Chernogor, L. F., Results of complex radiophysical observations of wave disturbances in the geo-cosmos, accompanied launchings and flights of rockets, *Kosmichna Nauka i Technologia,* Dodatok, 9, 51–61, 2003 (in Ukrainian).
13. Burmaka, V. P., Taran, V. I., and Chernogor, L. F., Wave disturbances in the ionosphere accompanied rockets' launchings at the background of natural transferred processes, *Geomagn. Aeronom.,* 44, 518–534, 2004.
14. Burmaka, V. P., Taran, V. I., and Chernogor, L. F., Complex radiophysical investigations of wave disturbances in the ionosphere accompanied launchings of rockets at the background of natural inhomogeneous non-stationary processes, *Radiophys. Radioastron.,* 9, 5–28, 2004 (in Russian).
15. Chernogor, L. F., Geo-cosmosphere—The open nonlinear dynamic system, *Radiophys. Electron.,* 570, 175–180, 2002 (in Ukrainian).

16. Chernogor, L. F., Physics of the Earth, atmosphere and geo-space in the lightness of system's paradigm, *Radiophys. Radioastron.*, 8, 59–106, 2003 (in Russian).

17. Chernogor, L. F., Earth–atmosphere–geo-space as an open nonlinear dynamic system, *Kosmichna Nauka i Technologia*, Dodatok, 9, 96–105, 2003 (in Ukrainian).

18. Garmash, K. P., Rozumenko, V. T., Tyrnov, O. F., Tsymbal, A. M., and Chernogor, L. F., Radio physical investigations of processes in the near-the-earth plasma disturbed by the high-energy sources. Part 2, *Foreign Radioelectronics: Success in Modern Radioelectronics*, 8, 3–19, 1999 (in Russian).

19. Garmash, K. P., Gokov, A. M., Kostrov, L. S., et al., Radiophysical investigations and modeling of ionospheric processes generated by sources of various nature. 2. Processes in a modified ionosphere. Signal parameters variations. Disturbance simulation, *Telecomm. Radioeng.*, 53, 1–22, 1999 (in Russian).

20. Burmaka, V. P., Taran, V. I., and Chernogor, L. F., Results of study of wave disturbances in the ionosphere by the method of incoherent scattering, *Foreign Radioelectronics: Success in Modern Radioelectronics*, 3, 4–35, 2005 (in Russian).

21. Burmaka, V. P., Chernogor, L. F., and Chernyak, Yu. V., Wave disturbances in Geo-space accompanied launchings and flights of rockets "Soyuz" and "Proton," *Radiophys. Radioastron.*, 10, 254–272, 2005 (in Russian).

22. Burmaka, V. P., Lisenko, V. N., and Chernogor, L. F., Results of study of wave processes in the ionosphere for various conditions of cosmic weather, *Kosmichna Nauka i Technologia*, 11, 37–57, 2005 (in Ukrainian).

23. Arendt, P. R., Ionospheric undulations following Apollo-14 launching, *Nature*, 231, 438–439, 1971.

24. Noble, S. T., A large-amplitude traveling ionospheric disturbance excited by the Space Shuttle during launch, *J. Geophys. Res.*, 95, 19037–19044, 1990.

25. Mendillo, M., Hawkins, G. S., and Klobuchar, J. A., An ionospheric total electron content disturbances associated with the launch of NASA's Skylab, *Tech. Rep. 0342*, Bedford, MA: Air Force Cambridge Res. Lab., 1974.

26. Mendillo, M., The effect of rocket launches on the ionosphere, *Adv. Space Res.*, 1, 275–290, 1981.

27. Sorokin, V. M. and Fedorovich, G. V., *Physics of Slow MGD-Waves in the Ionospheric Plasma*, Moscow: Energoatomizdat, 1982, 134.

28. Hantadze, A. G., On a new branch of own oscillations of the electro-conductive atmosphere, *Reports of Academy of Science*, Moscow, 376, 250–252, 2001 (in Russian).

29. Hantadze, A. G., Electromagnetic planetary waves in the Earth's ionosphere, *Geomagn. Aeronom.*, 42, 333–335, 2002.

30. Aburdgania, G. D. and Hantadze, A. G., Large-scale electromagnetic wave structures in *E*-region of the ionosphere, *Geomagn. Aeronom.*, 42, 245–251, 2002.

31. Vlasov, M. N. and Krichevsky, S. V., *Ecological Danger of Cosmic Activity: Analytical Issue*, Moscow: Science, 1999.

32. Burmaka, V. P., Lisenko, V. N., and Chernogor, L. F., Wave processes in the *F*-region of the ionosphere accompanied starts of rockets from cosmodrome Baikonur, *Geomagn. Aeronom.*, 46, 783–800, 2006.

33. Taran, V. I., Study of the ionosphere in the natural and artificially excited states by method of incoherent scattering, *Geomagn. Aeronom.*, 41, 659–666, 2001.

Chapter 6

Cluster Diagnostics of Plasma Disturbances

6.1 Overview

As mentioned in the previous chapters and as follows from numerous studies [1–11], in the present day, it can be taken into consideration that space vehicle (SV) launches and flights of rockets and spacecraft cause wave disturbances (WDs) in the ionosphere at distances greater than 2000 km [this is the distance between the Baikonur cosmodrome (Kazakhstan) and the observatory of Kharkov University located near Kharkov, Ukraine]. The authors of Refs. [1–11] carried out more than 40-year-long investigations into the ionospheric reaction on SV and rocket launches (RLs) and flights. For these purposes the whole gamut of radiophysical and magnetometric methods and techniques was used.

It is well known that radiophysical methods allow finding and studying of the WD electron concentration N. At the same time, for performance of the dynamic physical–mathematical model of the whole complex of processes caused by SV launches and flights, knowledge of spatiotemporal variations not only of N but also of other parameters of plasma is required. Complex diagnostics of the ionospheric plasma can be carried out using the results of incoherent scatter (IS) radar (ISR) measurement [12–15].

It is well known that the IS method allows determination of the parameters of ionospheric plasma such as concentration of electrons (ions; plasma is quasineutral), temperature of electrons and ions, velocity of movement of plasma, ion content of plasma, and so on, with sufficient height-temporal resolution and error. Using the major measured parameters of plasma, its secondary parameters can be derived that describe the dynamic processes in the medium. Collision frequencies of charged and neutral plasma components, ambipolar diffusion tensors, thermodiffusion, thermoconductivity, drift, plasma and heat flows, and so on are related to secondary parameters [16].

In geophysics, SV launches and flights can be considered as active experiments which allow tracking and watching the dynamic processes caused by rocket flight with the working engine. As mentioned in Chapter 3, study of these processes allows analyzing the interaction of the subsystems in the entire system of Earth–atmosphere–ionosphere–magnetosphere (EAIM) [17–20]. The main element of interaction of these subsystems is the waves of various physical natures.

Therefore, the study of the physical processes that accompany launches into geospace, diagnostics of the parameters of ionospheric plasma, and recording and identifying of wave types, which transfer disturbances excited by launches, all of these are actual tasks.

The aim of this chapter is to describe the results of the research and the analysis of spatiotemporal changes of the major parameters of ionospheric plasma caused by SV launches and flights [21,22].

6.2 Techniques and Methodology

The ISR was located at the Ionospheric Observatory of the Institute of the Ionosphere near Kharkov, Ukraine (see Appendix 3 and Refs. [3–8]). The results of the measurements were preprocessed based on the method proposed in Refs. [3,4]. This method is described in detail in Chapter 4. We will briefly repeat it here. After eliminating the average $\langle P_s \rangle$ in the temporal series of the signal power P_s calculated at the time interval of 180 min with a slicing step of $\Delta T_0 = 1$ min, the temporal series of magnitudes of changes in scattering power, $\delta P_s = P_s - \langle P_s \rangle$, were found, as well as the relative changes of this power $\delta_s = \delta P_s / \langle P_s \rangle$. For the chance of comparing the amplitude of WDs at different altitudes z, dependence $P_s \propto z^{-2}$ (or the dependence of SNR $q \propto z^{-2}$) was accounted for. In this case, the dependence of P_s on $(1 + T_e/T_i)$ was eliminated, where T_e and T_i are the temperatures of electrons and ions, respectively.

The time dependence of these temperatures was determined by the measured autocorrelation function (ACF) of the IS signal.

For given altitudes in this case, ACF is measured for a mixture of signal plus noise, from which the ACF of the noise can then be subtracted. The latter is defined at the end of the range coverage, which corresponds to altitudes higher than 2000 km where the useful IS signal is practically absent. As a result, the ACF of the IS signal is defined as distorted by the influence of the pulse character of sounding (due to "nonmonochromatic character" of sounding signal), peculiarities of the scattered medium (e.g., decrease of the correlated volume of plasma with growth of the time delay), and limitations of the bandwidth of the filter in the radio receiver. The obtained ACF is normalized as the subtraction of the mixed signal-plus-noise power at the corresponding altitudinal interval and the power of the noise.

The obtained coefficient of correlation dependence on the time delay then compares with *a priori* computed library dependences with a step of 10 K, which accounts for all mentioned earlier hardware and methodical distortions of ACT of the IS signal. As a result of selection of library functions (with decrease, step by step, in the process of definition of the minimal step of the temperature, from 250 to 50, and to 10 K), which is close to the measured one, is defined according to the criterion of minimization of the root-mean-square deviations. Then, the temperatures that correspond to the defined library dependence are found as those measured.

After definition of the height profile of the temperatures $T_e(t)$ and $T_i(t)$ of the charge particles, it is possible to define a nonnormalized profile of the electronic concentration $N(t)$ and the maximum magnitude of ionization. With the existence of the data on the critical frequency of the F-layer, the normalization of this profile can be done, as well as the definition of the absolute values of electron concentration.

Then, using the spatiotemporal variations of $N(t)$, $T_e(t)$, and $T_i(t)$ at a time interval that equals 180 min, average magnitudes $\bar{N}$, $\bar{T}_e$, and $\bar{T}_i$ were computed, as well as their deviations, $\delta N = N - \bar{N}$, $\delta T_e = T_e - \bar{T}_e$, $\delta T_i = T_i - \bar{T}_i$, and also relative amplitudes $\delta_N = \delta N/\bar{N}$, $\delta_{T_e} = \delta T_e/\bar{T}_e$, and $\delta_{T_i} = \delta T_i/\bar{T}_i$. For decrease of statistical broadening, the absolute and relative disturbances of these parameters were smoothened at the time interval of 5 min.

We note that at the diurnal period of April 28, 2001, the SNR q was not less than 10–40. This allows obtaining N, T_e, and T_i with absolute errors of about $10^{10}\,\mathrm{m}^{-3}$, $\pm(10–20)$ K, and $\pm(10–20)$ K, respectively, at a height range of 180–600 km.

At the diurnal period of May 25, 2004, with altitudal increase from 125 to 500 km the magnitudes of q were decreased from about 5 to 0.5, respectively. For such a situation, the errors of recording of N, T_e, and T_i increased, with respect to those obtained on April 28, 2001, by 3–6 times for altitudes of 125–500 km, respectively.

Spectral analysis was based on usage of the "window" and the "adaptive" Fourier's transform, as well as on wavelet transformation. Further, the diagrams of energy, as distribution of energy on periods, were calculated following Refs. [23,24]. The results of the computations were presented in the special format recommended in Refs. [23,24].

As an active source affecting the full system EAIM, two Soyuz-type launches were chosen. The first launch was done in the background of a moderate magnetic storm, the second one during quiet conditions.

6.3 Conditions of Space Weather

Let us describe the conditions of space weather during these two launches, which are convenient to present as temporal variations of the solar wind parameters, interplanetary magnetic field, density of the proton and electron flows, H_p-component of the geomagnetic field, as well as indexes of magnetic and auroral activity. Except the measured parameters, temporal variations of solar wind pressure, p_{SW}, and energy function of Akasofy, ε, were derived. The latter, as is known, describes the velocity of the solar energy transfer into the magnetosphere.

In the period before the launch, the condition of the cosmic plasma was characterized as nonperturbed (Figure 6.1). On the day of the launch, April 28, 2001, at around 03:00 UT (here and in further description the time is Universal Time or UT), the magnitude of concentration sharply increased (4 times), of the temperature (5–7 times), of the velocity (1.5–1.8 times), and of the solar wind pressure (10 times). The velocity of the transport of the energy of solar wind was increased up to 70–140 GJ s^{-1}. The sudden start of a moderate magnetic storm (according to the classification made in Refs. [17–19]) was observed with index D_{st}, which, from the beginning increased up to 45 nT, and then decreased to −50 nT at the main phase when the magnetic storm started. This phase continued until the beginning of the next day. The phase of the recovery took an additional 2 days. The change of sign of the interplanetary magnetic field component, B_z, which continued for about a day, corresponded to the main phase. Corresponding fluctuations were imposed on the regular variations of $H_p(t)$.

The sudden start of the magnetic storm was accompanied by a jump in proton density (by about 10 times) that continued for 15–20 h. The electron flow density, conversely, decreased several times. The index of auroral activity during the magnetic storm increased 5–10 times. A strong increase of the index K_p, without any reason, started on May 1, 2001, and continued for one more day.

We will now describe the conditions of space weather in the period May 22–28, 2004 (Figure 6.2). Before the RL, a weak magnetic storm (the index K_p changed from 1 to 3) was registered. Variations of other parameters were insufficient.

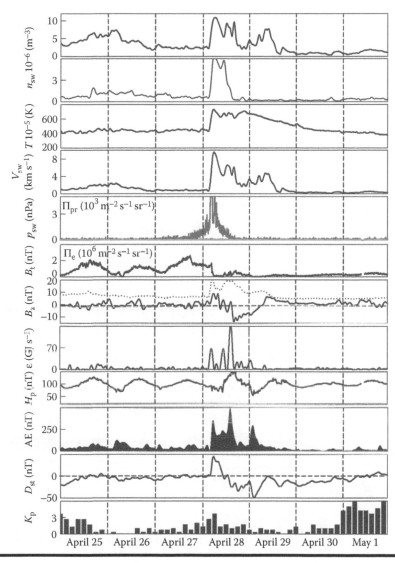

Figure 6.1 Temporal variations of solar wind parameters from April 25 to May 1, 2001: ε, function of Acasofu (calculations); Π_e, flow of electrons (GOES-12); Π_{pr}, flow of protons [GOES-8 (W75)]; AE, index of the aurora activity (WDC Kyoto); B_t and B_z, modules of the interplanetary magnetic field (ACE Satellite–magnetometer); D_{st}, index (WDC-C2); n_{sw}, concentration of particles; H_p, component of geomagnetic field; K_p, index (Air Force Weather Agency); p_{sw}, dynamic pressure (calculations); T, temperature; V_{st}, radial velocity (ACE Satellite—solar wind electron–proton alpha monitor).

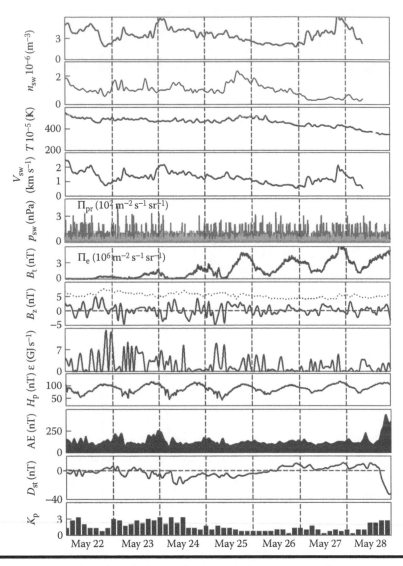

Figure 6.2 Temporal variations of solar wind parameters from May 22 to May 28, 2004: ε, function of Acasofu (calculations); Π_e, flow of electrons (GOES-12); Π_{pr}, flow of protons [GOES-8 (W75)]; AE, index of the aurora activity (WDC Kyoto); B_t and B_z, modules of the interplanetary magnetic field (ACE Satellite–magnetometer); D_{st}, index (WDC-C2); n_{sw}, concentration of particles; H_p, component of geomagnetic field; K_p, index (Air Force Weather Agency); p_{sw}, dynamic pressure (calculations); T, temperature; V_{st}, radial velocity (ACE Satellite—solar wind electron–proton alpha monitor).

Thus, electron concentration changed from 2×10^6 to 6×10^6 m^{-3}, temperature from 10^5 to 5.4×10^5 K, solar wind pressure from 1 to 2.5 nPa, velocities from 300 to 550 km s^{-1}, and magnitude of Akasofu's function from 1 to 14 GJ s^{-1}. Variations of proton density were insignificant. Conversely, the density of electron flow slowly increased according to the quasiperiodical process with a period of about 1 day.

The component B_z was changed chaotically at the limits of ± 5 nT. Variations of the H_p-component of the geomagnetic field were sufficiently ordered with the 1-day period. Magnitudes of the index D_{st} also fluctuated chaotically at the limits of ± 10 nT. Variations of the index AE were insignificant. The day of the launch, May 25, 2004, was quiet. This fact allows identifying those disturbances easily, which were related to the RL.

6.4 Results of Measurements and Signal Processing

6.4.1 Spatiotemporal Characteristics

Let us present, first of all, an example of the spatiotemporal variations of electron concentration (Figure 6.3) that accompanied Soyuz launch from the Baikonur cosmodrome. As mentioned earlier, the sudden beginning of the moderate magnetic storm registered at 03:00 h. From 05:00 to 12:00 h, three oscillations were observed with 180, 150, and 100 min duration. Their absolute and relative amplitudes achieve maximum values at 270 km altitude, which equal $\delta N \approx 2.5 \times 10^{11}$ m^{-3} and $\delta_N \approx 0.3$, respectively. It is interesting to note that the primary front of the second oscillation, which started 23 min after launch, was much more steep than its previous (e.g., secondary) front. We suggest that the nonlinear wave process with the clearly registered effects of the steepness of the wave front be observed.

The spatiotemporal variations of the electron temperature, $\delta T_e(t)$, as well as its relative amplitude, $\delta_{T_e}(t)$, are shown in Figure 6.4. The wave process in variations of $\delta T_e(t)$ and $\delta_{T_e}(t)$ are clearly seen before the RL. Following each other, the three oscillations were observed to achieve their maximal amplitudes of $\delta T_e \approx 200$ K and $\delta_{T_e} \approx 0.1$ at the time interval from 06:00 to 11:00 h, and were analogous to similar amplitudes of N. The main peculiarity between their behavior was that the oscillations of T_e shifted in phase by 180° relative to the oscillations of N.

Moreover, with increase in altitude shifting of minima and maxima of the oscillations was observed in the left direction along the temporal axis.

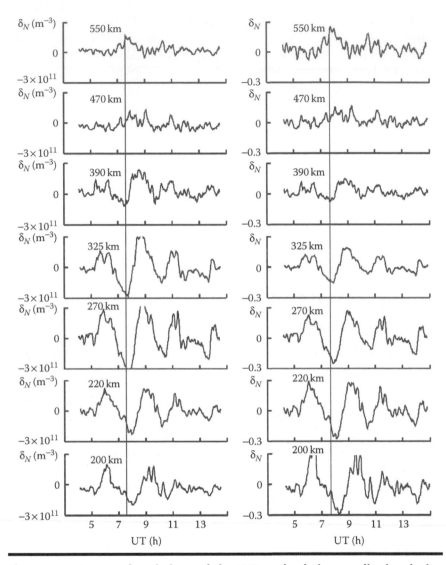

Figure 6.3 Temporal variations of the WD and relative amplitudes during Soyuz launch on April 28, 2001.

Thus, the difference in time at 220 and 470 km was of about an hour. We suppose that this shifting was caused by natural perturbation in the geospace environment that started at about 03:00 h. After launch, the period of oscillations also changed. These changes can be clearly seen at a height range of 200–600 km.

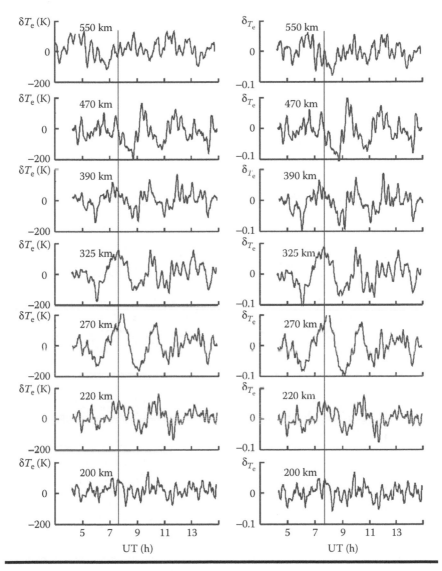

Figure 6.4 **Temporal variations of the WD and relative amplitudes of electron temperatures during Soyuz launch on April 28, 2001.**

The effect of increase in magnitudes of $\delta T_i(t)$ was observed at the narrow height range of 200–220 km (Figure 6.5). The absolute and the relative amplitudes of oscillations were estimated as $\delta T_i \approx 150$ K and $\delta_{T_i} \approx 0.15$, respectively. It is interesting to see that after the launch, the oscillations of T_e and T_i had practically similar phases (i.e., were sine-phased).

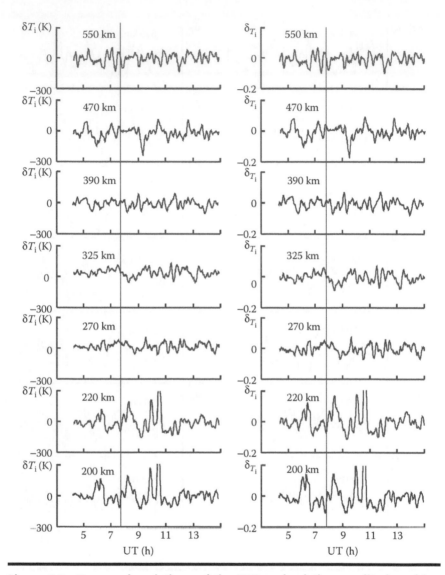

Figure 6.5   Temporal variations of the WD and relative amplitudes of ion temperatures during Soyuz launch on April 28, 2001.

The next example of spatiotemporal variations of the IS signal parameters that accompanied Soyuz launch at 12:24 h on May 25, 2004, from Baikonur cosmodrome, are presented in Figure 6.6. Before launch, the wave process was observed at a height range of 200–330 km, which then disappeared 15–20 min after the launch in dependences of $\delta N(t)$ and $\delta_N(t)$.

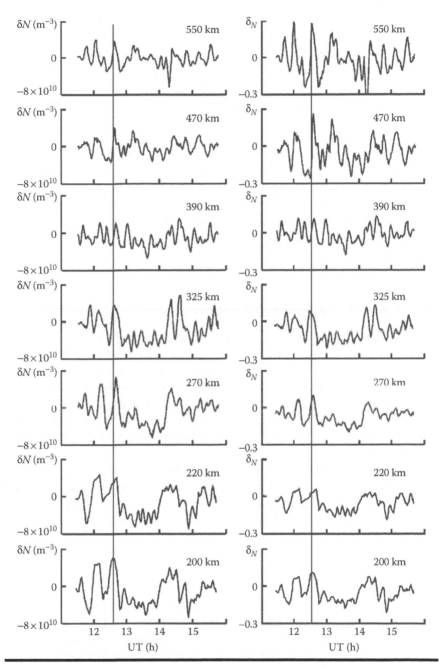

Figure 6.6 Temporal variations of the WD and relative amplitudes during Soyuz launch on May 25, 2004.

But, 70–80 min after the launch, oscillations with period $T \approx 50$–70 min occurred.

The dependences $\delta T_e(t)$ and $\delta_{T_e}(t)$ also had an oscillatory character (Figure 6.7). After the SV launch at the height range of 200–220 km, a WD occurred with a delay of $\Delta t_2 \approx 35$–40 min, duration $\Delta T \approx 90$ min, and corresponding absolute and relative amplitudes $\delta T_e \approx 100$–150 K and $\delta_{T_e} \approx 0.04$.

Let us describe the temporal variations of $\delta T_i(t)$ and $\delta_{T_i}(t)$. Before the "Soyuz" launch wave oscillations with amplitudes of $\delta T_i \approx 100$–150 K and $\delta_{T_i} \approx 0.1$ (Figure 6.8) were observed at all altitudes. $\delta T_i(t)$ and $\delta_{T_i}(t)$ decreased to $\delta T_i \approx 20$ K and $\delta_{T_i} \approx 0.02$ at altitudes of $z \approx$ 200–220 km 50 min after the launch. Then, after 70 min the amplitudes were reconstructed to their primary magnitudes. Thus, at 13:10 h increase of amplitudes to $\delta T_i \approx 100$ K and $\delta_{T_i} \approx 0.07$–0.08 was observed at altitudes of $z \approx$ 270–330 km.

6.4.2 Results of Spectral Analysis

Spectrograms and energy diagrams of the temporal dependences $\delta N(t)$ and $\delta T_e(t)$ for April 28, 2001, were practically the same (Figures 6.9 and 6.10). It is clearly seen from the presented spectrograms that these differ significantly before and after the launch. After launch oscillations with period $T \approx 150$–170 min appeared. It is clearly seen from the spectrograms and energy diagrams of the temporal dependences of $\delta T_i(t)$ that after launch, oscillations occurred with periods $T \approx 130$–150 min and $T \approx 30$–40 min (Figure 6.11).

The results of the spectral analysis of $\delta N(t)$ for observations carried out on May 24, 2004, are shown in Figure 6.12. It can be seen that before launch wave process with period $T \approx 25$–35 min existed; 15–20 min after the launch it disappeared, but 70–80 min after launch oscillations with period $T \approx 80$ min rose.

Spectral analysis of temporal dependences of $\delta T_e(t)$ indicates that 60 min after launch oscillations with period $T \approx 80$ min rose (Figure 6.13).

In Figure 6.14 the results of the spectral analysis of dependences $\delta T_i(t)$ are presented for May 25, 2004. It can be seen that before the launch a wave process with period $T \approx 15$–25 min was observed, but several minutes before the launch it disappeared. Only 100 min after the launch do oscillations with period $T \approx 30$–60 min rise.

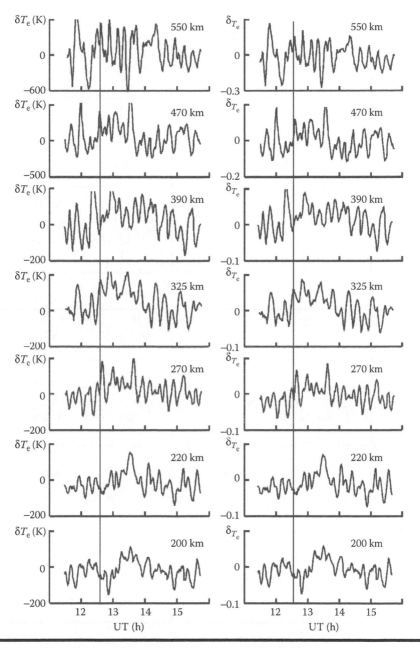

Figure 6.7 **Temporal variations of the WD and relative amplitudes of electron temperatures during Soyuz launch on May 25, 2004.**

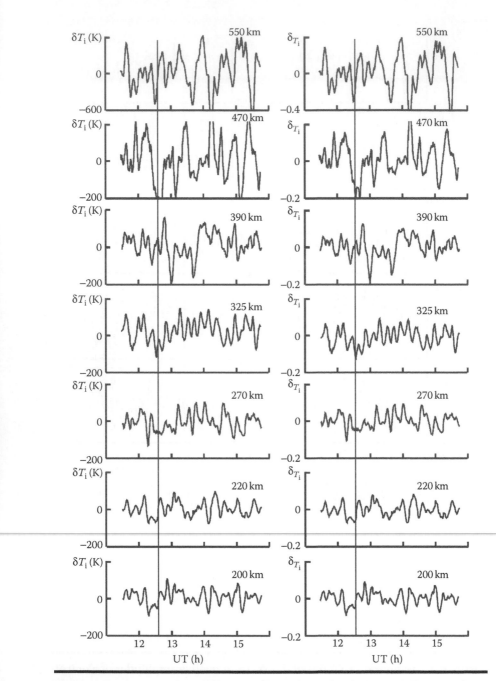

Figure 6.8 Temporal variations of the WD and relative amplitudes of ion temperatures during Soyuz launch on May 25, 2004.

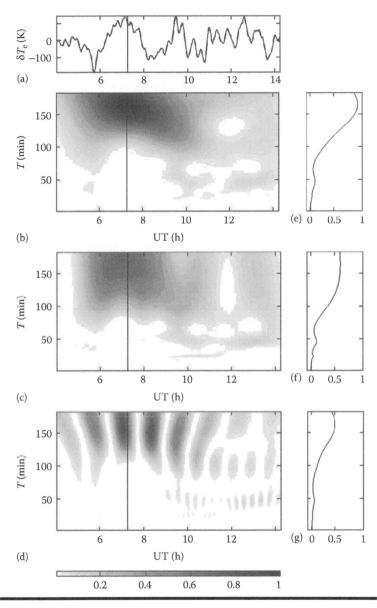

Figure 6.9 Results of the spectral analysis of temporal variations of δN on May 28, 2001 for an altitude of 325 km: (a) signal; (b–d) spectrograms (in relative units) for window, adaptive Fourier transform, and wavelet transform on the basis of Morlet wavelet signal, respectively; (e–g) energy diagrams for window, adaptive Fourier transform, and wavelet transform on the basis of Morlet wavelet signal, respectively.

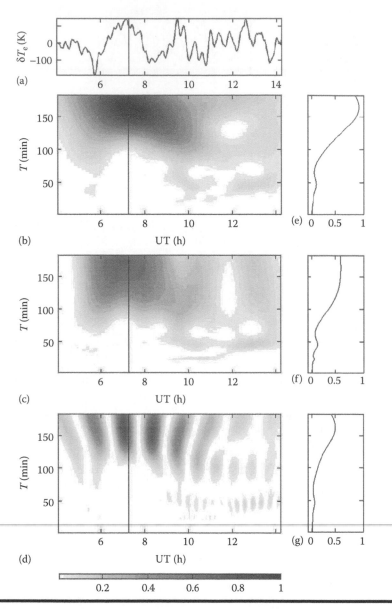

Figure 6.10   Results of the spectral analysis of temporal variations of δT_e at an altitude of 325 km: (a) signal; (b–d) spectrograms (in relative units) for window, adaptive Fourier transform, and wavelet transform on the basis of Morlet wavelet signal, respectively; (e–g) energy diagrams for window, adaptive Fourier transform, and wavelet transform on the basis of Morlet wavelet signal, respectively.

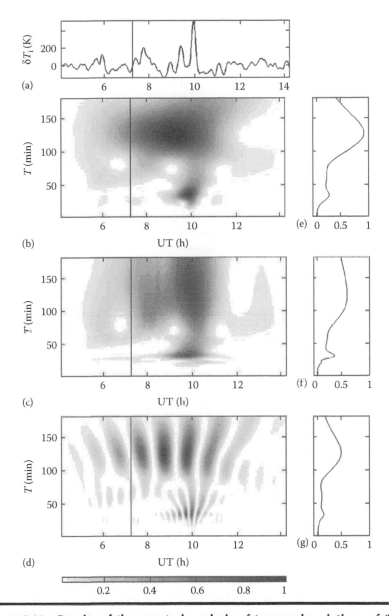

Figure 6.11 **Results of the spectral analysis of temporal variations of δT_i at an altitude of 220 km: (a) signal; (b–d) spectrograms (in relative units) for window, adaptive Fourier transform, and wavelet transform on the basis of Morlet wavelet signal, respectively; (e–g) energy diagrams for window, adaptive Fourier transform, and wavelet transform on the basis of Morlet wavelet signal, respectively.**

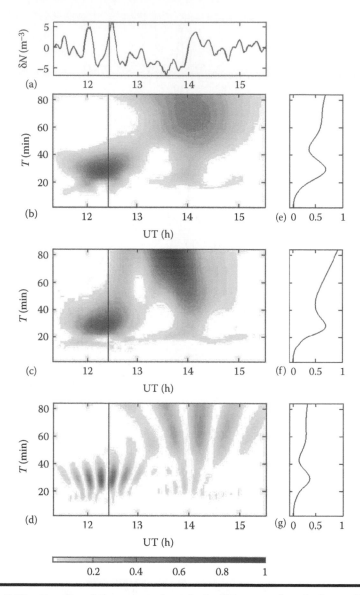

Figure 6.12 Results of the spectral analysis of temporal variations of δ*N* on May 25, 2004 at an altitude of 270 km: (a) signal; (b–d) spectrograms (in relative units) for window, adaptive Fourier transform, and wavelet transform on the basis of Morlet wavelet signal, respectively; (e–g) energy diagrams for window, adaptive Fourier transform, and wavelet transform on the basis of Morlet wavelet signal, respectively.

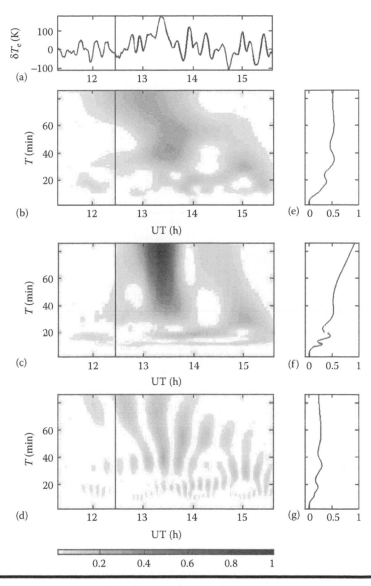

Figure 6.13 **Results of the spectral analysis of temporal variations of δT_e at an altitude of 220 km: (a) signal; (b–d) spectrograms (in relative units) for window, adaptive Fourier transform, and wavelet transform on the basis of Morlet wavelet signal, respectively; (e–g) energy diagrams for window, adaptive Fourier transform, and wavelet transform on the basis of Morlet wavelet signal, respectively.**

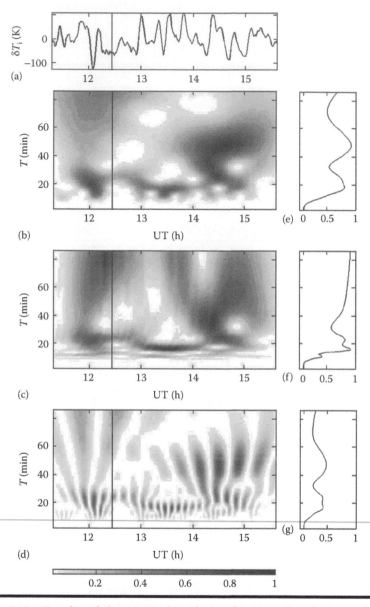

Figure 6.14 **Results of the spectral analysis of temporal variations of δT_i at an altitude of 270 km: (a) signal; (b–d) spectrograms (in relative units) for window, adaptive Fourier transform, and wavelet transform on the basis of Morlet wavelet signal, respectively; (e–g) energy diagrams for window, adaptive Fourier transform, and wavelet transform on the basis of Morlet wavelet signal, respectively.**

6.5 Results and Discussions

Let us discuss Figure 6.3 again. It can be seen that once the launch happened, a sharp and essential decrease of δN registered. This effect is evidently seen at the height range of 200–400 km, which is related to the delay of $\Delta t_1 \approx 20\,\text{min}$ that was defined by the minimum of δq. The time duration of this process eventually does not exceed 1 h.

The next change of signal characteristics, accompanied by strong oscillations with period 150 min and total duration about 5–6 h, have been recorded with a time delay of about 60 min (defined by the maximum of δN). During the flight of the rocket to the E-region of the ionosphere that takes a time of Δt_0, the corrected magnitudes of time delays $\Delta t_0 \approx 3\,\text{min}$ were estimated at $\Delta t_1' \approx 17\,\text{min}$ and $\Delta t_2' \approx 57\,\text{min}$. Accounting for a distance R from the place of WD generation to the place of their observation equaling ~2100 km, such delays relate to the estimated velocities of $v_1' \approx 2\,\text{km s}^{-1}$ and $v_2' \approx 600\,\text{m s}^{-1}$. We should add here that WDs caused by RLs were imposed on natural disturbances. This leads to amplification of the launch effect.

Now we will consider the launch that happened on May 25, 2004. For this launch the following delays, $\Delta t_1' \approx 23\,\text{min}$ and $\Delta t_2' \approx 64\,\text{min}$, were found, from which we get $v_1' \approx 1.5\,\text{km s}^{-1}$ and $v_2' \approx 550\,\text{m s}^{-1}$. It can be seen that there are two groups of velocities of propagating WDs, which depend on the height, daytime periods, and conditions in the atmosphere and the ionosphere. As mentioned in previous chapters and as shown in previous works [1–8,10], the velocities $v_1' \approx 1.5–2\,\text{km s}^{-1}$ have slow MHD (SMHD) waves [25], whereas the velocities $v_2' \approx 550–610\,\text{m s}^{-1}$ have intrinsic gravity (IG) waves [26]. Their propagation correlates with the rising of quasiperiodical changes not only N but also T_e and T_i (see Figures 6.4 through 6.9, respectively). At higher altitudes the velocity of propagating WDs was greater than that at the lower altitudes. Thus, for slower WDs transported by the help of IG waves, the time difference of the arrival of disturbance N, T_e, and T_i at altitudes of 1900 and 660 km, respectively, was achieved in 60 and 150 min, respectively. This allows us to estimate the vertical component of the velocity v_v for IG waves. It was close to that for all disturbances of plasma parameters (N, T_e, T_i) and was estimated at 105–130 m s^{-1}. The obtained magnitudes of v_v fully correspond to the vertical component of the velocity of IS waves [27].

Simultaneous use of the "window" and adaptive Fourier transform, which add to each other, and also the wavelet transform that together can allow watching the dynamics of WDs (e.g., quasiperiodical processes) in the ionosphere, can order their periods, amplitudes, and time duration, as also distribution of oscillation energy over periods (see Figures 6.11 and 6.12).

6.6 Major Results

1. Large-scale quasiperiodical variations of the main parameters of plasma that were caused by RLs ranging in distances more than 2000 km from the cosmodrome were analyzed for the first time with the help of ISR. It is interesting to note that the effects of similar RLs, which registered at the background of the natural magnetic perturbation, were eventually stronger than those that occurred in nonperturbed conditions.

2. It was confirmed that there are two groups of velocities of the propagating WDs of about $v_1' \approx 1.5-2$ km s^{-1} and of $v_2' \approx 550-610$ m s^{-1}. To these velocities the SMHD and the IS waves correspond, respectively.

3. It was shown that launches from Baikonur cosmodrome cause disturbances of N, T_e, and T_i with the absolute amplitudes of $(0.8-4) \times 10^{11}$ m^{-3} and 50–100 K, respectively. The relative amplitudes of these disturbances achieved 0.1–0.25 and 0.05–0.08.

4. Oscillations of N, T_e, and T_i in the general case phase-shifted relative to each other. Apparently, the phase-shifting differed during different observations.

5. It was confirmed that simultaneous usage of the "window" Fourier transform, wavelet transform, and the adaptive Fourier transform allows estimating the periods, amplitudes, and time duration of the wave packets (e.g., oscillation processes) in the ionospheric plasma more efficiently, as also the distribution of energy of the oscillations over periods.

References

1. Burmaka, V. P., Kostrov, L. S., and Chernogor, L. F., Statistical characteristics of signals of Doppler HF radar during sounding the middle ionosphere perturbed by rockets' launchings and solar terminator, *Radiophys. Radioastron.*, 8, 143–162, 2003 (in Russian).

2. Burmaka, V. P., Taran, V. I., and Chernogor, L. F., Results of complex radiophysical observations of wave disturbances in the geocosmos, accompanied launchings and flights of rockets, *Kosmichna Nauka i Technologia*, Dodatok, 9, 51–61, 2003 (in Russian).

3. Burmaka, V. P., Taran, V. I., and Chernogor, L. F., Wave disturbances in the ionosphere accompanied rockets' launchings at the background of natural transferred processes, *Geomagn. Aeronom.*, 44, 518–534, 2004.

4. Burmaka, V. P., Taran, V. I., and Chernogor, L. F., Complex radiophysical investigations of wave disturbances in the ionosphere accompanied launchings of rockets at the background of natural inhomogeneous non-stationary processes, *Radiophys. Radioastron.*, 9, 5–28, 2004 (in Russian).

5. Burmaka, V. P., Taran, V. I., and Chernogor, L. F., Results of study of wave disturbances in the ionosphere by the method of incoherent scattering, *Foreign Radioelectronics: Success in Modern Radioelectronics*, 4–35, 2005 (in Russian).

6. Burmaka, V. P., Chernogor, L. F., and Chernyak, Yu. V., Wave disturbances in geospace accompanied launchings and flights of rockets "Soyuz" and "Proton," *Radiophys. Radioastron.*, 10, 254–272, 2005 (in Russian).

7. Burmaka, V. P., Lysenko, V. N., and Chernogor, L. F., Results of study of wave processes in the ionosphere for various conditions of space weather, *Kosmichna Nauka i Technologia*, 11, 37–57, 2005 (in Russian).

8. Burmaka, V. P., Lysenko, V. N., and Chernogor, L. F., Wave processes in the *F*-region of the ionosphere accompanied starts of rockets from cosmodrome Baikonur, *Geomagn. Aeronom.*, 46, 783–800, 2006.

9. Garmash, K. P., Rozumenko, V. T., Tyrnov, O. F., Tsymbal, A. M., and Chernogor, L. F., Radiophysical investigations of processes in the near-the-Earth plasma disturbed by the high-energy sources. Part 2, *Foreign Radioelectronics: Success in Modern Radioelectronics*, 3–19, 1999 (in Russian).

10. Kostrov, L. S., Rozumenko, V. T., and Chernogor, L. F., Doppler radio sounding of perturbations in the middle ionosphere accompanied burns and flights of space vehicles, *Radiophys. Radioastron.*, 4, 227–246, 1999 (in Russian).

11. Garmash, K. P., Gokov, A. M., Kostrov, L. S., et al., Radiophysical investigations and modeling of ionospheric processes generated by sources of various nature. 2. Processes in a modified ionosphere. Signal parameter variations. Disturbance simulation, *Telecomm. Radioeng.*, 53, 1–22, 1999.

12. Rishbeth, H. and Williams, P. J. S., The EISCAT ionospheric radar: The system and its early results, *Quart. J. R. Astron. Soc.*, 26, 478–512, 1985.

13. Hysell, D. L., Incoherent scatter experiments at Jacamaca using alternating code, *Radio Sci.*, 35, 1425–1435, 2000.

14. Zhou, Q. J. and Morton, Y. T., A case study of mesosphere gravity wave momentum flux and dynamical instability using the Arecibo dual beam incoherent scatter radar, *Geophys. Res. Lett.*, 33, 1–4, doi:10.1029/2005GL025608, 2006.

15. Blaunstein, N. and Plohotniuc, E., *Ionosphere and Applied Aspects of Radio Communication and Radar*, Chap. 8, New York: Taylor & Francis, 2008.

16. Farley, D. T., Incoherent scatter radar probing, in *Modern Ionospheric Science*, Eds. Kohl, H., Ruster, R., and Schlogel, K., Katlenburg-Lindau: Max-Planck Institute of Aeronomy Press, 1996, 415–439.

17. Chernogor, L. F., Physics of the Earth, atmosphere and geospace in the lightness of system's paradigm, *Radiophys. Radioastronom.*, 8, 59–106, 2003 (in Russian).

18. Chernogor, L. F., Earth–atmosphere–geospace as an open nonlinear dynamic system, *Kosmichna Nauka i Technologia*, Dodatak, 9, 96–105, 2003 (in Russian).

19. Chernogor, L. F., Earth–atmosphere–ionosphere–magnetosphere, as the open dynamic nonlinear system (Part 1), *Nonlinear Univ.*, 4, 655–697, 2006 (in Russian).

20. Chernogor, L. F., Earth–atmosphere–ionosphere–magnetosphere, as the open dynamic nonlinear system (Part 2), *Nonlinear Univ.*, 5, 198–231, 2007 (in Russian).

21. Burmaka, V. P. and Chernogor, L. F., Complex diagnostics of parameters of the ionospheric plasma far from the trajectory of launching rockets, *Geomagn. Aeronom.* (in press).

22. Burmaka, V. P. and Chernogor, L. F., Complex diagnostics of the ionospheric plasma perturbed by far-located rockets' launching, *Radiophys. Radioastronom.*, 14, 26–44, 2009 (in Russian).

23. Burmaka, V. P., Panasenko, S. V., and Chernogor, L. F., Modern methods of spectral analysis of quasi-periodical processes in the geospace, *Foreign Radioelectronics: Success in Modern Radioelectronics*, 3–24, 2007 (in Russian).

24. Lasorenko, O. V., Panasenko, S. V., and Chernogor, L. F., Adaptive Fourier transform, *Electromagn. Waves Electron. Sys.*, 10, 39–49, 2006 (in Russian).

25. Sorokin, V. M. and Fedorovich, G. V., *Physics of Slow MGD-Waves in the Ionospheric Plasma*, Moscow: Energoatomizdat, 1982, 134 (in Russian).

26. Grigor'ev, G. I., Acoustic-gravity waves in the atmosphere of the Earth (issue), *Izv. Vuzov. Radiofizika*, 42, 3–25, 1999 (in Russian).

27. Gershman, B. N., *Dynamics of the Ionospheric Plasma*, Moscow: Nauka, 1974 (in Russian).

Chapter 7

System Spectral Analysis of Plasma and Geomagnetic Perturbations

7.1 Overview

As previously mentioned in Chapters 1 through 3, quasiperiodical processes play an essential role in the physics of the geospace environment (see also Refs. [1–4]). They play an important role in the interaction between subsystems in the system that comprises the Earth–atmosphere–ionosphere–magnetosphere [5–8].

For the study of the mechanisms of wave disturbance (WD) generation and their propagation, it is necessary to employ actual experiments and, particularly, to launch powerful rockets (which were indicated in the previous chapters as space vehicles or SVs) [9–17]. In the present case, the source of the WDs, time and place of their generation and energy of the source are known precisely, that is, the energy of the wave process, and so on.

As numerous observations showed, WDs in geospace environment have 1–2 oscillations, and sometimes, though rarely, 3–4 oscillations [18–20]. Moreover, such oscillations occur in the background of natural nonstationary processes. It means that for the detection and analysis of WDs, traditional theoretical framework based on one-dimensional (1D) Fourier transform is not effective and is not even valid in some cases [21]. The main objective for mitigating these difficulties is to find adequate mathematical tools for precise analysis of the non-stationary and the short temporal series that require frequency–temporal simultaneous resolution.

Wavelet analysis can be considered a very effective mathematical tool [22–26]. Recently, a modification of the *windowed* Fourier transform (WFT) was proposed, which is defined in Refs. [27,28] as *adaptive* Fourier transform (AFT).

The major goal of this chapter is to describe the results of a comparative analysis of the quasiperiodical processes and WDs occurring in ionospheric plasma in the presence of a geomagnetic field obtained by using the WFT, AFT, and wavelet transform (WT) as mathematical tools. The launch of powerful rockets along with a solar terminator (ST) was taken into consideration as a source of such quasiperiodical processes and WD generation.

7.2 Spectral Analysis of Temporal Series

7.2.1 General Presentation of the Adaptive Fourier Transform

It is well known that the traditional 1D Fourier transform, which was used for many years in science and technology for a wide variety of fields, can be presented as follows [21,22]:

$$\dot{F}[f(t)] \equiv \dot{F}(\omega) = \int_{-\infty}^{\infty} f(t) e^{-i\omega t} dt, \tag{7.1}$$

where:

$f(t)$ is the signal under analysis

$\dot{F}(\omega)$ is the spectral density function (SDF)

$\varphi_{\mathrm{m}}(t) = e^{-i\omega t}$ are the basic functions of the 1D Fourier transform

A minor drawback of such type of Fourier transform is the complete absence of information in $\dot{F}(\omega)$ on the position of the corresponding frequency components in time domain (e.g., along the time axis). Therefore, a secondary step in the development of the Fourier transform was the option of using WFTs [23,24]:

$$\dot{S}[f(t)] \equiv Sf(\omega, \tau) = \int_{-\infty}^{\infty} f(t) g(t - \tau) e^{-i\omega t} dt, \tag{7.2}$$

where:
  $g(t)$ is the real window function that has a property to be localized along the time axis t
  τ is the parameter that describes a shift in the window function with respect to the signal along the time axis

Displacement of the window function on this basis can be performed in such a manner for this case:

$$\varphi_{\omega,\tau}(t) = g(t-\tau)e^{-i\omega t}. \tag{7.3}$$

As a result, the SDF, $Sf(\omega,\tau)$, of the WFT transfers information simultaneously on both the temporal and frequency properties of the signal $f(t)$ during data processing.

The time and frequency (or period) dependence of the energy density for WFT is usually called a spectrogram and is defined as follows:

$$P_S f(\omega,\tau) = \left| Sf(\omega,\tau) \right|^2.$$

At the same time, the WFT mathematical framework has a drawback that relates to time-splitting on the frequency for a given width of the "window" function $g(t)$ and for a constant time of splitting that is equally distributed along all frequency components of the signal. This drawback is absent in the application of the WL that has adaptive properties (see, e.g., Refs. [23,24]). The continuous WT of the function $f(t)$ has the following form:

$$\dot{W}[f(t)] \equiv Wf(a,b) = \frac{1}{a^{1/2}} \int_{-\infty}^{\infty} f(t)\psi^* \left(\frac{t-b}{a} \right) e^{-i\omega t} dt. \tag{7.4}$$

Here the complex conjugate of the function $\psi(\bullet)$ is indicated by the symbol "*" and splitting is done on the basis of wavelets:

$$\varphi_{a,b}(t) = \frac{1}{a^{1/2}} \psi \left(\frac{t-b}{a} \right), \tag{7.5}$$

where:
  a is the parameter of scalability ($a > 0$)
  b is the parameter of shift in the time domain

Such basic real wavelets are particularly convenient to use for indicating sharp changes in the signal under investigation. The complex analytical wavelets allow differentiation of the amplitude and phase components of the desired signal. This shows their usefulness while estimating the immediately occurring frequencies.

A WT framework based on such kind of wavelets is called *analytical wavelet transform* (AWT) [24].

The dependence of energy density on *a* and *b* for WT is called scale diagram [24] and is defined as follows:

$$P_W f(a,b) = |Wf(a,b)|^2.$$

Another drawback of the WFT is the absence of the property of self-similarity of basic functions. The cause of this drawback is that in the constant window $g(t)$ (i.e., with the same width) in time domains for different magnitudes of frequency ω, different numbers of the harmonic functions $e^{-i\omega t}$ come into play.

We can follow both the drawbacks mentioned above with the help of the Fourier transform if it is used together with the window function introduced earlier and the idea of performance of self-similarity of basic functions taken from the WT theory. As the basis of a new transform, we choose the following [26]:

$$\varsigma_{a,\tau,v}(t) = Cg\left(\frac{t-\tau}{a}\right)\exp\left\{i\pi v\left(\frac{t-\tau}{a}\right)\right\}, \tag{7.6}$$

where:
  C is a constant
  a is the parameter of scalability that defines the width of the window function
    $g(t)$ $(a > 0)$
  τ is the parameter of the shifting of the window function in the time domain
  v is the coefficient $(v > 0)$ that defines the number of periods of the harmonic
    function that can be "hidden" onto the width of the window function
    for a given *a*

These basic functions have the property of self-similarity. The constant *C* is computed from the conditions of normalization of the basic function $\varsigma_{a,\tau,v}(t)$, that is,

$$\int_{-\infty}^{\infty} \varsigma_{a,\tau,v}(t)\varsigma_{a,\tau,v}(t)dt = 1, \tag{7.7}$$

from which it follows that $C = 1/\sqrt{a}$ for $\|g\| = 1$. Then, for a new transform we get

$$\varsigma_{a,\tau,v}(t) = \frac{1}{\sqrt{a}} g\left(\frac{t-\tau}{a}\right)\exp\left\{i\pi v\left(\frac{t-\tau}{a}\right)\right\}. \tag{7.8}$$

The transform by itself can be rewritten in this case as follows:

$$\dot{A}[f(t)] \equiv A_v f(a,\tau) = \int_{-\infty}^{\infty} f(t)\varsigma_{a,\tau,v}(t)dt$$

(7.9)

$$= \frac{1}{\sqrt{a}} \int_{-\infty}^{\infty} f(t)g\left(\frac{t-\tau}{a}\right) \exp\left\{i\pi v\left(\frac{t-\tau}{a}\right)\right\} dt.$$

In Equation 7.9, the letter A stands for the SDF and $A_v f(a,\tau)$ indicates *adaptive* transformation process. Because SDF $A_v f(a,\tau)$ of the AFT is a complex function, obviously its absolute value, $|A_v f(a,\tau)|$, phase of $A_v f(a,\tau)$, and the magnitude of

$$P_A f(a,\tau) = |A_v f(a,\tau)|^2$$

are to be considered separately to get the spectrogram of the AFT. The latter is simply a two-dimensional (2D) density of energy of the desired signal $f(t)$ under investigation. In Ref. [26], it was demonstrated that the AFT has properties of linearity, scalability, and shifting.

It is interesting to note that the AFT can equal the adaptive wavelet transform (AWT) in some definite conditions [25]. Using a window of the form of

$$g(t) = \exp\left(-\frac{t^2}{2\sigma^2}\right),$$

(7.10)

the AFT limits the AWT at the base of Gabor's wavelet presentation given as follows:

$$\psi(t) = \frac{1}{(\pi\sigma^2)^{1/4}} \exp\left(-\frac{t^2}{2\sigma^2}\right) \exp(i\eta t),$$

(7.11)

that for $\sigma = 1$ transfers to Morlet's complex wavelet (see all definitions, e.g., in Refs. [24–26]).

In our discussions further, we use the WFT and the AFT in the following forms:

$$Sf(t) = \sqrt{\frac{2}{t_{wS}}} \int_{-\infty}^{\infty} f(t)g\left(\frac{t-\tau}{t_{wS}/2}\right) \exp\left\{-i\frac{2\pi t}{T}\right\} dt,$$

(7.12a)

$$A_\nu f(T_\nu, \tau) = \sqrt{\frac{2}{\nu T_\nu}} \int\limits_{-\infty}^{\infty} f(t) g\left(\frac{t-\tau}{\nu T_\nu / 2}\right) \exp\left\{-i \frac{2\pi(t-\tau)}{T_\nu}\right\} dt, \quad (7.12b)$$

where:

t_{wS} is the window width for the WFT

$T = 2\pi/\omega$ and $T_\nu = 2a/\nu$ are the magnitudes of the periods of oscillation

Let us consider the relative periods, $\tilde{T} = T/t_{wS}$ and $\tilde{T}_\nu = T_\nu/t_{wS}$. For convenience, we can omit index ν in the future. All other magnitudes can be presented in relative units.

We will use Hemming's window function $g(t)$ that has well-known advantages [25,26] for the WFT and AFT in our discussions further, which can be presented as follows:

$$g_H(t) = \gamma\left[0.54 + 0.46\cos(\pi t)\right], \quad (7.13)$$

where:

$\gamma \approx 1.12$ is the normalized product factor

For WT, we use Morlet's wavelet for the basic function, $\psi(t)$, which can be presented as [23]:

$$\psi(t) = \exp\left(-\frac{t^2}{2}\right) \cos(5t). \quad (7.14)$$

Here, similar to that mentioned earlier, t is dimensionless time. The wavelet (given in Equation 7.14) is convenient for analyzing short wavelets that, as a rule, are observed in the ionosphere.

Except for the functions, $P_S(\tilde{T}, \tau)$, $P_A(\tilde{T}, \tau)$, and $P_W(\tilde{T}, \tau)$, we use the dependences of the following measures:

$$E_S(\tilde{T}) = \int\limits_{-\infty}^{\infty} P_S(\tilde{T}, \tau) d\tau, \quad E_A(\tilde{T}) = \int\limits_{-\infty}^{\infty} P_A(\tilde{T}, \tau) d\tau,$$

$$E_W(\tilde{T}) = \int\limits_{-\infty}^{\infty} P_W(\tilde{T}, \tau) d\tau.$$

It is known that the local maxima in the dependences $E_S(\tilde{T})$, $E_A(\tilde{T})$, and $E_W(\tilde{T})$ indicate the energy distribution on the spectrum (e.g., the interval of the periods) during the total time of the analysis, as well as indicate the existence of the defined harmonic components with the corresponding periods in the signal under investigation. These dependences are naturally called *energy diagrams*.

The major property of any transform is its ability to take necessary information from a signal during the limited period of its registration. Therefore, in our further explanation, we will discuss the possibilities of the AFT during the analysis of the harmonic signals with respect to the same possibilities of the WFT and the WT.

7.2.2 Resolution of the Signal Spectrum in the Period and Time Domains

Expressions for the median values of the absolute and relative resolution during estimations of the periods of oscillation were found in Refs. [25,26]. In the case of using the rectangular window, the absolute value of the WFT of signal complex function of $f(t) = \exp(i\omega_0 t)$ has the well-known form of *sinc* function:

$$\left| \hat{S}[f(t)] \right| = t_w \frac{\sin\left[(\omega - \omega_0)t_w/2\right]}{(\omega - \omega_0)t_w/2}, \tag{7.15}$$

where:

t_w is the width of the processing time interval

The absolute resolution for the frequency is determined as

$$\Delta\omega = \omega - \omega_0 = \frac{1}{2\pi\tilde{t}_w}, \quad \tilde{t}_w = \frac{t_w}{c_1}, \tag{7.16}$$

where:

$c_1 \le 1$ is the constant that depends on the level of estimation of the resolution (thus, for the resolution according to the level of zeros, $c_1 = 1$)

From Expression 7.16 it is simple to obtain relations on the absolute resolution for the period on the right side (ΔT_r) and on the left side (ΔT_l) from T_0 [26]:

$$\Delta T_r = \frac{T_0^2}{\tilde{t}_w - T_0}, \quad \frac{\Delta T_r}{T_0(T_0 + \Delta T_r)} = \frac{1}{\tilde{t}_w}, \tag{7.17a}$$

$$\Delta T_l = \frac{T_0^2}{\tilde{t}_w + T_0}, \quad \frac{\Delta T_l}{T_0(T_0 - \Delta T_l)} = \frac{1}{\tilde{t}_w}. \tag{7.17b}$$

After this procedure, the magnitude of the absolute mean time of such a division, $\Delta T_{av} = (\Delta T_r + \Delta T_1)/2$, is defined as follows:

$$\Delta T_{av} = \frac{\tilde{t}_w T^2}{\tilde{t}_w^2 + T^2}. \tag{7.18}$$

Dependences of the absolute separation, $\Delta T_{avS}(\tilde{T})$ and $\Delta T_{avA}(\tilde{T})$, for $c_l = 1$, $t_{wS} = 120$, and $v = 3$, which were taken from Ref. [27], are shown in Figure 7.1a for AFT (dashed line) and WFT (solid line). As is seen for the small normalized periods $\tilde{T}$ ($\tilde{T} \le 0.25$), the resolution for WFT is better than that for AFT; whereas for large periods $\tilde{T}$ ($\tilde{T} \ge 0.4$), the opposite is valid.

The situation is not the same regarding absolute resolution in time domain, Δt. It can be estimated approximately from the condition of vagueness: $\Delta t \Delta \omega \approx 2\pi$. Dependences of $\Delta t_S(T)$ and $\Delta t_A(T)$ for similar magnitudes of t_w and v are shown in Figure 7.1b.

In Figure 7.1c and d, the dependences of the relative resolutions on period $\Delta T_{av}/T_0$ and on time $\Delta t_{av}/T_0$ for AFT (dashed lines) and WFT (solid lines) are presented. It is seen that oscillations with small periods can be localized better

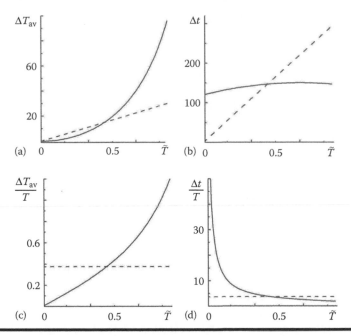

Figure 7.1   Resolution of AFT (dashed lines) and WFT (solid lines) vs. normalized periods $\tilde{T}$: (a) absolute resolution in period, (b) absolute resolution in time, (c) relative resolution on period, (d) relative resolution on time.

with the help of the AFT, but for large periods localization is better made by the WFT. For $\tilde{T} \to 1$, the magnitude is $\Delta T_{avS} \to \infty$. Hence, the same principles that apply for relative resolutions are valid for absolute resolution too. We notice that for the AFT, the relative resolutions in both period and time domains are equally constant. Only in this case, the adaptive properties are developed which is its major advantage.

Thus, it is necessary to jointly use the WFT and the AFT that add each other for the limited time of processing and allow obtaining full information on frequency–temporal (or temporal–periodical) localization of each component of the complicated signal.

As for the resolution regarding the use of the WT, it depends on the form of the "mother" wavelet and on the magnitudes of parameters "embedded" in its expression. For example, for WTs made on the basis of Morlet's wavelet, the resolutions via the periods are similar regarding the form of dependences; but when they are via the magnitude, they are comparable with the resolutions made by the AFT. This property is on account of similarity in functions $\varsigma_{a,\tau,v}$ and ψ.

7.2.3 Peculiarities of the Adaptive Fourier Transform

As mentioned before, it follows that AFT has a series of properties that are present in AWT. These are self-similarity, common form of the basic functions, compact carrier, and so on. Also, at the same time these transforms have significant differences.

As can be seen in Table 7.1, AFT can turn into AWT only for the given magnitudes of parameter v, whose minimum value, v_{min}, depends on the form of function $g(t)$. Particularly, for $v = 1$ there is no window, because of which the AFT turns into AWT. At the same time, for a v of such a magnitude, the basic functions of AFT are localized functions.

Table 7.1 Magnitudes of v_{min} for a Set of Windows $g(t)$ for Different χ

No.	Window	$\chi = 0.01$	$\chi = 0.005$	$\chi = 0.001$
1	Rectangular	16	31	159
2	Hemming	1.0	4.3	23
3	Gauss	1.5	1.6	1.8
4	Henning	1.4	1.9	3.3
5	Blackman	1.3	1.4	2.3

Therefore, the AFT can be taken as an effective technique for the identification of quasiperiodical and wave processes with small (from 1 to 3) number of periods.

One of the main properties, generally for WT, and particularly for AWT, is the equality of zero for the average magnitude of the "mother" wavelet. This condition limits the form of signals allocated efficiently with the help of the AWT. In Ref. [27], the following examples of signals are presented, for which use of the AFT is preferable:

1. A signal is presented as a wavelet of harmonic oscillations, whose duration is filled by a fractional number of periods. In this case, by variation of parameter ν, increase in the accuracy of estimations of the oscillation period can be achieved well. But, for the AWT we get $\nu \in N$, where N is the natural number. Therefore, such a possibility is absent for the latter transform.

2. The overall value of the harmonic signal is not a symmetrical function. Such kinds of signal are frequent and exist naturally. Such a kind of signal has oscillation with fast growth of the amplitude and slow decay, growth of signal amplitude with further achievement of its saturation, and so on. For that, it is possible to obtain full, with respect to AFT, information on the mentioned processes by choosing the corresponding form of $g(t)$.

Besides the advantages, the AFT has a series of drawbacks. Some of these were discussed earlier. Here, we should mention additionally that the AFT does not have the inverse transform, and therefore, does not allow us to reconstruct the initial signal in distinction to the WT and AWT. We should also note that an additional study is required for bringing out the other advantages and disadvantages as well as a definition of the quantitative indicators of its efficiency.

Hence, the AFT obviously can be used at least for one class of signals—harmonic or quasiharmonic processes—because such signals mostly describe the WDs in geospace. It is interesting to note that for defined forms of the signals, the efficiency of their parameter estimations with the help of the AFT is higher than that obtained with the help of the AWT.

7.3 Condition of the Space Weather

We now briefly describe the condition of the space weather during the periods of June 7–13, 2002, December 21–27, 2004, and September 28–October 4, 2005 (Figure 7.2a–c).

On the eve of the launch of the rocket Proton on June 10, 2002 (the time of the launch—01:04 UT) the geospace weather was relatively quiet and stable (see Figure 7.2a). It was observed to increase twice per particle concentration,

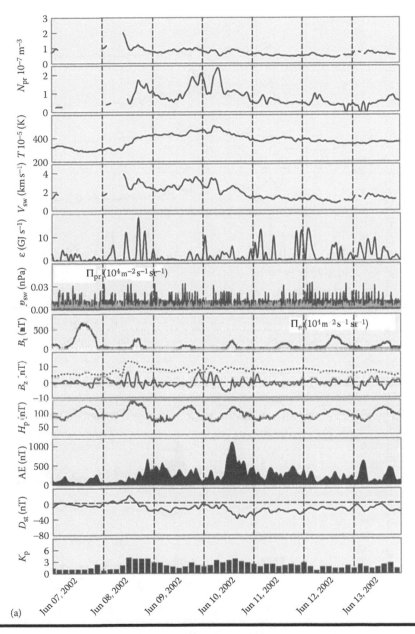

Figure 7.2 Temporal variations of the parameters of solar wind: (a) The start of Proton launch on June 10, 2002; (b) the start of Cyclone launch on December 24, 2004; and (c) the start of Soyuz launch on October 1, 2005. ε, function of Acasofu (computation); Π_e, flow of electrons (GOES-12); Π_{pr}, flow of protons

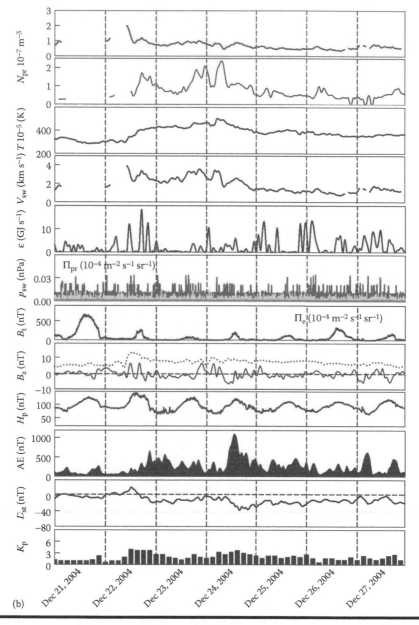

(b)

Figure 7.2 (Continued) [GOES-8 (W75)]; AE, index of the aurora activity (measured by WDC-C2 for geomagnetism of Kyoto University); B_t and B_z, components of vector modules of the interplanetary magnetic field (measured by ACE Satellite—magnetometer); D_{st}, index (WDC-C2); H_p, component of geomagnetic

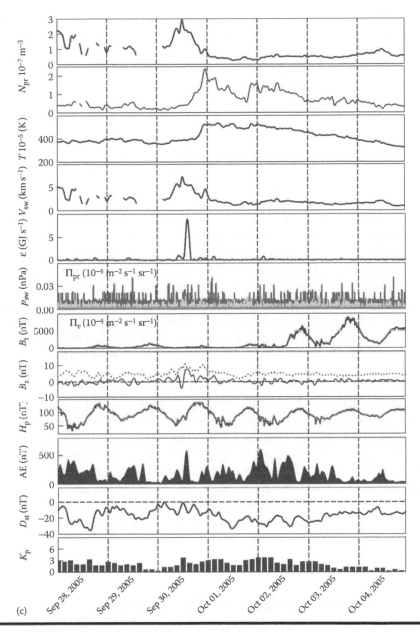

Figure 7.2 (Continued) field (measured by GOES-12); K_p, index (measured by Air Force Weather Agency); N_{pr}, concentration of particles (measured by ACE Satellite—solar wind electron–proton alpha monitor); p_{sw}, dynamic pressure (computations); T, temperature; V_{sw}, radial velocity accompanied.

temperature and velocity, as well as solar wind pressure from June 8 to 10. Power embedded in the magnetosphere and described by the function of Akasofu ε increased on the order of 1 episodically. The density of flow of the charge particles Π_{pr} and Π_e changed insignificantly. Sufficient disturbances of the interplanetary magnetic field were recorded (for the components, B_z and B_t), as well as for the component H_p. The index of aurora activity increased episodically 5–10 times. All of this was accompanied by the appearance of a geospace storm: the magnitude of the index D_{st} fell down to −40 nT, and the maximum value of the index K_p reached 4 on the day after the launch, $K_{p\,max} \approx 3$. Therefore, we can state that that day mostly relates to being a magnetically quiet day.

Before the launch of the rocket Cyclone 3 at 11:20 UT on December 24, 2004, December 21–23, 2004, was observed to increase the velocity of the solar wind approximately by 20%–25% as also the essential growth of concentration and temperature of particles inside the wind (see Figure 7.2b).

As a result, the pressure of solar wind and the velocity of penetration of energy into the magnetosphere both increased by 3–4 and 5–10 times, respectively. The density of protons flows in the radiation region changed insignificantly. At the same time, the density of electrons flow in this region decreased by 1–2 orders. The magnetic substorm was due to come. This indicates that the decrease of B_z was down to −7 nT, the growth of K_p was up to 4, and the fall of the index D_{st} was up to −30 nT. The next substorm, having approximately similar parameters, was observed in the period of December 26–26, 2004. December 24, 2004 was quiet.

Before the launch of the rocket Soyuz at 03:54 UT [here and in further descriptions time is Universal Time, denoted by the unit (UT)] on October 1, 2005, N_{pr} increased 3 times, temperature T, radial velocity V_{sw}, and dynamic pressure p_{sw}—increased 5, 15, and 2–3 times, respectively. On September 30, the function of Akasofu ε grew by an order of 1. Variations of the induction of the interplanetary magnetic field, the indexes AE and D_{st}, were relatively small. The magnitude of the index K_p did not exceed 3 (see Figure 7.2c).

As shown by the illustrations presented in Figure 7.a–c, all days that had rocket launches were magnetically quiet days. Therefore, they were chosen for further analysis and discussions.

7.4 Results of Modeling of the Processes

A signal $x(t) = s(t) + n(t)$ was used for detecting the good signal coming from the background of the noise, where $s(t)$ is the part of the sinusoid with a relative period of 0.25, period $T = 30$, and time of the beginning, $t_1 = 120$. In the end, $t_2 = 270$; $n(t)$ is the Gaussian noise (Figure 7.3a). The results of the

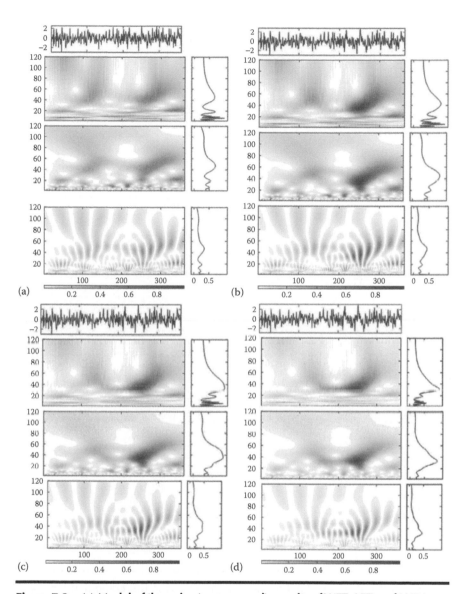

Figure 7.3 (a) Model of the noise (upper panel), results of WFT, AFT, and WT (panels from the top to the bottom) and the corresponding energy diagrams (right sides for each of them). Along the horizontal axis—the time, along the vertical axis—the periods (for the upper panel—the level of the noise). (b) The same notations, as in (a), computed for $q = 0.05$. (c) The same notations, as in (b), computed for $q = 0.1$. (d) The same notations, as in (b), computed for $q = 0.2$. (e) The same notations, as in (b), computed for $q = 0.5$. (f) The same notations, as in (b), computed for $q = 1$.

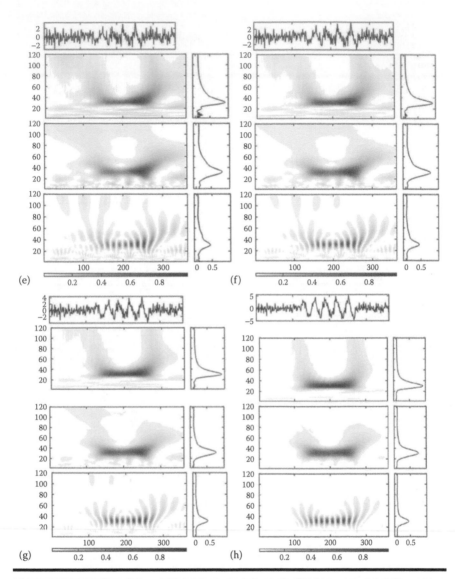

Figure 7.3 (Continued) **(g)** The same notations, as in (b), computed for $q = 2$. **(h)** The same notations, as in (b), computed for $q = 5$.

computer simulations for various magnitudes of the signal-to-noise (S/N) ratio, $q = A_s^2/2\sigma_n^2$, are shown in Figure 7.3b–h. Here, A_s^2 is the power of the good signal and σ_n^2 is the dispersion of the noise.

It is clearly seen from the figures that the AFT and WT evidently allow detection of the quasiperiodical signal, at least, for $q \geq 0.1$. The quasiperiodical signal is conveniently observed with the help of spectrograms as well as by energy diagrams. The first diagram allows us to search the development of the process in time domain, and the second one the energetic behavior of oscillations with the given periods.

7.5 Methods and Techniques of Experimental Studies

For obtaining information on WDs, observations using Doppler sounding (DS) and incoherent scattering (IS), as well as magnetometric measurements gave the results (see Appendix 3).

7.6 Results of Experimental Studies

For illustration of the practical application of the AFT, comparing its possibilities with those of the WFT and WT, let us describe the results of WDs observed during the diurnal and nocturnal time periods. Quasiperiodical processes occurred during the rocket launch and passing of the ST beside the background WDs.

7.6.1 Sounding of the Ionosphere by Doppler Radar

Temporal variations of the Doppler spectra (DSs), obtained with the help of Fourier transform at a time interval of 1 min, are shown in Figure 7.4a–c. The error in estimation of the Doppler frequency shift was about 0.01 Hz, which was about $T \approx 10$ min for the period of WDs and corresponded to a magnitude of the relative amplitude, δ_{Nm}, of perturbations of the electron concentration N of about 0.003 (i.e., ~0.3%). The supposed reaction on the launch of the rockets is searched better at a frequency of 4.0 MHz (Figure 7.4c).

From Figure 7.4c, it can be seen that the WDs were registered during the total time of the observation. Before the rocket's launch, the oscillations were predominant with periods of about 10–15, 20–25, and 40–45 min. The launch of the rocket was done at 11:20 h. After the launch, the character of the signal changed—oscillations with $T \approx 20$–60 min disappeared. The change in the character of the signal that followed took place at about 12:00 h when the components with $T \approx 20$–50 min began to appear gradually. The duration of this process was about 2 h. Approximately after 14:00 h, the spectral structure of the signal changed again—the effects of the evening terminator began to prevail. We note that sunset at the level of the Earth and in the ionosphere at an altitude of 450 km took place at about 13:40 and 16:00 h, respectively.

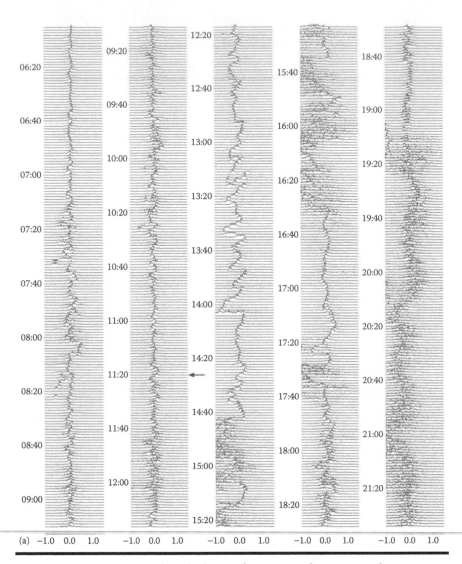

(a)   −1.0   0.0   1.0   −1.0   0.0   1.0   −1.0   0.0   1.0   −1.0   0.0   1.0   −1.0   0.0   1.0

Figure 7.4   (a) Temporal variations of DS at a frequency of 2.74 MHz during Cyclone 3 launch on December 24, 2004 (the effective altitudes are from 225 to 300 km). The moment of the rocket launch is denoted by an arrow.

Additionally, the effects at frequencies of $f_1 = 2.74\,\text{MHz}$ and $f_2 = 4.0\,\text{MHz}$ were often similar and the WDs were observed to be practically synchronous. The ratio of the amplitudes of the frequency shifts of Doppler that were related to WDs was $f_d(f_2)/f_d(f_1) = 1.43$. The ratio of frequencies f_1 and f_2: $f_2/f_1 \approx 1.46$ was approximately the same. This shows that during the propagation of the WDs

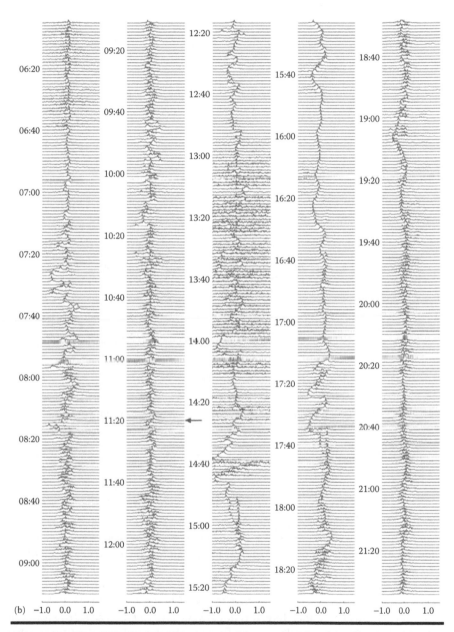

Figure 7.4 (Continued) **(b) The same, as in (a), but for the effective altitudes from 300 to 375 km.**

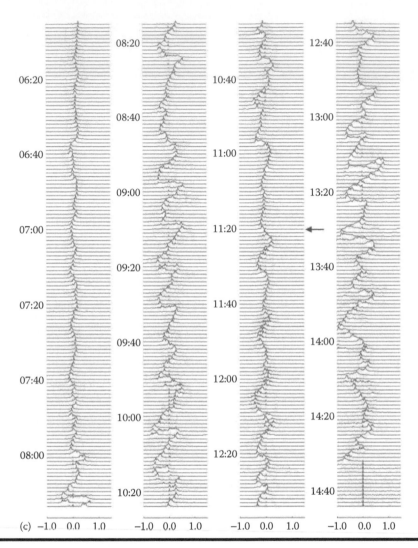

Figure 7.4 (Continued) (c) The same, as in (b), but for the frequency of 4.0 MHz.

the major contribution in the Doppler shift gave the height of the region that accompanies the region of radio-wave reflection. On the whole, the Doppler effect is an integral effect and, therefore, it is not always $f_d(f_2)/f_d(f_1) \approx f_2/f_1$. However, close to the region of the radio-wave reflection, the dependence $f_d(f) \propto f$ takes place [29]. In this case it is

$$f_d = -2\frac{f_2 v_r}{c} \tag{7.19}$$

where:

c is the speed of light

v_r is the velocity of movement in the region of reflection

For maximum amplitude of Doppler shift, $f_{dm}(f_2) \approx 0.6\,\text{Hz}$, we get $v_{rm} \approx 22.5\,\text{ms}^{-1}$. Such value of the amplitude v_{rm} for the period of WDs of $T = 12$ min corresponds to the amplitude, $\delta_{Nm} \approx 0.07$. Let us prove this fact.

Thus, the amplitude of oscillations of the height of the radio-wave reflection equals $\Delta z_{rm} = v_{rm}T/2\pi \approx 2.6\,\text{km}$. For the characteristic scale, the L_r of the concentration N changes in the proximity of the height of the radio wave z_r, so we get $\delta_{Nm} \approx \Delta z_{rm}/L_r \approx 0.07$. Here, $L_r \approx 40\,\text{km}$ and z_r were obtained by IS radar.

Sometimes, temporal variations of DS at frequencies f_1 and f_2 differed significantly. For example, in a time interval of 08:30–12:30 h, the magnitude $f_d(f_1) \approx 0$, whereas the magnitude, $f_d(f_2) \approx 0.6\,\text{Hz}$, that is, it should have experienced oscillations with an amplitude of about 0.6 Hz. This means that the WD traveled basically close to the maximum of ionization, in which a radio wave with $f_2 = 4.0\,\text{MHz}$ was reflected, for which $\delta_{Nm} \approx 0.07$ that fully corresponds to the results obtained by the IS radar (see the following paragraph).

After 14:30 h, that is 50 min after sunset at the ground level, the radio wave operating at f_1 frequency due to the decrease of N was shifted an hour up to 75 km (compare Figure 7.4a and b). In this case, $v_r \approx 20\,\text{ms}^{-1}$ and $f_d(f_1) \approx -0.35\,\text{Hz}$. The amplitude of the oscillations that followed the evening terminator after it passed was insufficient, $f_{dm} \approx 0.2–0.3\,\text{Hz}$, and the oscillations by themselves were irregular (see Figure 7.4a and b).

The effects that followed the launch of the rocket at frequency f_1 were expressed evidently. First, the character of the signal changed at the time interval of 11:40–12:10 h (see Figure 7.4a and b). In this case, $\Delta t_1 \approx 20\,\text{min}$. Second, the character of the signal changed during the interval of 12:10–14:00 h, when $\Delta t_2 \approx 50\,\text{min}$.

Much stronger oscillations of $f_d(f)$ were observed from 13:00 to 14:03 h (see Figure 7.4a) achieving $f_{dm}(f_1) \approx 0.5\,\text{Hz}$ with a period of $T \approx 6$ min. Let us estimate δ_{Nm} from the following expression:

$$\delta_{Nm} = \frac{cT}{4\pi L_d}\frac{f_{dm}}{f}, \tag{7.20}$$

where:

L_d is the height range giving necessary contribution to the Doppler effect

For $L_d \approx 30\,\text{km}$ estimations give $\delta_{Nm} \approx 0.02$ (or 2%).

7.6.2 Sounding of the Ionosphere by Incoherent Scatter Radar

The method of recording WDs based on measurements of the IS signal power, P_s, is the following. From the beginning, powers of the signal, P_s, and the noise, P_n, were averaged at the time interval of 1 min. Then, for the dependence $P_s(t)$ at the interval of 3 h, with the slicing step $\Delta T_t = 1 \min$, the mean $\bar{P}_s$ was found. Next, the series $\delta P_s(t) = P_s(t) - \bar{P}_s$, or $\delta q(t) = \delta P_s(t) / \bar{P}_n$, was analyzed, where $\bar{P}_n$ is the mean value of $P_n(t)$ obtained during the time of the observation (of over a day). It is important to note that magnitudes of $\delta P_s(t)$ and $\delta q(t)$ are proportional to the amplitudes of the electron concentration of WDs, $\Delta N(t)$. The series $\delta q(t)$ are more convenient than those of the $\delta P_s(t)$, because the former do not depend on the so-called radar constant. Moreover, the series $\delta q(t)$ are appropriate for comparing different altitudes and different sets of measurements. For computation of the coefficient of proportionality between $\delta q(t)$ and $\Delta N(t)$, corresponding calibration was done. In the experiments mentioned in this paragraph, $<\delta q> = 1$, and correspondingly, $<\Delta N> \approx 5 \times 10^{10} \, \mathrm{m}^{-3}$, the sign $<\bullet>$ indicates time-averaging procedure. We note the relative amplitude again $\delta_N = \Delta N/N \approx \delta P_s/P_s$, where N is the concentration of the background plasma electrons (or ions; plasma is quasineutral) in the absence of WDs.

To exclude high-frequency fluctuations in time dependences of $\delta P_s(t)$ and $\delta q(t)$, it is natural to perform a further averaging of these functions at the time interval of $T_{av} = 10-15 \min$. If so, the range of the WD periods under investigation is chosen to lie at the time interval defined by the bottom, $2T_{av}$, and the top, T_t, limits, that is, ranging from 10 to 20 min to 180 min. The relative error of δ_N estimation was on average 8%–26% for altitudes of $z \approx 150$–500 km, respectively. The relative errors in estimations of $<\Delta N>$ and $<\delta q>$ were about twice lesser.

The time dependences of $<\delta q>$ and δ_N, after averaging over 15 min with a slicing step of 1 min, are shown in Figure 7.5a and b. It is seen that WDs were observed during the whole period of their registration and also at all altitudes. In the diurnal and evening time periods with increase in altitude the magnitudes of $<\delta q>$ (as well as $<\Delta N>$) grew to the height of $z_m \approx 220$–240 km. Beyond it, these magnitudes have decreased to the height of $z_2 \approx 350$ km. At the height range of $z_2 \approx 350$–500 km, the height dependence of $<\delta q>$ was sufficiently weak. Here, the amplitude $<\delta q>$ is 5–6 times lesser than $<\delta q_m>$ at the height z_m. Approximately, the same relation took place among amplitudes at altitudes of about 150 and 220–240 km in diurnal time periods. At altitudes of $z_2 \approx 180$–400 km in nocturnal time periods, the amplitude $<\delta q_m>$ was approximately the same as in the diurnal time period. For $z_2 \approx 400$–500 km, the day dependence of $<\delta q_m>$ practically was not observed.

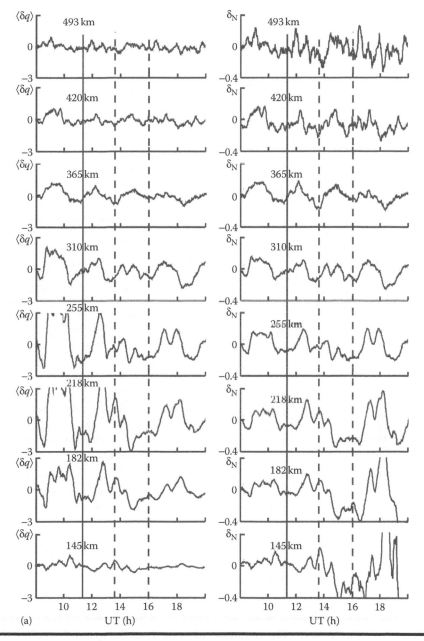

Figure 7.5   (a) Temporal variations of the amplitudes (left panel) and relative amplitudes (right panel) of electron concentration of WDs that accompanied Cyclone 3 launch on December 24, 2004 (denoted by solid vertical line).

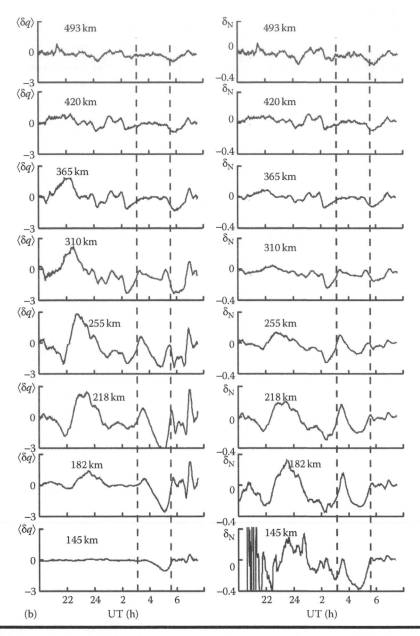

(b) UT (h)

Figure 7.5 (Continued)   The moments of sunset at the ground level at the observatory near Kharkov (Ukraine) and at an altitude of 450 km are denoted by dashed vertical lines. (b) Continuation in time of (a).

The behavior of the amplitude δ_{Nm} in the time dependence $\delta_N(t)$ was not similar. Thus, at altitudes of $z_2 \approx 150–240$ km, the nocturnal magnitudes of δ_{Nm} were several times greater than those for the diurnal period. At the same time, the day dependence of δ_{Nm} was weak at the height range of 250–500 km.

The height dependence of δ_{Nm} was also interesting. Thus, at the height range of 150–350 km the diurnal magnitudes $\delta_{Nm} \approx 0.05–0.1$ practically did not depend on height. For $z_2 \approx 350–500$ km, the amplitude δ_{Nm} increased continuously with the altitude from 0.05 to 0.12. During the nocturnal time period, the magnitudes of δ_{Nm} decreased by 0.3–0.4 to 0.1–0.2 with an increase in altitude from 150 to 500 km, respectively.

Let us examine the WDs that accompanied the rocket launch. The first essential change of the signal character (distortions of the quasiperiodical process) occurred soon after the launch of the rocket, for which $\Delta t_1 \approx 15–20$ min. This effect was registered in dependences of $<\delta q(t)>$ and of $\delta_N(t)$ at altitudes of $z_2 \approx 150–300$ km. The second essential change of the signal character (amplification of oscillations with the period of $T \approx 50–60$ min) was recorded with a time delay $\Delta t_2 \approx 55–65$ min. This effect was observed better at the height range of 150–380 km. It was expressed weakly when it was higher than 440 km. The durations of the first and the second effects were $\Delta T_1 \approx 20–30$ min and $\Delta T_2 \approx 2–3$ h.

After the time interval of 16:00–17:00 h, the growth of amplitudes $<\delta q_m>$ and δ_{Nm} occurred approximately twice. It was expressed more evidently at the height range of 150–290 km. The duration of such a process was not less than 2–3 h, and the magnitude of quasiperiod of the main oscillation was $T \approx 55–65$ min.

7.6.3 Magnetometric Observations of WDs in the Ionosphere

Generally, more than 50 rocket launches are analyzed at cosmodromes in Baikonur and Plesetsk. The results of the observations on the geomagnetic effects of rocket launches will be described in more detail later in Part III of this book. For example, we will describe only two of them here.

The launch of the rocket Proton happened on June 10, 2002 in the morning at 01:14 UT soon after sunrise at an altitude of $z \sim 100$ km (Figure 7.6a and b). The passing of the morning terminator was accompanied by a sharp (up to twice) increase in the amplitude of oscillations of the signal, denoted as $H(t)$ here, and by further decrease in their periods in 1.5–2 times (from 8–12 to 6 min). Approximately 5–7 min after launch, the amplitude of oscillations increased by 2–4 times and 1.5–2 times of their periods. These features continued for 20–30 min for the D- and H-components of the geomagnetic field, correspondingly. A further essential change of the signal character was observed approximately from

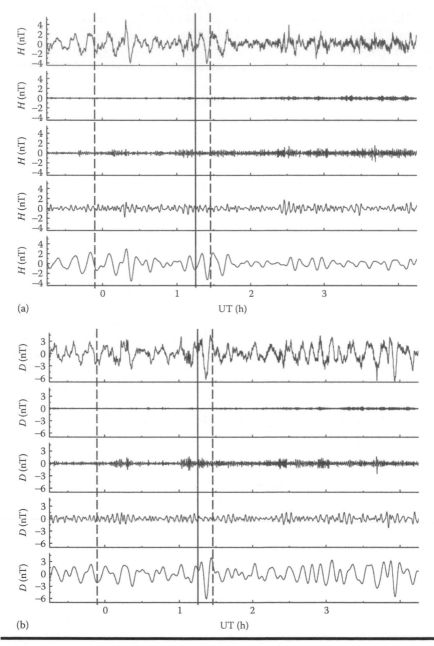

Figure 7.6 **(a) Temporal variations of the *H*-component of the geomagnetic field that accompanied Proton launch on June 10, 2002 at period intervals of 1–1000, 1–20, 20–100, 100–300, 100–300, and 300–1000 s (panels from**

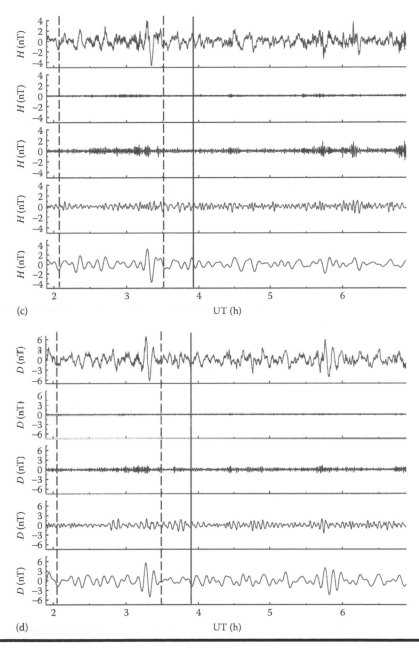

Figure 7.6 (Continued) top to bottom, respectively). The moment of launch is denoted by solid vertical line. The moments of sunrise at an altitude of 100 km and at the Earth's level at the place of location of observatory are denoted by

01:50 to 02:20–02:40 h for the *D*- and *H*-components, respectively. After that, the additional change of the signal character was registered having duration of about 30 min.

Example of variations of signals $H(t)$ and $D(t)$, their spectra, and energy diagrams in the morning is shown in Figure 7.6c and d. The passing of the morning time was accompanied by short-time growth of the amplitude $H(t)$ and $D(t)$ approximately twice. The duration of this process was about 15 min (approximately from 03:15 to 03:30 h). The launch of the rocket Soyuz happened on October 1, 2005 at 03:54 h.

As shown in Figure 7.6c and d, the first short-time (~10 min) and weakly expressed changes of signal character were observed at the time interval of 04:15–04:30 h. Further change of the signal character appeared at the interval from 04:30 to 05:00 h and from 04:20 to 05:15 h for signals $H(t)$ and $D(t)$, respectively. Approximately from 05:30 to 06:10 h, significant (in 1.5–2 times) growth of the amplitude of oscillations was recorded with the prevailed periods of 6–7 min.

Statistical analyses of the time delays (Δt), reaction of the magnetic field on the rocket launch, duration of this reaction (ΔT), and amplitudes of the prevailed oscillations showed that at the diurnal time period, as a rule, three instances of disturbances occurred with Δt that lies at limits of 6–7, 35–45, and 90–130 min, respectively. The first occurrence was observed not very clearly, and was fully absent even at the nocturnal periods. The durations of the magnetic field reactions of 17–27, 9–11, and 9–11 min were attributed to these three instances of disturbances. The amplitude of the magnetic disturbances usually was 3–6 nT.

7.7 Results of the Spectral Analysis of Experimental Data

7.7.1 Method of Doppler Sounding

The results of the spectral estimation carried out on the basis of three integral transforms of temporal variations of the main mode in the Doppler spectrum, $f_d(t)$, are shown in Figure 7.7a (before the launch of the rocket) and in Figure 7.7b (after the launch of the rocket). First, for the dependence $f_d(t)$ the

---

Figure 7.6 (Continued) dashed lines. **(b) The same, as in (a), but for the *D*-component. (c) The same, as in (a), but for the Soyuz rocket launch on October 01, 2005. (d) The same, as in (a), but for the Soyuz rocket launch and for the *D*-component.**

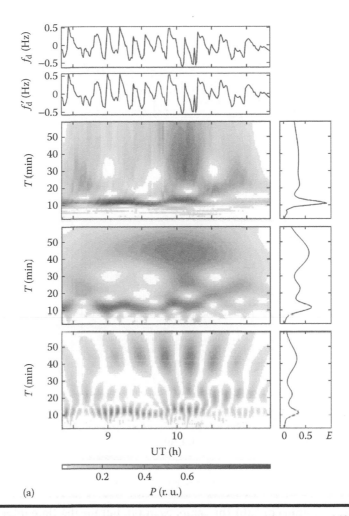

(a)

Figure 7.7 **(a) The temporal variations of the Doppler frequency shift for the main mode of the reflected radio wave recorded at a frequency of 4.0 MHz before the rocket launch on December 24, 2004 (top graph) and the same, but with deduction of the mean value (second graph from the top); the results of WFT, AFT, and WT of data (panels from top to bottom) and the corresponding energy diagrams (right side).**

mean value was computed at the time interval of $T_t = 60$ min with slicing step $\Delta T_t = 1$ min. Then WFT, AFT, and WT were used. The initial periods of WDs were $T = 1$–60 min.

From Figure 7.7a (bottom panel, where the results of the WT is presented), it is seen that before the launch of the rocket oscillations with $T \approx 12$, 22, and

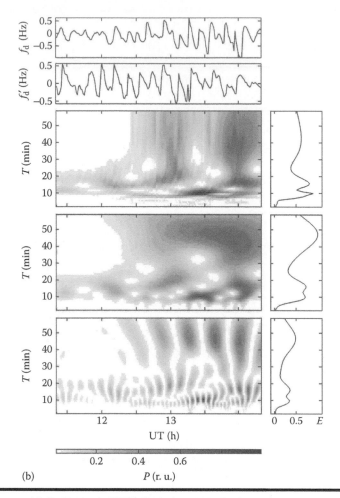

(b)

P (r. u.)

Figure 7.7 (Continued)   (b) The same, as in (a), but for the time interval from 11:20 to 14:20 h after rocket launch.

42 min were predominant in the spectrum, and their energy was recorded as 0.7, 0.5, and 0.9, respectively. The close results gave the AFT. As for WFT, it evidently provided only oscillations with $T \approx 12$ min. As was expected, WT gives the best frequency–time localization of WD spectra.

After the launch of the rocket (Figure 7.7b), as is seen in all panels, oscillations with $T \approx 20$–60 min disappeared from the spectrum of WDs. They occurred again after 12:00 h, that is, 40 min after the starting of the rocket. Further, the spectral content of WDs changed essentially, and almost importantly, after 14:00 h, when the effects of the evening terminator showed up vividly. After

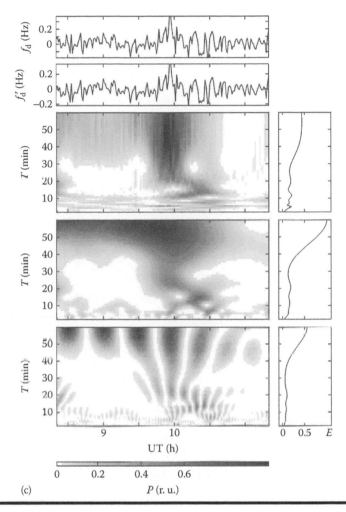

Figure 7.7 (Continued)   (c) The same, as in (a), but for the frequency $f_2 = 2.74$ **MHz.**

the rocket launch, the AFT and WT gave similar results that were essentially different from the results given by WFT.

Now, we will describe the results of the spectral analysis of the dependences $f_d(t)$ for a radio wave operated at the frequency of 2.74 MHz. From Figure 7.7c–f the dynamics of the spectral components is shown clearly. Thus, from about 08:30 to 09:30 h oscillations with $T \approx 50$–60 min were predominant. Then, at the time interval of 09:30–10:30 h a signal became broadband—the period of WDs changed from 5 to 60 min. Further, during 2–2.5 h signal amplitude was

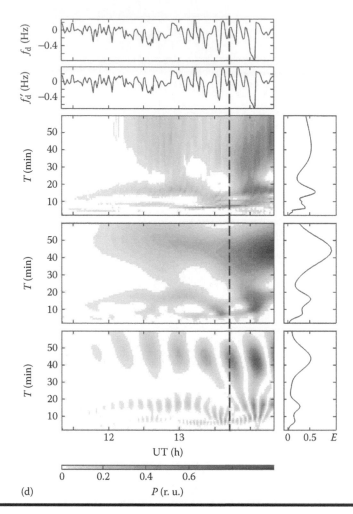

(d)

Figure 7.7 (Continued) (d) The same, as in (a), but for the frequency $f_2 = 2.74$ MHz and after rocket launch (11:20–14:20 h). The vertical dashed line denotes the moment of sunset at the ground level.

insignificant. The essential change in the signal character occurred at 12:30 h, when the oscillations with $T \approx 40$–50 min amplified. From about 13:40 to 16:00 h signal became broadband—oscillations with periods from 10 to 20 min and from 40 to 50 min were recorded.

The spectrum of the process became continuous and the oscillations with $T \approx 20$–60 min prevailed approximately after 15:00 h. A more sharp increase in the amplitude of oscillations with periods of 35–45 min was recorded from 14:30

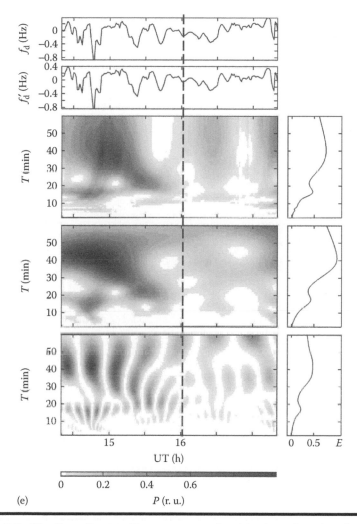

(e)

Figure 7.7 (Continued) (e) The same, as in (a), but for the frequency $f_2 = 2.74$ MHz and for the time interval after the launch from 14:20 to 17:20 h; the sunset is at an altitude of 450 km.

to 15:10 h. All variations of the main spectral mode described earlier are obviously related to the passing of the evening terminator.

At the time interval from 16:00 to 17:20 h signal amplitude was insignificant. The expected change in spectrum was observed from 17:20 to 19:30 h (Figure 7.7f). Oscillations with $T \approx 10$–60 min presented in this spectrum that continuously became narrow. It is interesting to note that from about 17:30 to 18:10 h and from 18:20 to 19:00 h a sharp increase in

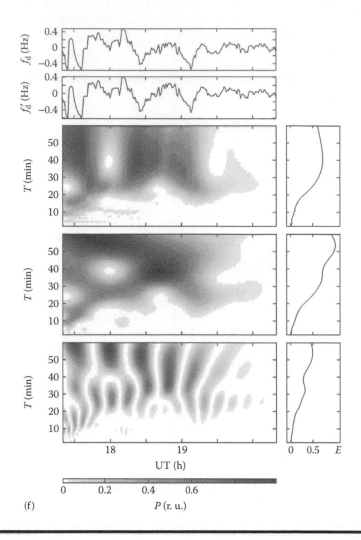

(f)

P (r. u.)

Figure 7.7 (Continued) **(f) The same, as in (a), but for the frequency $f_2 = 2.74$ MHz and for the time interval after launch from 17:20 to 19:30 h.**

the amplitude of oscillations occurred at periods of 50–60 min and about 35–45 min, respectively.

7.7.2 Method of Incoherent Scattering

Before the launch the intensive WDs, periods of 120–180 min, were registered (Figure 7.8a). The WDs with $T \approx 20$–60 min and $T \approx 70$-120 min

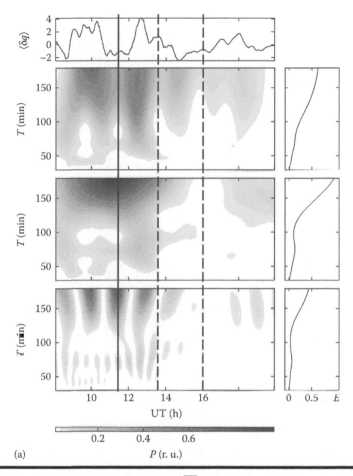

Figure 7.8 (a) Temporal variations of $\overline{\delta q}(t)$ for altitude of 200 km recorded on December 24, 2004 (upper panel); results of WFT, AFT, and WT (panels from top to bottom) and the corresponding energy diagrams (at the right side). Vertical dashed lines denote the moments of the sunset at the ground level and at an altitude of 450 km, the solid line—the moment of launch.

were weakly expressed. After the launch of the rocket, the WDs with $T \leq 60$ min disappeared fully. This continued for not less than 2 h. After about 14:00 h new changes occurred in the dynamics of the WDs spectra (see Figure 7.8a and b).

Sufficiently intensive WDs appeared with the continuously decreased period—from 90–140 to 20–60 min and these were clearly caused by the passing of the ST. A picture described above was observed that was similar to the spectra of the WFT, as well as the spectra of AFT and WT. Evidently, the dynamics of

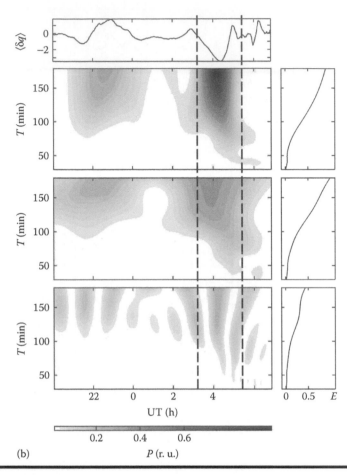

(b)

P (r. u.)

Figure 7.8 (Continued) **(b) The same, as in (a), but for the rocket on December 24 and 25, 2004.**

the WDs is presented by wavelet spectrograms. The dynamics of the WDs at the nocturnal time period and during the passing of the morning terminator can be seen from Figure 7.8b.

7.7.3 Spectral Analysis by Using a Magnetometer

The results of the spectral analysis by using the WFT, AFT, and WT for variations of the H- and D-components of the geomagnetic field at the period interval of 300–1000 s, observed in nocturnal time period of June 9 and 10, 2002, are shown in Figure 7.9a and b. It is seen that the amplification of the quasiperiodical

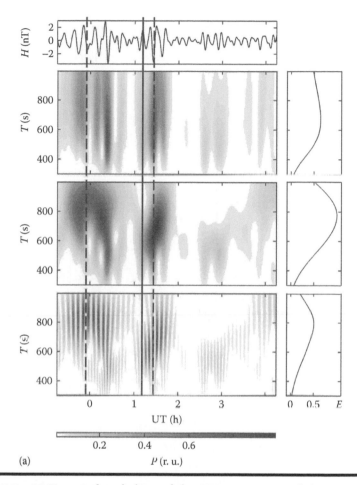

Figure 7.9 (a) Temporal variations of the *H(t)*-component of the geomagnetic field at the periods' interval of 300–1000 s during Proton launch on June 9 and 10, 2002. Other notations are as given in Figure 7.8a.

processes with $T \approx 600{-}1000$ s took place at the time interval of 23:45–00:35 h which accompanied the passing of the morning terminator. Further amplification of the geomagnetic field variation level was observed from about 01:20 to 01:30 h, from 02:30 to 03:15 h, and from 03:40 to 04:00 h. These processes accompanied the Proton rocket launch.

The results of the spectral analysis of temporal variations of the *H-* and *D*-components of the geomagnetic field in the nocturnal and morning time periods on October 1, 2005 at 300–1000 s interval periods are shown in Figure 7.10a and b.

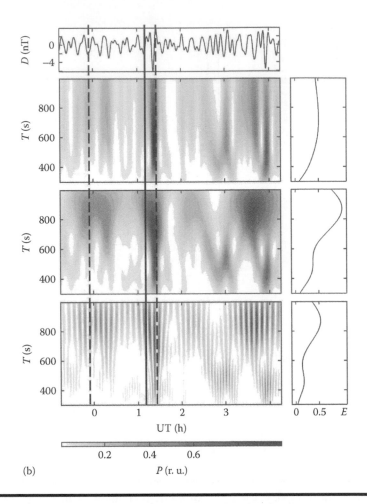

Figure 7.9 (Continued) **(b) The same, as in (a), but for the $D(t)$-component.**

It can be seen from the figures that before sunrise at the ground level, a WD packet of 15–20 min duration was observed, in which oscillations with periods of 350–450 s prevailed. After the launch of the rocket Soyuz, there are two wavelets of oscillations arising with durations of 20–30 min. In the first one, oscillations with periods of 600–800 s prevailed, and in the second one with periods of 350–450 s.

From Figures 7.9 and 7.10 it can be seen that energy diagrams that correspond to AFT and WT are similar to each other. Only the close lining maxima in it differ. At the same time, the mentioned energy diagrams differ significantly from those of the WFT that has the worst resolution on the periods. Only one maximum is seen in the later energy diagrams.

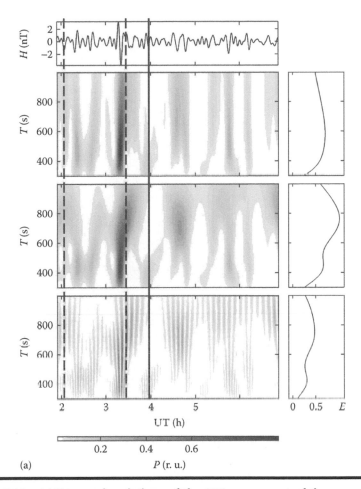

Figure 7.10 (a) Temporal variations of the *H(t)*-component of the geomagnetic field at the periods' interval of 300–1000 s during Soyuz launch on June 9 and 10, 2002. Other notations are as given in Figure 7.8a.

7.8 Discussion of the Results

Using WFT, AFT, and WT for experimentally recording the temporal series showed that each of them, with its own degree of accuracy, can identify WDs occurring in the ionosphere and correct the parameters of the WDs. Naturally, the above-indicated transformations have different frequency–temporal resolutions. As was expected, traditional WFT loses in resolution with respect to AFT and WT. The possibilities of the two latter transforms are close to each other. Despite the sufficiently high resolution of these latter transforms in the joint frequency–time

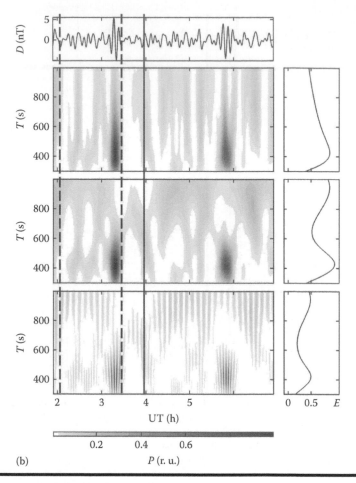

Figure 7.10 (Continued) **(b) The same, as in (a), but for the $D(t)$-component.**

domain, the moment of rise of the reaction of the medium (e.g., the moment of arrival of the WDs) often can be defined more correctly by using the temporal dependence of the output signal, in this case—by use of $f_d(t)$, $\langle \delta q(t) \rangle$, and $\delta_N(t)$. It is natural to note that the AFT and WT cannot be replaced during definition of the spectral content while tracing its dynamics, and so on.

7.8.1 Effects of the Rocket Launches

7.8.1.1 Radiophysical Methods

The methods of DS and IS already show that for $\Delta t_1 \approx 15–20\,min$ after the launch of the rocket, the essential changes of the signal occur that can be recorded as the temporal variations $f_d(t)$, $\langle \delta q(t) \rangle$, and $\delta_N(t)$, or as the result of spectral estimations. Let us derive the velocity of the propagation of WD of the first type, $v_1 = R/\Delta t_1$, where $R = 1500$ km is the distance between the cosmodrome Plesetsk and the place of registration, and $\Delta t_1 \approx 20$ min. It was proved that $v_1 \approx 1.25 \, \text{km s}^{-1}$. If we assume that the time of rocket movement, Δt_0, to altitude $z \approx 100$ km, where generation of WDs takes place, is about $\Delta t_0 \approx 3$ min, then the corrected value of the time delay is $\Delta t_1' = \Delta t_1 - \Delta t_0 \approx 17$ min. In this case, the corrected magnitudes are $R' \approx R$ and $v_1' \approx 1.5 \, \text{km s}^{-1}$. This velocity is close to the velocity of slow MHD waves [14,30].

For the WDs of the second type, the time delay interval Δt_2 is about 40 min obtained by the DS technique and about 50–65 min obtained by the IS technique. The difference in magnitudes of Δt_2 can be related to the following given below.

First, the DS technique gives Δt_2 only for altitudes close to the height of radio-wave reflection. In the experiments under consideration, the frequency of the Doppler radar was $f_2 \approx 4 \, \text{MHz}$ and the radio wave at diurnal time was reflected in the altitude of 200 km, where $N \approx 2 \times 10^{11} \, \text{m}^{-3}$. At this altitude, essential growth of $\langle \delta q \rangle$ began with the delay $\Delta t_2 \approx 50$ min, that is, the IS technique gives the magnitude of $\Delta t_2 \approx 10$ min larger than that obtained by the DS technique.

Second, in the DS technique f_d is proportional to the temporal derivative $d(\Delta N)/dt$, or to $d\delta_N/dt$, whereas in the IS, it is proportional to $\langle \delta q \rangle \propto \Delta N$. As is known, the derivative of the harmonic function forestalls the function on the phase at the quarter of the period. As can be seen from the results of the AFT and WT, at close to 12:00 h DS radar mainly WDs with the periods from 10 to 60 min were observed. If that were so, then $T/4 \approx 2.5$–15 min. The mean value of this correction equals to about 9 min. With enough confidence, we can assume that for to this reason Δt_2 is estimated by the DS technique at approximately at 10 min less than that estimated by the IS technique.

Because the differential device allows calculation of the temporal intervals exactly, it can be stated that the DS technique is preferable compared to the IS technique during measurement of the magnitudes of the time delay. For $\Delta t_2' \approx 37$ min and $v_2' \approx 675 \, \text{m s}^{-1}$. Such a value of v_2' is less than that for the sound velocity v_s at an altitude of about 200 km. It is known that (see also Chapter 2)

$$v_s = \left(\gamma \frac{k_B T_n}{M} \right)^{1/2}, \tag{7.21}$$

where:
    $\gamma = 1.4$ is the ratio of the specific thermocapacities
    $k_B = 1.38 \times 10^{-23} \, \text{J K}^{-1}$ is the Boltzmann's constant

$M = 3.8 \times 10^{-26}$ kg is the mean mass of the neutral particle

$T_n \approx 1000$ K is the magnitude of the neutral component of plasma that was supposed to be equal to the magnitude of the temperature of ions (see Chapter 1), defined by the use of the IS technique

For the above-indicated parameters at an altitude of 200 km at the diurnal time period, the sound velocity $v_s \approx 725 \text{ m s}^{-1}$. The velocity magnitude close to 675 m s^{-1} has an internal gravity wave (IGW) that has earlier been mentioned in Chapters 2 and 3. For this wave, the maximum magnitude of the group velocity equals $v_g \approx 0.9 \times v_s \approx 650 \text{ m s}^{-1}$ [31]. The fact is that $v_{g\,max} < v'_2$ witnesses a part of the path wave traveling as a shock wave for which $v_g > v_s$ [13].

Disturbances in the ionosphere following the rocket launch, according to their magnitudes of velocities (of about 1.5 and 0.6–0.7 km s^{-1}), durations, and periods are similar to those parameters that were observed and registered by other authors [9–13] as well as by one of the authors of this book along with his colleagues [5,12,14,16–19].

7.8.1.2 Magnetometric Technique

Let us consider disturbances of the geomagnetic field that accompany the launch of the rockets. The effects obtained from the launch of the rocket Soyuz and from the launch of the rocket Proton were seen to be finally similar.

Three groups of disturbances were usually presented in the magnetometric signals having different times of delay: $\Delta t_1 \approx 6$–7 min, $\Delta t_2 \approx 35$–45 min, and $\Delta t_3 \approx 90$–130 min.

The first instance of disturbances was not always observed—it happened mostly in the nocturnal time periods. Often these had an irregular (i.e., noisy) character and relatively small durations (17–20 min). The characteristic level of these fluctuations was 2–4 nT. The periods of the predominant pulses were about 5–8 min.

Disturbances with $\Delta t_2 \approx 35$–45 min were observed systematically. They presented themselves as packets of oscillations with periods of 9–11 min and amplitudes of 3–5 nT.

The packets of oscillations with periods of 9–11 min, amplitudes 3–6 nT, and duration of 40–70 min were presented with more slower disturbances, and had a time delay of $\Delta t_3 \approx 90$–130 min. The parameters of the second and the third instances were close, and, therefore, it can be assumed that they were a part of the same disturbances.

Let us estimate the characteristic velocities of propagation of the disturbances caused by the launch and flight of the rockets. Generally speaking, these disturbances are generated directly in the ionosphere when the rocket achieves an altitude of $z \sim 100$ km. From the action of the reactive jet (due to a torch of the rocket) the conductivity of plasma is changed essentially in the dynamo region of the ionosphere (at altitudes of $z \approx 100$–200 km) and waves of electromagnetic nature (mostly magnetohydrodynamic, MHD) are generated [14]. At the same altitudes shock waves are generated effectively in gas, which propagate from the rocket with a velocity that exceeds sound velocity in gas (see, e.g., Refs. [13,14]). At distances sufficiently far from the rocket trajectory, a shock wave is transformed to an acoustic gravity wave (AGW). As shown in Ref. [13], this wave corresponds to a period of 4–5 min.

A rocket achieves an altitude of $z \approx 100$–200 km during time Δt_0 that depends on the type of rocket and equals about 2–3 min. In this case, the time of propagation of the disturbance is $\Delta t' = \Delta t - \Delta t_0$. Then, $\Delta t_1' \approx 4$, $\Delta t_2' \approx 32$–42, and $\Delta t_3' \approx 90$–130 min. As was shown in Ref. [13], the region of generation of disturbance is located approximately 500 km to the east from the place of the launch of the rocket. Accounting for these facts, the corrected value of the distance of $R' \approx 2500$ km can be estimated. These magnitudes of $\Delta t'$ and R' correspond to velocities $v_1' \approx 10$, $v_2' \approx 1.0$–1.3, and $v_3' \approx 0.3$–$0.5 \, \mathrm{km \, s^{-1}}$.

The waves of $v_1' \approx 10 \, \mathrm{km \, s^{-1}}$ obviously relate to the electromagnetic (possibly to MHD) waves [14,30]. The velocity of $v_2' \approx 1.0$–$1.3 \, \mathrm{km \, s^{-1}}$ relates to the slow MHD waves [14,30], as well as to the shock waves at altitudes of the *F*-region of the ionosphere [13]. However, the velocity of the later waves exceeds the velocity of AGWs at these altitudes, which is close to 0.7–0.8 km s^{-1}. The velocity of $v_3' \approx 0.3$–$0.5 \, \mathrm{km \, s^{-1}}$ corresponds approximately to the velocity of AGWs at altitudes of the *E*-region of the ionosphere.

Let us compare the above-presented results of our estimations with those obtained from Ref. [15]. Thus, for short-periodical ($T \approx 2$–3 min) oscillations of the geomagnetic field strength, $\Delta t_1 \approx 3$–4 min, $\Delta t_1' \approx 1$–2 min, and $v_1' \approx 7$–$13 \, \mathrm{km \, s^{-1}}$ (the mean is $v_1' \approx 10 \, \mathrm{km \, s^{-1}}$). This value is very close to that which we obtained and described above. This is something unexpected because the velocity of MHD waves depends essentially on the orientation of the direction of wave propagation with respect to the geomagnetic field. In the case of our research, it was nearly across the magnetic field $\mathbf{B}_0$ wave propagation; whereas in Ref. [15], it was nearly along $\mathbf{B}_0$ wave propagation was observed.

The amplitude of the short-periodic disturbances according to Ref. [15] was about 6–20 nT, whereas in our case it was from 2 to 4 nT, which is several times lesser. Accounting for distances R of 800 km (their case) and 2100 km (our case), such a correspondence between the amplitudes was expected.

Slow disturbances in Ref. [15] had $\Delta t_2 \approx 10$–15 min, $\Delta t_2' \approx 7.5$–12 min, and $v_2' \approx 1.1$–$1.8 \mathrm{km \, s^{-1}}$. These magnitudes are close to those obtained by us and described above (of 0.8–1.1 km s^{-1}). The amplitude of slow disturbances according to Ref. [15] was about 10 nT, that is, decreased with the decrease of distance R. It is evident that these disturbances have the same nature.

We add that the observed magnitudes of amplitudes of the geomagnetic field disturbances corresponded well to the derived values that follow from the performed theoretical model regarding the geomagnetic field disturbances. According to this model, the observed disturbances are the result of modulation of the ionospheric conductivity and of the capture of electrons and ions by neutral particles affected by the AGWs caused by the rocket launches.

7.8.2 Effects of the Solar Terminator

Using the three types of spectral analyses described above allows detailed study and observance of the dynamics of the WDs generated by launch of rockets (see Figures 7.7b and 7.8). As was proved, its passage resulted in the redistribution of energy between the spectral components of WDs. These WDs began soon after the moment of passing of the evening terminator at the ground surface level and continued until the moment of its passing through an altitude of $z \approx 450$ km. Periods of about 15, 45, and 60 min prevailed in the WDs. The duration of the WD was achieved in 3 h. The picture of the dynamics of the WD during its passage through the evening and the morning terminator was different (compare Figure 7.8a and b). The parameters of the WD in the ionosphere described above were observed earlier and are described in Refs. [14,18–20,29,31–37], as well as by other authors too (see, e.g., Ref. [33]).

Let us briefly consider the variations of the geomagnetic field accompanied when it passed the ST. The corresponding reaction appeared during passage by the terminator at the height range of 100–150 km, where the dynamo region of the ionosphere is usually located. In this case, the WDs can be amplified (sometimes they became too weak). The duration of the reaction, as a rule, did not exceed 1–2 h (see Figures 7.6, 7.9, and 7.10).

Therefore, the main advantages of the modern methods of spectral analysis can be related to the possibility of obtaining a vivid picture of the evolution of WDs in the ionosphere with the high-frequency time (or period-time) resolution.

7.9 Major Results

1. For the first time the experiment was performed to consider a comparative analysis of the possibilities of the WFT, AFT, and WT for the detection and estimation of the spectral density of the WDs in the ionosphere and in the geomagnetic field with a high-frequency time (or period-time) resolution. It was shown that the AFT, because of its possibilities, exceeds the possibilities of the WFT and limits those of

the WT. Therefore, the AFT can be used as an effective mathematical tool for the detection and estimation of the parameters of WDs in geospace. The AFT and WT allow detection of the quasiperiodical processes clearly at the noisy background for a signal-to-noise ratio (SNR) of not less than 0.1.

2. The WFT, AFT, and WT can be recommended to be used jointly. As a basis, the format presented in this chapter can be used.

3. The WFT, AFT, and WT were used for the detection of the WDs in the ionosphere and in the geomagnetic field related to the rocket launches and to the passing of the ST at the background of significant wave activity. It was stated that AFT and WT allowed the WDs accompany the rocket launch or the passing of the ST well localization in the joint period-time domain.

4. It was shown that the rocket launch in the ionosphere caused WDs of two types with velocities not more than 1.5 and 0.4–0.7 km s^{-1}. These disturbances had durations of 20–30 min and 2–3 h, respectively. They can be expressed more strongly at the height range of 150 to 300–380 km. Faster traveling disturbances were observed with confidence than those with weaker ones that were less faster.

5. In the magnetometric signals, the three groups of WDs were detected with a time delay of 6–7, 35–45, and 90–130 min. The following durations of the processes corresponded to these delays: 17–27, 45–80, and 40–70 min and the predominant periods of 5–8, 9–11, and 9–11 min.

6. It was found that the fast disturbances (velocity of 10 km s^{-1}) were expressed with less evidence and were observed in approximately 78% cases, and, as a rule, at diurnal time periods. They have been transferred with the help of MHD waves. Disturbances with velocities of 1.0–1.3 and 0.3–0.5 km s^{-1} were always clearly seen in about 70% of the cases, and, as a rule, at diurnal time period.

7. Appearance of the much slower waves was caused by movements of plasma in the magnetic field that was affected by AGW. The nature of the waves having velocities between 1.0 and 1.3 km s^{-1} can be related to the propagation of either a packet of the captured AGW, created by the shock wave in the atmosphere, ionosphere, and geomagnetic field, or the slow MHD waves.

8. It is stated that the amplitude of all kinds of disturbances was from 3 to 6 nT.

9. It was demonstrated that the passing of the ST causes essential change of the spectral content of WDs in the ionosphere and the geomagnetic field. The duration of the effects from the evening terminator in the ionosphere was not less than 2–3 h. The effects of the ST depended significantly on altitude. The duration of the effects of ST in the geomagnetic field was of 1–2 h.

References

1. Gossard, E. E. and Huk, Y. X., *Waves in the Atmosphere*, 1978. 532c (in Russian).
2. Grigor'ev, G. I., Acoustic-gravity waves in the earth's atmosphere (issue), *Izv. Vuzov. Radiofizika*, 42, 3–25, 1999 (in Russian).
3. Hocke, K. and Schlegel, K., A review of atmospheric gravity waves and traveling ionospheric disturbances: 1982–1995, *J. Ann. Geophys.*, 14, 917–940, 1996.
4. Williams, P. J. S., Tides, atmospheric gravity waves and traveling disturbances in the ionosphere, in *Modern Ionospheric Science*, Kohle, H., Ruster, R., and Schlegel, K., Eds., Katlenburg-Lindau: Max-Planck Institute of Aeronomy Press, 1996, 136–180.
5. Chernogor, L. F., The Earth–atmosphere–geospace as an open dynamic nonlinear system, *Cosmic Sci. Technol.*, 9, 96–105, 2003 (in Russian).
6. Chernogor, L. F., Physics of the Earth, atmosphere and geospace in the lightening of the system paradigm, *Radiophys. Radioastron.*, 8, 59–106, 2003 (in Russian).
7. Chernogor, L. F., Tropical cyclone as an element of the system: Earth–atmosphere–ionosphere–magnetosphere, *Cosmic Sci. Technol.*, 12, 16–36, 2006 (in Russian).
8. Chernogor, L. F., The Earth–atmosphere–ionosphere–magnetosphere as an open dynamic nonlinear system (Part 1), *Nonlinear Univ.*, 4, 655–697, 2006 (in Russian).
9. Arendt, P. R., Ionospheric undulations following Apollo 14 launching, *Nature*, 231, 438–439, 1971.
10. Noble, S. T., A Large-amplitude travelling ionospheric disturbance excited by the space shuttle during launch, *J. Geophys. Res.*, 95, 19037–19044, 1990.
11. Nagorsky, P. M., Inhomogeneous structure of the F-region of the ionosphere generated by the rockets, *Geomagn. Aeronom.*, 38, 100–106, 1998.
12. Kostrov, L. S., Rozumenko, V. T., and Chernogor, L. F., Doppler radio sounding of disturbances in the middle ionosphere accompanied launchings and flights of space vehicles, *Radiophys. Radioastron.*, 4, 227–246, 1999 (in Russian).
13. Afra'movich, E. L., Perevalova, N. P., and Plotnikov, A. V., Registration of ionospheric responses on the shock-acoustic waves generated during launchings of the rocket-carriers, *Geomagn. Aeronom.*, 42, 790–797, 2002.
14. Burmaka, V. P., Kostrov, L. S., and Chernogor, L. F., Statistical characteristics of signals of the Doppler HF radar during sounding of the middle ionosphere perturbed by the rockets' launchings and by the solar terminator, *Radiophys. Radioastron.*, 8, 143–162, 2003 (in Russian).
15. Sokolova, O. I., Krasnov, V. M., and Nikolaevsky, N. F., Changes of geomagnetic field under influence of rockets' launch from the cosmodrome Baykonur, *Geomagn. Aeronom.*, 43, 561–565, 2003.
16. Burmaka, V. P., Taran, V. I., and Chernogor, L. F., Wave disturbances at the background of the natural transfer processes, *Geomagn. Aeronom.*, 44, 518–534, 2004.
17. Burmaka, V. P., Lisenko, V. N., Chernogor, L. F., and Chernyak, Yu. V., Wave processes in the F-region of the ionosphere accompanied starts of rockets from the cosmodrome Baykonur, *Geomagn. Aeronom.*, 46, 6, 783–800, 2006.
18. Burmaka, V. P., Taran, V. I., and Chernogor, L. F., Results of study of wave disturbances in the ionosphere by method of incoherent scattering, *Foreign Radioelectronics: Success in Modern Radioelectronics*, 3, 4–35, 2005 (in Russian).

19. Burmaka, V. P., Taran, V. I., and Chernogor, L. F., Wave processes in the ionosphere during quiet and perturbed conditions. 1. Results of observations at the Kharkov's radar of incoherent scattering, *Geomagn. Aeronom.*, 46, 193–208, 2006.

20. Burmaka, V. P., Taran, V. I., and Chernogor, L. F., Wave processes in the ionosphere during quiet and perturbed conditions. 2. Analysis of results of observations and modeling, *Geomagn. Aeronom.*, 46, 209–218, 2006.

21. Bracewell, R., *The Fourier Transform and Its Applications*, 3rd edn., New York: McGraw-Hill, 1999.

22. Holschneider, M., *Wavelets: An Analysis Tool*, Oxford: Calderon Press, 1995.

23. Diakonov, V. P., *Wavelets: From Theory to Practice*, Moscow: Solon-R, 2002.

24. Mallat, S., *A Wavelet Tour in Signal Processing*, 2nd edn., San Diego, CA: Elsevier, 1999, 489.

25. Burmaka, V. P., Panasenko, S. V., and Chernogor, L. F., Modern methods of the spectral analysis of quasi-periodical processes in the Geospace, *Foreign Radioelectronics: Success in Modern Radioelectronics*, 11, 3–24, 2007 (in Russian).

26. Lazorenko, O. V., Lazorenko, S. V., and Chernogor, L. F., Usage of wavelet analysis to the problem of detection of super-wideband signals at the background of obstructions, *Radiophys. Radioastron.*, 7, 46–63, 2002 (in Russian).

27. Lazorenko, O. V., Panasenko, S. V., and Chernogor, L. F., Adaptive Fourier transform, *Electromagn. Waves Electron. Sys.*, 10, 39–49, 2005 (in Russian).

28. Chernogor, L. F., Modern methods of spectral analysis of quasi-periodical and wave processes in the ionosphere. Peculiarities and results of experiments, *Geomagn. Aeronom.*, 48, 681–702, 2008.

29. Namazov, S. A., Novikov, V. D., and Khmel'nitsky, I. A., Doppler shift of frequency during ionospheric propagation of the decametric waves, *Izv. Vuzov. Radiofizika*, 7, 3–15, 1999; 8, 3–19, 1999.

30. Sorokin, V. M. and Fedorovich, G. V., *Physics of Slow MHD-waves in the Ionospheric Plasma*, Moscow: Energoizdat, 1982 (in Russian).

31. Grigirenko, E. I., Lazorenko, S. V., Taran, V. I., and Chernogor, L. F., Wave disturbances in the ionosphere accompanied the flash at the Sun and the strong magnetic storm occurred on September 25, 1998, *Geomagn. Aeronom.*, 43, 770–787, 2003.

32. Garmash, K. P., Rozumenko, V. T., Tyrnov, O. F., Tsymbal, A. M., and Chernogor, L. F., Radiophysical investigations of the processes in the near-Earth plasma perturbed by the high-energy sources, *Foreign Radio Electronics: Successes of Modern Radio Electronics*, 3–15, 1999; 3–19, 1999 (in Russian).

33. Somsikov, V. M., Waves in the atmosphere stipulated by the solar terminator (issue), *Geomagn. Aeronom.*, 31, 1–12, 1991.

34. Akimov, L. A., Grigorenko, E. I., Taran, V. I., Tyrnov, O. F., and Chernogor, L. F., Complex radiophysical and optical studies of dynamic processes in the atmosphere and Geospace caused by solar eclipse occurred on August 11, 1999, *Foreign Radio Electronics: Successes of Modern Radio Electronics*, 25–63, 2002 (in Russian).

35. Kostrov, L. S., Rozumenko, V. T., and Chernogor, L. F., Doppler radio sounding of disturbances in the naturally-perturbed middle ionosphere, *Radiophys. Radioastron.*, 4, 206–226, 1999 (in Russian).

36. Taran, V. I., Study of the ionosphere in the natural and artificially-perturbed conditions by the method of incoherent scattering, *Geomagn. Aeronom.*, 41, 659–666, 2001.
37. Garmash, K. P., Leus, S. G., Pazura, S. A., Pohil'ko, S. N., and Chernogor, L. F., Statistical characteristics of fluctuations of the Earth's electromagnetic field, *Radiophys. Radioastron.*, 8, 163–180, 2003 (in Russian).

DIAGNOSTICS OF GEOMAGNETIC DISTURBANCES

Chapter 8

Geomagnetic Effects of Rocket Launches from Russian Cosmodromes

8.1 Rocket Launches from Baikonur Cosmodrome

8.1.1 Overview

As mentioned in the Introduction and summarized in a special issue [1], the effects caused by the launch and flight of rockets and space vehicles (SVs) in the atmosphere, ionosphere, and geospace were studied for more than 50 years. In most of the works published in this period, the effects that come up along the trajectory of the SV flight were described (see, e.g., Refs. [2–4]). The study of processes that ranged at large distances from the SV trajectory has been discussed only in several papers [5–9]. One of the authors of this book along with his colleagues investigated the possibilities of the appearance of large-scale and global effects caused by rocket launches (RLs) [10–21]. In this case, as was shown by researchers, the existence of such disturbances is a problematic aspect. For their search and detection, it is obvious from the beginning to carry out observations of the possible effects occurring during the launches of sufficiently powerful rockets from cosmodromes close to the places of registration (in our case—from Kharkov Observatory in Ukraine).

Such cosmodromes, as previously mentioned in Chapters 1 and 2, are in Plesetsk and Baikonur (Russia), which are close to Kharkov Observatory. In the case of the former cosmodrome, such SV launches were comparably rarely declared, and the power of the starting rockets was sometimes less than those of rockets that started from Baikonur cosmodrome.

Therefore, in this chapter we discuss and describe the results of observations of ionospheric disturbances that accompany Soyuz and Proton launches from Baikonur cosmodrome during the period 2002–2006. Main attention was paid to the study of geomagnetic field reaction on such launches and wave disturbances (WDs) in the medium that are responsible for the transport of WDs to distances of about 2100 km.

As mentioned in the previous chapters, the study of WDs in the atmosphere, ionosphere, and geospace has its own additional interests (see, e.g., Refs. [22,23]). This is because WDs can always be observed in the near-the-Earth environment. They play an essential role in the transport of energy and momentum from some desired regions of the atmosphere and the geospace to others, and vice versa. They are also responsible for the interactions of subsystems in the system Earth–atmosphere–ionosphere–magnetosphere (EAIM) [24–27]. The WDs significantly affect the working functions and characteristics of radio and telecommunication systems (see Chapters 11 through 13), which essentially limit their potential possibilities.

On the other hand, by themselves the WDs can be used for distant diagnostics of the near-the-Earth environment. All that mentioned earlier *a priori* define the increasing demand and interests of the wave processes occurring in the ionosphere. The idea to study WDs caused by the SV launches and flights was proved to be sufficiently useful. In this case, it is well known that at the place and time of yield of energy, the source power reaches 10^{11} W, and its energy becomes 10^{13} J. About 0.1%–1% of the total energy of the reactive jet [16,19] can be transformed into wave energy.

As follows from the previous chapters and the references mentioned earlier, for observation of the near-the-Earth and space media reaction on RLs and flights, ionosondes, Doppler radar systems, incoherent scatter radars, and so on were used. Only in individual works [4,8], the magnetometric technique was used based on observations of low-frequency changes (in units of tens and hundreds of minutes) of the geomagnetic field. In Ref. [4], the distance between the cosmodrome and the place of observation R was about 800 km. From the results of Ref. [8], it is impossible to adequately conclude on the nature of the long period (of about 80 min) geomagnetic field disturbances. It is assumed that they could be caused by natural sources as well as by RLs.

The aim of this chapter is to describe the observational results and theoretical modeling of sufficiently high-frequency (with periods from 1 to 1000 s) geomagnetic field fluctuations that accompanied 65 Soyuz and Proton launches from Baikonur cosmodrome ($R \approx 2100$ km) close to Kharkov Observatory [28].

8.1.2 Method of Analysis

A no-serial highly sensitive magnetometer is described in Appendix 3. From the beginning, the magnetometric signals that correspond to *H-* and *D*-components of the geomagnetic field, which account for the amplitude–frequency characteristics of the magnetometer/fluxmeter, were transformed into signals *H*(*t*) and *D*(*t*) of geomagnetic field fluctuations. The first are measured in relative units, but the latter are measured in absolute units (in nanotesla or nT).

Then, signals *H*(*t*) and *D*(*t*) were processed by digital filtering with the filter bandwidths corresponding, for example, to periods of 1–20, 20–100, 100–300, and 300–1000 s, by windowed Fourier transform (WFT), adaptive Fourier transform (AFT), and wavelet transform (WT) as described in Chapter 4 (see also Ref. [29]).

WT, as shown in Chapter 4, has the following form:

$$\hat{W}\left[s(t)\right] \equiv Ws(a,b) = \frac{1}{a^{1/2}} \int_{-\infty}^{\infty} s(t)\psi^*\left(\frac{t-b}{a}\right)dt, \qquad (8.1)$$

where:
s(*t*) is a signal [*H*(*t*) or *D*(*t*)]
a and *b* are the parameters of scalability and shifting, respectively
ψ is the wavelet basis
* symbol indicates complex conjunction
t is dimensionless time

To express the basis ψ(*t*), Morlet's wavelet was used (see, e.g., Ref. [29])

$$\psi(t) = \exp\left\{-\frac{t^2}{2}\right\}\cos\tilde{\omega}t, \qquad (8.2)$$

which represents the wave packet with dimensionless frequency $\tilde{\omega}$ (usually select $\tilde{\omega} = 5$). This basis is convenient for detection and further analysis of wavelets (i.e., geomagnetic pulses of type *Pc*).

Time delays of the possible geomagnetic field reaction on RLs, their duration, amplitudes, and spectral content of the occurred disturbances were then determined. Further, by using the averaging procedure, the statistical characteristics of these disturbances were calculated. From the beginning, the averaging procedure was carried out separately for Soyuz and Proton for the diurnal and nocturnal time periods, and for the *H-* and *D*-components. If histograms of any parameter were similar for both the components or for both types of rockets, the initial amounts of data were combined into one.

8.1.3 Level of Magnetic Perturbations

The level of magnetic perturbations was controlled with the help of indexes A_p and K_p (which were introduced in Chapter 3) taken from the Internet. The magnitudes of these indexes were analyzed during the day of the RL, a day before, and 2 days after this event. It was shown that all launches happened in magnetically quiet conditions. Only the launch on February 2, 2003, took place in some perturbed conditions in which index K_p achieved figure 5).

8.1.4 Results of Observations

Totally 65 RLs were analyzed of which 38 were Soyuz and 27 Proton. These took place at different periods of the day, different seasons, and different levels of geomagnetic activity.

As proven, variations of the geomagnetic field that followed RLs were principally not sufficiently different from natural variations. Therefore, for selection of disturbances that accompanied RLs, the following approaches were used. First of all, the disturbances relating to well-known sources (such as the solar terminator, ST) were fully eliminated. Then, changes in the character of the signal compared to the background period were studied (e.g., several hours prior to RL). Furthermore, changes in the signal character were compared for the same diurnal time intervals of RL and for the same control days (both prior to and after launches). Finally, the effects of launches on the H- and D-components of the geomagnetic field were compared with each other.

Analysis of separate events could not give full evidence that the observed changes in the signal character were related to SV launches. Such evidence could give the corresponding statistical analysis of registrations for the 65 launches.

Examples of temporal variations of geomagnetic field horizontal component, their wavelet spectra, and energy diagrams (i.e., distribution of the mean energy of oscillations on the periods) that accompanied SV launches at different diurnal and nocturnal time periods and for different seasons of the year are described below. It should be noted from the beginning that the supposed geomagnetic field reaction on SV launches in its registrations was different. Only the common change in the characteristic of the magnetometric signal remained that followed the SV launch.

8.1.4.1 Soyuz Launch

The Soyuz launch happened at the diurnal time period (at 12:59 UT—here and in later occurrences time is expressed as Universal Time or UT) on February 2, 2003 (Figure 8.1a). Prior to the launch, the level of geomagnetic fluctuations did not exceed 3–4 nT.

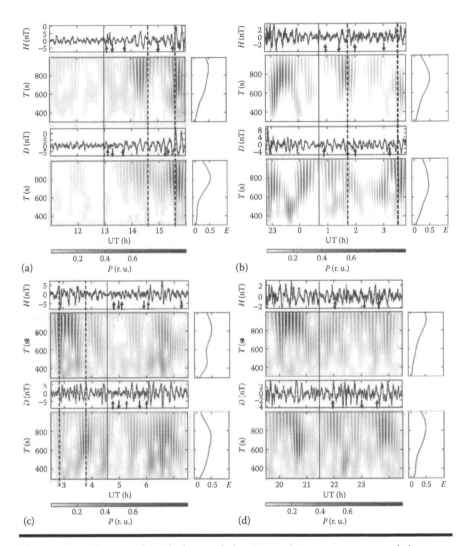

Figure 8.1 Temporal variations of the *H*- and *D*-components of the geomagnetic field, wavelet spectra corresponding to them, and energy diagrams at 300–1000 s interval period. The solid vertical line indicates the moments of RL; the dashed lines indicate the moments of the sunset at the ground surface level (at the location of the observatory and at an altitude of 100 km. Arrows show the moment of the beginning (↑) and the end (↓) of changes in the signal character: (a) Soyuz RL on February 2, 2003; (b) Soyuz RL on April 15, 2005; (c) Proton RL on October 17, 2002; and (d) Proton RL on October 14, 2004.

The period of the prevailed oscillations was $T \approx 8–10$ min. The first change in the character of the signal for the duration of $\Delta T_1 \approx 10–15$ min happened with a delay of $\Delta t_1 \approx 5–6$ min after the SV launch. Oscillations of the H-component became higher in frequency ($T \approx 3–5$ min) and their amplitude increased from 0.5–1 to 2.0–2.5 nT. The amplitudal oscillations of the D-component, conversely, decreased approximately twice. A further change in the character of the signal was observed 45 min after the launch. The amplitude of oscillations with $T \approx 12–15$ min increased from 1–2 to 5–7 nT. This process continued for 80–110 min for the H- and D-components, respectively. The next change of signal character registered close to about 140 min after the launch. The amplitude of the H- and D-components increased up to 10–15 nT, and the period up to 13–15 min. This continued for not less than 40–50 min. Let us note that the last change in the signal character registered at the same time as that of the passing of the evening terminator at altitudes of 100–150 km. Therefore, it is evident that cumulative effects of the two sources were observed.

At the nocturnal time period (00:46 h), the RL rose on April 4, 2005 (Figure 8.1b). Sunrise at altitudes of 200, 100, and 0 km took place approximately at 01:45, 02:15, and 03:20 UT, respectively. Prior to the RL, strong variations of the geomagnetic field were observed, the amplitude of the H- and D-components attained 2–4 nT, and the periods of the prevailed oscillations changed from 5 to 12 min. Prior to about 20 min of the RL, the oscillation amplitude decreased up to 1 nT, and then increased up to 2–3 nT. The essential change in the signal character occurred 10–15 min after the RL for the D- and H-components. It continued for about 40 min with prevailing oscillation periods of 11–13 min. In a time interval of 02:00–03:00 h, the signal character changed again. We assume that variations caused by the passing of the morning terminator (observed from 02:05 to 02:35 h) were added on to these changes. After 02:35 h, oscillations with $T \approx 10–15$ min and amplitude of about 2–3 nT were observed for the H- and D-components, respectively.

8.1.4.2 Proton Launches

Diurnal RL (at 04:41 h) occurred on October 17, 2002 (Figure 8.1c). Sunrise at altitudes of 100 and 0 km (ground surface level) was observed at about 03:00 and 04:00 h, respectively. During this time, an increase of geomagnetic pulsation amplitude from 2 to 6 nT was registered. The prevailing oscillation periods were 6–10 min for the D-component and 10–20 min for the H-component. We can state that during this registration, the effect of the morning terminator was evidently observed.

Approximately 6 min after the RL, a chaotic change in the level of both components, H and D, was observed for 10 min. The second change in the signal character came at about 25–30 min after the RL and continued for 40–50 min. The level of fluctuations increased up to 3–4 nT. Oscillations with periods of

7–8 min were predominant. A further change in the signal character was observed from about 06:00 to 07:30 h, that is, 80 min after the RL. The amplitude of oscillations attained 6 nT.

Nighttime RL (at 21:23 h) happened on October 14, 2004 (Figure 8.1d). Close to 40 min after the RL, the level of geomagnetic pulsations increased from 2 to 4 nT for the *D*-component. The duration of such oscillations was about 1 h. After that, for about 20 min the level of pulsations was lesser—about 1 nT. It increased again to 3 nT for the *D*-component after 110 min of the RL. The prevailing oscillation period was 12–14 min. For the *H*-component, the effects were expressed weakly.

8.1.5 Results of the Statistical Analysis

Results of the analysis of all 65 RLs are presented in Figure 8.2. During the processing, a large amount of data from the signals $H(t)$ and $D(t)$ combined, because, as shown during the preliminary analysis, this amount is related to the homogeneous one.

Separately for the light and dark time of a day for the two signal components, the maximal number of realizations for estimation of the duration and period of disturbance could be $N_{1,2max} = n_1 N_{1,2}$, where n_1 is the number of components. During estimation of the time delay for the three groups of disturbances (see below) $N_{1,2max} = n_1 n_2 N_{1,2}$, where n_2 is the number of groups. Here, $N_{1,2}$ is the number of RLs at the diurnal and nocturnal time periods.

The real number of selections was illustrated in the corresponding histograms. The number of realizations N is less than $N_{1,2max}$ in these illustrations. In the remaining cases, its number is $N_{1,2max}-N$, which complicated to outline concerning any decision on the existence of geomagnetic field reaction on the RL. From Figure 8.2, it is seen that disturbances appearing after the RL had three groups of delay: $\Delta t_1 \approx 6$ min, $\Delta t_2 \approx 30–50$ min, and $\Delta t_3 \approx 70–110$ min at the diurnal time period and $\Delta t_1 \approx 7$ min, $\Delta t_2 \approx 30–70$ min, and $\Delta t_3 \approx 100–130$ min at the nocturnal time period. The durations of the first, second, and third groups usually did not exceed 10–20, 30–60, and 45–65 min, respectively, during the diurnal time period and 10–20, 40–70, and 40–70 min, respectively, during the nocturnal time period.

The periods of oscillation usually were in limits of 6–12 min. Only for the first group of disturbances did the period T usually make up 3–6 min. Analogous statistical processing was done for the background days (i.e., a day before and a day after the RL). It was assumed that on these days "RL took place" at the same time as for the actual launches. The results of data processing are presented in Figure 8.3. As expected, the laws of the time-delay distributions from imaginary RLs and their durations were close to the uniform law.

Some deviations from the uniform law are related to the limitation of data selection as well as to the influence of disturbances of other nature

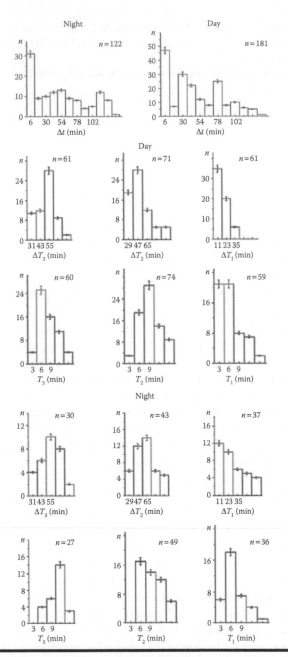

Figure 8.2 **Histograms of the time delays of geomagnetic field disturbances with respect to the moment of the RL (Δt), their durations (ΔT), and periods T of the prevailing oscillations (65 events in all).**

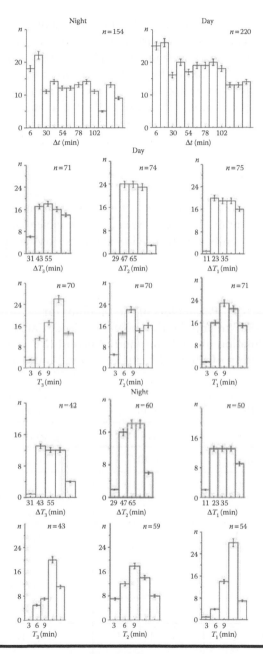

Figure 8.3 Histograms of the time delays of geomagnetic field disturbances with respect to the moment of the RL (Δ*t*), their durations (Δ*T*), and periods *T* of the prevailing oscillations (65 events in all) only for the control days (130 days).

(caused, namely, by the passing of the ST, magnetic activity, variations in space and atmospheric weather, etc.). It turned out that the magnitudes of the oscillation periods were close to those that took place on the days of the RLs. This means that disturbances caused by the RLs and other natural sources on the periods do not differ from each other.

All the above-mentioned facts testify to the fact that the RLs can amplify or weaken (if generation of WDs happens in the antiphase regime) the existing quasiperiodical processes with periods from units to 10–15 min in near-the-Earth environment.

Estimations of the main disturbance parameters obtained by statistical data processing for 65 RLs are presented in Table 8.1. All parameters presented in the table are defined and described in a further section on the process of theoretical model description. It is clearly seen from the table that the parameters of the geomagnetic disturbances that follow the RLs vary comparatively weakly during daytime.

Table 8.1　Parameters of the Medium for $z \approx 150$ km and the Geomagnetic Effect Caused by Modulation of the Ionospheric Current ($\Delta B_{x,y}$) and Excitation of Charged Particles by the Wave ($\Delta B_{xv,yv}$)

Night	Day	Parameter
2×10^{10}	2×10^{11}	N (m^{-3})
3.2×10^{-7}	3.2×10^{-6}	j_c (A m^{-2})
5×10^{-2}	5×10^{-2}	δ_{Nn}
3.3×10^{-2}	3.3×10^{-3}	δ_N
-2.7×10^{-2}	-5.7×10^{-2}	$\delta_{\sigma P}$
3.3×10^{-2}	3.3×10^{-3}	$\delta_{\sigma H}$
50	50	Δz_x (km)
25	25	Δz_y (km)
-0.2	-0.2	ΔB_x (nT)
1.1	11	ΔB_y (nT)
-1.8	-17.6	ΔB_{xv} (nT)
7.1	71	ΔB_{yv} (nT)

8.1.6 Theoretical Models

8.1.6.1 General Information

Movement of the rocket with switching-off of its engines affects the atmosphere and the geospace in a complex manner. This action is related to the cumulative effect of thermal, chemical, acoustic, dynamic, electromagnetic, and other types of factors. Particularly, ultrasound plasma spreading of the reactive jet and ultrasound movement of plasma accelerated by the SV in the gyrotropic ionosphere leads to the generation of electromagnetic and magnetohydrodynamic (MHD) waves. These waves have a characteristic velocity of ~10–100 and 10 km s^{-1}, respectively [16,17].

As mentioned previously in Chapter 3, the atmosphere at altitudes up to 120–150 km is comparably dense. Here, the work of the reactive engines and the rocket movement is accompanied by generation of shock-acoustic waves (SAWs). As estimated in Chapter 3, the power of the rocket engines, P, can achieve ~10^{11} W, and because only 0.1%–1% of this power is transferred to SAW power, P_a, we get that $P_a \approx 10^8$–10^9 W. The optimal height range for SAW generation is 100–130 km. On the one hand, the atmosphere is sufficiently dense, and the rockets moves with ultrasound velocity, that is, the condition of satisfactory generation of SAWs is quite correct. On the other hand, this height range lies above the mesospheric waveguide which effectively captures and channels waves of the acoustic type generated on the surface of the ground and in the near-the-Earth atmosphere. Furthermore, at the indicated range of height a rocket trajectory is close to horizontal one that, finally, increases the efficiency of SAW generation. This is confirmed from the result of computations [30,31], as well as by observation data [3].

8.1.6.2 Mechanisms of Geomagnetic Disturbance Generation

Magnetometers record a magnetic component of the electromagnetic or MHD waves. Obviously, the disturbances with a time delay Δt_1 correspond to those waves on the order of several minutes. What is the nature of disturbances having time delays of $\Delta t_2 \sim 40$ min and $\Delta t_3 \sim 100$ min. To answer this question, we should account for the fact that the SAW is transformed into a power acoustic pulse the farther it is from the rocket. Further, a nonlinear broadening of this pulse takes place accompanied by generation of a wide spectrum of acoustic gravity waves (AGWs), internal gravity waves (IGWs), and captured AGW (CAGW) in the thermospheric waveguide. A spectrum of these waves in the real medium, accounting for stratification of the atmospheric temperature on altitude, is discrete (see, e.g., Ref. [32]). As estimations show, periods of the prevailed waves are from 6 to 15 min. For the observation of waves with such periods, it is convenient, as was mentioned earlier, to use a magnetometer "flux-meter." The absolutely minimal attenuation the CAGW has is by propagating in the thermospheric waveguide. These waves can reach distances of several thousands of kilometers without any attenuation. CAGWs have maximum amplitude at a height of about 150 km, that is, at the upper boundary of the dynamo region of the ionosphere.

With period T, CAGWs modulate parameters of the atmosphere as the concentration of neutrals (N_n), and their temperature (T_n), and parameters of the ionosphere (N), such as the concentration of electrons (or ions, plasma is quasineutral [33]), their temperature ($T_{e,i}$), frequencies of electron–neutral collisions (ν_{en}), ion–neutral collisions (ν_{in}), and so forth. As a result, the components of the conductivity tensor of the ionospheric plasma ($\hat{\sigma}$), will also be modulated. In the presence of the ambient (of ionospheric and magnetospheric origin) electric field ($\mathbf{E}_0$), modulation of the tensor $\hat{\sigma}$ yields to the modulation of the density of ionospheric currents with amplitude

$$\Delta \mathbf{j} = \Delta \hat{\sigma} \mathbf{E}_0. \tag{8.3}$$

On the other hand, movements of neutral particles in the CAGW with velocity $\mathbf{v}_n$ lead to movement of ions and electrons with velocities $\mathbf{v}_{i,e}$ due to collisions. Finally, the induced current with density $\mathbf{j}_1 = eN(\mathbf{v}_i - \mathbf{v}_e)$ occurs, where e is the charge of plasma electrons (we assume here, following Ref. [33], that ions, like electrons, have unit charge ($Z = 1$) and only differ from electrons by the sign of the charge). The appearance of currents with densities $\Delta \mathbf{j}$ and $\mathbf{j}_1$ stipulates generation of an electromagnetic field with induction $\mathbf{B}$, which can be calculated from the second Maxwell's equation [33]:

$$\nabla \times \mathbf{B} = \mu_0 \Delta \mathbf{j} + \mathbf{j}_1, \tag{8.4}$$

where:

μ_0 is the magnetic permeability of the background ionospheric plasma
$\nabla = (\partial/\partial x)\mathbf{i}_x + (\partial/\partial y)\mathbf{i}_y + (\partial/\partial z)\mathbf{i}_z$ is the differential vector operator

Equation 8.4 does not account for the time-varied electrical induction $\mathbf{D}$ or the displacement current that is not actual for the wave periods of $T \sim 1$–10 min.

Considering this problem we direct vector $\mathbf{B}$ along the z-axis, and the x-axis orients from the place of observation to the place of generation of CAGWs (close to Baikonur cosmodrome). In this case, the y-axis forms the right Cartesian 3D coordinate system. At large (~2000 km) distances far from the region of wave generation, vector $\mathbf{v}_n$ is conversely directed parallel to the *x*-axis. In this case, from Equation 8.4 it follows that

$$-\frac{\partial B_y}{\partial z} = \mu_0 (\Delta j_x + j_{1x}), \tag{8.5a}$$

$$-\frac{\partial B_x}{\partial z} = \mu_0 (\Delta j_y + j_{1y}). \tag{8.5b}$$

In Equation 8.5a and b $|\partial/\partial z| \gg |\partial/\partial x|, |\partial/\partial y|$ is also accounted for. In fact, for the first derivative the characteristic scale is on the order of several tens of kilometers, that is, close to the thickness of the dynamo region filled with a height range of 100–150 km, whereas for the other derivatives this scale is on the order of the wavelength $\lambda = v_{ph} T \approx 300$–1000 km. Here, v_{ph} is the phase velocity of the wave, which is close to the speed of sound c_s in the atmosphere ($c_s \approx 0.3$–0.8 km s^{-1}).

8.1.6.3 Modulation of the Ionospheric Currents

Using a well-known approach (see, e.g., Ref. [33]), it is easy to derive the density of the generated electric current in nonperturbed ionosphere under the action of atmospheric wind:

$$\mathbf{j}_0 = eN(\mathbf{v}_i - \mathbf{v}_e). \tag{8.6}$$

Velocities of ions and electrons can be found from quasistationary equations:

$$v_{in}(\mathbf{v}_i - \mathbf{v}_n) = \frac{e}{m_i}(\mathbf{v}_i \times \mathbf{B}_0), \tag{8.7a}$$

$$v_{en}(\mathbf{v}_e - \mathbf{v}_n) = \frac{e}{m_e}(\mathbf{v}_e \times \mathbf{B}_0). \tag{8.7b}$$

Here, as above, $\mathbf{B}_0$ is the induction of the geomagnetic field and $m_{i,e}$ is the mass of ions and electrons. The quasistationary condition in the ionosphere has the appearance: $2\pi/T \ll v_{in}, v_{en}, \omega_{Bi}, \omega_{Be}$, where ω_{Bi} and ω_{Be} are the gyrofrequencies of ions and electrons, respectively.

For the western (or eastern) wind direction, solutions of Equation 8.7a and b have the appearance:

$$v_{ix} = \frac{v_n}{1 + \Omega_i^2}, v_{iy} = -v_n \frac{\Omega_i}{1 + \Omega_i^2}, \tag{8.8a}$$

$$v_{ex} = \frac{v_n}{1 + \Omega_e^2}, v_{iy} = -v_n \frac{\Omega_e}{1 + \Omega_e^2}, \tag{8.8b}$$

where:

$$\Omega_{i,e} = \omega_{Bi,Be}/\nu_{in,en}$$

For the northern (or southern) wind direction, the projects of $v_{ix,ex}$ and $v_{iy,ey}$ are changed in their "role." Further, we account for the fact that in the dynamo region of the ionosphere (see Table 8.1) $\omega_{Bi} \approx 180$ s^{-1}, $\omega_{Be} \approx 8.8 \times 10^6$ s^{-1}, $\nu_{in} \approx 10^4 - 10^2$s^{-1}, and $\nu_{en} \approx 10^5 - 10^3$s^{-1} for $z \approx 100$–150 km, respectively.

If so, $\Omega_i^2 \ll \Omega_e^2$ and $v_{ix} \gg v_{ex}$, $v_{iy} \gg v_{ey}$. Hence,

$$j_{0x} \approx eNv_{ix} \approx eNv_n, \tag{8.9a}$$

$$j_{0y} \approx eNv_{iy} \approx eNv_n\Omega_i. \tag{8.9b}$$

Due to the fact that electrons are much more magnetized compared to ions, the current in the dynamo region is fully determined by the movement of ions everywhere (except the upper boundary of the dynamo region) $j_{0x} \approx -j_{0y}$. Close to this boundary $\Omega_i \approx 1$

$$j_{0x} \approx \frac{1}{2}eNv_n, \tag{8.10a}$$

$$j_{0y} \approx \frac{1}{2}eNv_n. \tag{8.10b}$$

The characteristic density of the current in the ionospheric dynamo region $j_c = eNv_n$, where v_n is defined by the velocity of the neutral wind v_w. For $v_w = 100$ m s^{-1} and $N = 2 \times 10^{10}$ m^{-3} and 2×10^{11} m^{-3} at an altitude of 150 km in the nocturnal and diurnal time periods, we get $j_c = 3.2 \times 10^{-7}$ A m^{-2} and 3.2×10^{-6} A m^{-2}, respectively.

It is well known (see, e.g., Ref. [33]) that in the dynamo region of the ionosphere, conductivity of plasma is determined by Pedersen's (σ_P) and Hall's (σ_H) conductivities. It is important to note that the main contribution in the first conductivity is to give ions, whereas in the second it is electrons, that is,

$$\sigma_P \approx \sigma_{Pi} = \varepsilon_0 \frac{\omega_{pi}^2 \nu_{in}}{\omega_{Bi}^2 \nu_{in}^2}, \tag{8.11a}$$

$$\sigma_H \approx \sigma_{He} = -\varepsilon_0 \frac{\omega_{pe}^2}{\omega_{Bi}}, \tag{8.11b}$$

where:
 ε_0 is the electric permittivity
 ω_{pi} and ω_{pe} are the plasma frequencies of ions and electrons, respectively

For small deviations of electron concentration ($\Delta N \ll N$), temperature of ions ($\Delta T_i \ll T_i$) and concentration of neutrals ($\Delta N_n \ll N_n$) correspond to small deviations of $\Delta \sigma_P$ and $\Delta \sigma_H$:

$$\Delta \sigma_P = \sigma_P \delta_{\sigma_P} = \sigma_P \left(\delta_N \frac{1+\gamma}{2} \delta_{N_n} \right), \tag{8.12a}$$

$$\Delta \sigma_H = \sigma_H \delta_{\sigma_H} = \sigma_H \delta_N, \tag{8.12b}$$

where:

δ_{σ_P} and δ_{σ_H} are the relative changes of the Pedersen and Hall conductivities
$\delta_N = \Delta N/N$, $\delta_{N_n} = \Delta N_n/N_n$, γ is the ratio of the specific thermocapacities

It is taken into account in Equation 8.12 that $v_{in} \propto N_n T_n^{1/2}$ and $\Delta T_n/T_n = (\gamma - 1)\delta_{N_n}$. The magnitude of δ_N can be estimated following the results obtained in Ref. [34], according to which

$$\delta_N = \frac{k_z v_z}{(\omega_N^2 + \omega^2)^{1/2}}, \tag{8.13}$$

where:

$\omega_N = t_N^{1} = 2\alpha_r N, t_N$ is the time required for achieving a standing regime of N due to recombination
α_r is the coefficient of the dissociative recombination (i.e., of electrons with molecular ions)
k_z is the vertical component of the wave number k of CAGW
v_z is the vertical component of the particle velocity within a wave

Thus, for derivation of a magnitude of disturbance of the conductivity tensor component, it is necessary to know the magnitudes of δ_N, or k_z and v_z. Strict derivation of v_z is a complicated and independent task that is out of the scope of our book. Such a task was solved, for example, in Refs. [31,32].

Here we used two approaches. In the first, we used magnitudes of δ_N obtained in Ref. [32] for the same RLs using vertical Doppler sounding and incoherent scattering methods employing the methodology described in Refs. [17–24,34]. From the obtained experimental data magnitudes of ω_N, k_z, and, which means, v_z, were computed. Magnitudes of v_z at altitudes of the E-region were in limits of 1–4 m s^{-1} for different RLs.

In the second approach, the magnitude of v_z was derived analytically from hydrodynamic suggestions. The obtained magnitudes were compared with the results of detailed numerical computations [38]. These computations also gave the magnitudes v_z on the order of several meters per second at altitudes of the E-region of the ionosphere. Therefore, in further discussions we will focus on magnitude

$v_z = 1-4 \text{ m s}^{-1}$. For the estimation of δ_N let us suppose that $k_z = 3k$, $v_z = 0.1v$, and $v = c_s \gamma \delta_{N_n}$. At an altitude of $z \approx 150$ km, we get $\alpha_r \approx 3.5 \times 10^{-13} \text{ m}^3 \text{ s}^{-1}$, $c_s \approx 500 \text{ m s}^{-1}$, and $\delta_{N_n} = 0.05$. Then $T = 1000$ s, $v = 35 \text{ m s}^{-1}$, $v_z = 3.5 \text{ m s}^{-1}$, $k \approx 1.3 \times 10^{-5} \text{ m}^{-1}$, and $k_z \approx 1.3 \times 10^{-4} \text{ m}^{-1}$; ω_N equals 0.14 and 0.014 s^{-1} for the diurnal and nocturnal time periods, respectively. This is seen usually close to the altitude of 150 km for $\omega^2 \ll \omega_N^2$. Then, $\delta_N = k_z v_z / \omega_N$ which equals 3.3×10^{-3} and 3.3×10^{-2} at the diurnal and nocturnal time periods, respectively. For the same cases, $\delta_{\sigma P}$ equals -5.7×10^{-2} and -2.7×10^{-2}, and $\delta_{\sigma H}$ equals 3.3×10^{-3} and 3.3×10^{-2}.

From Equation 8.5a and b, it follows for $j_1 = 0$ that the amplitudes of disturbance of geomagnetic field components equal

$$\Delta B_y \approx \mu_0 \int \Delta j_x dz \approx \mu_0 \Delta j_x \Delta z_x, \tag{8.14a}$$

$$\Delta B_x \approx \mu_0 \int \Delta j_y dz \approx \mu_0 \Delta j_y \Delta z_y. \tag{8.14b}$$

Here, we accounted for the fact that the contribution in ΔB_y gives all in the dynamo region, where $\Delta z_x \approx 50$ km, whereas ΔB_y is the only region close to the upper boundary of the dynamo region ($z \approx 150$ km), having $\Delta z_y \approx 25$ km. In the first case, Δj_x is the deviation of the Pedersen current, that is,

$$\Delta j_x \approx \delta_{\sigma P} j_c. \tag{8.15}$$

In addition, in the second case Δj_y is the deviation of the Hall current, that is,

$$\Delta j_y \approx \delta_{\sigma H} j_{0y} \approx -\frac{1}{2} \delta_{\sigma H} j_c. \tag{8.16}$$

Introducing Equations 8.15 and 8.16 into Equation 8.14a and b, respectively, we finally get:

$$\Delta B_y \approx \mu_0 \delta_{\sigma P} j_c \Delta z_x, \tag{8.17a}$$

$$\Delta B_x \approx -\frac{1}{2} \mu_0 \delta_{\sigma H} j_c \Delta z_y. \tag{8.17b}$$

The results of estimations of the geomagnetic effect according to Equation 8.17a and b are presented in Table 8.1.

8.1.6.4 Excitation of Charged Particles by Waves

The components of the induction current density can also be described by expressions such as Equations 8.9 and 8.10, in which velocity of the neutrals v_n

is defined by the velocity of the particles in the wave, $v = c_s \gamma \delta_{N_n}$. The expressions for ΔB_y and ΔB_x are similar to Equations 8.14 and 8.17 and can be expressed as

$$\Delta B_y \approx \mu_0 j_{1x} \Delta z_x, \tag{8.18a}$$

$$\Delta B_x \approx \mu_0 j_{1y} \Delta z_y, \tag{8.18b}$$

where in the dynamo region of the ionosphere, as in Equation 8.9a and b

$$j_{1x} \approx eNv, \tag{8.19a}$$

$$j_{1y} \approx - eNv\Omega_i. \tag{8.19b}$$

At the upper boundary of the dynamo region ($z \approx 150$ km), the same expressions as Equation 8.10, are used

$$j_{1x} \approx \frac{1}{2} eNv, \tag{8.20a}$$

$$j_{1y} \approx -\frac{1}{2} eNv. \tag{8.20b}$$

Accounting for Equations 8.19 and 8.20, we can rewrite a system of Equation 8.18 as follows:

$$\Delta B_{yv} \approx \frac{1}{2} \mu_0 eNv\Delta z_x, \tag{8.21a}$$

$$\Delta B_{xv} \approx -\frac{1}{2} \mu_0 eNv\Delta z_y. \tag{8.21b}$$

The results of estimations of the geomagnetic effect for $v = 35$ m s^{-1} ($\delta_{N_n} = 0.05$) are presented in Table 8.1 as well.

8.1.7 Results and Discussions

All the considered RLs related to notable or essential variations of the level of geomagnetic field fluctuations and their spectral content. Often, the three groups of disturbances were presented as having different time delays: 6, 30–50, and 70–110 min at daytime and 7, 30–70, and 100–130 min at nighttime.

The first group of disturbances was not always observed (in about 60% of the cases). Often, this group had an irregular character, was noisy, and had more high-frequency pitch and relatively small duration, $\Delta t_1 \approx 10$–20 min. The characteristic level of these fluctuations was 2–4 nT with 5–8 min periods of prevailing pulsations.

Disturbances with $\Delta t_2 \approx 30-70$ min were observed systematically. They took the form of oscillation wavelets with 30–70 min duration, 6–12 min periods, and 3–5 nT amplitudes.

Most slow disturbances also took the form of wavelets of oscillations with 6–12 min periods, up to 3–6 nT amplitudes, and 45–70 min duration, as also $\Delta t_3 \approx 30-130$ min time delay. The parameters of the second and third groups are quite close, therefore, we can assume that they have a similar nature or they are part of the same disturbance.

Let us estimate the characteristic velocities of disturbances propagating on account of RLs and flights. We can notice with a high probability that these disturbances are generated in the ionosphere when the rocket achieves $z \sim 100$ km altitude. Under the affect of the rocket jet (e.g., plume of the rocket), plasma conductivity is essentially changed in the dynamo region of the ionosphere (at altitudes of $z \approx 100-150$ km) and waves having an electromagnetic (magneto-dynamic) nature are generated [16]. Shock waves are also generated in gas at the same altitudes, which propagate from the rocket with a velocity that exceeds the speed of sound in gas (see, e.g., Refs. [3,16]). At distances sufficiently far from a rocket trajectory, the shock wave is transformed into an AGW.

Following Ref. [3], it was shown in Chapter 3 that this wave responds to 4–5 min periods. The time, Δt_0, during which the rocket attains $z \approx 100-120$ km altitude depends on the type of rocket, and equals 3–4 min. In this case, the time of propagation of the disturbance equals $\Delta t' = \Delta t - \Delta t_0$ (see Chapter 3). These are $\Delta t_0 \approx 3$ min, $\Delta t_1' \approx 3$ min, $\Delta t_2' \approx 27-47$ min, and $\Delta t_3' \approx 67-107$ min in diurnal time conditions. The region of wave generation, as shown in Ref. [3], is placed 400–500 km eastward from the cosmodrome. In this case, the corrected magnitude of the distance from the place of wave generation to the observatory was $R' \approx 2500$ km. The corrected magnitudes $\Delta t'$ and R' correspond to velocity $v' = R'/\Delta t'$, that is, $v_1' \approx 14$ km s^{-1}, $v_2' \approx 0.9-1.6$ km s^{-1}, and $v_3' \approx 390-630$ m s^{-1}. In the nocturnal time conditions, we get $\Delta t_1' \approx 4$ min, $\Delta t_2' \approx 27-67$ min, and $\Delta t_3' \approx 97-127$ min. In this case, we get that $v_1' \approx 11$ km s^{-1}, $v_2' \approx 0.6-1.6$ km s^{-1}, and $v_3' \approx 330-430$ m s^{-1}.

The waves having $v_1' \approx 10-14$ km s^{-1} evidently relate to electromagnetic waves (strictly speaking, to MHD waves). The velocity $v_2' \approx 0.6-1.6$ km s^{-1} corresponds to slow MHD waves, as well as to shock waves at altitudes of the F-region of the ionosphere and their cause—the CAGW described above and in Ref. [3]. The velocity of shock waves higher than the velocity of AGWs at these altitudes, and their magnitudes do not exceed 0.7–0.8 km s^{-1}. The velocity $v_3' \approx 330-430$ m s^{-1} approximately relates to that of IGWs described in Chapter 3. Let us compare results described above with those obtained in Ref. [4]. For short-periodical oscillations (with $T \approx 2-3$ min) of the geomagnetic field, $\Delta t_1 \approx 3-4$ min, and for $\Delta t_0 \approx 2$ min we get $\Delta t_1' \approx 1$ min. Then for $R' \approx 1000$ km, we get $v_1' \approx 17$ km s^{-1}.

This magnitude of v_1' obtained in Ref. [4] is close to that was obtained by us. The amplitude of the short-periodic disturbances in Ref. [4] was of 6–20 nT, whereas we obtained above 2–4 nT, that is, several times lesser. Accounting for the fact that distances R were of 800 and 2100 km (but $R' = 100$ and 2500 km), such differences between amplitudes was predictable. The slow disturbances in Ref. [4] have $\Delta t_2 \approx 10$–15 min. Then for $\Delta t_0 \approx 3$ min, $\Delta t_2' \approx 7$–12 min, and $v_2' \approx 1.4$–2.4 km s^{-1}. These magnitudes of v_2' are close to those obtained by us (of 0.6–1.6 km s^{-1}, see earlier). The amplitude of slow disturbances in Ref. [4] was about 10 nT, whereas in our observations it was 3–5 nT, that is, it decreased with increase of distance R from the rocket. We can suppose with a great probability that these disturbances have the same nature. Generally speaking, the effects of Soyuz and Proton type RLs were similar.

Let us describe in more detail the waves responsible for the transfer of disturbances and having the time delays Δt_2 and Δt_3. In contrast to most fast disturbances of the geomagnetic field that are generated in the proximity of the rocket and then propagate as electromagnetic or MHD waves to the place of observation, the magnetic disturbances with the time delays Δt_2 and Δt_3 are generated, first, at the place of observation.

These magnetic disturbances are caused by traveling of CAGWs and IGWs, whose velocities equal 0.6–1.6 and 0.3–0.6 km s^{-1}, respectively. The magnetic effect occurred due to propagation of CAGW- and IGW-stimulated simultaneous appearance of two mechanisms: (1) modulation of the tensor of conductivity of the ionospheric plasma (e.g., modulation of the natural ionospheric currents) and (2) generation of the induced current in the dynamo region of the ionosphere as the result of capture of charged particles by the wave in the neutral gas.

As computations show, for sufficiently intensive CAGWs, a second mechanism is more effective. Amplitudes of magnetic disturbances in the ionosphere (as well as in the magnetosphere) in daytime conditions can achieve tens of nanotesla, and at nighttime periods of 2–7 nT. Only magnetic effects of the magnetosound can be registered at the Earth's surface or transformed in them in the dynamo region the Alfven's waves. It is natural that at the planet's surface the amplitude of magnetic disturbances that depends on λ, is 1.5–2 times less than that in the ionosphere or the magnetosphere, but is sufficient for detection (not less than 1 nT [34]).

Thus, the theoretical model and the results of computations of the geomagnetic disturbances qualitatively correspond to the results of observations. But this correspondence is not full. According to computations, the magnetic effect at nighttime conditions should be one order less compared to that at daytime conditions. Observational data show that these differences are sufficiently small. Not full correspondence, in our opinion, can be explained by the absence of taking the wave attenuation into account, which at conditions of daytime and nighttime ionosphere, differ significantly.

We add that the magnetic effects of RLs should be observed at the magnetic conjugate regions where, according to our estimations, $\Delta B_{x,y} \geq 1$ nT.

Hence, repetition of the main features during observations of RLs of various types at different daytime periods and different yearly periods for different levels of magnetic activity, as well as the realistically obtained magnitudes of waves parameters and the nonconflicted results of measured data at the close zone ($R \approx 800$ km), coincide with the results of observations of those obtained by the theoretical modeling—all these facts allow us to state that for the first time significant variations were found in the level of geomagnetic field at distances of more than 2000 km from the place of the RL.

8.2 Geomagnetic Effects of RLs from Plesetsk Cosmodrome

8.2.1 Overview

As mentioned in Section 8.1 and summarized in a book [1], despite the fact that during the past decades a great interest was observed on effects of the RLs, only effects caused in the ionosphere were mainly investigated (see, e.g., Refs. [2–10]). Possible "reaction" of the ionosphere on the RLs, accounting for the effects in the geomagnetic field, was described only in several separate works [4,8,9]. In the last work, the distance between the cosmodrome and the place of observation R was about 800 km. It is quite possible that in this work only relatively low-frequency disturbances related to the RL were selected. Unfortunately, this cannot be stated following only the results of Ref. [8,35].

Following the results obtained by one of the authors in Ref. [36], Section 8.1 describes the results of observations and theoretical modeling of sufficiently high-frequency fluctuations of the geomagnetic field (measurements were carried out for the periods of 1–1000 s) close to the Kharkov city (Ukraine), which accompanied launches and flights of 65 rockets of Soyuz and Proton type from Baikonur cosmodrome ($R \approx 2100$ km). As mentioned in Chapter 7, based on the corresponding statistical analysis, the three groups of geomagnetic field disturbances were identified, whose main parameters, generally speaking, depended on the daytime conditions and conditions in the geospace. The characteristic velocities of wave propagation and their periods allow show the nature of the wave, which transfers such kinds of disturbances, nature.

We mentioned in Chapters 1 and 2 about Plesetsk and Baikonur cosmodromes (Russia), which are close to Kharkov Observatory (Ukraine). Here, it is important to note that compared to Kharkov, Baikonur cosmodrome is located nearly perpendicular in direction to the magnetic strength lines, whereas the direction of Plesetsk cosmodrome is close to the direction of the magnetic meridian. Therefore, it is reasonable to compare the effects of RLs from Baikonur cosmodrome, described in Chapter 7, with those carried out in Plesetsk cosmodrome. Ranges from the cosmodromes to the place of observation are comparable—of about 2100 and 1500 km, respectively.

Therefore, the aim of this chapter is to describe the results of the analysis of the quasiperiodical geomagnetic field oscillations that accompanied 22 RLs of four types from Plesetsk cosmodrome and their comparison with similar results obtained during launches from Plesetsk cosmodrome, following Ref. [37]. Measurements were carried out at the observatory of Kharkov University, whose geomagnetic coordinates are 45°20′ northern latitude and 119°20′ eastern longitude. The experimental setup is described in Appendix 3. The method of measurements and signal and data processing are the same as were mentioned in Section 8.1 (see also Ref. [36]). As mentioned in Chapter 4, usage of the WT for spectral analysis based on the Morlet's wavelet was sufficiently convenient for studying the separate spectral components dynamics that form the short wave packets.

8.2.2 The Level of Magnetic Perturbation

For the general control of the level of magnetic perturbation the indexes A_p and K_p taken from the Internet were used (see Table 8.2). In this table, the magnitudes of indexes A_p and ΣK_p are presented during the day of RLs, a day, and 2 days prior to this event. As a rule, ΣK_p presents the sum of K_p indexes observed for 3 h per day. From Table 8.2, it can be clearly seen that launches occurred on magnetically quiet days as well as in the weakly and moderately magnetoperturbed days.

Only the launch that happened on October 29, 2003, was done during a strong magnetic storm (with $A_p = 189$ and K_p that achieved of 9).

8.2.3 Results of Observations

Out of 22 RLs, 5 launches of the heavy rockets of Soyuz and Molnia type, 16 launches of the rockets of mean weights (Kosmos and Rokot), and one launch of the light rocket Topol with the initial mass not less than 50 t were analyzed according to the classification given in Chapters 1 and 2 (see also Ref. [37]). The RLs happened at different daytime periods, at different seasons of the year, and for different levels of geomagnetic activity (see Table 8.2). General information on all types of rockets is given in Appendix 1.

Let us describe temporal variations of the horizontal components of the geomagnetic field, their wavelet spectra, and energy diagrams (i.e., distribution

Table 8.2   The Level of Magnetic Perturbation Prior to and during an RL Day

Data	Time (UT)	Rocket Type	Two Days Prior to RL		One Day Prior to RL		Day of RL	
			A_p	$\sum K_p$	A_p	$\sum K_p$	A_p	$\sum K_p$
June 20, 2002	09:34	Rokot	11	18	16	26	10	22
July 8, 2002	06:36	Cosmos-3M	23	30	11	25	10	19
September 26, 2002	15:30	Cosmos-3M	6	16	6	17	8	17
November 28, 2002	06:07	Cosmos-3M	14	23	21	29	15	24
April 2, 2003	01:25	Molnia	31	33	12	24	20	28
June 4, 2003	19:23	Cosmos-3M	39	36	26	32	21	29
June 19, 2003	20:01	Molnia	50	39	54	40	18	27
June 30, 2003	14:15	Rokot	32	34	26	32	20	29
August 19, 2003	10:50	Cosmos-3M	15	21	86	48	21	27
September 27, 2003	06:12	Cosmos-3M	28	33	17	26	9	18
October 29, 2003	13:43	Rokot	15	22	20	29	189	55
November 2, 2004	09:52	Topol	10	18	5	11	4	9
May 31, 2005	12:00	Soyuz	22	22	67	42	17	26
August 27, 2005	18:34	Rokot	24	29	11	21	7	14
October 8, 2005	15:02	Rokot	4	8	11	18	22	28
October 27, 2005	06:52	Cosmos-3M	19	26	8	13	6	10

(*Continued*)

Table 8.2 Continued

Data	Time (UT)	Rocket Type	Two Days Prior to RL		One Day Prior to RL		Day of RL	
			A_p	ΣK_p	A_p	ΣK_p	A_p	ΣK_p
July 28, 2006	07:05	Rokot	6	11	9	12	29	26
December 19, 2006	14:00	Cosmos-3M	4	6	8	15	14	21
June 7, 2007	18:00	Soyuz	3	4	2	3	3	6
July 2, 2007	19:38	Cosmos-3M	6	11	5	11	3	6
November 1, 2007	19:00	Cosmos-3M	10	17	4	7	3	6

of the average energy of oscillations on the periods). Generally speaking, the supposed "reaction" of the geomagnetic field on RLs in different registrations was different. The common changes of the character of the magnetometric signal, such as spectral content, amplitude of the spectral components, and so on still remain.

8.2.3.1 Rokot Launch

Launch of this rocket was done on June 20, 2002 in daytime conditions (at 09:34 UT), as seen from Figure 8.4a. The first change in the signal character (where they appeared as high-frequency components with the period $T \approx 20$–100 s) was registered about 18 min after RL and continued for about 15 min. Further change in the signal character, when T was increased up to 10–15 min and the amplitude—at 1–2 nT, took place at the time interval approximately from 10:25 to 11:00 h. The next change in the signal character (also increase of the periods and the amplitudes of oscillations) was observed from about 11:20–11:30 to 12:20 h.

8.2.3.2 Cosmos-3M Launch

The nocturnal (at 19:00 h) RL occurred on November 1, 2007 (Figure 8.4b). At the moment of launch, the reaction on the passing of the evening terminator was completed. This reaction was vividly expressed at the time interval of 17:00–18:00 h. The first change in the signal character, when the period of oscillations was increased up to 12–15 min and the amplitude at about 1 nT, occurred at the

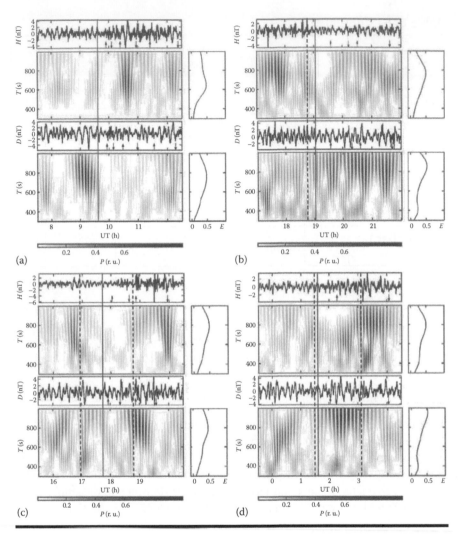

Figure 8.4 Temporal variations of the *H*- and *D*-components of the geomagnetic field, the corresponding wavelet spectra, and the energy diagrams in the 300–1000 s interval period. The solid vertical line indicates the moments of RL; the dashed lines indicate at the moments of sunrise, the sunset at ground surface level at the location of the observatory, and at an altitude of 200 km. Arrows show the moments of the beginning (↑) and the end (↓) of changes in the signal character: (a) Rokot RL on June 20, 2002; (b) Cosmos-3M RL on November 1, 2007; (c) Soyuz RL on May 3, 2006; and (d) Molnia RL on April 2, 2003.

time interval of 19:30–20:10 h. The second change in the signal character (also increase of the period and the amplitude of oscillations) took place from about 20:30 to 21:40–21:45 h.

8.2.3.3 Soyuz Launch

The described event occurred at evening time (at 17:38 h) on May 3, 2006 (Figure 8.4c). The action on the passing of the evening terminator registered at the time interval from 18:00 to 19:00 h. For this reason, the first reaction on RL was nonvividly searched. Perhaps it started at about 18.00 h and continued 30–40 min, respectively, for the D- and H-components of the geomagnetic field (see Figure 8.4c). The second change in the signal character began at about 72 min after the RL and continued for 40–60 min for the D- and H-components of the geomagnetic field, respectively.

8.2.3.4 Molnia Launch

The nocturnal (at 01:35 h) RL occurred on April 2, 2003. The reaction on passing of the morning ST was awaited at the time interval from about 01:30 to 03:00 h. The change in the signal character (increase of the period and the amplitude of oscillations) for the H-component of the geomagnetic field began at about 40 min after the RL and continued for 110 min (Figure 8.4d). In this case, the amplitude of the H-component of the geomagnetic field was increased at 1–2 nT.

As for the D-component, increase of the amplitude of oscillations began 30 min after the RL and continued for not less than 120 min. Such large signal character duration (of 110–120 min) for both components of the geomagnetic field can be explained by combining its "reaction" on the morning terminator and the effects caused by the RL. Due to the morning ST, it was impossible to observe separately different groups of disturbances of the geomagnetic field.

We note that study of the reaction of the magnetic field on separate RLs does not give full confidence that the observed variations of the level of geomagnetic field relates to RLs. As was shown in Ref. [36], such a confidence can give statistical processing of sufficiently large numbers of RLs.

8.2.4 Results of Statistical Analysis

The results of such a processing are shown in the form of histograms in Figure 8.5. From 22 RLs, 13 happened in the morning, 6 in the evening, and 3 at night. The number of realizations is shown in the corresponding histograms. For the

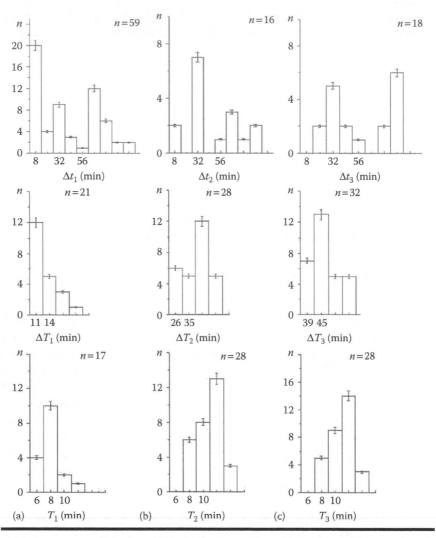

Figure 8.5 **Histograms of the time delays of geomagnetic field disturbances with respect to the moment of the RL (Δt), their durations (ΔT), and periods T of the prevailing oscillations (22-day events in all): (a) for daytime, (b) for evening time, and (c) for nighttime.**

estimation of time delay Δt, the maximum number of realizations is $N = N_1 N_2 N_3$, where N_1 is the number of launches in the observed period of a day, $N_2 = 3$ is the number of possible disturbances, and $N_3 = 2$ is the number of geomagnetic field components. For estimating the duration ΔT of disturbances and their periods T, the number of realizations $N = N_3 N_4$, where N_4 is the number of all launches.

From Figure 8.5, it can be concluded that after launches in daytime, and possibly, in the evening, three groups of disturbances appeared with delays: $\Delta t_1 \approx 5.6 \pm 0.5$ min, $\Delta t_2 \approx 32.1 \pm 1.3$ min, and $\Delta t_3 \approx 79.5 \pm 4.5$ min. The corresponding durations of disturbances were $\Delta T_1 \approx 13.0 \pm 0.6$ min, $\Delta T_2 \approx 36.3 \pm 1.9$ min, and $\Delta T_3 \approx 46.7 \pm 1.2$ min. Usually, the periods of oscillations are $T_1 \approx 8.0 \pm 0.4$ min, $T_2 \approx 10.9 \pm 0.4$ min, and $T_3 \approx 10.9 \pm 0.3$ min.

At nighttime, disturbances with time delay Δt_1 did not register. Results of the statistical data processing for all 22 RLs are summarized in Table 8.3. Analogous statistical processing that was carried out found that the maximal number of realization in this case doubled for the days prior to and after the RL, which are called background days. Identification of similar effects was done relative to the times of the real RLs (i.e., which occurred during the day of their start). The results of data processing in this case are presented in Figure 8.6.

Laws of the time delay distribution and the duration of disturbances caused by "virtual" RLs were found to be regular. Some deviation from the regular law is related to the limitation of realization as well as to the influence of disturbances of another nature. The passing of the ST, variation of space and atmospheric weather, changes in magnetic activity, and so forth relate to them.

Table 8.3 Statistical Characteristics of the Main Parameters of Disturbances That Accompany RLs

Magnitudes (min)	Parameter
Δt_1	5.6 ± 0.5
Δt_2	32.1 ± 1.3
Δt_3	79.5 ± 4.5
ΔT_1	13.0 ± 0.6
ΔT_2	36.2 ± 1.9
ΔT_3	46.7 ± 1.2
T_1	8.0 ± 0.4
T_2	10.9 ± 0.4
T_3	109.0 ± 0.3

Note: Δt are the delays of disturbances; ΔT and T are their durations and periods, respectively.

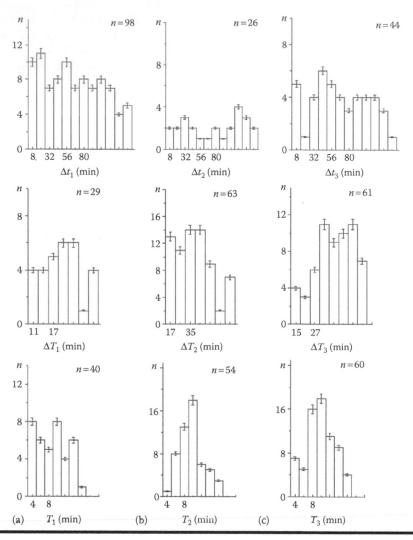

Figure 8.6 **Histograms of the time delays of geomagnetic field disturbances with respect to the moment of the RL (Δt), their durations (ΔT), and periods T of the prevailing oscillations (22-day events in all): (a) for daytime, (b) for evening time, and (c) for nighttime for 44 control days.**

8.2.5 Results and Discussions

Practically all RLs accompanied essential or significant changes in their signal character. In most cases, increase in pulsation amplitude registered, but in four cases depression of oscillations took place, which can be explained by summation of the wave processes with the opposite phases. In addition, changes in the signal character were practically always accompanied by changes in the period of quasiperiodical processes. The time evolution of the prevailing oscillation periods and their energy is seen clearly from the wavelet spectra and energy diagrams shown in Figure 8.4.

During daytime, three groups of disturbances with time delays of $\Delta t_1 \approx 4$–8 min, $\Delta t_2 \approx 20$–40 min, and $\Delta t_3 \approx 65$–90 min were often presented. At nighttime, only two groups of reactions were observed with some larger delays of time equaling 25–75 and 75–100 min, respectively, for the second and third disturbances.

The first group of disturbances was absent at nighttime periods, and during daytime and evening periods it appeared in 50% of cases, having a familiar noise (high-frequency) character, and its duration on the order of 10 min, with small amplitudes (of 2–4 nT) and 7–9 min periods.

The second and third groups were recorded in the form of wave packets (i.e., wavelets) of oscillations with 40–50 min duration and 10–12 min periods. The parameters of disturbance in the second and the third groups are close. Therefore, it is obvious to suppose that they can be considered as part of the same wave process.

Disturbances of the magnetic field are generated at altitudes of 100–150 km. This was substantiated from Ref. [36]. On the one hand, at these altitudes there are sufficiently high magnitudes of electron concentration and current densities, therefore, waves of electromagnetic (MHD) nature can be generated. On the other hand, at these altitudes the density of neutral particles is still quite high. Therefore, shock waves are generated here in gas, which are transported to large distances as AGWs.

As mentioned in Chapter 7, the time Δt_0, during which the rocket achieves altitudes of $z \approx 100$–150 km, equals 3–4 min, and depends on the type of rocket. In this case, the time of propagation of the disturbance that was observed close to Kharkov Observatory, equals $\Delta t_2 = \Delta t - \Delta t_0$. Then, taking the mean values Δt_1, Δt_2, and Δt_3 from Table 8.3, and for $\Delta t_0 \approx 3$ min, we get $\Delta t'_1 \approx 2.1$–3.1 min, $\Delta t'_2 \approx 27.8$–31.4 min, and $\Delta t'_3 \approx 72$–81 min. For $R \approx 1500$ km these magnitudes correspond to velocities $v'_1 \approx 8.3$–12.5 km s^{-1}, $v'_2 \approx 800$–900 m s^{-1}, and $v'_3 \approx 310$–350 m s^{-1}. These velocities of disturbance propagation are close to those obtained in Ref. [36] and described in Chapter 7 for the RL from Baikonur cosmodrome, that is, $v'_1 \approx 9$ k m s^{-1}, $v'_2 \approx 800$–1100 m s^{-1}, and $v'_3 \approx 30$–400 m s^{-1}.

Therefore, we can state that the parameters of three types of disturbances for 22 RLs from Plesetsk cosmodrome coincide well with the analogous parameters for 65 RLs from Baikonur cosmodrome—described in Ref. [36] and summarized in Section 8.1. The minimal value from the velocities $v'_3 \approx 310$–350 m s^{-1} is the velocity of AGWs in the ionosphere. The velocities $v'_2 \approx 800$–900 m s^{-1} are associated with the velocity of the shock waves, which can be transformed into AGWs due to their continuous extension for some distance from the source.

Most fast waves (with $v_1' \approx 8.3-12.5\,\mathrm{km\,s^{-1}}$) supposedly relate to waves of the nature of MHD [23].

In the earlier days (prior to and after the RL), as was expected, histograms are essentially different from the analogous parameters obtained during the days of the real RLs for the time delays of disturbances from the "virtual" RLs and their durations. Groups of time delay for disturbances and their durations are absent, which were clearly expressed earlier. The periods of quasiperiodical disturbances in the earlier days are larger than in RL days.

Hence, the existence of RL effect in geomagnetic field variations was confirmed in this chapter. These effects were determined as the main parameters and the disturbance propagation velocities coincide with the results from other works [10–21]. It is important to note that the parameters of magnetic and ionospheric disturbances are close to each other. This increases the reliability of the results obtained by different authors, as well as witnessed to exist through common links in the chain of generation and transfer of disturbances in the ionosphere (atmosphere) and the geomagnetic field. It is thus reasonable to assume that some links can be related only to plasma processes while others to only electromagnetic processes.

8.3 Main Results

1. It was stated that all 65 Soyuz and Proton launches and flights were accompanied by notable variations in the level and the spectral content of geomagnetic field, mainly in the 300–1000 s interval period.
2. For the first time, three groups of disturbances having delays of 6, 30–40, and 70–110 min in the daytime conditions as well as delays of 7, 30–70, and 100–130 min in the nighttime conditions were identified. The following durations of the processes 10–20, 30–70, and 45–70 min and the prevailing periods of 6, 6–12, and 6–12 min corresponded to these data.
3. It was noted that the fastest disturbances (with a velocity of about 10 km s^{-1}) were weakly expressed and observed only in 60% of cases. We suggest that these disturbances are transferred with the help of MHD waves. Disturbances with velocities of 0.6–1.6 and 0.3–0.4 km s^{-1} were always observed confidently. The appearance of very slow waves was caused by the movement of plasma in the magnetic field under the action of IGWs. The nature of waves having a velocity of 0.6–1.6 km s^{-1} can be related either to the propagation of CAGW that are generated by SAWs in the atmosphere or to the propagation of slow MHD waves having a magnitude close to the velocity.

4. It was found that the amplitude of all kinds of disturbances does not exceed 3–6 nT.

5. It was noted that reaction of the geomagnetic field on Soyuz and Proton RLs is generally similar.

6. A theoretical model was proposed describing magnetic disturbances that have a characteristic velocity of 0.6–1.6 and 0.3–0.6 km s^{-1}. For finding the difference with the fastest disturbances that are generated close to the rocket, slow disturbances are transferred with the help of CAGWs and IGWs that modulate the ionospheric current density and lead to generation of the injection electric current. The latter by itself excites variations of the geomagnetic field above the ground-based observatory with periods of about 5–17 min.

7. The theoretical model proposed in this chapter corresponds to the results of observations on the whole.

8. The reaction of the geomagnetic field on 22 RLs from the Plesetsk cosmodrome was studied. It was found that in all cases of RL essential or significant changes in the character of oscillations and their spectral content and energy of the prevailing oscillations were observed.

9. Three groups of disturbances having delays of 4–8, 30–35, and 70–90 min were identified. The following durations of the processes 12–14, 25–45, and 40–55 min and the prevailing periods of 7–9, 8–12, and 10–12 min corresponded to these data.

10. It was noted that fast disturbances (with a velocity of about 8–12 km s^{-1}) were weakly expressed and observed at daytime and evening time periods only in 11 cases from 19 (in about 58%). The disturbances with velocities of 800–900 and 310–350 m s^{-1} were indicated practically during all times of observation and at any time of the day. There is no doubt that they are a part of the same process.

11. It was established that the results of the investigations of geomagnetic field reaction on RLs from Baikonur and Plesetsk cosmodromes coincided well with each other, and the parameters were generally similar.

References

1. Adushkin, V. V., Kozlov, C. I., and Petrov, A. V., Eds., *Ecological Problems and Risks of the Effects of Rocket-Cosmic Technique on the Environment: Handbook*, Moscow: Ankil, 2000, 640.

2. Nagorsky, P. M., Inhomogeneous structure of the *F* region of the ionosphere generated by the rockets, *Geomagn. Aeronom.*, 38, 100–106, 1998.

3. Afra'movich, E. L., Perevalova, N. P., and Plotnikov, A. V., Registration of ionospheric responses on the shock-acoustic waves generated during launchings of the rocket-carriers, *Geomagn. Aeronom.*, 42, 790–797, 2002.

4. Sokolova, O. I., Krasnov, V. M., and Nikolaevsky, N. F., Changes of geomagnetic field under influence of rockets' launch from the cosmic port Baikonur, *Geomagn. Aeronom.*, 43, 561–565, 2003.

5. Arendt, P. R., Ionospheric undulations following Apollo 14 Launching, *Nature*, 231, 438–439, 1971.

6. Zasov, G. F., Karlov, V. D., Romanchuk, T. E., et al., Observations of disturbances in the low ionosphere during experiments on the program Soyuz-Apollo, *Geomagn. Aeronom.*, 17, 346–348, 1977.

7. Noble, S. T., A large-amplitude traveling ionospheric disturbance excited by the space shuttle during launch, *J. Geophys. Res.*, 95, 19037–19044, 1990.

8. Foster, J. C., Holt, J. M., and Lanzerotti, L. J., Mid-latitude ionospheric perturbation associated with Spacelab-2 plasma depletion experiment at Millstone Hill, *J. Ann. Geophys.*, 18, 111–120, 2000.

9. Deminov, M. G., Oraevsky, V. N., and Ruzhin, Yu. Ya., Ionospheric-magnetospheric effects from launchings of rockets to the direction of high latitudes, *Geomagn. Aeronom.*, 41, 772–781, 2001.

10. Chernogor, L. F., Garmash, K. P., Kostrov, L. S., et al., Perturbations in the ionosphere following U.S. powerful space vehicle launching, *Radiophys. Radioastron.*, 3, 181–190, 1998.

11. Garmash, K. P., Gokov, A. M., Kostrov, L. S., et al., Radiophysical investigations and modeling of processes in the ionosphere perturbed by sources of different nature. 2. Processes in the artificially perturbed ionosphere. Variations of characteristics of radio signals. Modeling of disturbances, *Radiophys. Electron.*, 427, 3–22, 1999 (in Russian).

12. Garmash, K. P., Kostrov, L. S., Rozumenko, V. T., et al., Global perturbations of the ionosphere caused by launchings of rockets at the background of the magnetic storm, *Geomagn. Aeronom.*, 39, 72–78, 1999.

13. Garmash, K. P., Rozumenko, V. T., Tyrnov, O. F., Tsymbal, A. M., and Chernogor, L. F., Radiophysical investigations of the processes in the near-the-Earth plasma perturbed by the high-energy sources, *Foreign Radio Electronics: Successes of Modern Radio Electronics*, 7, 3–15, 1999; 8, 3–19, 1999 (in Russian).

14. Kostrov, L. S., Rozumenko, V. T., and Chernogor, L. F., Doppler radio sounding of disturbances in the middle ionosphere accompanied launchings and flights of space vehicles, *Radiophys. Radioastron.*, 4, 227–246, 1999 (in Russian).

15. Garmash, K. P., Gokov, A. M., Kostrov, L. S., et al., Radiophysical investigations and modeling of ionospheric processes generated by sources of various nature. 2. Processes in a modified ionosphere. Signal parameters variations. Disturbance simulation, *Telecomm. Radioeng.*, 53, 1–22, 1999.

16. Burmaka, V. P., Kostrov, L. S., and Chernogor, L. F., Statistical characteristics of signals of the Doppler HF radar during sounding of the middle ionosphere perturbed by the rockets' launchings and by the solar terminator, *Radiophys. Radioastron.*, 8, 143–162, 2003 (in Russian).

17. Burmaka, V. P., Taran, V. I., and Chernogor, L. F., Results of complex radiophysical observations of wave disturbances in geo-space accompanied launchings and flights of the rockets, *Cosmic Sci. Technol.*, 9, 57–61, 2003 (in Russian).

18. Burmaka, V. P., Taran, V. I., and Chernogor, L. F., Results of complex radiophysical investigations of wave disturbances in the ionosphere accompanied launchings of the rockets at the background of the natural non-stationary processes, *Radiophys. Radioastron.*, 9, 5–28, 2004 (in Russian).

19. Burmaka, V. P., Taran, V. I., and Chernogor, L. F., Wave processes in the ionosphere accompanied launchings and flights of rockets at the background of natural transmission processes, *Geomagn. Aeronom.*, 44, 518–534, 2004.

20. Burmaka, V. P., Chernogor, L. F., and Chernyak, Yu. V., Wave disturbances in geo-space accompanied launchings and flights of the rockets "Soyuz" and "Proton," *Radiophys. Radioastron.*, 10, 254–272, 2005 (in Russian).

21. Burmaka, V. P., Lisenko, V. N., Chernogor, L. F., and Chernyak, Yu. V., Wave processes in the *F*-region of the ionosphere accompanied starts of rockets from the cosmodrome Baikonur, *Geomagn. Aeronom.*, 46, 783–800, 2006.

22. Gossard, E. E. and Hooke, W. H., *Waves in the Atmosphere: Atmospheric Infrasound and Gravity Waves: Their Generation and Propagation*, Amsterdam: Elsevier, 1975, 476.

23. Sorokin, V. M. and Fedorovich, G. V., *Physics of Slow MHD-Waves in the Ionospheric Plasma*, Moscow: Energoizdat, 1982 (in Russian).

24. Chernogor, L. F., Physics of the Earth, atmosphere and geospace in the lightening of the system paradigm, *Radiophys. Radioastron.*, 8, 59–106, 2003 (in Russian).

25. Chernogor, L. F., The Earth–atmosphere–geospace as an open dynamic nonlinear system, *Cosmic Sci. Technol.*, 9, 96–105, 2003 (in Ukrainian).

26. Chernogor, L. F., The Earth–atmosphere–ionosphere–magnetosphere as an open dynamic nonlinear system (Part 1), *Nonlinear Univ.*, 4, 655–697, 2006 (in Russian).

27. Chernogor, L. F., The Earth–atmosphere–ionosphere–magnetosphere as an open dynamic nonlinear physical system (Part 2), *Nonlinear Univ.*, 5, 198–231, 2007 (in Russian).

28. Garmash, K. P., Leus, S. G., Chernogor, L. F., and Shamota, M. A., Variations of the geomagnetic field accompanied launchings and flights of the space vehicles, *Cosmic Sci. Technol.*, 13, 87–98, 2007 (in Russian).

29. Lazorenko, O. V., Panasenko, S. V., and Chernogor, L. F., Adaptive Fourier transform, *Electromagn. Waves Electron. Sys.*, 10, 39–49, 2005 (in Russian).

30. Li, Y. Q., Jacobson, A. R., Carlos, R. C., et al., The blast wave of Shuttle plume at ionospheric heights, *Geophys. Res. Lett.*, 21, 2737–2740, 1994.

31. Ahmedov, R. R. and Kunitsyn, V. E., Simulation of generation and propagation of acoustic-gravity waves in the atmosphere during a rocket flight, *Int. J. Geomagn. Aeron.*, 5,GI2002, doi:10.1029/2004GI000064, 2004.

32. Ahmedov, R. R., Numerical modeling of generation of the acoustic-gravity waves and ionospheric disturbances from the ground-based and atmospheric sources, PhD Dissertation, Moscow: Moscow State University Press, 2004, 131.

33. Blaunstein, N. and Plohotniuc, E., *Ionosphere and Applied Aspects of Radio Communication and Radar*, New York: CRC Press/Taylor & Francis, 2008, 577.

34. Burmaka, V. P., Taran, V. I., and Chernogor, L. F., Wave processes in the ionosphere at quiet and perturbed conditions. 2. Analysis of results of observation and modeling, *Geomagn. Aeronom.*, 46, 209–218, 2006.

35. Nagorsky, P. M. and Tarashuk, Yu. E., Artificial modification of the ionosphere during launchings of rockets leading out at the orbit the space vehicles, *Izv. Vuzov. Radiofizika*, 10, 94–106, 1993 (in Russian).
36. Chernogor, L. F., Geomagnetic field fluctuations near Kharkov, which accompanied rocket launches from the Baikonur site, *Geomagn. Aeronom.*, 49, 384–396, 2009.
37. Chernogor, L. F. and Shamota, M. A., Wave disturbances of the geomagnetic field accompanied launchings of the rockets from the cosmodrome Plesetsk, *Cosmic Sci. Technol.*, 14, 29–38, 2008 (in Ukrainian).
38. Pogorel'tsev, A. I. and Bidlingma'er, E. R., Numerical modeling of structure of electromagnetic disturbances caused by acoustic-gravity waves, *Geomagn. Aeronom.*, 32, 131–141, 1992.

Chapter 9

Geomagnetic Effects of Rocket Launches from Cosmodromes Worldwide

9.1 Geomagnetic Effects of Rocket Launches from Chinese Cosmodromes

9.1.1 Overview

As mentioned in the previous chapters, space vehicle (SV) launches and flights significantly affect all subsystems of the system that is the Earth–atmosphere–ionosphere–magnetosphere (EAIM), creating a spectrum of phenomena (see also Refs. [1–3]). Going through the previous chapters and utilizing corresponding literature [1,3–8], it is clear now that rocket launch (RL) is accompanied by thermal, dynamic, chemical, acoustic, and electromagnetic action on media.

Local high-energy action must affect the ionospheric current system due to change in conductivity as a result of injection of a large amount (in hundreds of tons) of burnt fuel into the environment as high-temperature (~3000 K) ionized gas.

It is possible that there is another mechanism of action on the subsystems of EAIM. As described earlier, during the super-high velocity of rocket motion shock waves are generated, which modulate atmospheric density, electron concentration, and electrical field in the ionosphere. At large distances from the place of action, the shock waves are transformed into acoustic gravity waves (AGWs) that draw the ionized component of plasma into motion. As a result, geomagnetic pulsations, particularly, with the same periods as those of AGWs [9,10] should appear at the place of registration. Therefore, the ionospheric and magnetic effects of RLs are related to each other. For more than three decades of registration, the study and modeling of RL effects were done and a huge volume of experimental data obtained by using radiophysical techniques was selected and analyzed (see, e.g., Refs. [7,8,10–13]). The researches were mainly carried out at close distances (i.e., along the trajectory of the SV flight) [4,8,14]. However, a special interest was evinced in the registration of the effects for large distances [2,6,7,9,14–23]. At the same time, only a few separate works [9,10,22] described the observations of the RL effects by using the magnetometric method.

Following from Refs. [9,10], the results of investigation into the reaction of the geomagnetic field in the proximity of Kharkov city for 43 RLs from Baikonur cosmodrome and for 16 RLs from Plesetsk cosmodrome were described in Chapter 8. As mentioned there, three groups of quasiperiodical disturbances with velocities of 5–9 km s^{-1}, 800–1100, and 300–400 m s^{-1}, respectively, were identified. The amplitudes of oscillations varied, depending on the level of geomagnetic activity, and their average magnitude was 3–6 nT. At the same time, a great interest arose to study the possibility of detecting the effects from RLs at distances significantly exceeding those of Plesetsk cosmodrome (1500 km), or to Baikonur cosmodrome (2100 km). For this purpose, Chinese cosmodromes are more convenient because distances from them to the place of registration are $R \approx 5000$–6000 km.

The aim of this chapter is to describe the results of observation of variations in the level and the spectral components of the geomagnetic field near Kharkov city. These accompanied 33 RLs of different types from Chinese Zeuan', Sichan, and Taiuan cosmodromes following the results obtained in Ref. [24]. The procedure of the program apparatus that measured the setup used for observations is described in Ref. [21] (see also Appendix 3). The method of measurement and analysis are analogous to those described in Ref. [9].

9.1.2 Level of Magnetic Perturbation

The level of magnetic perturbation was controlled by recording indexes A_p and K_p taken from the Internet (see Table 9.1). In the table, the magnitudes of indexes A_p and $\sum K_p$ are presented for the day of the RLs, a day prior, and 2 days prior to this event. From the table, it is clearly seen that launches occurred as in magnetically quiet days, as well as in the magnetoperturbed days.

Table 9.1 Levels of Magnetic Perturbation before RL and on Day of RL

Date	Time (UT)	Cosmodrome	Rocket Type	Two Days Prior to Launch		One Day Prior to Launch		The Day of Launch	
				A_p	ΣK_p	A_p	ΣK_p	A_p	ΣK_p
March 25, 2002	14:15	Zeuan'	2F	8	17	9	15	47	37
May 15, 2002	01:50	Ta'uan'	4B	12	22	32	35	16	25
October 26, 2002	03:17	Ta'uan'	4B	47	40	40	37	27	32
December 29, 2002	16:40	Zeuan'	2F	37	35	19	28	13	23
May 24, 2003	16:34	Sinchan	3A	25	30	21	27	22	30
October 15, 2003	01:00	Zeuan'	2F	13	22	48	40	42	36
October 21, 2003	03:16	Ta'uan'	4B	32	34	30	34	39	38
November 3, 2003	19:20	Zeuan'	2D	21	28	18	27	10	21
November 14, 2003	16:01	Sinchan	3A	26	32	42	37	37	35
December 29, 2003	19:06	Sinchan	2C	12	24	12	22	6	14
April 18, 2004	15:59	Sinchan	2C	12	22	9	18	11	18
July 25, 2004	07:31	Ta'uan'	2C	47	38	27	33	122	54

(Continued)

Table 9.1 (Continued) Levels of Magnetic Perturbation before RL and on Day of RL

Date	Time (UT)	Cosmodrome	Rocket Type	Two Days Prior to Launch		One Day Prior to Launch		The Day of Launch	
				A_p	ΣK_p	A_p	ΣK_p	A_p	ΣK_p
August 29, 2004	07:50	Zeuan'	2C	8	17	12	23	8	17
September 8, 2004	23:14	Ta'uan'	4B	14	23	16	25	9	20
September 27, 2004	08:00	Zeuan'	2D	5	12	4	12	5	11
October 19, 2004	01:20	Sinchan	3A	3	6	4	8	4	8
November 6, 2004	08:10	Ta'uan'	4B	7	14	4	7	3	4
November 18, 2004	10:45	Sinchan	2C	8	15	6	11	3	5
July 5, 2005	22:40	Zeuan'	2D	11	21	7	15	5	10
October 12, 2005	01:00	Zeuan'	2F	10	17	6	10	1	1
July 26, 2006	22:48	Ta'uan'	4B	7	14	5	10	5	11
September 9, 2006	07:00	Zeuan'	2C	8	11	4	8	2	3

Date	Time	Site	Type						
September 12, 2006	16:02	Ta'uan'	3A	6	11	8	15	4	8
October 24, 2006	11:34	Ta'uan'	4B	13	21	3	6	4	8
December 8, 2006	00:53	Ta'uan'	3A	28	31	25	31	25	28
February 2, 2007	16:28	Sinchan	3A	16	24	8	17	2	6
April 11, 2007	03:27	Ta'uan'	2C	9	14	7	13	4	5
April 13, 2007	20:11	Sinchan	3A	4	5	8	16	2	3
May 14, 2007	16:01	Sinchan	3B	3	3	3	7	3	5
May 25, 2007	07:12	Zeuan'	2D	42	36	28	32	16	25
May 31, 2007	16:08	Sinchan	3A	4	8	4	8	4	7
September 19, 2007	03:26	Ta'uan'	4B	2	3	2	5	3	8
October 24, 2007	10:05	Sinchan	3A	5	11	7	13	3	6

9.1.3 Results of Observation

In this chapter, we consider the reaction of the geomagnetic field on 33 Long March RLs having the following modifications: "2C" (seven starts), "2D" (four starts), "2F" (four starts), "3A" (nine starts), "3B" (one start), and "4B" (eight starts). The main characteristics and parameters of the rockets are presented in Appendix 1. It is clearly seen from Appendix 1 that rockets of 2F and 3B modifications relate to the class of heavy rockets, but the rockets of 2C, 2D, 3A, and 4B modifications relate to median weight rockets [11]. A short description of the cosmodromes is presented in Appendix 2. It is seen that the number of the RLs from different cosmodromes is approximately the same. The distances from the observatory (Kharkov city, Ukraine) to the cosmodromes are $R \approx 5000$–6000 km (see Table 9.2).

Figure 9.1 shows the variations of geomagnetic field components, $H(t)$ and $D(t)$, which accompanied 3A RL on November 14, 2003, from Sinchan cosmodrome that took place at 16:01 UT (here and in further instances time is expressed as Universal Time or UT).

At about 15:25 h, sunset was observed at an altitude of 200 km near Kharkov. Probably, the effects relating to the evening solar terminator (ST) ended at the moment of the RL. The first change in the character of the oscillation process was registered at the time interval from 16:40 to 16:45 h and continued for 30–40 min for both components, $H(t)$ and $D(t)$. An increase of the amplitude from 2 to 4 nT was registered in the D-component, whereas, conversely,

Table 9.2 Corrected Directions from the Region of Generation of Disturbances to the Observatory

Corrected Distance from the Place of Generation to the Observatory (km)	Time of Movement of the Rocket to Altitude of Generation (min)	Name of Cosmodrome
1650	3–5	Plesetsk
2300	3–5	Baikonur
5400	3–5	Zeuan'
6100	3–5	Ta'uan'
6300	3–5	Sichan
8000	9	Cape Canaveral
9000	5	Kourou

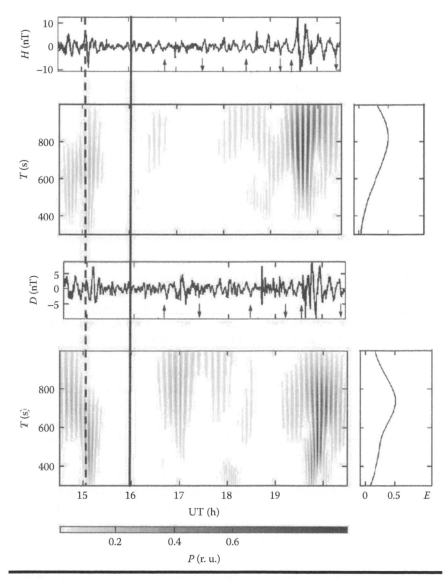

Figure 9.1 Temporal variations of the *H*- and *D*-components of the geomagnetic field and the corresponding wavelet spectra and energy diagrams at time interval 300–1000 s accompanying 3A RL from Sinchan cosmodrome on November 14, 2003. Here and in further presentations the solid vertical line indicates the moment of RL (at 16:01 UT); the dashed line indicates the moment of sunset at an altitude of 200 km close to Kharkov Observatory.

a repression of the wave process was observed in the H-component. The periods of oscillations were from 10 to 14 min.

Launch of the Chinese 3A rocket from Sinchan cosmodrome happened on October 24, 2007, at 10:05 h. Figure 9.2 shows temporal dependences of the H- and D-components of the geomagnetic field. The first, vividly registered change in the character of oscillations [when the high-frequency component in variations of $H(t)$ was totally repressed] had a delay of 23–25 min, and its duration was 58–63 min. The period of the prevailing oscillations for the H- and D-components equaled 7 and 12 min, respectively.

A third disturbance of 52 min duration appeared after 180–200 min. It was accompanied by significant increase in the period, from 7 to 14 min, and in the amplitude, from 1 to 3.5 nT in the H-component. For the D-component, the effects were expressed weakly. All three processes were accompanied by an essential increase in the wave process period that was achieved in 12–14 min.

9.1.4 Results of Statistical Analysis

Histograms of the time delays, durations, and periods of the prevailing pulsations of the geomagnetic field are shown in Figure 9.3. Processing was carried out for 33 RLs. The number of realizations is denoted in the corresponding histograms. It can be concluded from the results presented in Figure 9.3 that after the RL three groups of disturbances with delays took place: $\Delta t_1 \approx 39$–$53\,\text{min}$, $\Delta t_2 \approx 100$–$129\,\text{min}$, and $\Delta t_3 \approx 140$–$180\,\text{min}$. The periods of oscillation were 9–12 min.

The results of statistical processing of the large number of delays, durations, and periods of oscillation for all RLs, as well as the parameters of the root-mean-square deviations of the parameters of geomagnetic disturbances are presented in Table 9.3. The analogous statistical processing procedure was carried out for the background days, that is, a day prior to and a day after the RL. Recording of similar effects was done in relation to the time of the actual launches. The results of statistical processing for the background days are shown in Figures 9.4 and 9.5. Analyzing the illustrations in the figures, it can be concluded that the experimental law of distribution of the magnitudes of the wave disturbance (WD) delays is close to regular, which indicates the random nature of disturbances during the background days.

9.1.5 Results of Spectral Analysis

The spectra of the wavelet also indicate the existence of the three groups of delays (Figure 9.1). Those disturbances that have delays of about 200–230 min are revealed most evidently in both spectrograms.

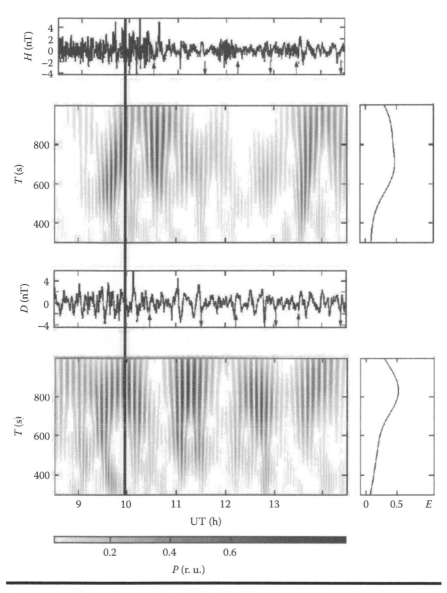

Figure 9.2 Temporal variations of the *H*- and *D*-components of the geomagnetic field and the corresponding wavelet spectra and energy diagrams at time interval 300–1000 s accompanying 3A RL from Sinchan cosmodrome on October 24, 2007. Here and in further presentations the solid vertical line indicates the moment of RL (at 10:05 UT); the dashed line indicates the moment of sunset at an altitude of 200 km close to Kharkov Observatory.

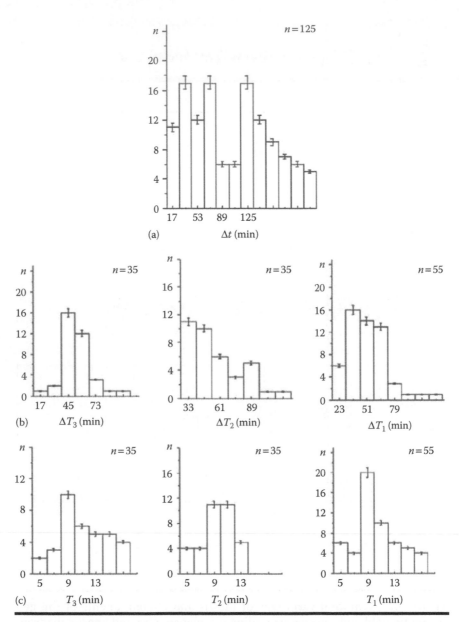

Figure 9.3 Histograms of (a) the time delays (Δt); (b) their durations (ΔT_i), $i = 1$, 2, 3; and (c) the periods (T_i), $i = 1, 2, 3$, of the prevailed oscillations of three types of disturbances of the geomagnetic field, which accompanied the RL.

Table 9.3 Statistical Characteristics of the Main Parameters of Disturbances Accompanying RLs

Magnitudes (min)	Parameters
Δt_1	46.0 ± 7.1
Δt_2	116.8 ± 11.9
Δt_3	166.0 ± 37.1
ΔT_1	51.6 ± 6.8
ΔT_2	56.9 ± 15.5
ΔT_3	53.2 ± 8.03
T_1	10.4 ± 0.2
T_2	10.3 ± 0.3
T_3	11.6 ± 0.3

Note: Δt is the time delay; ΔT and T are their durations and periods, respectively.

Three groups of disturbances were also observed on October 24, 2007, after the RL (Figure 9.2). The amplitude of oscillations was larger for the D-component, with respect to H-component, of the signal.

9.1.6 Results and Discussions

Essential or significant changes in the spectral content of the wave processes accompanied practically all considered RLs. In most cases, increase of the amplitude of pulsations registered, but in some cases the repression of the signal was indicated, which can be explained by summation of wave processes with opposite phases. The variations in the signal character were accompanied by changes in the periods of the quasiperiodical processes.

Disturbances occurring after RLs are not different principally compared to the natural disturbances. Therefore, the analysis of geomagnetic field variations after separate launches does not give us full confidence about the relation of the observed disturbances to the RLs. Such confidences give the results of statistical analysis. As later indicated, more often the three groups of disturbances

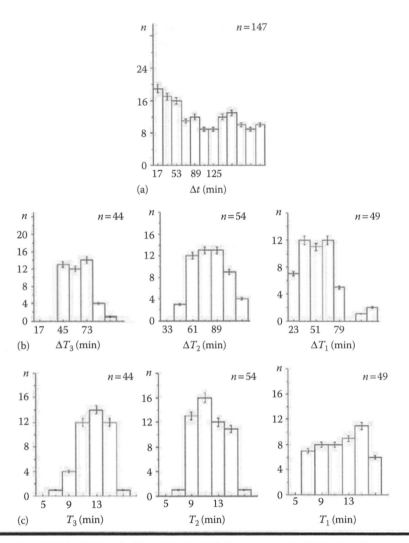

Figure 9.4 Histograms of (a) the time delays (Δt); (b) their durations (ΔT_i), $i = 1, 2, 3$; and (c) the periods (T_i), $i = 1, 2, 3$, of the prevailed oscillations of three types of disturbances of the geomagnetic field for the background days (a day prior to the RL).

registered with $\Delta t_1 \approx 39$–53 min, $\Delta t_2 \approx 100$–129 min, and $\Delta t_3 \approx 140$–180 min. These disturbances have the form of wave packets (e.g., wavelets) with durations of about 30–50 min and periods of 9–12 min. Knowledge of signal delays, Δt_1, Δt_2, and Δt_3 allows calculation of the velocities of the corresponding wave processes.

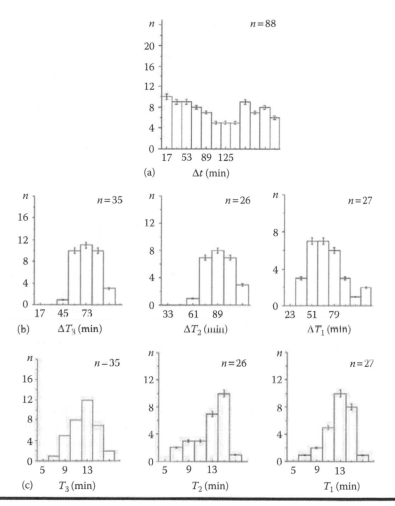

Figure 9.5 Histograms of (a) the time delays (Δt); (b) their durations (ΔT_i), $i = 1, 2, 3$; and (c) the periods (T_i), $i = 1, 2, 3$, of the prevailed oscillations of three types of disturbances of the geomagnetic field for the background days (a day after the RL).

We can declare that geomagnetic field disturbances are generated at altitudes of the current jet maximum ($z \approx 100$–150 km). As mentioned in Chapter 8, the time Δt_0, during which the rocket attains an altitude of $z \approx 100$–150 km, depends on the rocket type and equals 3–4 min. In this case, the propagation time of the disturbance observed close to Kharkov Observatory is $\Delta t' = \Delta t - \Delta t_0$. Then, taking the mean values Δt_1, Δt_2, and Δt_3 from Table 9.3, we get the corrected time delays: $\Delta t'_1 \approx 37$–49 min, $\Delta t'_2 \approx 102$–124 min, and $\Delta t'_3 \approx 149$–199 min.

For the ranges $R \approx 5000-6000$ km (see Table 9.2), the velocities $v_1' \approx 2.0-3.7\,\mathrm{km\,s^{-1}}$, $v_2' \approx 670-980\,\mathrm{m\,s^{-1}}$, and $v_3' \approx 420-670\,\mathrm{m\,s^{-1}}$ correspond to such delays. Slow MHD waves have the highest velocity, which was mentioned in Chapter 3 following Ref. [25]. Such wave velocities that follow RLs were observed many times (see, e.g., Refs. [5,11,15–19]). Velocities of the propagation of second group disturbances are close to those obtained and described in Refs. [9,10]: $v_2' \approx 800-1100\,\mathrm{m\,s^{-1}}$ and $v_2' \approx 880-910\,\mathrm{m\,s^{-1}}$, respectively. These waves are associated with the shock wave velocities, which due to their extension far from the source can continuously be transformed into AGWs (we will call these formally as "shock-wave" AGW). Internal gravity waves (IGWs) in the *F*-region of the ionosphere [26,27] have a velocity of $v_3' \approx 420-670\,\mathrm{m\,s^{-1}}$.

The natural question that arises is: Can the above-mentioned waves pass distances greater than 5000–6000 km and in this case conserve the registered amplitude? Concerning gravity waves AGWs and IGWs, they can be said to propagate to the aforementioned distance [26]. The following circumstances indicate to this. First of all, waves of such type propagate inside the waveguide with an energy loss proportional to R^{-1}, but an amplitude of the disturbance of pressure and density of particles as $R^{-1/2}$. Therefore, in the absence of attenuation of waves, their amplitude would be only 1.5–1.7 times lesser than it is for RLs from the Baikonur cosmodrome for the same types of rockets and RLs from the Chinese cosmodromes ($R = 2300$ km, see Table 9.2). The decrement of attenuation is not more than $3 \times 10^{-4}\,\mathrm{km^{-1}}$, as our estimations show.

The situation is more complicated in the case of slow MHD waves. This question, as we know, has not been discussed in the literature. We can only suggest that their attenuation does not change from that of AGW and IGW. The reason for this statement can be the result of observations carried out by several authors on waves of such velocities [5,7,11,15–19].

9.2 Geomagnetic Effects of RLs from Cape Canaveral and Kourou Cosmodromes

9.2.1 Overview

As mentioned in the Introduction, and we point it out again, rocket and SV launches and flight are the most powerful sources of anthropogenic action on the atmosphere and the geospatial environment. In our discussions further in Chapter 8, it was shown that a degree of action on the near-the-Earth medium is determined by the rocket type, the power of its engine, and the number of its launches. A special research carried out in Ref. [1] found that disturbances of the electron concentration and ambient electric and magnetic fields arise with magnitudes of 50%–100%, 5–10 nR, and 1–5 mV m^{-1}, respectively, in the proximity of the SV.

These disturbances can move to great distances, that is, the local processes can yield large-scale or even global disturbances in the ionosphere at the farthest regions from the rocket trajectory. It should be realized that the EAIM system is an open, dynamic, nonlinear system and a triggering mechanism for liberation of energy is possible in it [2–4]. Hence, because it is now clear to the reader familiar with the gist of the previous chapters, the effects of RLs may have a response in all subsystems of the entire EAIM system and can propagate great distances, having a later effect not only close to the place of action but also sufficiently far enough from it.

A set of works by different authors [1,5–8,11–13,15–23] investigates iono-spheric disturbances and those in the geospatial environment. The experimental data have mainly been analyzed by radiophysical methods, and the ionospheric effects of RLs have been studied [1,5–13,17–21]. As mentioned in Chapter 8, investigations related to the effects of RLs on the magnetic field are lesser known nowadays [15,16,22,23].

Based on the work in Refs. [16,22,23] carried out by one of the authors of this book along with his colleagues, in Chapter 8, and in Section 9.1, the results of the analysis of temporal variations of the Earth's magnetic field in the diapason of geomagnetic pulsations accompanying RLs from Baikonur, Plesetsk, and Chinese cosmodromes were outlined. Registration of pulsations was done close to Kharkov city (Ukraine) located about 1500, 2300, and 5000–6000 km (see Table 9.2), respectively, away from these cosmodromes.

Practically, all considered launches were accompanied by essential and significant changes in the character of horizontal component oscillations of the geomagnetic field at the diapason periods of $T = 1$–1000 s.

An interesting investigation is undertaken on the possibility of the rising effects caused by RLs at distances of about 10,000 km. For this purpose, American and European cosmodromes are convenient, such as the location of Cape Canaveral (FL) and Kourou (French Guiana), ranging from the place of registration to about 10,000 km (see Table 9.2).

The aim of this chapter is to describe the results of analysis of variation of the level and the spectral content of geomagnetic pulsations measured at Kharkov Observatory, which accompanied 11 Space Shuttle and 15 Ariane-5 launches (see Appendix 1) from Cape Canaveral and Kourou cosmodromes, respectively (see Appendix 2). In our explanation below, we will explain the obtained results described in Ref. [28].

The programming apparatus set up for measurements, used for observations, is fully described in Appendix 3 and in Ref. [21]. The technique and the method of measurements and analysis are analogous with those described in Ref. [9].

9.2.2 Level of Magnetic Perturbations

For common control of the level of magnetic perturbations, A_p and K_p indexes were used, which are taken from the Internet (see Table 9.4). In this table, the magnitudes of indexes A_p and $\sum K_p$ are presented as on the day of RL, as well as on a day and 2 days prior to this event. It is clearly seen from the table that launches occurred on magnetically quiet ($\sum K_p < 24$) and magnetoperturbed ($\sum K_p \geq 24$) days. The first were 18 days, and the second were 8 days.

9.2.3 Results of Observation

9.2.3.1 Ariane-5 Launch

For example, let us consider the two Ariane-5 RLs. Figure 9.6 shows the temporal variations and wavelet spectra of the horizontal components of the geomagnetic field that accompanied the RL, which took place on February 12, 2005, at 19:49 UT.

Currently, a cumulative index equals $\sum K_p = 10$. The first notable change in the character of oscillations for both components, $H(t)$ and $D(t)$, registered 47 min after the launch. It had duration of 34 min and period of 11 min. The next quasiperiodical process accompanied by increase of amplitude from 1 to 2 nT has delays of 222 and 226 min, duration of 41 min, and periods of about 14–15 min for the H- and D-components of the signal, respectively. The third disturbance of duration 82–84 min registered 300–306 min after RL. The periods of pulsations were 11–15 min for both the components.

Further, launch of Ariane-5 happened on October 13, 2006, at 20:56 h (Figure 9.7). This day was related to the magnetoperturbed day because $\sum K_p = 28$. After about 34–41 min a change in the character of the signal was observed, which continued for about 45–50 min and was accompanied by an increase of the period in the D-component (from 7 to 15 min).

A second type of disturbance had delays of 139–158 min, with duration 45–56 min. The appearance of the high-frequency component and decrease in the period from 15 to 8 min was observed in the H-component. The period of oscillations increased from 11 to 15 min, and the amplitude increased from 2 to 3.5 nT in the D-component.

9.2.3.2 Space Shuttle Launch

For example, let us consider two Space Shuttle launches. Figure 9.8 presents the variations of $H(t)$ and $D(t)$ that accompanied Atlantis launch on October 7, 2002, at 19:45 h.

On the day of the RL, a magnetic storm took place with a maximum of $K_p = 6$ and $\sum K_p = 36$. Starting from 20:29 to 20:32 h depression of the

Table 9.4 Levels of Magnetic Perturbation before RL and on Day of RL

Rocket	Type	Date	Time (UT)	Two Days Prior to RL		One Day Prior to RL		Day of RL	
				A_p	$\sum K_p$	A_p	$\sum K_p$	A_p	$\sum K_p$
Ariane	5	June 5, 2002	23:22	7	18	7	16	13	24
Ariane	5	August 28, 2002	22:45	18	26	15	25	10	21
Ariane	5G	April 9, 2003	22:52	6	14	20	31	25	29
Ariane	5	June 11, 2003	22:38	28	31	27	32	15	25
Ariane	5	September 27, 2003	13:14	28	33	17	26	9	18
Ariane	5	March 2, 2004	07:17	21	28	18	27	17	27
Ariane	5G	July 18, 2004	00:44	12	19	24	29	9	18
Ariane	5	February 12, 2005	19:49	17	25	11	19	5	10
Ariane	5G	October 13, 2005	22:32	6	10	1	1	4	8
Ariane	5	March 11, 2006	22:32	4	8	12	20	12	20
Ariane	5	October 13, 2006	20:56	3	7	5	11	24	28
Ariane	5	December 8, 2006	22:08	28	31	25	35	25	28
Ariane	5	March 11, 2007	22:03	2	3	4	7	8	14

(Continued)

Table 9.4 (Continued)   Levels of Magnetic Perturbation before RL and on Day of RL

Rocket	Type	Date	Time (UT)	Two Days Prior to RL		One Day Prior to RL		Day of RL	
				A_p	ΣK_p	A_p	ΣK_p	A_p	ΣK_p
Ariane	5	April 7, 2007	22:29	3	4	4	9	2	3
Ariane	5	November 14, 2007	22:06	2	5	8	17	9	16
Space Shuttle	Atlantis	April 8, 2002	20:44	5	13	7	17	5	11
Space Shuttle	Endeavour	June 5, 2002	21:22	12	23	17	26	10	19
Space Shuttle	Atlantis	October 7, 2002	19:45	29	32	19	25	39	36
Space Shuttle	Endeavour	November 24, 2002	00:49	24	30	19	27	16	27
Space Shuttle	Discovery	July 26, 2005	14:39	5	12	6	13	6	13
Space Shuttle	Columbia	July 4, 2006	18:37	2	5	4	8	13	19
Space Shuttle	Atlantis	September 9, 2006	16:29	8	11	4	8	2	3
Space Shuttle	Discovery	December 10, 2006	01:47	2	3	3	6	5	11
Space Shuttle	Atlantis	June 8, 2007	23:38	2	3	3	6	8	16
Space Shuttle	Endeavour	August 8, 2007	22:38	12	17	23	28	6	11
Space Shuttle	Discovery	October 23, 2007	15:38	4	8	5	11	7	13

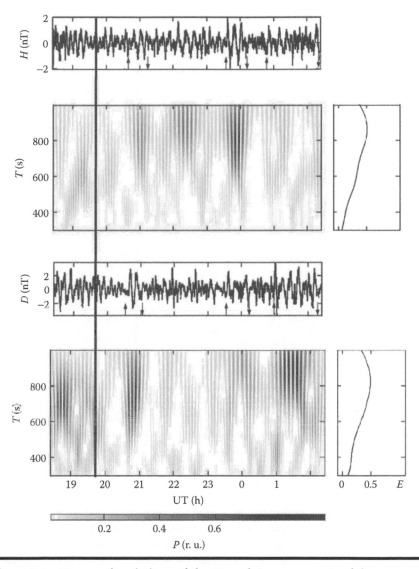

Figure 9.6 Temporal variations of the *H*- and *D*-components of the geomagnetic field and the corresponding wavelet spectra and energy diagrams at the time interval of 300–1000 s accompanied by Ariane-5 RL from Kourou cosmodrome on February 12, 2005. Here and in further presentations the solid vertical line indicates the moment of RL (at 19:49 UT); the arrows (↑) and (↓) indicate a moment of the beginning and the end of measurements, respectively, of the signal character.

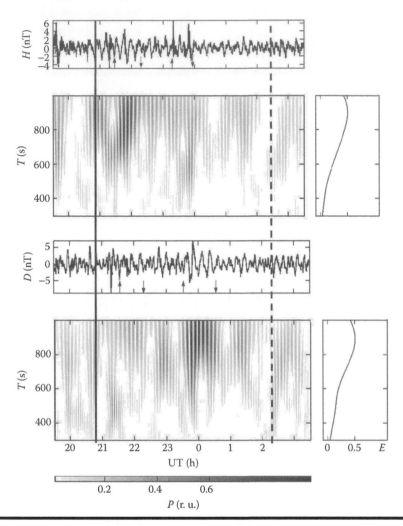

Figure 9.7 Temporal variations of the *H*- and *D*-components of the geomagnetic field and the corresponding wavelet spectra and energy diagrams at the time interval of 300–1000 s accompanied by Ariane-5 RL from Kourou cosmodrome on October 13, 2006 (20:56 UT). The dashed line indicates the moment of sunrise at an altitude of 200 km close to Kharkov (Ukraine).

wave process with the periods of 7 and 13 min in the *H*- and *D*-components were observed and continued for 38 min. The next disturbance occurred 135–137 min after the RL and continued for 68–84 min. The periods were 13–15 min. The third change of signal character registered at the time interval from 00:00 to 00:10 h for the *H*-component and 00:19 h for the

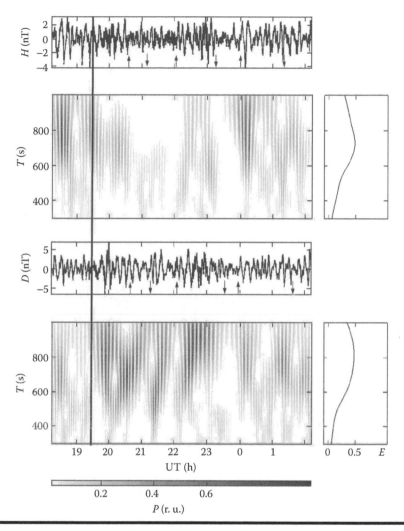

Figure 9.8 **Temporal variations of the *H*- and *D*-components of the geomagnetic field and the corresponding wavelet spectra and energy diagrams at the time interval of 300–1000 s accompanying Space Shuttle launch from Cape Canaveral on October 7, 2002 (19:45 UT).**

D-component. The periods of oscillations of the level of both components equaled 11–12 min.

Discovery launch on October 23, 2007, at 15:38 h practically was at the same time of the passing of the ST at Kharkov Observatory (Figure 9.9). The RL was on a relatively magnetically quiet day (with $\sum K_p = 13$). This fact makes it difficult to

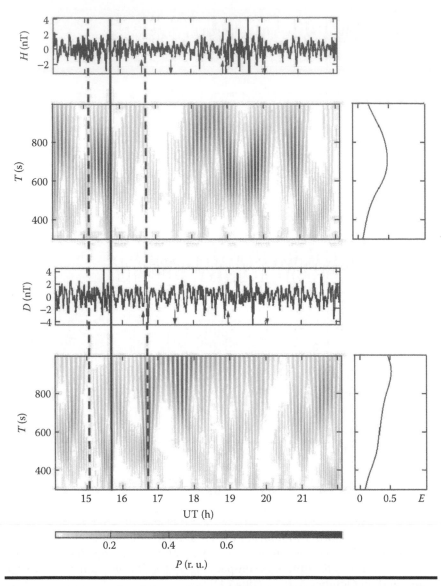

Figure 9.9 Temporal variations of the *H*- and *D*-components of the geomagnetic field and the corresponding wavelet spectra and energy diagrams at the time interval of 300–1000 s accompanying Space Shuttle launch from Cape Canaveral cosmodrome on October 23, 2007 (15:38 UT). The dashed line indicates the moment of sunrise at an altitude of 200 km and at ground surface close to Kharkov (Ukraine).

decide on the existence of the RL, because in our opinion the wave processes from these two sources were added to one another.

Significant increase in the period (from 7 to 15 min) and increase in the amplitude (from 1.5 to 4 nT) in the D-component were recorded approximately after 53 min and these continued for 60 min. At the same time interval, insufficient depression of the wave process was observed with a period of 7 min in the H-component.

The next disturbance appeared at about 18:55–19:00 h, and had a vividly expressed character in the H- and D-components related to the occurrence of high-frequency signal component. This disturbance continued for about 65–67 min and had periods of 10–12 min for both components.

9.2.4 Results of Spectral Analysis

The wavelet spectra testify to the existence of three groups of disturbances with periods of 10–14 min (Figure 9.6). Figure 9.7 shows the second type of disturbances having delays of 139–158 min expressed more clearly.

Spectrograms of temporal variations of the level of geomagnetic pulsations that accompanied Space Shuttle launches are shown in Figures 9.8 and 9.9. Three groups of disturbances are also seen here. After SV launch on October 7, 2002, increase of the period of the wave process was basically observed that is seen clearly in the wavelet spectra (Figure 9.8). On October 23, 2007, for the Space Shuttle launch, the spectrogram of temporal variations at the H-component level especially expressed disturbances with delays of about 200 min and periods of 10–12 min clearly (Figure 9.9). At the same time, amplification of the level of pulsations in the D-component took place after about 60 and 120 min of the SV launch. The amplitude of disturbances was larger for the D-component of all RLs.

9.2.5 Results of Statistical Analysis

Histograms of the time delays, durations, and periods of the prevailing geomagnetic field oscillations for Ariane-5 launches are shown in Figure 9.10. Processing was carried out for 15 RLs. The number of realizations is denoted in the corresponding histograms.

From the results presented in Figure 9.10, it can be concluded that after the RL three groups of magnetic disturbances with delays of $\Delta t_1 \approx 13$–$88\,$min, $\Delta t_2 \approx 193$–$238\,$min, and $\Delta t_3 \approx 300$–$330\,$min took place. The durations of disturbance were the following: $\Delta T_1 = 30$–$70\,$min, $\Delta T_2 = 33$–$63\,$min, and $\Delta T_3 = 33$–$70\,$min. The periods of oscillation were 9–12 min.

Histograms of the parameters of the three groups of disturbances for Space Shuttle launches are shown in Figure 9.11. Histograms testify for the

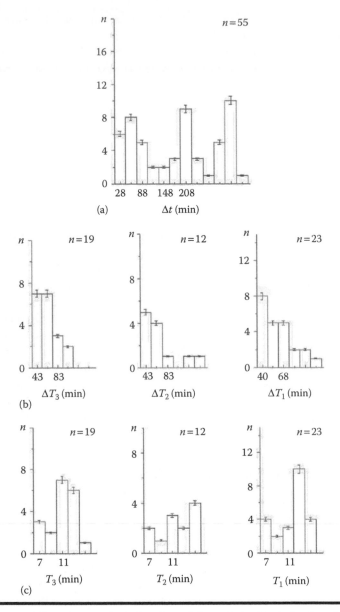

Figure 9.10 Histograms of (a) the time delays (Δt); (b) their durations (ΔT_i), $i = 1, 2, 3$; and (c) the periods (T_i), $i = 1, 2, 3$, of three types of geomagnetic field disturbances, accompanying Ariane-5 launch.

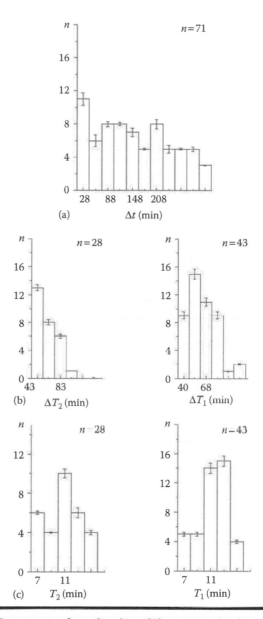

Figure 9.11 Histograms of (a) the time delays (Δt); (b) their durations (ΔT_i), $i = 1, 2, 3$; and (c) the periods (T_i), $i = 1, 2, 3$, of three types of geomagnetic field disturbances.

existence of three groups of disturbances with delays: $\Delta t_1 \approx 13\text{--}58\,\text{min}$, $\Delta t_2 \approx 133\text{--}163\,\text{min}$, and $\Delta t_3 \approx 193\text{--}253\,\text{min}$. The durations of disturbance were the following: $\Delta T_1 = 36\text{--}60\,\text{min}$, $\Delta T_2 = 54\text{--}90\,\text{min}$, and $\Delta T_3 = 63\text{--}93\,\text{min}$. The periods of oscillation were the following: $T_1 = 6\text{--}10\,\text{min}$, $T_2 = 10\text{--}14\,\text{min}$, and $T_3 = 10\text{--}14\,\text{min}$. The results of statistical processing of the massive magnitude of delays, durations, and periods of oscillation for all RLs are presented in Tables 9.5 and 9.6 for Ariane-6 and Space Shuttle, respectively.

The analogous statistical processing procedure was carried out for the background days, that is, for a day prior to and a day after the RL. Definition of the parameters of changes in the signal character was carried out relative to the time of the actual launches. The results of statistical processing of registrations for the background days for the two types of rockets are shown in Figures 9.12 and 9.13. Analyzing the figures, it can be concluded that the experimental law of distribution of magnitude for the delays of WD is close to regular, which testifies to the random nature of disturbances during a day prior and after the RL.

9.2.6 Results and Discussions

Practically, all RLs registered fluctuations in the horizontal components of the geomagnetic field and changes were observed in the signal character. Usually,

Table 9.5 Statistical Characteristics of the Main Parameters of Disturbances Accompanying Ariane-5 Launches

Magnitudes (min)	Parameters
Δt_1	59.4 ± 4.9
Δt_2	214.0 ± 5.9
Δt_3	29.1 ± 9.9
ΔT_1	75.1 ± 8.3
ΔT_2	60.9 ± 5.7
ΔT_3	62.5 ± 4.6
T_1	11.7 ± 0.7
T_2	12.1 ± 0.8
T_3	11.9 ± 0.7

Note: Δt is the time delay; ΔT and T are their durations and periods, respectively.

Table 9.6 Statistical Characteristics of the Main Parameters of Disturbances Accompanying Space Shuttle Launches

Magnitudes (min)	Parameter
Δt_1	36.1 ± 3.3
Δt_2	139.3 ± 5.9
Δt_3	245.6 ± 9.9
ΔT_1	50.8 ± 2.9
ΔT_2	66.9 ± 3.9
ΔT_3	66.8 ± 3.8
T_1	11.1 ± 0.9
T_2	12.4 ± 0.4
T_3	12.0 ± 0.4

Note: Δt is the time delay; ΔT and T are their durations and periods, respectively.

the amplitude and the period of geomagnetic pulsations increased. Sometimes the existing oscillations were repressed. We suggest that this may be due to the incoming wave process with a phase shift of 180° with respect to the existing wave process. Their summation was observed as repression of the signal.

It is important to note that pulsations of the geomagnetic field, caused by RLs, were insignificantly different from natural pulsations [9,10,24]. Therefore, changes in the signal character occurring after RLs do not confidently assure that these changes are caused by RLs. This is proof of the results of the statistical analysis (Figures 9.10 and 9.11).

As shown in Figure 9.10, there are three groups of time delays of geomagnetic field disturbances. Of course, the number of Ariane-5 launches during 5 years was comparatively small (only 15 launches). Despite this, all three groups of disturbances were evidently observed. The duration of the three types of disturbances were close to each other and equal 30–70 min. The periods of the prevailing oscillations usually were at limits from 12 to 14 min.

Accounting for the fact that the mentioned groups of disturbances were caused by RLs, let us estimate the velocities of their propagation. As shown earlier, we will assume that the region of effective generation of waves is located at altitudes of 100–150 km [1,16–23], and that the direction of disturbance propagation is

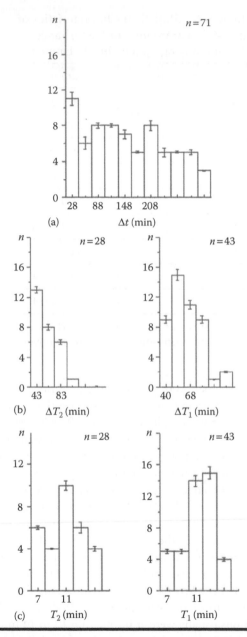

Figure 9.12 Histograms of (a) the time delays (Δt); (b) their durations (ΔT_i), $i = 1, 2, 3$; and (c) the periods (T_i), $i = 1, 2, 3$, of three types of geomagnetic field disturbances for the background days of Ariane-5 launch.

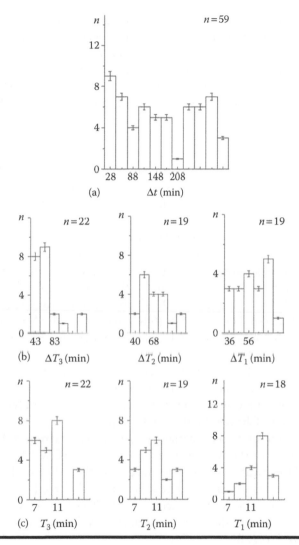

Figure 9.13 **Histograms of (a) the time delays (Δt); (b) their durations (ΔT_i), $i = 1, 2, 3$; and (c) the periods (T_i), $i = 1, 2, 3$, of three types of geomagnetic field disturbances for the background days of Space Shuttle launch.**

close to horizontal. If so, we see that at the time of movement of Ariane-5 to the middle of this region is about $\Delta t_0 \approx 5$ min. In this case, the corrected delays are $\Delta t' = \Delta t - \Delta t_0$, and the corrected distances are $R' = R - \Delta R \approx 9000$ km. The following magnitudes of the corrected velocities: $v'_1 \approx 2.5–3.1\,\mathrm{km\,s^{-1}}$, $v'_2 \approx 680–740\,\mathrm{m\,s^{-1}}$, and $v'_3 \approx 500–530\,\mathrm{m\,s^{-1}}$ correspond to the three time delays.

During Space Shuttle launch also three groups of delays were observed (see Figure 9.11). As for Ariane-5 RLs, the duration of disturbances in all groups was 40–90 min, and the periods of prevailing oscillations were changed from 10 to 14 min. Let us now estimate the velocities of propagating disturbances. We account for the fact that the trajectory of Space Shuttle flight essentially differs from the trajectories of Soyuz, Proton, and Ariane-5. The SVs have a trajectory that is more horizontal. Space Shuttle achieves altitudes for the effective generation of waves at about $\Delta t_0 \approx 9$ min after launch. During this time, a rocket goes away from the cosmodrome in the eastward direction at 1000–1500 km s^{-1}, so let us suppose that $\Delta R \approx 1300$ km on average. Then, $\Delta R' \approx 8000$ km. For the corrected time delays, according to the data in Table 9.4, $\Delta t'$ corresponds to the corrected velocities: $v_1' \approx 4.4$–5.5 km s^{-1}, $v_2' \approx 980$–1070 m s^{-1}, and $v_3' \approx 540$–590 m s^{-1}.

It is seen that the magnitude of the velocities for Ariane-5 RLs and for Space Shuttle RLs are sufficiently close. They are close to the velocities observed during RLs from Plesetsk cosmodrome ($v_1' \approx 4.9$–6.1 km s^{-1}, $v_2' \approx 880$–910 m s^{-1}, and $v_3' \approx 325$–400 m s^{-1}), from Baikonur ($v_1' \approx 7.0$–12.0 km s^{-1}, $v_2' \approx 800$–1100 m s^{-1}, and $v_3' \approx 300$–440 m s^{-1}), and from Chinese cosmodromes ($v_1' \approx 2.0$–3.7 km s^{-1}, $v_2' \approx 800$–970 m s^{-1}, and $v_3' \approx 500$–600 m s^{-1}) (see Chapter 8 and Section 9.1, which follow the results obtained in Refs. [5,9,10,15–19,24]).

We suggest that the magnitude of velocity v_1 that depends on the conditions of ionospheric plasma and, generally speaking, on the direction of propagation of the geomagnetic field, is related to slow magnetohydrodynamic (MHD) waves described in Chapter 3 following Refs. [25,28].

The magnitude of the second velocity v_2 that changes from 700 to 1100 m s^{-1} relates, probably, to shock-acoustic gravity waves (s-AGWs) (nonlinear at the beginning of the path) as waves of neutral gas density. Close to the place of generation of waves their relative amplitude is sufficiently large (~0.5–1). Therefore, we call them shock waves. With increase in the length of the wave's path from the place of generation, they are transformed, step by step, into regular AGW (see, e.g., Refs. [26,27]).

The velocity $v_3 \approx 300$–600 m s^{-1} corresponds to linear intrinsic gravity waves (IGWs). There is no doubt that delays Δt_2 and Δt_3 are related to the same wave, but also to two of their neighboring periods. In this case, $\Delta t_3 - \Delta t_2 \approx T$, where T is the period of AGW. If so, the magnitude of $T \approx 50$–60 min was obtained for the cosmodromes (such as Plesetsk and Baikonur) located not far from Kharkov Observatory. For Cape Canaveral and Kourou cosmodromes that are located far from the observatory, namely, about 10,000 km away, the magnitude of T changes from 90 to 100 min.

Now we will try to answer more complicated questions: Can the above-mentioned type of waves propagate to global distances? How fast do they attenuate? Can they cause disturbances to register close to Kharkov city (Ukraine)?

We will start by considering waves that correspond to v_2 and v_3 values. According to our estimates, a decrement of attenuation $\gamma \approx (2.5/3.0) \times 10^{-4}\,\mathrm{k\,m^{-1}}$ happens for these waves. This means that for RLs from Cape Canaveral and Kourou cosmodromes, the wave amplitude decreases according to law $\beta = \exp(\gamma R')$, for $R' \approx 7000$ km, $\beta \approx 5.8$–8.2 and for $R' \approx 9000$ km, $\beta \approx 9.5$–14.9.

Further, AGWs (or VGWs) attenuate due to spatial spreading of waves [28]. During channeling of these waves in the thermosphere waveguide, disturbance of density ΔN and pressure Δp in the atmosphere decreases as $1/\sqrt{R'}$, for Baikonur, $R' \approx 2500$ km; for Cape Canaveral, $R' \approx 7000$ km; and for Kourou, $R' \approx 9000$ km. In this case, magnitude Δp decreases by 1.7 and 1.9 times, respectively, compared to the magnitude for RLs from the Baikonur cosmodrome. Generally speaking, for RLs from the American continent waves attenuate by 10–14 and 19–28 times. If we suggest that $\Delta p/p = 1$ at the place of generation of a GW (VGW), then in the proximity of Kharkov city $\Delta p/p \approx 0.07$–0.1 (launch from Cape Canaveral cosmodrome) and $\Delta p/p \approx 0.03$–0.05 (launch from Kourou cosmodrome).

Disturbances of pressure Δp in the wave yield to WDs of the electron concentration N and the geomagnetic file. For the above-mentioned magnitudes $\Delta p/p$, relative density disturbances are $\Delta N/N \approx 0.01$–0.1. Such effects were found by radiophysical methods [4–8,11,14–20,23].

As our estimates show, for $\Delta N/N \approx 0.01$–0.1, the amplitude of geomagnetic pulsations, generated by WDs N at the diapason of periods 10–20 min, is on the order of several units of nanotesla, which is in good agreement with the measured data observed in Refs. [4,15–19].

Unfortunately, attenuation of small MHD waves is lightened weakly as given in the literature. We can assume that it is not greater than the attenuation of AGWs.

The above-mentioned data allows us to conclude that disturbances caused by RLs from Cape Canaveral and Kourou cosmodromes can attain the standards set at Kharkov Observatory and can eventually be registered. The results of investigations carried out many years ago testify to this, in which one of the authors of this book also took an active part, which were published in Refs. [4–7,11,14–20]. It is important to note that the velocities of propagation of disturbances given by radiophysical and magnetometrical techniques are close to each other. This means that geomagnetic pulsations are generated by disturbances N close to the place of registration of pulsations. Due to this factor, velocities of propagation obtained by geomagnetic observations do not depend (or depend weakly) on the orientation of the track of propagation to geomagnetic field lines.

An additional argument for the usefulness of geomagnetic pulsation generation by RLs is the comparison of histograms of the time delay for the days with RLs and those for the background days (i.e., a day prior to and after RL). As was expected, in the background days a law of distribution of the time delays was close to regular.

It should also be noted that the power of the Space Shuttle is about 3–7 times greater than that of Proton and Soyuz, respectively. This means that the magnitude of the created disturbances at the place of generation is 3–7 times greater than at the start of rockets with lesser power.

As for Ariane-5, its power is only 2.2 times greater than the power of Soyuz. The power of Proton and Ariane-5 is about the same.

9.3 Geomagnetic Effects of RLs from Various Cosmodromes Worldwide

9.3.1 Overview

As mentioned in previous chapters, for more than 40 years researchers gathered their efforts for detecting and studying disturbances in the atmosphere and geospace that accompanied the RLs. This question is illuminated in a set of works [4–12,14–24,28,29]. Some of these works were related to the investigation into the effects that occur in regions close to the rocket trajectory, that is, to the place of its action on the medium [4,8,14]. More interesting, however, is to study the effects that occur far from the place of launch. This question was investigated more precisely by one of the authors of this book along with his colleagues and published recently in Refs. [4–6,10,15–24,28,29]. As a rule, most of these works dealt with ionospheric effects caused by RLs [4–6,10,15–21,23,24,29].

For effects caused by RLs on the geomagnetic field close to the cosmodrome, we are familiar with only one such research (see Ref. [22]), where the scope of geomagnetic disturbance generation was shown with periods of about 140 min and amplitudes of 10 nT. The disturbances appeared 10–15 min after the RL. In this work, the repeated character of geomagnetic field disturbances in the time interval of periods of 2–3 min with the amplitude of 5–20 nT was assumed, occurring 2–3 min after the RL, and showing the ability to generate quasiperiodical disturbances by RLs. All data were obtained by Alma-Ata Observatory (Kazakhstan) from 23 RLs during the period 1999–2001 [22]. As mentioned in Chapter 8, following the results obtained in Ref. [9], we also investigated the reaction of sufficiently high-frequency fluctuations of the geomagnetic field (with the frequencies from 10^{-3} to 1 Hz) on 43 powerful RLs from Baikonur cosmodrome ($R \approx 2100$ km).

As outlined in this chapter, statistical analysis identified three groups of disturbances and determined their main parameters—delays, durations, and periods. The time delays were equal to 6–7, 35–45, and 90–130 min, the durations to 17–27, 45–80, and 40–70 min, and the periods to 5–8, 9–11, and 9–11 min, respectively. To these three groups of disturbances, velocities of 9 km s^{-1}, 800–1100, and 300–400 m s^{-1} corresponded, respectively. In Chapter 8 and Sections 9.1 and 9.2, using the same method, the reaction of the Earth's magnetic field that accompanied RLs at the closest cosmodromes to the place of registration ($R \approx 2100$–2300 km), as well as to far cosmodromes ($R \approx 5000$ km and $R \approx 9500$ km) was analyzed, following Refs. [10,24,28,29]. For this purpose, those cosmodromes were selected which were actively functional and from which sufficiently powerful rockets had been launched. Such cosmodromes as mentioned earlier are Plesetsk (Russia); Sichan, Zeuan', and Ta'uan' (China); Cape Canaveral (FL); and Kourou (French Guiana). Registration of the geomagnetic field temporal variations was done close to Kharkov city (Ukraine) that is ranged from these cosmodromes at distances from 1500 to 9500 km (see Table 9.2). Practically, all RLs were accompanied by essential changes in the character of horizontal component oscillations of the geomagnetic field at the range of geomagnetic pulsations.

Also identified were three groups of disturbances for all cosmodromes. For Plesetsk cosmodrome, the velocities: 4.9–6.1 km s^{-1}, 880–910, and 325–400 m s^{-1} corresponded (see Chapter 8 and Ref. [10]). For the cosmodromes in China—2.0–3.7 km s^{-1}, 800–970, and 500–600 m s^{-1} (see Section 9.1 and Ref. [24]). For Cape Canaveral cosmodrome—4.4–5.5 km s^{-1}, 980–1070, and 540–590 m s^{-1} (see Section 9.2 and Ref. [28]). For the European rockets from Kourou cosmodrome, the magnitudes of velocities were the following: 2.5–3.1 km s^{-1}, 680–740, and 480–540 m s^{-1} (see Section 9.2 and Ref. [28]).

There is a set of questions for which we need to provide an evident answer: What tendencies of changes occur in the parameters of disturbances with increase of distance from the cosmodrome? How far can waves generated by the RLs propagate? What is the magnitude of such geomagnetic disturbances? and so on.

The aim of this chapter is to study and analyze dependences of the delays, durations, and periods of disturbances accompanying RLs from various cosmodromes worldwide (ranged from 1500 to 9500 km) on distance to the desired cosmodrome, using the results of observation of the sufficiently high-frequency fluctuations (from 10^{-3} to 1 Hz) of the geomagnetic field close to Kharkov Observatory, following Ref. [29].

9.3.2 Methods of Analysis

For the investigation, we used data from the magnetometer, which is described in detail in Appendix 3. From the beginning, magnetometric signals corresponding to the *H*- and *D*-components of the geomagnetic field and accounting for the amplitude–frequency characteristics of the magnetometer fluxmeter were transformed into signals *H*(*t*) and *D*(*t*) of geomagnetic field fluctuations. The former are measured in relative units and the latter in absolute values (in nanotesla).

Thereafter, signals *H*(*t*) and *D*(*t*) were passed for further preprocessing through digital filtering with the filter's bandwidths corresponding to, for example, periods of 1–20, 20–100, 100–300, and 300–1000 s and, finally, the wavelet transform (WT) described in Chapter 7 (see also Ref. [22]).

The WT, as shown in Chapter 7 (we repeat it here for recall by changing some notations), can be presented as [22]

$$\hat{W}\left[s(t)\right] \equiv Ws(a,b) = \frac{1}{a^{1/2}} \int_{-\infty}^{\infty} s(t)\psi^*\left(\frac{t-b}{a}\right) e^{-i\omega t} dt, \qquad (9.1)$$

where:

s(*t*) is a signal *H*(*t*) or *D*(*t*)

a and *b* are the parameters of scalability and shifting, respectively

ψ(*) is the basis function of the wavelet, where the symbol "*" denotes a complex conjugate

t is the dimensionless time

As for ψ(*t*), it was used for Morlet's wavelet [22]

$$\psi(t) = \exp\left(-\frac{t^2}{2}\right)\cos\omega t, \qquad (9.2)$$

which is the wave packet with the dimensionless frequency ω [usually select ω = 5 (see Chapter 7)]. This basis is convenient for the detection and analysis of wavelets that follow (i.e., geomagnetic pulsations of type *Pc*).

The time delays were determined for the possible reaction of the geomagnetic field on RLs, its duration, and the spectral content of rising disturbances. Further, using a procedure of averaging, the statistical characteristics of these disturbances for different cosmodromes were calculated and dependences of delays, durations, and periods versus distances to the cosmodromes were compared.

9.3.3 Results of Observations

The reaction of the geomagnetic field on the start of 21 rockets from Plesetsk cosmodrome, 69 rockets from Baikonur cosmodrome, 33 rockets from Chinese cosmodromes, 11 rockets from Cape Canaveral cosmodrome, and 15 rockets from Kourou cosmodrome was studied in this chapter. All RLs took place in the period from 2002 to 2008. The characteristics and main parameters of the corresponding rockets are presented in Appendix 1, and brief information on the cosmodromes is presented in Appendix 2. For the common control of the level of magnetic perturbations, indexes A_p and K_p were used prior to and on the day of the RL. This information was taken from the Internet. The RLs were done on magnetically quiet and magnetically perturbed days.

For example, we will consider five RLs from different cosmodromes worldwide. In Figure 9.14, variations of $H(t)$ and $D(t)$ are presented that accompany median-weight rocket Rokot launch on June 30, 2003, at 14:15 UT from Plesetsk cosmodrome on June 30, 2003. On this day $\sum K_p = 22$. Heavy Russian rocket Proton launch from Baikonur cosmodrome took place on April 9, 2007, at 22:54 h (Figure 9.15). The day of the RL was magnetically perturbed weakly with $\sum K_p = 14$.

The first disturbance (significant decrease of the period from 9–10 to 3–4), clearly fixed the two components, with a delay of 4–5 min and continued for about 42 min for the *H*- and *D*-components, respectively. The next disturbance came 60–64 min later, continued for 34–42 min and had periods of 9 and 13 min for the *H*- and *D*-components, respectively. The third change in the character of the wave process had delays of 108 and 112 min, duration of 38 min, and periods of 11 and 12 min for the *H*- and *D*-components, respectively.

Chinese heavy rocket 2F launch took place on December 29, 2002, at 16:40 h from Zeuan' cosmodrome. The day of the RL was at times magnetically perturbed with $\sum K_p = 23$ (Figure 9.16). The first change in the character of oscillations was observed 35 and 40 min after the RL for the *H*- and *D*-components, respectively. In the *H*-component, repression of the wave process and decrease of the period took place twice (from 14 to 7 min). In the *D*-component, decrease of the period registered. This process continued for 54 min and had periods of 7 min for both the components of the signal. The next disturbance was of 49–60 min duration with delays of 99–107 min. The periods were 8 and 13 min for the *H*- and *D*-components, respectively. A third process appeared as an increase of the period of prevailing oscillations in the *H*-component up to 13 min, and in the *D*-component up to 15 min. It had delays of 161–191 min and duration of 51–57 min for the *H*- and *D*-components of the signal.

Figure 9.17 shows variations of the horizontal components of the geomagnetic field that accompanied Atlantis launch on June 8, 2007, at 23:38 h. On the day

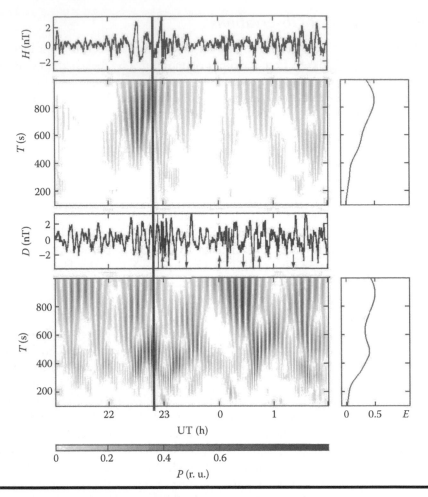

Figure 9.14 Temporal variations of the *H*- and *D*-components of the geomagnetic field, the corresponding wavelet spectra, and the energy diagrams in the range of 100–1000 s accompanying Rokot launch on June 30, 2003. The vertical solid line indicates the moment of the RL (14:15 UT).

of the launch, $\sum K_p = 16$. The SV launch happened at the same time as the passing of the ST close to Kharkov Observatory; therefore, it was difficult to identify the first type of disturbances for this start. Change in the character of oscillations accompanied by increase of the amplitude twice, took place at time interval from 02:10 to 03:25 UT. Other types of disturbances were observed from 03:50 to 05:13 UT for the *H*-component and from 04:15 to 05:15 UT for

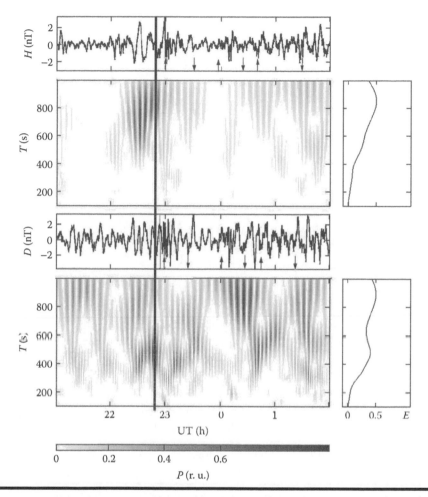

Figure 9.15 **Temporal variations of the *H*- and *D*-components of the geomagnetic field, the corresponding wavelet spectra, and the energy diagrams in the range of 100–1000 s for Proton launch from Baikonur cosmodrome on April 9, 2007 (22:54 UT is indicated by a solid vertical line).**

the *D*-component. The geomagnetic pulsation periods were about 13 min for both types of disturbances.

Fluctuations of the *H*- and *D*-components of the magnetic field accompanied Ariane-5 launch on July 18, 2004, at 00:44 h and are shown in Figure 9.18. On the day of the launch, $\sum K_p = 18$. In the time interval from about 23:55 to 01:40 h, sunrise was observed at an altitude of 200 km and on the ground surface level close to Kharkov Observatory. The purported effects related to the RL

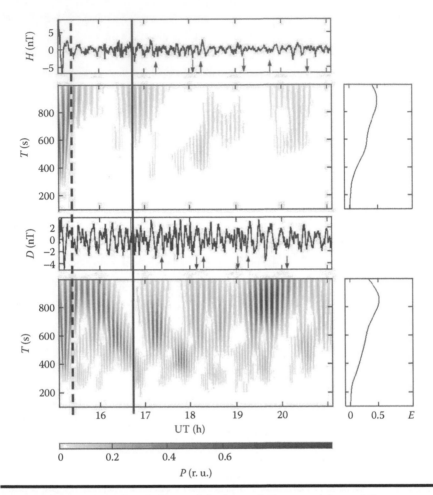

Figure 9.16 Temporal variations of the *H*- and *D*-components of the geomagnetic field, the corresponding wavelet spectra, and the energy diagrams in the range of 100–1000 s for the launch from Zeuan' cosmodrome on December 29, 2002 (16:40 UT). The dashed line indicates the moment of sunrise at an altitude of 200 km close to Kharkov (Ukraine).

start after about 103 and 93 min of launch for the *H*- and *D*-components of the signal, respectively. They were accompanied by an increase in the period (from 6 to 11 min) and the amplitude (twice) in the *H*-component and insufficient repression of the wave process in the *D*-component with periods of 9 min. The duration of disturbance was from 60 to 66 min. Further change in the character of the oscillations was registered 215–228 min after the RL, which continued for

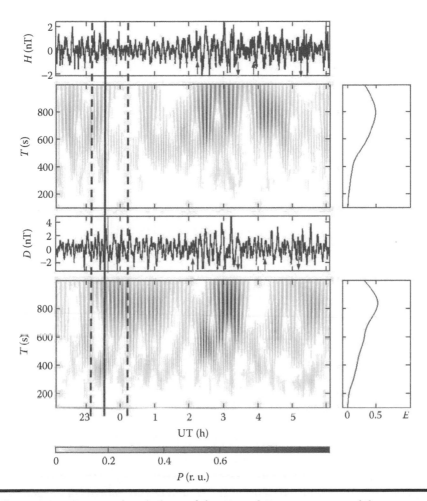

Figure 9.17 **Temporal variations of the *H*- and *D*-components of the geomagnetic field, the corresponding wavelet spectra, and the energy diagrams in the range of 100–1000 s for Space Shuttle launch from Cape Canaveral on June 8, 2007 (23:38 UT is indicated by solid vertical line). The dashed lines indicate the moment of sunrise at an altitude of 200 km and at the ground surface level close to Kharkov (Ukraine).**

about 67 min, in the *D*-component with an increase in the period of prevailing oscillations from 9 to 15 min being fixed, and in the *H*-component decrease in the period from 11 to 8 min. A third disturbance, observed in the both components, had delays of 315–340 min. It continued for 93–108 min, and the periods of oscillations were about 14–15 min for both the components of the signal.

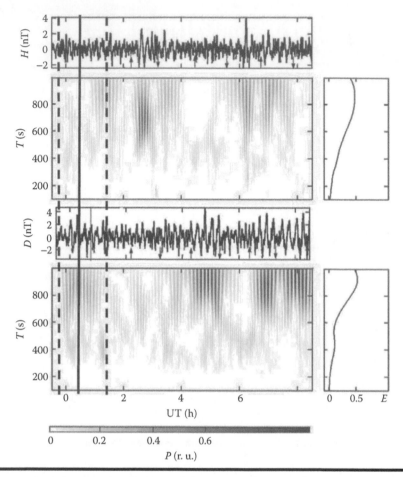

Figure 9.18 Temporal variations of the *H*- and *D*-components of the geomagnetic field, the corresponding wavelet spectra, and the energy diagrams in the range of 100–1000 s for Ariane-5 launch from Kourou cosmodrome on July 18, 2004 (00:44 UT is indicated by a solid vertical line). The dashed lines indicate the moment of sunrise at an altitude of 200 km and at the ground surface level close to Kharkov (Ukraine).

9.3.4 Results of Statistical Analysis

A statistical preprocessing was carried out separately for each cosmodrome as given in Chapter 8 and Sections 9.1 and 9.2, as well as in previous works [9,10,24,28]. Three groups of disturbances were identified and their main parameters defined (see Table 9.7). It is seen from the table that an average magnitude of delays, durations, and periods increased with increase in range from the cosmodrome.

Table 9.7 Statistical Characteristics of the Main Parameters of Disturbances Accompanying the RLs

Parameter (min)	Cape Canaveral	Sichan	Ta'uan'	Zeuan'	Baikonur	Plesetsk	Kourou
Δt_1	36.1 ± 3.3	42.3 ± 3.6	53.5 ± 5.6	40.2 ± 4.1	8.7 ± 0.7	10.1 ± 1.3	59.4 ± 4.9
Δt_2	139.3 ± 5.9	121.1 ± 8.1	117.4 ± 4.2	10.2 ± 6.9	41.6 ± 1.4	30.5 ± 1.1	147.3 ± 14.9
Δt_3	245.9 ± 9.9	153.6 ± 12.0	190.4 ± 4.5	161.7 ± 7.4	103.4 ± 2.3	78.7 ± 5.2	300.6 ± 8.9
ΔT_1	50.8 ± 2.9	45.5 ± 3.0	53.4 ± 5.6	56.7 ± 5.2	22.1 ± 1.8	23.0 ± 1.7	51.8 ± 4.4
ΔT_2	66.9 ± 3.9	45.6 ± 5.0	67.3 ± 8.1	56.1 ± 5.6	53.6 ± 2.2	42.3 ± 1.9	70.5 ± 5.3
ΔT_3	66.8 ± 3.8	62.3 ± 5.0	47.9 ± 4.8	49.6 ± 2.2	52.0 ± 1.5	46.8 ± 1.4	62.5 ± 4.6
T_1	11.1 ± 0.9	10.1 ± 0.9	9.6 ± 0.8	11.0 ± 0.9	6.8 ± 0.7	6.7 ± 0.6	11.7 ± 0.7
T_2	12.4 ± 0.4	9.4 ± 1.1	10.7 ± 0.9	10.0 ± 0.9	9.7 ± 0.3	11.1 ± 0.4	12.1 ± 0.8
T_3	12.0 ± 0.4	13.2 ± 0.8	10.7 ± 1.4	11.0 ± 1.0	10.0 ± 0.4	11.9 ± 0.5	11.9 ± 0.7

As given in Chapter 8 and the earlier sections, we consider the region of effective generation of waves placed at an altitude of 100–150 km and the direction of disturbance propagation is close to horizontal (see also Refs. [5,9,10,15–19,24]). It is necessary to note that during its flight the rocket moves aside from the vertical direction, and the distance from the place of registration to the region of disturbance generation differs from the geographic distance between the observatory and the cosmodrome. For this purpose, we introduce corrections to the delay and to distance from the place of RL. The magnitudes of the corrected distances R' are presented in Table 9.2. Corrections of the distance were introduced accounting for the deviation of the trajectory of the given types of the rockets during time Δt_0 to attain the middle region for effective generation of waves.

Figure 9.19 shows the dependences of the median magnitudes of the delays, durations, and periods along with their deviation intervals. The regression analysis

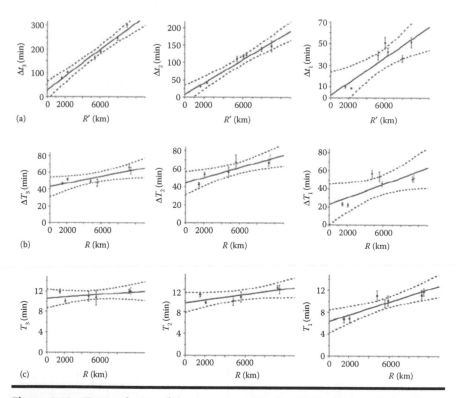

Figure 9.19 Dependence of the mean magnitudes of (a) delays, (b) durations, and (c) periods. A straight solid line indicates the lines of regression, the dashed lines, and the deviation intervals.

of the results of observations was also made. Lines of regression for the delays, durations, and periods of the three groups of disturbances were drawn. The lines of regression for delays of disturbances are described by the following expressions:

$$\Delta t_1 = 2.36 + 0.006 \cdot R', \tag{9.3}$$

$$\Delta t_2 = 7.82 + 0.016 \cdot R', \tag{9.4}$$

$$\Delta t_3 = 28.83 + 0.028 \cdot R', \tag{9.5}$$

for the first, second, and third types of disturbances, respectively. Here, as for other regressions, the temporal characteristics are changed in minutes, and the distances in kilometers.

It is evident from Equations 9.3 through 9.5 that $\Delta t_1(R' = 0) \approx 2.4\,\text{min}$, $\Delta t_2(R' = 0) \approx 8.0\,\text{min}$, and $\Delta t_3(R' = 0) \approx 29\,\text{min}$. The magnitude of $\Delta t_1(0)$ estimated earlier proves that geomagnetic disturbances arise when the rocket achieves altitudes of $z \approx 100$ km. The disturbances of the second type are generated with delays of about 8 min, when the rocket moves nearly horizontally at altitudes of $z \approx 120$–150 km. We note that the difference $\Delta t_3(0) - \Delta t_2(0) \approx 20\,\text{min}$ is close to the period of IGWs generated by the RLs.

From Equations 9.3 through 9.5, it follows that the time delays of the three groups are linearly increased with the distance from the observatory to the place of disturbance generation. The magnitudes of durations and the periods are also increased practically according to linear law.

9.4 Results and Discussions

The regressive analysis of dependences of the main parameters versus distances described in Equations 9.3 through 9.5 showed that the time delays of the three groups of disturbances linearly increased with the increase of distance from the place of observation to the place of disturbance generation (Figure 9.19a).

Using Equations 9.3 through 9.5 for the lines of regression, it is possible to calculate the velocities of waves which correspond to the three groups of delays. The magnitude of the velocity v_i' ($i = 1, 2, 3$) can be described by the following expression:

$$v_i' = \left[\frac{d(\Delta t_i)}{dR'} \right]. \tag{9.6}$$

From Equation 9.6, it is estimated that wave velocity of the three groups of disturbances are $v_1' \approx 2.7$, $v_1' \approx 1.04$, and $v_3' \approx 0.6\,\text{km s}^{-1}$. The magnitude of velocity v'_1 corresponds to slow MHD waves. Such velocities have often been observed through radiophysical techniques [4–6,11,15–21,23]. The velocity v'_2 is associated with the velocity of shock waves, and more precisely to shock-wave AGWs. Close to the source, these waves are nonlinear and their velocity exceeds the velocity of linear waves. For the movement of rocket away from the place of generation, these waves are continuously transformed into regular AGWs [26,27]. The velocity v'_3 has linear IGW, which also was observed earlier (see, e.g., Refs. [5,6,11,15–21]). The proximity of the main parameters of the second and third groups allows us to suppose that they can be parts (two neighboring periods) of the same wave process.

We now raise a question: Can the amplitudes of the above-mentioned types of waves be registered at the place of the observatory? To answer this question, we should note that the obtained magnitudes of velocities are in good agreement with the magnitude of velocities obtained by radiophysical methods [4–6,11,15–21,23]. Hence, geomagnetic field disturbances, in all probability, are generated by disturbances of the electron concentration in the ionosphere close to the location of the observatory and do not depend on the orientation of the wave propagation trace.

As our estimations show, for cosmodromes that are the farthest [28], the amplitude of geomagnetic pulsations generated by the WDs of plasma concentration in the range of periods of 10–20 min is on the order of several nanotesla for waves of the second and the third groups. Just these magnitudes of the amplitude of disturbances were observed during our experimental sessions.

Attenuation of slow MHD waves in the literature was not explained until now. We suggest that the attenuation of these waves is not greater than those of AGWs and IGWs. Regarding their durations and periods, these changed practically according to linear law depending on the distance from the cosmodrome.

Lines of regression for durations of the three types of WDs are shown in Figure 9.19b. As shown in the figure, the magnitudes of ΔT_i ($i = 1, 2, 3$) increased from 20–30 to 60–70 min with increase of the distance from the cosmodrome during the rocket flight. The characteristic distance, at which durations were increased twice, was about 5,000, 15,000, and 21,000 km for the three types of disturbances, respectively. If so, the expressions for regressions can be expressed as follows:

$$\Delta T_1 = 22.81 + 0.004R, \tag{9.7}$$

$$\Delta T_2 = 44.11 + 0.003R, \tag{9.8}$$

$$\Delta T_3 = 42.98 + 0.002R. \tag{9.9}$$

Regarding the mean magnitudes of the prevailing pulsation periods, these increase linearly in the narrow interval (7–12 min) with increase in the distance from the cosmodrome during the rocket flight. The magnitude of the periods is limited from the bottom by the period of Brunt–Väisälä value (see Chapter 3 and Refs. [26,27]), but from the top by the magnetometer fluxmeter frequency characteristic ($T_{max} = 17$ min). The lines of regression for the periods of the three types of disturbances are described by the following expressions:

$$\Delta T_1 = 6.32 + 5.87 \times 10^{-4} R, \tag{9.10}$$

$$\Delta T_1 = 9.65 + 2.46 \times 10^{-4} R, \tag{9.11}$$

$$\Delta T_1 = 6.32 + 5.87 \times 10^{-4} R. \tag{9.12}$$

We assume that the physical cause for the increase in duration and period is because of the existence of dispersion of the phase wave velocity in the medium.

Finally, we raise another question: Are the observed pulsations the reaction of the medium on RLs? The following arguments can prove this statement. First of all, the law of distribution of time delays has expressed the maxima well. In the background days (prior to and after RLs), this law is close to regular [9,10,24,28]. Second, the computed magnitudes of velocities are in a good agreement with the velocity magnitudes obtained by other methods [4–6,11,12, 15–21,23]. Third, the lines of regression for delays of the three groups of disturbances are represented by straight lines.

All of these allow us to state that the observed changes in the signal character evidently relate to the RLs.

9.5 Main Results

1. Variations of the geomagnetic field following 33 RLs from Sinchan, Zeuan', and Ta'uan' cosmodromes in China. It was stated that practically in all cases after RLs essential or significant changes in the character of oscillations were observed.

2. As a result of the statistical analysis, three groups of disturbances were identified, having time delays of 39–53, 100–129, and 150–180 min. To these data, process durations of 30–50, 37–45, and 38–57 min and the periods of these processes 9–12 min corresponded.

3. In assuming that the disturbances were caused by RLs, the velocities of propagation of the three groups of disturbances $v'_1 \approx 2.0$–2.7, $v'_2 \approx 0.8$–0.97, and $v'_3 \approx 0.5$–$0.6 \, \mathrm{km \, s}^{-1}$ were determined. Such velocities have slow MHD waves, shock-wave AGWs, and IGWs, respectively.

4. Noncontradiction of the obtained parameters of disturbances with the results of our own research and with the results of studies by other authors allows us to state with a sufficient degree of confidence that the methods

described in this chapter on disturbances can be related to the RLs from Chinese cosmodromes.

5. The characteristics of geomagnetic pulsations that accompanied 11 Space Shuttle and 15 Ariane-5 RLs from Cape Canaveral and Kourou cosmodromes during the period of 2002–2007 were studied. It was stated that practically all RLs were accompanied by generation of pulsations at the period diapason of 10–15 min. The amplitude of pulsations was of some nanotesla.

6. As a result of the statistical analysis of observed data, three groups of disturbances having delays of 13–58, 133–163, and 193–253 min during RL from Cape Canaveral cosmodrome and delays of 13–88, 193–238, and 300–330 min for RL from Kourou cosmodrome were identified.

7. By assuming that geomagnetic field disturbances were caused by RLs, the velocities of propagation that correspond to these three groups of delays were determined. For the American rockets, these velocities were close to 4.4–5.5, 0.98–1.07, and 0.54–0.59 km s^{-1}. For the European rockets, the velocities were 2.5–3.2, 0.68–0.74, and 0.48–0.54 km s^{-1}.

8. Noncontradiction of the obtained parameters of disturbances with the results of our research and with the results of studies by other authors allows us with a sufficient degree of confidence to outline the method described in this chapter that geomagnetic pulsations can be generated during the flight of the rockets launched from Cape Canaveral and Kourou cosmodromes.

9. Statistical and spectral analysis of the level of geomagnetic pulsations accompanied 149 RLs from cosmodromes far away from the place of registration at distances of ~1500–9500 km showed that after the RL it changed the character of pulsations notably, that is, it increased (rarely was decreased) their level and essentially changed their spectral content.

10. Three groups of disturbances were identified. Their time delays increased linearly with increase of distance from the place of disturbance generation to the place of observations. These disturbances corresponded to the mean velocities of about 2.2, 1.0, and 0.6 km s^{-1}. The first group of disturbances was not registered so confidently. The second and the third disturbances were able to present the same group of disturbances.

11. The duration of disturbances was usually changed from 20–40 to 60–70 min with increase of distance from the observatory to the cosmodrome, R, from about 1500 to 9500 km, respectively. The duration of disturbances increased according to nearly linear law. The characteristic distance at which the duration of disturbances doubled was about 5,000, 15,000, and 21,000 km for the three groups of disturbances mentioned earlier.

12. The magnitude of the prevailing periods of pulsations practically changed linearly with increase in distance R from 1500 to 9500 km, but these changes were insignificant.

References

1. Chernogor, L. F., Physics of the Earth, atmosphere and geospace in the lightening of the system paradigm, *Radiophys. Radioastron.*, 8, 59–106, 2003 (in Russian).
2. Chernogor, L. F., The Earth–atmosphere–ionosphere–magnetosphere as an open dynamic nonlinear system (Part 1), *Nonlinear Univ.*, 4, 655–697, 2006 (in Russian).
3. Chernogor, L. F., The Earth–atmosphere–ionosphere–magnetosphere as an open dynamic nonlinear physical system (Part 2), *Nonlinear Univ.*, 5, 55–97, 2007 (in Russian).
4. Afra'movich, E. L., Perevalova, N. P., and Plotnikov, A. V., Registration of ionospheric responses on the shock-acoustic waves generated during launchings of the rocket-carriers, *Geomagn. Aeronom.*, 42, 790–797, 2002.
5. Burmaka, V. P., Kostrov, L. S., and Chernogor, L. F., Statistical characteristics of signals of the Doppler HF radar during sounding of the middle ionosphere perturbed by the rockets' launchings and by the solar terminator, *Radiophys. Radioastron.*, 8, 143–162, 2003 (in Russian).
6. Garmash, K. P., Gokov, A. M., Kostrov, L. S., et al., Radiophysical investigations and modeling of processes in the ionosphere perturbed by sources of different nature. 2. Processes in the artificially perturbed ionosphere. Variations of the characteristics of radio signals. Modeling of disturbances, *Radiophys. Electron.*, 427, 3–22, 1999 (in Russian).
7. Zasov, G. F., Karlov, V. D., Romanchuk, T. E., et al., Observations of disturbances in the low ionosphere during experiments on the program Soyuz–Apollo, *Geomagn. Aeronom.*, 17, 346–348, 1977.
8. Nagorsky, P. M. and Tarashuk, Yu. E., Artificial modification of the ionosphere during launchings of rockets leading out at the orbit the space vehicles, *Izv. Vuzov. Radiofizika*, 94–106, 1993 (in Russian).
9. Garmash, K. P., Leus, S. G., Chernogor, L. F., and Shamota, M. A., Variations of the geomagnetic field accompanied launchings and flights of the space vehicles, *Cosmic Sci. Technol.*, 13, 87–98, 2007 (in Russian).
10. Chernogor, L. F. and Shamota, M. A., Wave disturbances of the geomagnetic field accompanied launchings of the rockets from the cosmodrome Plesetsk, *Cosmic Sci. Technol.*, 14, 29–38, 2008 (in Ukrainian).
11. Kostrov, L. S., Rozumenko, V. T., and Chernogor, L. F., Doppler radio sounding of disturbances in the middle ionosphere accompanied launchings and flights of space vehicles, *Radiophys. Radioastron.*, 4, 227–246, 1999 (in Russian).
12. Arendt, P. R., Ionospheric undulations following Apollo 14 launching, *Nature*, 231, 438–439, 1971.
13. Foster, J. C., Holt, J. M., and Lanzerotti, L. J., Mid-latitude ionospheric perturbation associated with Spacelab-2 plasma depletion experiment at Millstone Hill, *J. Ann. Geophys.*, 18, 111–120, 2000.
14. Nagorsky, P. M., Inhomogeneous structure of the *F* region of the ionosphere generated by the rockets, *Geomagn. Aeronom.*, 38, 100–106, 1998.
15. Burmaka, V. P., Taran, V. I., and Chernogor, L. F., Results of complex radiophysical observations of wave disturbances in geospace accompanied launchings and flights of the rockets, *Cosmic Sci. Technol.*, 9, 57–61, 2003 (in Russian).

16. Burmaka, V. P., Taran, V. I., and Chernogor, L. F., Wave processes in the ionosphere accompanied launchings and flights of rockets at the background of natural transmission processes, *Geomagn. Aeronom.*, 44, 518–534, 2004.

17. Burmaka, V. P., Taran, V. I., and Chernogor, L. F., Complex radiophysical investigations of wave disturbances in the ionosphere accompanied launchings of the rockets at the background of the natural non-stationary processes, *Radiophys. Radioastron.*, 9, 5–28, 2004 (in Russian).

18. Burmaka, V. P., Chernogor, L. F., and Chernyak, Yu. V., Wave disturbances in geo-space accompanied launchings and flights of the rockets "Souz" and "Proton," *Radiophys. Radioastron.*, 10, 254–272, 2005 (in Russian).

19. Burmaka, V. P., Lisenko, V. N., Chernogor, L. F., and Chernyak, Yu. V., Wave processes in the F-region of the ionosphere accompanied starts of rockets from the cosmic port Baikonur, *Geomagn. Aeronom.*, 46, 783–800, 2006.

20. Garmash, K. P., Kostrov, L. S., Rozumenko, V. T., et al., Global perturbations of the ionosphere caused by launchings of rockets at the background of the magnetic storm, *Geomagn. Aeronom.*, 39, 72–78, 1999.

21. Garmash, K. P., Leus, S. G., Pazura, S. A., Pohil'ko, S. N., and Chernogor, L. F., Statistical characteristics of fluctuations of the Earth's electromagnetic field, *Radiophys. Radioastron.*, 8, 163–180, 2003 (in Russian).

22. Sokolova, O. I., Krasnov, V. M., and Nikolaevsky, N. F., Changes of geomagnetic field under influence of rockets' launch from the cosmic port Baikonur, *Geomagn. Aeronom.*, 43, 561–565, 2003.

23. Chernogor, L. F., Garmash K. P., Kostrov L. S., et al., Perturbations in the iono-sphere following U.S. powerful space vehicle launching, *Radiophys. Radioastron.*, 3, 181–190, 1998 (in Russian).

24. Chernogor, L. F. and Shamota, M. A., Geomagnetic pulsations accompanied the rocket launchings from cosmodromes of China, *Cosmic Sci. Technol.*, 14, 92–101, 2008 (in Russian).

25. Sorokin, V. M. and Fedorovich, G. V., *Physics of Slow MHD Waves in the Ionospheric Plasma*, Moscow: Energoizdat, 1982 (in Russian).

26. Gershman, B. N., *Dynamics of the Ionospheric Plasma*, Moscow: Nauka, 1974, 256 (in Russian).

27. Gossard, E. E. and Hooke, W. H., *Waves in the Atmosphere: Atmospheric Infrasound and Gravity Waves: Their Generation and Propagation*, Amsterdam: Elsevier Scientific Publishing Co., 1975, 476.

28. Chernogor, L. F. and Shamota, M. A., Geomagnetic pulsations accompanied the rocket launchings from cosmodromes at Cape Canaveral and Kuru, *Cosmic Sci. Technol.*, 14, 89–98, 2008.

29. Garmash, K. P., Leus, S. G., Chernogor, L. F., and Shamota, M. A., Geomagnetic pulsations accompanied starts of the rockets from various cosmodromes of the world, *Cosmic Sci. Technol.*, in press.

Chapter 10

Peculiarities of Magneto-Ionospheric Effects during Launch of a Group of Rockets

10.1 Overview

As mentioned in the Introduction, active experiments (AEs) are convenient for the study of dynamic processes in the system that is the Earth–atmosphere–ionosphere–magnetosphere (EAIM) and mutual interactions of its subsystems (see also Refs. [1–4]). In this case, it is possible to select a place and time of AE, control, dozing energy of the source, and so on. From the beginning, we have pointed out that one of the convenient sources of energy allocation can be rocket-space systems with working engines that are launched from different cosmodromes worldwide. It is important to note that for the observer this source of the whole complex of dynamic processes in all subsystems of the system EAIM is "cost-effective." Various ranges of cosmodromes, different powers of rockets, their fuels, and various trajectories guarantee variations of the reaction of each subsystem as also the whole system of EAIM on the action of such source.

We mentioned in the previous chapters that the reaction of the near-the-Earth medium on rocket launches (RLs) was studied for a sufficiently long time (see also Ref. [5] and the references therein). The described effects in the medium caused by the launch and flight of only a single rocket is given in all well-known works. Wave disturbances (WDs) occur in the ionosphere at distances more than 2000 km from the cosmodrome, having velocities of hundreds of meters per second, and sometimes, even tens of kilometers per second with a full duration of up to 203 h.

The aim of this chapter is to describe the results of observations of ionospheric plasma and the geomagnetic field reactions that accompany the launches of four powerful rockets, each after the other, from different ground-based launch facilities (this chapter is written on the basis of Ref. [6]).

10.2 Description of Active Experiments

A global multiaimed campaign "Security 2004" [7] was carried out in Russia on December 18, 2004. During this campaign, three ballistic missiles were launched from a ground-based facility and one ballistic missile from a sea-based facility with a time interval from 07:05 to 10:28 UT (here and in further instances time is expressed as Universal Time or UT). For geophysics, this was a unique opportunity to investigate such a global program. One of the authors of this book took part in this experimental campaign.

The main parameters and characteristics of the rocket, whose effects are analyzed in this chapter, are presented in Appendix 1, from where it can be seen that the Molnia rocket is a heavy type of rockets, Rokot rocket is of median weight, and Topol and Sineva rockets are of light weight.

The power and energy flowing out from the engines of straits of the rocket are presented in Appendix 1 and can be estimated from the following relationships:

$$P_f = q \frac{dm}{dt}, \tag{10.1}$$

$$E_f = qm, \tag{10.2}$$

where:

$q \approx 10$ MJ kg^{-1} is the dozen energy-consisting fuel (i.e., fuel mixture and ozone), which for simplification was taken as equal to liquid-fueled and solid-fueled rockets

dm/dt is the output of the fuel

m is the mass of the fuel

The last can be calculated according to the formula:

$$m = \frac{dm}{dt} \Delta t_f, \tag{10.3}$$

where:

Δt_f is the working time of the engine of the corresponding strait

10.3 Methods of Analysis

Following the results obtained in Refs. [7,8], in this chapter we use a special algorithm of data preprocessing and format of data presentation for analysis of temporal variations of the DSs central frequency, obtained with an error of about 0.017 Hz, as well as for the study of the horizontal component level of the geomagnetic field, $H(t)$ and $D(t)$ recorded by the magnetometer. Temporal variations of the signal parameter, results of spectral analysis with the help of the window Fourier transform (WFT), wavelet transform (WT), and adaptive Fourier transform (AFT) described in detail in Chapter 7 are presented in the format. Later, we will repeat some formulas needed for further explanation of the concept.

Thus, for computation of signal spectra $f(t)$ based on WFT, WT, and AFT, the corresponding relationships were used (see also Refs. [7–12]):

$$S[f(T,\tau)] \equiv A_v f(a,\tau) = \sqrt{\frac{2}{t_{Ws}}} \int_{-\infty}^{\infty} f(t)g\left(\frac{t-\tau}{t_{Ws}/2}\right)\exp\left\{-i\frac{2\pi t}{T}\right\}dt, \quad (10.4)$$

$$\ddot{W}[f(t)] \equiv Wf(a,b) = \frac{1}{a^{1/2}} \int_{-\infty}^{\infty} f(t)\psi^*\left(\frac{t-b}{a}\right)\exp(-i\omega t)dt, \quad (10.5)$$

$$A_v f(T_v,\tau) = \sqrt{\frac{2}{vT_v}} \int_{\infty}^{\infty} f(t)g\left(\frac{t-\tau}{vT_v/2}\right)\exp\left\{-i\frac{2\pi(t-\tau)}{T_v}\right\}dt, \quad (10.6)$$

where:

$g(*)$ is the window function
t_{wS} is the window width for the WFT
$T = 2\pi/\omega$ and $T_v = 2a/v$ are the magnitudes of the periods of oscillations
ψ^* is the complex conjugate of the function ψ

The latter function is the basic wavelet function on which WFT from Equation 10.5 has been expanded. We present it in the following form (see also Chapter 7):

$$\psi_{a,b}(t) = \frac{1}{a^{1/2}}\psi\left(\frac{t-b}{a}\right), \quad (10.7)$$

where:

a is the parameter of scalability ($a > 0$)
b is the parameter of shifting

It is convenient in computation to treat the relative periods $\tilde{T} = T/t_{wS}$ and $\tilde{T}_v = T_v/t_{wS}$, and the index v will be omitted in future. All other magnitudes can be presented in relative units.

As shown in Chapter 7, for the WFT and AFT we will use Hemming's window function $g(t)$ that has well-known advantages (see, e.g., Ref. [10]) which can be presented as follows:

$$g_H(t) = \gamma\left[0.54 + 0.46\cos(\pi t)\right], \tag{10.8}$$

where:

$\gamma \approx 1.12$ is the normalized product factor

For the WT, we use $\psi(t)$ for the basic function, Morlet's wavelet, which can be presented by the well-known expression [12,30]:

$$\psi(t) = \exp\left(-\frac{t^2}{2}\right)\cos(5t). \tag{10.9}$$

Here, as shown earlier, t is dimensionless time. The wavelet (Equation 10.9) is convenient for the analysis of short wavelets, which, as a rule, are observed in the ionosphere and in geomagnetic field variations.

For the study of the dynamic processes in the near-the-Earth medium and their relative energies, together with the spectra described in Equations 10.4 through 10.6, the following spectrograms were analyzed (see also Chapter 7):

$$P_S f(\omega, \tau) = |Sf(\omega, \tau)|^2, \tag{10.10}$$

$$P_W f(a, b) = |Wf(a, b)|^2, \tag{10.11}$$

$$P_A f(a, \tau) = |A_v f(a, \tau)|^2. \tag{10.12}$$

As is known, a spectrogram presents by itself the two-dimensional density of energy of the signal $f(t)$ under investigation. Except the functions, $P_S(\tilde{T}, \tau)$, $P_A(\tilde{T}, \tau)$, and $P_W(\tilde{T}, \tau)$, we use the dependences of the following measures in our further discussions:

$$E_S(\tilde{T}) = \int_{-\infty}^{\infty} P_S(\tilde{T}, \tau)d\tau, E_A(\tilde{T}) = \int_{-\infty}^{\infty} P_A(\tilde{T}, \tau)d\tau, \tag{10.13a}$$

$$E_{\mathrm{W}}(\tilde{T}) = \int\limits_{-\infty}^{\infty} P_{\mathrm{W}}(\tilde{T},\tau)\mathrm{d}\tau. \qquad (10.13\mathrm{b})$$

It is known that the local maxima in dependences $E_{\mathrm{S}}(\tilde{T})$, $E_{\mathrm{A}}(\tilde{T})$, and $E_{\mathrm{W}}(\tilde{T})$ indicate the energy distribution on the spectrum (e.g., on the interval of the periods) during the entire time of analysis, and thereby indicate the existence of the definite harmonic components with corresponding periods in the signal under investigation. These dependences are naturally called *energy diagrams* [8,9].

The combined use of the WFT, WT, and AFT is done because for the time limit of the processing they add to each other and allow obtaining much more specific information on the frequency–temporal (i.e., temporal–periodical) localization of the complicated signal components, as well as to watch the dynamics (i.e., evolution) of its components.

10.4 Condition of the Space Weather

For estimation of the condition of the space weather, as given in Chapter 7, the temporal weekly varying solar wind parameters were analyzed. the concentration of particles, n_{sw}; temperature, T; the radial velocity, V_{sw}; dynamic pressure, p_{sw} (computations); concentration of particles, N_{pr} (measured by ACE Satellite—solar wind electron proton alpha monitor); densities of proton flows, Π_{pr} [measured by GOES-8 (W75)] and electrons, Π_{e} (measured by GOES-12); components of the vector modules, B_{z} and B_{t}, of the interplanetary magnetic field (measured by ACE Satellite—magnetometer); the computed magnitudes of the energetic Akasofu function, ε; a component of geomagnetic field, H_{p} (measured by GOES-12); an index D_{st} (measured by WDC-C2 at Kyoto University for geomagnetism); an index of the aurora activity AE (measured by WDC at Kyoto University); and an index K_{p} (measured by Air Force Weather Agency) (Figure 10.1).

On February 18, 2004, the concentration, temperature, velocity of particles, and pressure of the solar wind happened to be typical for nonperturbed conditions. Charge particle flows also corresponded to the quiet conditions. Disturbances between the interplanetary and the geomagnetic fields were insignificant.

Particularly, prior to 14:00 h indexes $K_{\mathrm{p}} \approx 1\text{–}2$ and $D_{\mathrm{st}} \approx -(10\text{–}20)\mathrm{nT}$. The power of the solar wind that entered the magnetosphere and described by the average function of Akasofu was about 5 GJ s^{-1} (gigajoules per second). Therefore, we can state that the natural disturbance of the geospace on the day of the AE was weak.

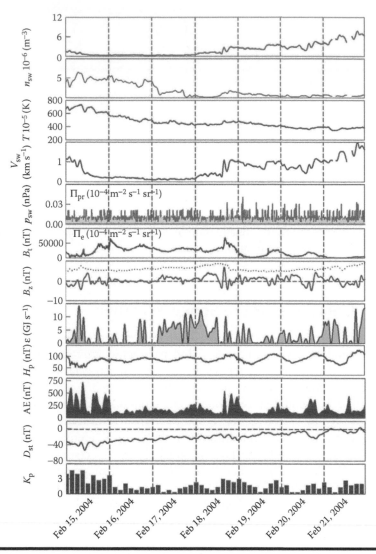

Figure 10.1 Temporal variations of solar wind parameters: ε, function of Acasofu (computation); Π_e, flow of electrons (GOES-12); Π_{pr}, flow of protons [GOES-8 (W75)]; AE, index of the aurora activity (measured by WDC-C2 for geomagnetism of Kyoto University); B_t and B_z, components of vector modules of the interplanetary magnetic field (measured by ACE Satellite—magnetometer); D_{st}, index (WDC-C2); H_p, component of geomagnetic field (measured by GOES-12); K_p, index (measured by Air Force Weather Agency); n_{sw}, concentration of particles (measured by ACE Satellite—solar wind electron–proton alpha monitor); p_{sw}, dynamic pressure (computations); T, temperature; V_{sw}, radial velocity.

10.5 Results of Observations of Launches in Russia

10.5.1 Molnia-M Launch

The task of this rocket that started from the Plesetsk cosmodrome at 07:05 h was placement of a communications satellite of Cosmos series into orbit. Let us describe the temporal variegations of DS presented in Figure 10.2, and the results of the spectral analysis shown in Figure 10.3. In the morning at about 06:40 h, WDs with periods of about 15–20 and 50–70 min took place in the ionosphere. Approximately, 20 min prior to the RL, variations of DS were insignificant.

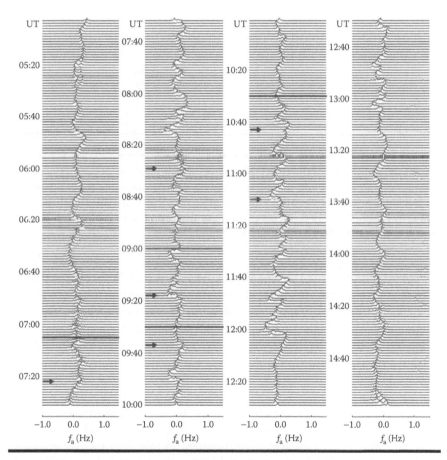

Figure 10.2 Temporal variations of DS at the frequency 2.4 MHz during Molnia-M launch on February 18, 2004, for the strobe, corresponding to the diapason of altitudes from 250 to 325 km. The moment of the RL is denoted by solid lines. The arrows indicate the moments of the beginning of changes in the signal character.

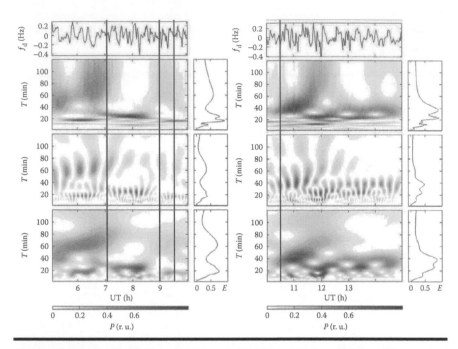

Figure 10.3 Results of the spectral analysis of temporal variations of the central frequency $f_d(t)$ in DSs and the corresponding energy diagrams (panels from top to bottom): temporal variations of $f_d(t)$ after moving away the trend results of WFT, WT, and AFT. Here, *a* corresponds to the time interval of 05:00–10:00 h, and *b* corresponds to 10:00–15:00 h.

Seventeen minutes after the RL, WD activity increased: WDs appeared corresponding to maximal variations of the central frequency of DS, $f_d \approx 0.2$ Hz. The periods from 5 to 10 min and from 20 to 27 min were predominant in the spectrum of WDs. In the time interval of 08:10 to 08:15 h, the magnitude of f_d decreased by about 0.5 Hz and returned to its steady-state condition after 2 min.

For the next 15 min, $f_d \approx 0$. A change in the character of DS variations was observed from 08:30 to 09:18 h. Let us describe the temporal variations of the geomagnetic field presented in Figure 10.4 and the results of the spectral analysis presented in Figure 10.5. As can be seen from the presented illustrations, the levels of both the components changed according to quasiperiodical law with period $T \approx 7$–12 min prior to the RL. Approximately 24 min after the RL, repression of the existing oscillation registered, which continued for 7–11 min. The next change in the signal character began 42 min after the RL and continued for 15 and 30 min for the *D*- and *H*-components, respectively. In the time interval of 08:05 to 09:05 h, oscillations with period $T \approx 10$–15 min were observed in

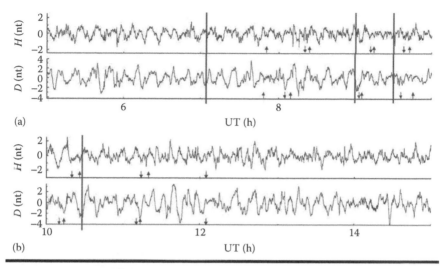

Figure 10.4 Temporal variations of the *H*- and *D*-components of the geomagnetic field that occurred on February 18, 2004. The moments of RLs are denoted by vertical solid lines. Arrows ↑ and ↓ indicate moments of the beginning (a) and ending (b) of changes in the signal character, respectively.

the *D*-component. The two groups of rising disturbances are seen well in the spectrograms (mostly on spectrograms obtained with AFT).

10.5.2 Rokot Launch

This rocket started at 09:00 h from the mine-ring setup located at Baikonur cosmodrome. The first change in the character of DS temporal variations occurred approximately 18 min after the RL (Figures 10.2 and 10.3). For the next 17 min, $f_d \approx 0$. Then, WDs appeared with $T \approx 10$ min that were observed up to 10:27 h. The behavior of geomagnetic field temporal variations, as can be seen from Figures 10.4 and 10.5, was as follows. At 7–10 min after Rokot launch, the character of the signal variations changed: for the *D*-component the signal level decreased approximately twice, and it took a high-frequency form. This behavior went on for up to 10:15 h. For the *H*-component, in the time interval from 09:10 to 09:40 h, the signal level also decreased twice and took a high-frequency form.

10.5.3 Sineva Launch

The Sineva RL occurred at 09:30 h from Karelia submarine which was at its acvatory in the Barents Sea. The distance *R* from the observatory to the place

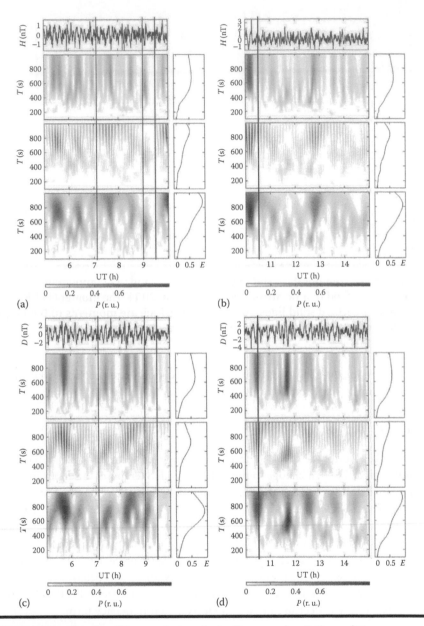

Figure 10.5 Results of the spectral analysis of temporal variations of *H*-component (a, b) and *D*-component (c, d) of the geomagnetic field and the energy diagrams corresponding to them (panels from top to bottom): temporal variations of *H*(*t*) and *D*(*t*) results of WFT, WT, and AFT. Here, a and c correspond to time interval 05:00–10:00 h, and b and d correspond to time interval 10:00–15:00 h.

of RL was about 3000 km. The rocket flight passed through a nonstate regime, because at 98 s after the RL, it registered a deviation from the given trajectory when the rocket had attained an altitude of 10 km. A command for liquidation by the corresponding system was automatically created and the rocket exploded. About 15 t of fuel remained in the rocket along with a dozen flammable substance close to 10 MJ kg^{-1} in weight. If all these exploded, the energy of explosion would be about 150 GJ, which is equivalent to an explosion of 40 t of TNT.

The first change in the character of DS variations after the RL, which correspond to WD with $T \approx 10-12$ min, rose at about 09:37 h and continued up to 10:44 h (see Figures 10.2 and 10.3). After that, the WD period decreased up to 5–6 min. This was observed for about 50 min.

Let us describe variations in the magnetometric signals (Figures 10.4 and 10.5). A signal remained in a high-frequency state up to 10:15 h and its amplitude did not exceed 1–1.5 nT for the D-component. In the time interval 10:15–11:15 h, oscillations with $T \approx 14-18$ min prevailed in the signal. The level of the H-component was also in the limits of 1–1.5 nT. From 10:00 to 10:17 h, its level increased sharply—an M-type response with a quasi-period magnitude of about 8–9 min arose.

The most intensive wave packet that was seen well in spectrograms was observed in the time intervals from 09:50 to 10:15 h and from 10:20 to 10:35 h for the H- and D-components, respectively. Its period of oscillation was 9–13 min.

10.5.4 Topol Launch

The RL happened at 10:28 h from Plesetsk cosmodrome. The first change in the character of the DS variations took place at 10:44 h (see Figures 10.2 and 10.3). In this case, instead of one oscillation with $T \approx 5$ min another one came with $T \approx 9$ min. Further, the amplitude of the oscillation process sharply increased with a period of 35–40 min. This status continued for 40–60 min.

The second change in the character of the DS variations registered in the time interval of 11:10–11:30 h. In this case, the oscillation process had a period of about 5–6 min. Further changes in the character of DS variations were observed from 11:30 to 12:10 h. The amplitude of the oscillations essentially increased by 10–20 min period and then by 25–30 min period.

Relatively short-period ($T \approx 5$ min) oscillations of f_d also took place from 12:30 to 13:04 h. Later, variations of f_d with 15–20 min period registered. We add that the sunset at the ground surface level near the observatory took place at about 15:00 h. Spectrograms for D-component showed the process with $T \approx 7-12$ min clearly protruding, which was observed in the time interval 11:40–11:50 h.

10.6 Results and Discussions

Before entering into discussions on the results of observations, let us note from the outset that the magneto-ionospheric disturbances relating to RLs principally do not differ from natural disturbances at distances exceeding $R \sim 1500$–2000 km from the cosmodromes. If they are close to the location of the RL the amplitude of disturbances can essentially exceed the level of natural disturbances far (for $R > 2000$ km) from the cosmodromes and the artificially excited disturbances can be weaker in comparison to the natural ones. This is the main problem with their selection. Therefore, to obtain a mechanism of precise prediction of the processes caused by RLs, researchers should increase the number of observations and try to find repeated features and principles. The latter consist of a similar character of rising signal characteristic variations, arrangement of time delays of disturbances around the definite temporal magnitudes into groups, their durations and periods, and so on during RLs carried out in similar ambient conditions from the same cosmodrome and for the same type of rockets, and so on. Expert analysis of RLs in our previous works [1,2,4,13–20] was done in such a manner, which will be summarized later in this section.

10.6.1 Effects of the Ionosphere

The analysis of the temporal DS variations showed that a wave activity increased notably in the time interval from about 07:20 to 12:10 h. The periods of oscillation became shorter (up to 5 min), and the amplitude of the Doppler frequency shifting (DFS), f_{da}, increased up to 0.2–0.3 Hz (in separate time intervals—up to 0.4 Hz). A well-known formula for DFS can be used (see also Chapter 7) for estimation of the relative disturbance amplitude of electron concentration, δ_N, in WDs.

$$f_d = -2\frac{f}{c}\frac{d}{dt}\int_0^{z_0} n(t,z)dz, \tag{10.14}$$

where:
   z_0 is the height of radiowave reflection
   n is the index of refraction of the wave
   z is the current height

For vertical sounding of the ionosphere by a radiowave with ordinary polarization, the main contribution of the Doppler effect gives a sufficiently narrow region of altitude in the proximity of height z_0. Then, the DFS amplitude caused by passing of the harmonic WD (close to the altitude of radiowave reflection) is true according to the following equation [16,20]:

$$f_{\mathrm{da}} \approx \frac{4\pi}{T}\frac{f}{c}\delta_N L, \qquad (10.15)$$

where:

f is the working frequency of the radiowave

T is the period of the prevailing oscillation

L is the thickness of the layer providing the main contribution to the DFS

Thus, for $T \approx 8$ min, $f \approx 3$ MHz, and $f_{\mathrm{da}} \approx 0.4$ Hz, we get $\delta_N \approx 7\%$, whereas for a typical magnitude of $f_{\mathrm{da}} \approx 0.2$–$0.3$ Hz, observed during this AE, $\delta_N \approx 3\%$–5%.

Let us now estimate the characteristic velocities of disturbance propagation caused by RLs and flights. After Molnia RL, the first change in the character of DS variations took place after $\Delta t_1 \approx 18$ min. If we account for the plasma disturbance that is generated only after the rocket attains ionospheric altitudes, the corrected time $\Delta t_1' = \Delta t_1 - \Delta t_0 \approx 15$ min, where $\Delta t_0 \approx 3$ min is the time of movement of the rocket to an altitude of $z \approx 100$ km. Then for the distance of the rocket from the place of observation $R \approx 1500$ km, $v_1' \approx 1.7$ km s^{-1}. The second change in the character of DS temporal variations registered after a time of $\Delta t_2 \approx 47$ min post RL. In this case $v_2' \approx 0.57$ km s^{-1}.

After Topol launch, we get $\Delta t_2 \approx 60$ min and $v_2' \approx 0.44$ km s^{-1}. The magnitude Δt_1 was impossible to estimate definitely because of disturbance caused by the rocket Topol; probably the disturbance was caused by the explosion of Sineva. We will consider the explosion of Sineva more in detail later.

The energy output during the explosion of Sineva probably did not exceed the equivalent of 40 t of TNT. It is important to observe that the explosion happened at an altitude of about 10 km where density of the air is approximately 4–5 times less than that near the ground surface. Therefore, according to its parameters and spatiotemporal scales, this explosion was nearly equivalent to a ground explosion with an energy output corresponding to 160–200 t of TNT. For comparison, we note that in the classical project MASSA (explosion was made under the ground), for which complex experimental and theoretical investigations of the dynamic processes in EAIM were carried out, the energy output was equivalent to 251 t of TNT (see Refs. [13,21]).

Therefore, the explosion of Sineva was sufficiently powerful to affect the whole EAIM system. For example, following Ref. [5], we will estimate the characteristic spatiotemporal scales of the shock wave caused by such an explosion. Thus

$$R_{\mathrm{sw}} = \left(\frac{E}{p_0}\right)^{1/3}, \ \tau_{\mathrm{sw}} = \frac{R_{\mathrm{sw}}}{v_s}, \qquad (10.16)$$

where:

$E = 150$ GJ (gigajoule) is the upper estimation of the energy of explosion

$p_0 \approx 2 \times 10^4$ Pa is the pressure of the air at the altitude of explosion

$v_s \approx 300$ m s^{-1} is the sound velocity at this altitude

Calculations show that $R_{sw} \approx 200$ m and $\tau_{sw} \approx 0.67$ s. At a distance of $R_0 = 5R_{sw} = 1$ km, the relative disturbance of the gas pressure $\Delta p/p_0 = 0.07$. At ionospheric altitudes, this disturbance increases with increase in altitude according to the law:

$$\frac{\Delta p}{p_0} = \left(\frac{\Delta p}{p_0}\right)_{R = R_0} \frac{R_0}{z} \exp\left\{\frac{z}{2H}\right\}, \tag{10.17}$$

where:

$H \approx 7$–8 km is the standard height of the neutral atmosphere introduced in Chapter 7

Then at an altitude of $z \approx 100$ km, we get $\Delta p/p_0 = 0.7$. According to such values of relative disturbances of air pressure, acoustic gravity waves (AGWs) are usually generated by filling relative amplitude of ~0.01. Such magnitudes are sufficient for the observation of AGWs by magnetometric and DS methods.

After the Sineva explosion time $\Delta t_1 \approx 5$ min the character of temporal variations of DS changed. Second, this occurred after time $\Delta t_2 \approx 71$ min. We note that for attaining ionospheric altitudes ($z \approx 100$ km), the AGW pulse needs about 5 min, which is close to the time delay of $\Delta t_1 \approx 5$ min. This pulse could generate an electromagnetic disturbance, which can reach the observation place and variations of SD at the time interval of 09:37–10:20 h can be recorded in the form of MHD waves via the ionosphere. The delay $\Delta t_2' = \Delta t_2 - \Delta t_1 \approx 66$ min for $R = 3000$ km corresponded to $v_2' \approx 0.76$ km s^{-1}. We note that changes in the character of DS variations also took place at the time interval of 11:10–11:30 h. In this case, $\Delta t_3 \approx 100$ min, $\Delta t_3' = \Delta t_3 - \Delta t_1 \approx 95$ min, and $v_3' \approx 0.53$ km s^{-1}.

Doppler measurements during this AE campaign were carried out at three frequencies: 2.4, 3.8, and 4.105 MHz. Qualitatively, temporal variations of Doppler shift frequency (DSF) for these frequencies were similar. But there were also differences. In DS at a frequency of 4.105 MHz after Molnia-M, Rokot, and Topol RLs, a second component appeared whose amplitude was 2–3 times lesser than the amplitude of the main component, whereas the magnitude of DFS did not exceed 0.1–0.2 Hz.

With increase in the frequency of radio sounding, time delay Δt_2 sometimes decreased. For example, for Molnia-M RL it was about 47, 43, and 40 min, respectively, for 2.4, 3.8, and 4.105 MHz. Thus, for smaller magnitudes of Δt_2 larger magnitudes of velocity v'_2 corresponded. The velocities of AGWs at ionospheric altitudes corresponded to this law.

It is important to note that approximately the same velocities were observed during studies by other authors on the effects of RLs [22–26,31] and by us in our previous research [13,17–20]. The correspondence of the velocities and the amplitudes of WDs show the usefulness during the AE campaign carried out on February 18, 2004, and the effects related to RLs and flights (including the explosion of one of them) were actually observed.

10.6.2 Effects in the Geomagnetic Field

If the first change in the signal character that has a delay of $\Delta t_1 \approx 24\,\text{min}$ was related to Molnia-M RL, then we get $v'_1 \approx 1.2\,\text{km s}^{-1}$. For $\Delta t_2 \approx 42\,\text{min}$, we get $v'_2 \approx 0.64\,\text{km s}^{-1}$. The possible reaction on Rokot launch started after $\Delta t_2 \approx 42\,\text{min}$. In this case, $v'_2 \approx 0.9\,\text{km s}^{-1}$. If the effect of RLs relates to a significant growth of the H-component amplitude of the signal, then $\Delta t_2 \approx 60\,\text{min}$ and $v'_2 \approx 0.61\,\text{km s}^{-1}$. The last velocity seems to be more true.

If there was a reaction to Sineva explosion, then it could have started after time $\Delta t \approx 43$ min that is clearly seen from the change in the character of D-component variations of the signal. For $\Delta t_0 \approx 5\,\text{min}$, $\Delta t' \approx 38$ min and $v' \approx 1.3\,\text{km s}^{-1}$. The duration of variations, ΔT, was about 60 min. A value of Δt for the H-component was about 50 min, and $\Delta T \approx 50$ min. Topol RL could be related to the change in the signal character having a delay of $\Delta t_2 \approx 42\,\text{min}$, for which $v'_2 \approx 0.64\,\text{km s}^{-1}$.

It can be affirmed that during AE, the changes in the character of magnetometric signal variations with high probability were related to RLs, and disturbances were transported with velocities $v'_1 \approx 1.2{-}1.3\,\text{km s}^{-1}$ and $v'_2 \approx 0.61{-}0.64\,\text{km s}^{-1}$. It is interesting to note that plasma and magnetic disturbances had a common nature. These low-MHD waves have the same magnitude as the first velocity [27]; the second velocity corresponds to AGW velocities at ionospheric altitudes [28,29]. Such velocities were often observed by us in previous observations [13,17,20] as also by other authors [22–26].

We add that during the AE campaign, temporal changes of the magnetometric signals were quicker at the time interval from 07:20 to 12:00 h on the whole. From these features, the registrations differed from registrations obtained during the background days of February 17 and 19, 2004

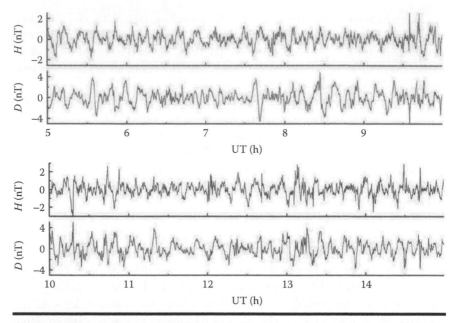

Figure 10.6 Temporal variations of the *H*- and *D*-components of the geomagnetic field on the background day of February 17, 2004 (a day prior to the RLs).

(Figures 10.6 and 10.7). Increase of geomagnetic pulsation frequency at time interval 07:20–12:00 h was also confirmed by the results of the spectral analysis (compare Figure 10.5 with Figures 10.8 and 10.9). Amplification of wave activity observed from DS variations was also recorded at the same time interval of 07:20–12:00 h.

Hence, the following facts prove our suggestion that continuous launches of a group of rockets cause disturbances in the ionosphere and the magnetic field:

1. The parameters of disturbances are that the time delays, periods, and durations of disturbances are close to those observed by other authors and by us in previous observations.
2. Reasonable magnitudes of disturbance propagation velocities corresponded to well-studied types of waves, such as slow MHD and AGW.
3. Proximity of the disturbance parameters that accompanied Molnia-M and Topol launches from Plesetsk cosmodrome.
4. Nearly the same synchronism of disturbances was there in the ionosphere and the geomagnetic field.

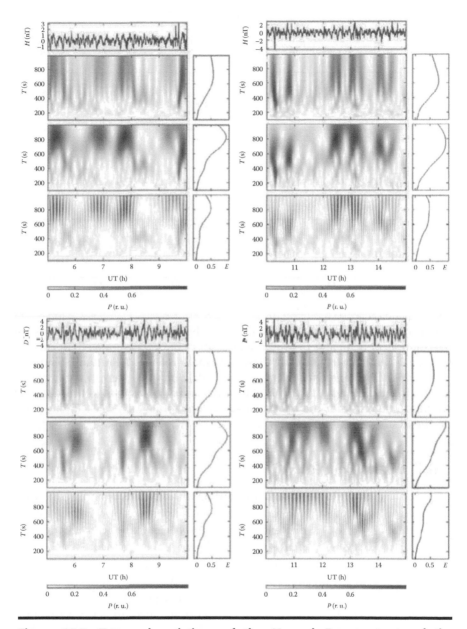

Figure 10.7 Temporal variations of the *H*- and *D*-components of the geomagnetic field on the background day of February 19, 2004 (a day after the RLs).

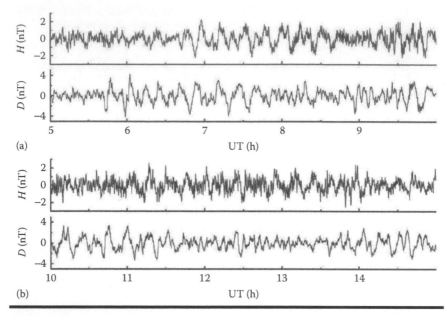

Figure 10.8 Results of the spectral analysis of temporal variations of *H*-component (a, b) and *D*-component (c, d) of the geomagnetic field and the energy diagrams corresponding to them (panels from top to bottom): temporal variations of *H*(*t*) and *D*(*t*), results of WFT, WT, and AFT. Here, a and c correspond to time interval 05:00–10:00 h, and b and d correspond to time interval 10:00–15:00 h but for the background day of February 17, 2004 (a day prior to the RLs).

10.7 Main Results

1. Considering the AE campaign, during which four RLs and flights were made as well as an explosion of one of the rockets took place, which caused amplification of wave activity in the ionosphere and the geomagnetic field. Reactions from different sources (different rockets) added to each other yielding a serious problem for their identification.

2. The amplitude of DFS variations during AE usually was 0.2–0.3 Hz, sometimes attaining 0.4 Hz. For the average period of $T \approx 8$ min, the magnitudes $\delta_N \approx 2\%$–5% and 7% corresponded to such variations, respectively.

3. During AE campaign, WDs with periods of infrasound of 5–6 min, as well as intrinsic gravity waves with $T \geq 10$ min were observed.

4. Explosion of the rocket accompanied by energy output on the order of 150 MJ, probably was the cause of appearance of WDs in the ionosphere,

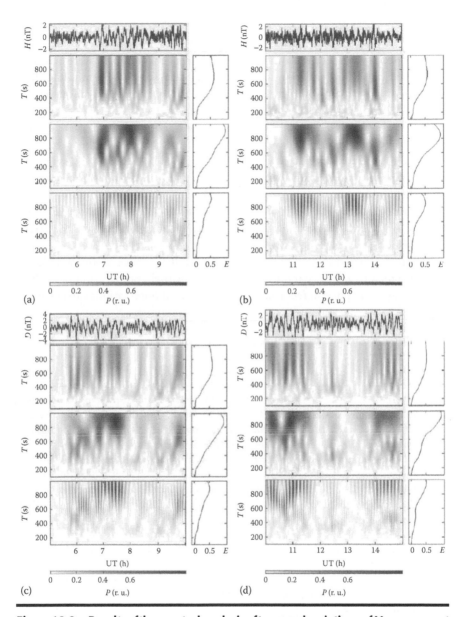

Figure 10.9 Results of the spectral analysis of temporal variations of *H*-component (a, b) and *D*-component (c, d) of the geomagnetic field and the energy diagrams corresponding to them (panels from top to bottom): temporal variations of *H*(*t*) and *D*(*t*), results of WFT, WT, and AFT. Here, a and c correspond to time interval 05:00–10:00 h, and b and d correspond to time interval 10:00–15:00 h but for the background day of February 19, 2004 (a day after the RLs).

having velocities of propagation of about 760 km s^{-1}, period 6–8 min, and duration close to 55 min.

5. Molnia-M and Topol launches from Plesetsk cosmodrome were accompanied by similar variations of DS, having delays of 50–60 min and durations close to 25–30 min, respectively. Disturbance propagation velocity of 0.44–0.539 km s^{-1} corresponded to such delays.

6. On the day of AE (more precisely, from 07:20 to 12:00 h), variations of the geomagnetic field level were more irregular (with higher frequency) compared to background variations.

7. The disturbances of the parameters of the ionosphere and the geomagnetic field that accompanied the AE campaign corresponded to closed velocities, equaling 1.7 and 440–760 m s^{-1}—for the ionosphere, and 1.2–1.3 and 0.61–0.64 km s^{-1}—for the geomagnetic field. The largest velocity relates to slow MHD waves, the slowest to the AGWs.

8. The estimated magnitudes of the value of disturbance of ionospheric and geomagnetic field parameters, velocities of their propagation, and durations are in good agreement with the results of observations carried out by other authors and with our previous investigations.

References

1. Chernogor, L. F., Physics of the Earth, atmosphere and geospace in the lightening of the system paradigm, *Radiophys. Radioastron.*, 8, 59–106, 2003 (in Russian).
2. Chernogor, L. F., The Earth–atmosphere–geospace as an open dynamic nonlinear system, *Cosmic Sci. Technol.*, 9, 96–105, 2003 (in Ukrainian).
3. Chernogor, L. F., The Earth–atmosphere–ionosphere–magnetosphere as an open dynamic nonlinear system (Part 1), *Nonlinear Univ.*, 4, 655–697, 2006 (in Russian).
4. Chernogor, L. F., The Earth–atmosphere–ionosphere–magnetosphere as an open dynamic nonlinear physical system (Part 2), *Nonlinear Univ.*, 5, 55–97, 2007 (in Russian).
5. Adushkin, V. V., Kozlov, C. I., and Petrov, A. V., Eds. *Ecological Problems and Risks of the Effects of Rocket-Cosmic Technique on the Environment: Handbook*, Moscow: Ankil, 2000, 640.
6. Chernogor, L. F., Effects in geospace accompanied starts of the groups of rocket, *Radiophys. Radioastron.*, 13, 39–53, 2008.
7. www.itar-tass.com.
8. Lozorenko, O. V., Panasenko, S. V., and Chernogor, L. F., The adaptive Fourier transform, *Electromagn. Waves Electron. Sys.*, 10, 39–49, 2005 (in Russian).
9. Burmmaka, V. P., Panasenko, S. V., and Chernogor, L. F., Modern methods of the spectral analysis of quasi-periodical processes in the geospace, *Foreign Radioelectronics: Success in Modern Radioelectronics*, 11, 3–24, 2007 (in Russian).

10. Bracewell, R., *The Fourier Transform and Its Applications*, 3rd edn., New York: McGraw-Hill, 1999.

11. Holschneider, M., *Wavelets: An Analysis Tool*. Oxford: Calderon Press, 1995.

12. D'akovnov, V. P., *Wavelets: From Theory to Practice*, Moscow: Solon-R, 2002.

13. Kostrov, L. S., Rozumenko, V. T., and Chernogor, L. F., Doppler radio sounding of disturbances in the middle ionosphere accompanied launchings and flights of space vehicles, *Radiophys. Radioastron.*, 4, 227–246, 1999 (in Russian).

14. Garmash, K. P., Rozumenko, V. T., Tirnov, O. F., Tsimbal, A. M., and Chernogor, L. F., Radiophysical investigations of the processes in the near-the-Earth plasma perturbed by the high-energy sources, *Foreign Radio Electronics: Successes of Modern Radio Electronics*, 3–15, 1999; 3 19, 1999 (in Russian).

15. Garmash, K. P., Leus, S. G., Pazura, S. A., Pohil'ko, S. N., and Chernogor, L. F., Statistical characteristics of fluctuations of the Earth's electromagnetic field, *Radiophys. Radioastron.*, 8, 163–180, 2003 (in Russian).

16. Burmmaka, V. P., Kostrov, L. S., and Chernogor, L. F., Statistical characteristics of signals of the Doppler HF radar during sounding of the middle ionosphere perturbed by the rockets' launchings and by the solar terminator, *Radiophys. Radioastron.*, 8, 143–162, 2003 (in Russian).

17. Burmaka, V. P., Taran, V. I., and Chernogor, L. F., Wave processes in the ionosphere accompanied launchings and flights of rockets at the background of natural transmission processes, *Geomagn. Aeronom.*, 44, 518–534, 2004.

18. Burmmaka, V. P., Taran, V. I., and Chernogor, L. F., Results of study of wave disturbances in the ionosphere by method of incoherent scattering, *Successes in Modern Radio Electronics*, 4–35, 2005 (in Russian).

19. Burmmaka, V. P., Taran, V. I., and Chernogor, L. F., Wave processes in the ionosphere during quiet and perturbed conditions. 1. Results of observations at the Kharkov's radar of incoherent scattering, *Geomagn. Aeronom.*, 46, 193–208, 2006.

20. Burmaka, V. P., Lisenko, V. N., Chernogor, L. F., and Chernyak, Yu. V., Wave processes in the F-region of the ionosphere accompanied starts of rockets from the cosmodrome Baikonur, *Geomagn. Aeronom.*, 46, 783–800, 2006.

21. Alperovich, L. S., Gokhberg, M. B., Drobgev, V. I., et al., Project MASSA—investigation of the magnetospheric-ionospheric communications during seismo-acoustic phenomena, *Phys. Earth*, 5–8, 1985.

22. Arendt, P. R., Ionospheric undulations following Apollo 14 launching, *Nature*, 231, 438–439, 1971.

23. Noble, S. T., A large-amplitude travelling ionospheric disturbance excited by the Space Shuttle during launch, *J. Geophys. Res.*, 95, 19037–19044, 1990.

24. Nagorsky, P. M., Inhomogeneous structure of the F region of the ionosphere generated by the rockets, *Geomagn. Aeronom.*, 38, 100–106, 1998.

25. Afra'movich, E. L., Perevalova, N. P., and Plotnikov, A. V., Registration of ionospheric responses on the shock-acoustic waves generated during launchings of the rocket-carriers, *Geomagn. Aeronom.*, 42, 790–797, 2002.

26. Sokolova, O. I., Krasnov, V. M., and Nikolaevsky, N. F., Changes of geomagnetic field under influence of rockets' launch from the cosmic port Baikonur, *Geomagn. Aeronom.*, 43, 561–565, 2003.

27. Sorokin, V. M. and Fedorovich, G. V., *Physics of Slow MHD Waves in the Ionospheric Plasma*, Moscow: Energoizdat, 1982 (in Russian).

28. Gossard, E. E. and Hooke, W. H., *Waves in the Atmosphere: Atmospheric Infrasound and Gravity Waves: Their Generation and Propagation*, Amsterdam: Elsevier Scientific Publishing Co., 1975, 476.

29. Grigor'ev, G. I., Acoustic-gravity waves in the Earth's atmosphere (issue), *Izv. Vuzov. Radiofizika*, 42, 3–25, 1999 (in Russian).

30. Mallat, S., *A Wavelet Tour in Signal Processing*. 2nd edn., San Diego, CA: Elsevier, 1999, 489.

31. Alperovich, L. S., Ponomarev, E. A., and Fedorovich, G. V., Geophysical phenomena modeling by explosion (issue), *Phys. Earth*, 9–20, 1985.

ROCKET BURN AND LAUNCH AND RADIO COMMUNICATION

IV

Chapter 11

Modification of the Ionosphere Caused by Rocket and Space Vehicle Burn and Launch

11.1 Overview

As mentioned in the previous chapters via intensive discussions regarding numerous experimental and theoretical investigations, the era of burn, launch, and flight of rockets, spacecrafts, and artificial satellites started more than 60–70 years ago, conjoined with the era of "active space experiments" carried out in the near-the-Earth environment, including the atmosphere; the low-, middle-, and upper ionosphere; and the magnetosphere, that is, in the subsystems of the whole system of the Earth–atmosphere–ionosphere–magnetosphere (EAIM). The main goal of most of these active-space experiments, in which both the authors of this book were, and currently are, involved, was modification of the atmosphere, ionosphere, and magnetosphere by artificially induced heating, caused by the pumping of radio waves generated from ground-based radars, injection of neutral and ionic clouds, and beams from special geophysical rockets and space vehicles (SVs), chemical releases and explosions, and so forth.

The main attention of the scientific society dealing with the study of the whole EAIM system accumulated from investigations into the processes and phenomena occurring in the near-the-Earth environment and its "reaction" on different kinds of modifications. A major task was to obtain more information on the structure and morphology of each subsystem of the EAIM system, as well as to understand and explain specific mechanisms of their coupling (see Refs. [1–22] and the references therein).

Moreover, drastic intensification of the complex experimental campaigns was observed from the middle of the twentieth century (specifically from 1950s to 1960s), in which ground-based radiophysical, magnetometric, and optical facilities (incoherent and coherent radars, Doppler radars, ionosondes, magnetometers, optical infrared and ultraviolet cameras, etc.), together with satellite observations (using the same or more precise instrumentations) were combined [23–62]. The main task of these investigations was the precise study of natural phenomena in each of the nonperturbed or perturbed portion of the subsystem of EAIM, namely, those affected by magnetic storms, streams and currents in the auroral and subauroral (also called *polar*) regions of the ionosphere [63–81], and solar eclipses [82–88], as well as by plasma bubbles in the equatorial ionosphere [89–93] and meteor trails in the middle-latitude ionosphere [94–103]. Simultaneously, the active experiments for modification of the ionosphere by powerful pump waves from ground-based facilities [50–62], injection of ion clouds and beams from special geophysical rockets [104–121], and special rocket releases and rocket engine and plume exhausts [122–172], which generate the so-called dusty or dirty plasmas [173–178], until now have been the key objects of investigation by the geophysical and radiophysical scientific society.

It was found that most of the naturally and artificially induced processes and phenomena are similar, because rocket pollutants, such as exhaust gases, engine fuel, and engine plume, as well as fuel fluxes, jets, and flashes, can be considered additional sources of ionospheric plasma modification. Moreover, as mentioned in the previous chapters, these active experiments are cost-effective and do not require additional expenses for their organization and performance.

Of course, we understand that the scenario for the future readers can be far from geophysics, cosmic space, radiophysics, or optical physics. Many of them, such as ecologists, mechanical and rocket engineers, and so forth, who are not familiar, partly or extensively, with atmospheric, ionospheric, and magnetospheric physics, as well as with the processes of radio propagation in such media, would be interested to understand the effects of rocket burns, launches, and flights not only for wireless communications but also for ecological and geophysical problems arising from such "cost-effective active experiments."

To understand such difficulties and overcome many "classical" physical phenomena, recalling the limited length of this book, we recommend to the reader some special monographs [1–22], where all atmospheric, ionospheric, and magnetospheric phenomena, observed in the regular nonperturbed and perturbed near-the-Earth media that are caused by natural and artificially induced phenomena, are fully presented.

Therefore, to be more precise on the wide spectrum of our future explanation of the subject to the readers, we present the most evident observations of rocket and SV burns and launches in Section 11.1. Then, in Section 11.2, we compare the results of observations with similar results observed in active-space experiments, such as heating of plasma [50–62] and injection of ion beams and clouds into plasma [104–121]. Section 11.3 compares the processes observed in plasma during natural phenomena caused by magnetic storms in the polar ionosphere [63–81], bubbles in the equatorial ionosphere [89–93], and meteor trails in the middle ionosphere [94–103]. In Section 11.4, we especially bring the reader's attention to these aspects, because most of the phenomena and features observed during these experimental campaigns are similar to the effects of rocket and SV burns, launches, and flights [122–172]. The main results that come from the described experimental observations are summarized in Section 11.5.

In addition, it should be mentioned that the natural and artificially induced effects observed during the "ionosphere modification" campaigns are widely given in the literature and have been proved by numerous experiments and observations, and nowadays have a clear physical explanation by using advanced theoretical models and frameworks [108–121].

11.2 Geophysical Effects of Rocket Burn and Launch Compared to Those Observed in Active-Space Experiments and Associated with Natural Phenomena

The pioneering experimental observations of plasma behavior in SV, during its flight through the ionosphere, were started at the end of 1960s to the beginning of 1970s. Such "cost-effective" active-space experiments were carried out simultaneously with active experiments and campaigns using the same kind of instrumentations and techniques. Both types of "active experiments" have shown that they are roughly the same and have similar features and peculiarities that are observed in the near-the-Earth media.

In the first and second sections, we briefly describe the most interesting experiments done in both kinds of active-space experiments (the full information can be seen in Refs. [122–172]), and compare the corresponding features and observed effects, illuminating their similarity and specific peculiarities. Some similar natural ionospheric phenomena observed during the corresponding worldwide experimental campaigns, from the equatorial to the polar ionosphere, is discussed briefly in the third section.

11.2.1 Observations of Effects Caused by Rocket Burn and Launch

In most rocket and SV burn, launch, and flight observations performed in the 1960s, there is a dramatically sudden increase of total plasma content (called the *total electron content* (TEC), in the pioneering investigations, which accounts for plasma quasineutrality; see Chapter 1), the significant ionization of the background ambient plasma in the lower ionosphere (from 50 to 120–140 km), and generation of plasma depletion regions (called *plasma holes*) in the upper ionosphere (from 150 to 350–400 km). Moreover, inside the observed global plasma holes (ranged approximately from 100 to 1000 km), the irregular (e.g., *sporadic*) structures created by small- and moderate-scale plasma turbulences filling the depletion areas, have been found experimentally using ground-based and satellite-arrangement radiophysical and optical instrumentations and techniques [122–172]. To avoid entering into numerous publications, we will describe below only those that evidently and adequately describe the observed phenomena and the corresponding rocket-induced effects.

11.2.1.1 Ionospheric Plasma Perturbations Caused by Rocket Burn and Launch

The first pioneering observations of the SV Vanguard II launch from Cape Canaveral started shortly prior to 11:00 h (local time, LT) on February 17, 1959, which were done by monitoring the ionosphere with two ionospheric sounders: one was located close to the site of the rocket's launch and the other located on Grand Bahama Island, that is about 300-km downrange and closer (with respect to the first sounder) to the place of the rocket entry into the *F*-region of the ionosphere, as is seen from the geometry of active experiment presented in Figure 11.1 (extracted from Ref. [122]).

The corresponding ionograms obtained by radio sounders, operating up to 18 MHz frequency band, are shown in Figure 11.2a–d (extracted from Ref. [122]) for each rocket's firing after the launch (at about 11:02, 11:09, 11:15, and 11:17 h, respectively).

In Figure 11.2, the letter F indicates a radio echo from the regular *F*-region, whereas the letter M indicates an echo associated with the effects of rocket burnout and ignition stages. During the first 15 min of the launch (Figure 11.2a–c), the first and the second multiple radio echoes produced

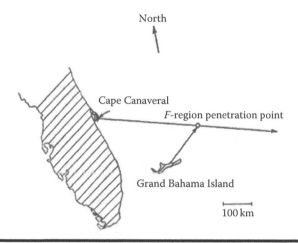

Figure 11.1   Mapping of the geometry of RL from Cape Canaveral.

by rocket exhausts were clearly seen in the form of loops, in addition to the regularly observed *F*-echoes. Seventeen minutes after the rocket start, the primary "anomalous" echo associated with the effects of the rocket's exhaust had disappeared, whereas the other anomalous multiple radio echoes could still be seen from Figure 11.2d. Observers had lost the first multiple echoes at 11:30 h, and after that all anomalous echoes associated with the rocket were no longer recorded. What is interesting to note is that the penetration frequencies for the *F*-layer are identical to those associated with the echoes produced by the rocket's exhaust gases.

Supposing that the rocket exhaust can be modeled as an ionized cloud surrounding the missile and expecting to strike heating-induced irregularities that are produced in addition to the increase of ionization in the preexisting ionization of the *F*-layer and yielding the corresponding scattering effects, the researchers could not explain the phenomenon illustrated in Figure 11.2, because the anomalous radio echoes have traces almost the same as those obtained from the regular *F*-region radio echoes.

Therefore, finally it is supposed that such traces as are clearly seen from the rocket-induced regions can be explained only by "a local decrease in ionization density," that is, by the existence of a deep depletion plasma region [122,123], which was also observed experimentally by other researchers [124–126].

We do not enter into detailed explanation of the observed phenomena now based on existing current theoretical models. This will be done in Chapter 12. At the same time, we should note that during the time of observation, researchers found multiple radio echoes produced by various-scale plasma irregularities

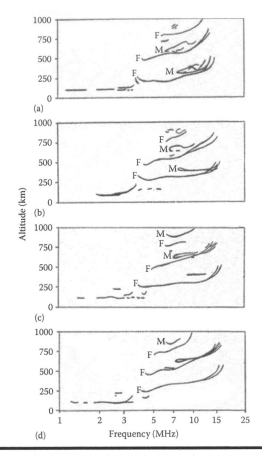

Figure 11.2 Ionograms (a–d) recorded by ionosonde during rocket burn and launch.

filling an artificially induced plasma hole in the *F*-region of the ionosphere produced by the passage of a rocket, in the same manner as usually happened during radio-star scintillation and the spreading of the *F*-layer phenomenon. We will return to these aspects in our theoretical analysis further described in Chapter 12.

The other experimental investigations, based on the Faraday rotation technique of the radio signal operating at 73.6 MHz sent by a special NASA rocket launched at 12:38 h (LT) from Wallops Island, VA, were carried out to check the effects of the rocket exhaust on the background ionosphere (see Ref. [129]). Such a technique is based on measuring the total electron content (TEC) along a path of radio signals from the rocket's transmitter to the ground-based receiver (radio station or ionospheric station). Thus, during the fourth-stage ignition of the solid-fuel rocket engine, creation of a deep ionospheric disturbance along its trajectory has been observed. The first two ignitions were observed in the lower ionosphere (below 90 km), whereas the third- and the fourth-stage ignitions were observed in the ionosphere at altitudes from the upper *E*-layer (~150 km) to the upper *F*-layer (~350 km).

The signal radiating at the carrier frequency of 73.6 MHz was recorded by the ground-based station very close to the launch site. Therefore, during the rocket's flight, the radio path of the receiving signal was almost the same as the rocket flight path. At the same time, ionosonde records, which selected signals reflected from nondisturbed ionospheric plasma (the so-called regular ionospheric profile), were compared with those obtained from a local sounding station that recorded the signal after using the Faraday rotation technique. Using these comparative results simultaneously, the authors of Ref. [127] performed a complex ionogram, shown in Figure 11.3, where on its left side, the rocket trajectory and the typical ray paths of the recorded signal are also indicated for

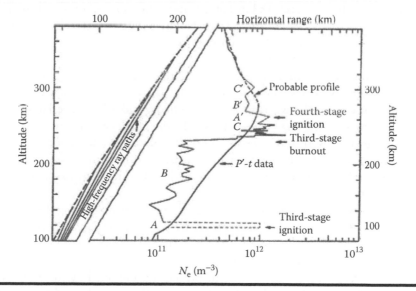

Figure 11.3 **Trajectory of the rocket and the corresponding radio path between the transmitter and the receiver (left panel); the ionogram of the undisturbed profile of the ionospheric plasma profile (bold-dashed curve); the perturbed ionospheric plasma profile using Faraday rotation technique (solid curve).**

further discussions made in further subsections. To differentiate the observed results shown on the right side of Figure 11.3, the researchers have divided the ionosphere into three typical regions:

1. Region *A* corresponds to the third stage of rocket's engine ignition that occurs in the *E*-layer. This evidently indicates enhancement (e.g., increase) of plasma density with respect to that in the background nondisturbed ionospheric plasma.
2. Region *B* corresponds to deep depletion region (*cavity* or *hole*) expanded from the upper *E*-layer to the *F*-layer.
3. Region *C* indicates plasma density enhancements (compared to background plasma) before the fourth-stage ignition and after the third stage of rocket burnout.

The additional letters *A'*, *B'*, and *C'* in Figure 11.3 indicate some local depletion and enhancement regions occurring after the fourth stage of rocket ignition. Because the authors of Ref. [127] present only their observed results without intensive physical explanation, we will return to the results during our discussions on the observed rocket's burn and launch effects based on the theoretical models and frameworks presented later. Now we note that the observed ionograms described in Ref. [127] fully correspond to the ionograms observed in Ref. [127] and presented in Figure 11.2.

The same technique of Faraday rotation of the transmitted radio signal via ionosphere was described in Ref. [130]. These measurements have been carried out during the RL from Cape Kennedy, and all the five launches of the large rocket Atlas have been analyzed. The effect of Faraday rotation of the signals passing through the length of the plume of a large rocket during its burning stage was undertaken for the five similar launches by using an experimental site located in Yankeetown, FL (Figure 11.4, extracted from Ref. [130]).

All launches were done during late afternoon or evening along the azimuth within a few degrees of the extension of the line connecting Yankeetown and Cape Kennedy. The corresponding Atlantic rocket range trajectory map, shown in Figure 11.4, indicates the launch azimuth line.

The corresponding plasma content profile was obtained by the authors of Ref. [130] from the data of Faraday rotation for the ionospheric *F*-region perturbed by the rocket plume. The disturbed plasma profiles were obtained by measuring with Faraday rotation. Information on the plasma profile of the nonperturbed background ionosphere was obtained from the ground-based ionospheric station that mapped the corresponding ionograms. Due to their similarity, we present only the two (from five) curves that correspond to two

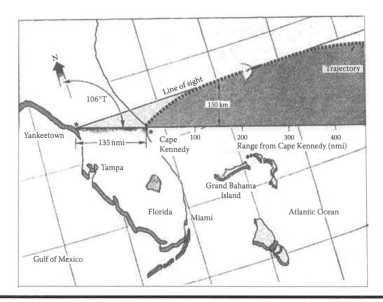

Figure 11.4   Mapping of the rocket launch experiment from Cape Canaveral; observations were done at Yankeetown place and Grand Bahama island.

different tests of Atlas launches, obtained from ionograms, ambient or non-disturbed, and disturbed conditions in Figure 11.5a and b (rearranged from Ref. [130]). The results of the observations clearly indicate that the plasma (e.g., electron and ions, because plasma is quasineutral) density in the rocket plume region begins to drop steeply below ambient plasma at the altitudinal range from the upper *E*-layer to the lower *F* layer. It can be clearly seen that the observed plasma density decreases with altitude, whereas the ambient plasma profile increases toward the *F*-layer maximum.

Moreover, the resulting disturbed plasma profiles obtained from the soundings at 97.7 MHz (Figure 11.5a) and at 259 MHz (Figure 11.5b) indicate increase in the plasma densities in the wake of the rockets with respect to the ambient plasma density profile at altitudes of the lower *E*-region of the ionosphere clearly and consistently.

As for higher altitudes, significant reduction of plasma density below the ambient value by a factor of 10–20 has been clearly observed in the wake of large rockets, such as Atlas, burning as it travels from the lower *E*-region to the *F*-region of the ionosphere.

The same observations of the decrease of the plasma density profile from ionosonde records obtained on February 16, 1965, at Cape Kennedy and from ionospheric stations at Grand Bahama Island and San Salvador were

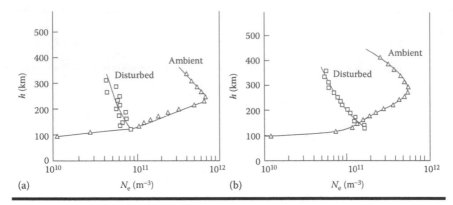

Figure 11.5 (a) Ambient ionospheric plasma altitudinal profile compared to the disturbed one caused by Atlas burn and launch in the *E*- and *F*-regions of the ionosphere for the first test. (b) Ambient ionospheric plasma altitudinal profile compared to the disturbed one caused by Atlas burn and launch in the *E*- and *F*-regions of the ionosphere for the second test experiment.

discussed in Ref. [132]. It was clearly and consistently indicated by both stations that the decrease of the electron content in the *F*-region of the ionosphere was 20%–25% with respect to the ambient TEC.

To explain these effects, the authors of Refs. [130–135] have proposed simplest models of adiabatic continuum expansion of the rocket exhaust gases (with a fixed total number of electrons) to the steady-state regime having ambient pressure surrounding the rocket. Additional hypotheses were proposed regarding the role of shock wave propagation and the thermal expansion of rocket exhaust gases. Based on these assumptions, the electron density in the expanded plume was derived in Refs. [130–135] for various altitudes. We will briefly describe their assumption in further subsections, where we will present the corresponding explanation of the phenomena, observed clearly in Figures 11.3 and 11.5a and b.

Detailed experiments based on radio and optical observations of the evolution of the rocket's exhaust and plume gas extent during the rocket's burn and launch through the cold quasineutral ionospheric plasma were carried out from the beginning to the end of 1970s [136–141]. The corresponding theoretical approaches and models created for explaining the observed phenomena caused by rocket burn, launch, and flight, as well as the corresponding active experiments, using rocket releases of chemical compositions of various gaseous and liquid clouds, were presented and described in Refs. [142–160].

Thus, on May 14, 1973, using the geostationary satellite ATS-3, whose trajectory is shown in Figure 11.6 (extracted from Ref. [140]), Faraday rotation

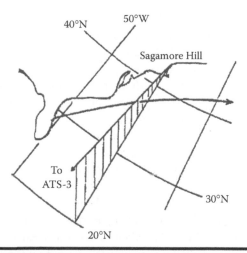

Figure 11.6 The trajectory of geostationary satellite ATS-3 and mapping of the experimental campaign by ground-based facilities located at the Sagamore Hill Radio Observatory.

observations were carried out from Sagamore Hill Radio Observatory in Hamilton, MA (43°N and 70°W), to obtain the TEC along the rocket travel path, whose launch was started close to 12:30 h local east time (LET).

This was NASA's Skylab 1 rocket launch (RL) from Kennedy Space Center (28.4°N and 80.6°W), Cape Canaveral, FL. Figure 11.7, extracted from Ref. [130], shows a sudden decrease of the TEC (e.g., the plasma profile) with respect to the

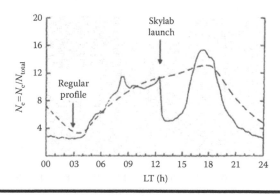

Figure 11.7 The disturbed relative plasma density before, during, and after the RL vs LT at the experimental site; the dashed curve represents undisturbed ambient plasma profile.

nondisturbed plasma profile (dashed curve) plotted for a period of 15 min [in units of plasma density of 10^{18} (m^{-3} × $column^{-1}$)].

It is clearly seen that the TEC value decreased from 11.3 to 5.8 units, that is, approximately twice and remained below 5 units during a 13:00–14:00 h time period with further slow growth of about 15 units. The same overall effect of plasma depletion region (e.g., plasma "hole") generation, with an essential decrease of TEC of ~50%, was observed for nearly 4 h.

Selecting not less than 31 observations (during the month of May 1973) of TEC along ATS-3 satellite trajectory made by the ground-based radio facilities at Sagamore Hill Radio Observatory prior to, during, and after the rocket's launch, full statistics is summarized and presented in Figure 11.8, according to Ref. [138] (that was repeated in Refs. [139–143]).

It is consistently seen that only during midday on May 14, 1973, when the RL was made, a dip in the TEC, called *cavity*, was clearly distinguished from the normal variability of the ionosphere itself caused by the specific changes in solar activity (during sunrise and sunset diurnal periods) and geomagnetic activity (during nocturnal periods). To differentiate the effects of the geomagnetic activity as a natural phenomenon described in Chapters 2 and 3, with those caused by the rocket burn and launch, both effects were compared, combining the corresponding data in Figure 11.9 (rearranged from Refs. [138,139]).

Now, the TEC obtained monthly during May 1973 was computed with data obtained during the magnetic storm period of May 13 and 14, 1973. It is clearly seen that despite the essential decrease in ionospheric plasma content during the

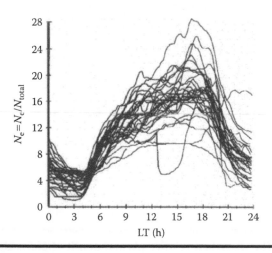

Figure 11.8 The monthly recorded profile of the ionospheric plasma density with deep depletion region during midday on May 14, 1973, of the RL.

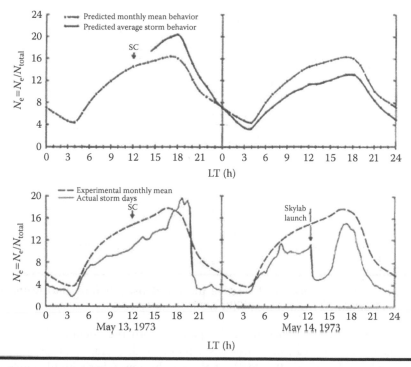

Figure 11.9 The monthly recorded relative plasma density profile compared to the predicted average magnetic activity monthly behavior (top panel) and the same perturbed profile during magnetic storm days on May 13 and 14, 1973, with respect to ambient measured profile (dashed curve in the bottom panel).

magnetic storm period, the TEC data decrease is manifested again on May 14 from 12:00 to 15:00 h LT at the Sagamore Hill Radio Observatory.

All illustrations allow us to conclude that only one evident decrease of TEC curve and generation of the cavity region arises during the whole month of May 1973, which closely related to the RL and was recorded both from the satellite and Sagamore Hill Radio Observatory based on Faraday rotation method.

The same observations were also carried out from Sagamore Observatory during the flight of the other satellite, ATS-5, located approximately 5° longitude westward to the ATS-3 satellite position. Figure 11.10 (rearranged from Ref. [138]) shows the corresponding results that were observed by ATS-5 plotted together with those observed by ATS-3.

It is seen that both radio observations of TEC (via computation of the TEC of the surrounding area of Skylab-1 launch) show the same tendency of TEC for

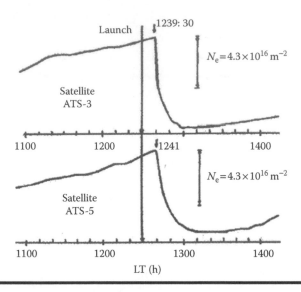

Figure 11.10 **The disturbed relative plasma density before and during the RL vs LT obtained from two satellites, ATS-3 and ATS-5.**

the essential decrease along the radio path of both the satellites through the plasma-disturbed area caused by rocket plume and exhaust. The same effect, but weaker (of only in 10%–15%), of the regular ionospheric profile depletion was also observed and followed from radio measurements carried out by other observatories in the United States, Canada, and Greenland, such as in London, ON; Urbana, IL; Goose Bay, Newfoundland and Labrador; and Narsarsuaq, Greenland.

Six years later, during a large RL on September 20, 1979, a deep cavity in the plasma profile (with respect to the background nonperturbed one, denoted by a "zero line" in units of 10^{11} m^{-3}) was measured by the Millstone Hill radar based on Faraday rotation technique.

This cavity was created for 15 s at the altitudinal range of about 400–430 km, with positive enhancements (with respect to "zero level," Figure 11.11; rearranged from Ref. [138]) of plasma density below (from 360 to 400 km) and above (from 430 to 450 km) the depletion zone, where the rocket's exhaust caused strong perturbation of the regular ionospheric plasma.

The experimentally observed and theoretically predicted effects of the RLs and flights are summarized by Mendillo in Ref. [143]. We take only some experimental results mentioned there relating to a special "Logopedo" experiment to discover the effects of the Skylab spacecraft. Thus, by using various weather satellite launching missions (NOAA-A, NOAA-B, NOAA-C), observation of large rocket engine exhaust and plume effects were carried

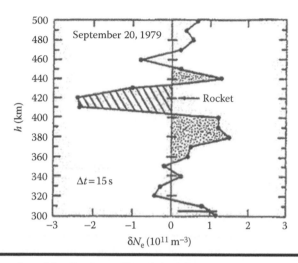

Figure 11.11 **The altitudinal profile of the disturbed relative plasma density before, during, and after the RL (total time of observation was 15 s) surrounding a "zero level" obtained during RL experiment carried out on September 20, 1979.**

out by Millstone Hill incoherent scatter radar (ISR). First, this was done for HEAD-C RL. Large-scale depletions observed during HEAD-C launch in the form of ionospheric cavities (holes) are clearly seen from Figure 11.12, rearranged according to Refs. [137,143]. Moreover, these depletions are filled with small-scale plasma irregularities (called in the special literature *plasma turbulences*).

The observation of the other Spacelab-2 RL during nighttime on July 29, 1985, were carried out at the same Millstone Hill observatory by not only ISR, whose results are the same as those presented in Figure 11.2.

The imaging of burn release of neutral gases by the rocket's engine and their dynamics and penetration through the background ionospheric plasma behind the rocket engine (Figure 11.13; extracted from Ref. [143]). A peak of the cloud brightness (called "airglow") was imaged 2 min after the burn from the exhaust gas that was emitted at an altitude of 320 km. The location of the 47-s engine burn is indicated by a white line ending near 40°N and 70°W.

As was found experimentally, the brightened cloud (e.g., airglow) was created by the CO_2 component of the rocket exhaust gases via some special chemical reactions that we will discuss in further subsections.

The same Spacelab-2 atmospheric experiments have been reported in Refs. [161,118], where "reaction" of the background ionosphere on the Space Shuttle engine burn plumes was investigated quite precisely. The researchers have mostly studied neutral exhaust gas dynamics streaming from rocket

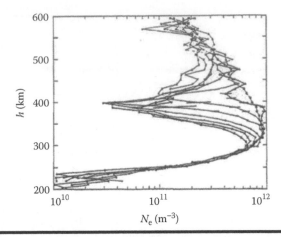

Figure 11.12 **The depletion region temporal dynamics during HEAD-C RL in a special Logopedo Space campaign. Each profile was obtained for each continuously changed time of recording.**

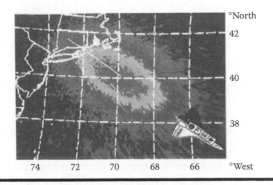

Figure 11.13 **The cloud brightness (airglow) generated by Spacelab-2 rocket engine exhausts 2 min after its burn and launch.**

engine and plume, as well as the dynamics of the so-called dusty plasma created by the hypersonic exhaust molecules during their interactions with background plasma electrons and ions (see Chapter 12). Thus, it was reported in Ref. [118] that a deep hole at the altitudinal range of 320–333 km was found by Arecibo, PR, backscatter incoherent radar operated at 430 MHz. This deep depletion of the background plasma was recorded close to 14 s of the rocket burn plumes and was observed for several minutes. We present here only four moments after the rocket burn, depicted in Figure 11.14, according to Ref. [118].

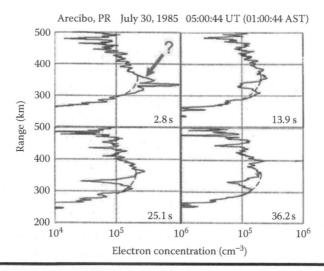

Arecibo, PR   July 30, 1985   05:00:44 UT (01:00:44 AST)

Figure 11.14   Altitudinal distribution of the TEC after the rocket burn: 2.8, 13.9, 25.1, and 36.2 s. The dashed curves depict nondisturbed ionospheric plasma density.

It is interesting to note that immediately after the rocket burn the exhaust gas flows and penetrates the background plasma and a huge chunk of plasma enhancement (with respect to background ionospheric plasma) was observed, which was indicated by a question by the authors of Ref. [118] (see top left panel in Figure 11.14). Later, as the researchers suggested in Ref. [172], the hypersonic molecules of the exhaust gas during the fast aeronomic and chemical reactions may generate hypersonic molecular plasma ions, which may produce a huge plasma density enhancement due to the appearance of a highly dense, "dusty" plasma surrounding the rocket inside the rocket region.

We will try to explain this statement also in Chapter 12 taking into consideration the formation of a dense multi-ion plasma surrounding the rocket after its engine burn from the beginning that in literature is called a "dusty" or "dirty" plasma [173–178]. The same result is clearly seen in the illustration presented in Figure 11.3, in which a huge enhancement is observed at an altitude of 110–115 km after the third ignition of the rocket motor. We will discuss this aspect in a phenomenological manner in the next section and give a quantitative theoretical description of this phenomenon in Chapter 12.

Altitudinal distribution at the bottom and top sides of the global cavity was observed after about 14 s, with large-scale cavities (called "holes") filled by small- and moderate-scale irregularities for complicated plasma density, which had the same features that were observed in previous experiments and shown in

Figures 11.11 and 11.12. All these cavities were fully explained by researchers following the theoretical results presented in Ref. [118]. Actually, as was shown by the authors of this work, hypersonic exhaust gases (molecules) transform their energy to the "hypersonic ions" after interaction with background plasma (definition taken from Refs. [118,172]). Finally, these hypersonic ions produce kinetic instabilities that excite low-frequency acoustic waves and low hybrid-ionic waves, that is, they generate small-scale and moderate-scale plasma irregularities (e.g., turbulences). These aspects will be illuminated in Chapter 13 based on the corresponding quantitative theoretical analysis that, finally, will be in satisfactory agreement with the results obtained in Ref. [118].

Next, to explain the results related to the neutral gas dynamics observed experimentally in Ref. [161], a 2D numerical code based on Monte Carlo simulation algorithm for explanation of the key physical features observed experimentally during exhaust gases transport from the rocket engine or plume to the background plasma was created in Ref. [169], accounting for molecules such as N_2, H_2O, CO_2, and H_2, the main molecular species of the exhaust stream (see the next section).

Using the corresponding aeronomic reactions with the background molecules O_2, O, and N_2, distributed in the wide range of the ionospheric altitudes (from 200 to 400 km), their rate, density, and temperature have been modeled as in Ref. [169].

Evolution of the H_2O-molecular cloud in the space and time domains accounting for the above-mentioned parameters was obtained in the specific numerical experiment using the initial parameters of the real rocket experiment. Thus, the initial rate of the exhaust plume was taken to be ~3 km s^{-1}, with the exhaust molecules rate of 5×10^{26} mole s^{-1}, observed during the processing time of 10 s.

As an example of a water vapor cloud spatial distribution at 10 s after the rocket burn (moving initially with velocity of 7.7 km s^{-1}), as well as its temperature and axial velocity at the moment of "observation" of 30 s is shown in Figure 11.15, extracted from Refs. [118,169].

It is clearly seen from the given illustrations that a huge exhaust of molecule transport and the corresponding heat transport are observed during the rocket burn plume, as well as the corresponding penetration of neutral cloud into the background cool ionospheric plasma. Thus, if the background plasma has temperature of around 2,000–3,000 K, the exhaust hot gas stream produced by the plume or engine can achieve 8,000–10,000 K. The kinetic energy, determined by the axial velocity of the neutral cloud, decreases fast (see the top panel) due to transfer of the energy of the exhaust "high-energy" molecular species to "low-energy" background molecules. These features will be discussed qualitatively in the next section and more precisely in Chapter 12.

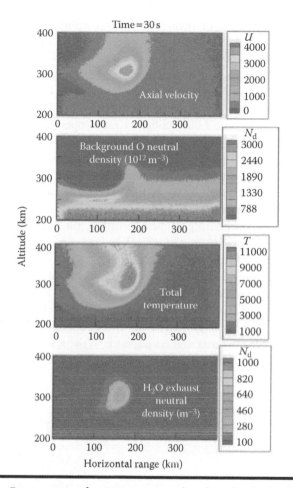

Figure 11.15 Parameters of water vapor molecular cloud 2D altitude–range spatial distribution obtained in numerical simulations of a 10-s rocket engine burn observed 30 s later.

11.2.1.2 Specific Features of Plasma Disturbance Generation by Rocket Burn and Launch

To observe and understand the so-called thin effects that accompanied the rocket engine or plume burn, complex experimental campaigns have been carried out during the recent past decade by using mutual space satellite and ground-based facilities, such as those of high-frequency (HF) and ultrahigh-frequency (UHF) range, incoherent and coherent radars, and optical systems combined with special spectral instrumentation [163–172].

We will present those of them that, in our opinion, can illuminate and explain some specific features and processes associated with rocket burns and launches. Thus, in Refs. [163–166], a mutual satellite–ground radar as well as optical spectral measurements has been carried out using incoherent UHF radar, located at Millstone Hill radio observatory and HF radar located over Jicamarca radio observatory. A deep hole of the TEC in the ionosphere was observed by the Millstone Hill incoherent radar operated in the UHF-range band and transmitting signals with radiated pick power of 2.5 MW generated by a steerable antenna 46-m high. Additionally, strong turbulent structures were observed during the rocket engine or plume burn in the perturbed ionospheric region above the Millstone Hill radar (the reader can find all details in Refs. [163–166]). It is important to note for our further discussions of the observed effects that according to measurements, the exhaust cloud from the rocket engine was directed across the UHF radar beam, as is clearly seen from Figure 11.16, extracted from Refs. [166,172].

Results of the spectral analysis made by the orbiting satellite ISS spectroscope (Figure 11.16) is presented in Figure 11.17, according to the results obtained in the earlier mentioned works (extracted from Refs. [166,172]). As is shown in Figure 11.17a, before the engine exhaust release, a standard emission ion-line in thermal spectra is observed that corresponds to the resonance of the ions O^+, a main ion consisting of the background ionospheric plasma at the F-region.

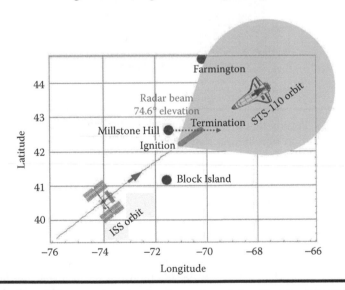

Figure 11.16 Exhaust plasma cloud produced by Shuttle STS-110 engine burn that is directed across the field of view from the Millstone Hill Radio Observatory.

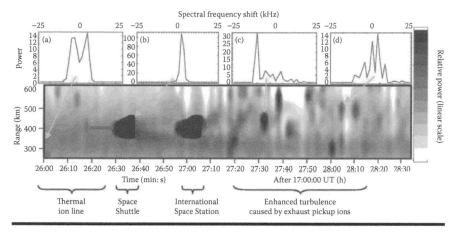

Figure 11.17 **The DS shape and the echoes from Space Shuttle before and after engine burn, and from the enhanced plasma turbulence.**

The panel denoted by (b) presents a δ-shaped spectral response of the echoes from a shuttle, whereas between 30 and 120 s after the rocket engine burn (see Figure 11.16c and d), a series of "side" maxima are clearly observed for more than 60 s in addition to the widespread Doppler power spectral pattern. The authors of Refs. [163–166] had explained the effect of broadening of the Doppler power spectral shape by introducing effects of strong plasma turbulent structures filling the global plasma holes produced by the dusty plasma exhaust surrounding the moving SV as well as in its wake behind the rocket.

At the same time, as was mentioned by the authors of Refs. [163–166], it is complicated to explain the effect of "splitting" of the wide DS shape into several maxima by the results obtained from incoherent radar only, using purely radio methods. In their opinion, this feature can be explained by the existence of different kinds of ions in the multicomponent exhaust plasma, which correspond to different ion-acoustic waves that excite the corresponding instabilities and generate small-scale plasma turbulences filling the large-scale plasma enhancements observed experimentally. These processes can be identified only by special optical spectroscopic instrumentation and techniques (see the corresponding results below). We will try to explain all these features based on the theoretical analysis described in Chapters 12 and 13, following the results obtained in Refs. [163–166].

It is interesting to discuss the observations recently carried out in the lower ionosphere, in which rocket engine burns were sometimes observed experimentally. Thus, in Ref. [168] a special experiment of the Space Shuttle that was launched at 18:36:42 h (eastern LT) on August 8, 2007, was reported.

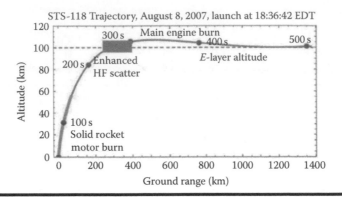

Figure 11.18 Trajectory of Shuttle STS-118 with the solid motor and main engine burns at altitudes of ~35 and ~100 km, respectively. EDT, eastern daylight time.

The main engine burn occurred at an altitude of about 199 km, as is shown in Figure 11.18 (extracted from Ref. [168]), where the trajectory of the SV is clearly presented. During the period of the main engine burn for 512 s, the exhaust gas with a mass of about 2.6×10^5 kg streamed at the rate of ~500 kg s^{-1}. The solid rocket motor burned at much lower part—at an altitude of 35 km, that is, before the entry of the SV into the lower ionosphere.

Using bistatic HF-scatter radar, a huge sporadic (e.g., turbulent) structure of small-scale turbulences filled the enhancement located at an altitude of about 100 km is evidently seen from Figure 11.19 (extracted from Ref. [168]). Each of the four panels is separated in Figure 11.19 by a 10-km range on the horizontal plane. The altitude of the scatter pattern is about 100 km, that is, it corresponds to the lower *E*-layer. This means that for 30–40 s, the dusty plasma exhaust spread axially about 30–40 km from the place of the engine burn. The corresponding recorded DS was about ±20 Hz (see bottom sentence along the horizontal axis). What is interesting to note is that this DS was not observed from the natural sporadic *E*-layer before the RL. After the RL, the observed "anomaly" was associated with the turbulent exhaust plasma structure created by the rocket engine burn.

Using the results of sporadic previously pioneering observations and systematic recent observations of the rocket burn and launch effects, researchers started to think of ways to repeat the observed phenomena, occurring in the background ionospheric plasma, using special active experiments based on special rocket releases of different kinds with special liquids and gases, which are similar to the main components or species of rocket engine and plume exhausts. Below, we present some of them that more adequately reproduce the effects and phenomena associated with SV burn and launch.

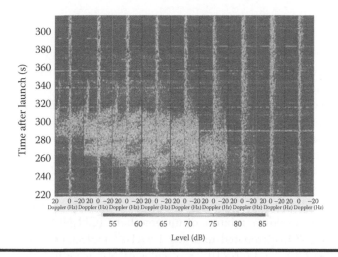

Figure 11.19 **Echo signals from the sporadic *E*-layer at an altitude of 100 km created by the Space Shuttle main engine burn on August 8, 2007, and recorded by bistatic HF radar. Each of the four panels is separated by 10-km range.**

11.2.2 Special Releases Related to Rocket Burns and Launches

The major purpose of the special "active-space" experiments was to check for those effects that can be created artificially by injecting different chemically reactive gaseous or liquid substances directly in the ionospheric plasma that yields the same effects observed during rocket burn and launch. During a series of previous experiments, it was found that engine fuel substances and other rocket exhausts consist of specific chemical elements, such as nitromethane, ammonium nitrate, nitrogen tetroxide (N_2O_4), liquid oxygen–hydrogen mixture, and so forth. These gaseous chemical mixtures during their reaction with the cold atmosphere release simpler molecules of gases, such as nitrogen, N_2; carbon dioxide, CO_2; hydrogen, H_2; oxygen, O_2; as well as molecules of water, H_2O [140–153]. The effects of rocket releases of more complex and heavier substances, such as sulfur hexafluoride (SF_6), were also investigated experimentally and theoretically [154–156].

It was shown experimentally and explained theoretically in Ref. [146] that the release of a chemically reactive gas, such as H_2, into the *F*-region of the ionosphere results in depletion of plasma density and significant plasma "hole" generation at altitudes from 300 to 350 km, depending on the weight of the initial releases. Thus, a 5-kg release at an altitude of 300 km over the magnetic equator at 19:30 LT produces a cavity region 40 km wide for about 1 min after the release. The TEC in this case reduced by 37% from its prerelease value. Increase of the weight of initial gas substance by four times (~20 kg) gives the releases, occurring at 300 km, depletion region 60 km wide with a 54% plasma density reduction. Release of the same gas of 10 kg weight at 350 km produces a plasma "hole" 65 km wide and a plasma density reduction of 51%.

Some specific active-space campaigns, called Logopedo, were designed to study the reaction of the ionosphere to the injection of large number of reactive molecules. The first report on preliminary results of such injection was published in Ref. [147], in which the effects of the special "Logopedo Uno" RL that happened on September 2, 1977, at 05:50:54 h (universal time, UT) from Kauai, HI, was described. In the 186th s of the rocket flight, when it was at an altitude of 261 km, an explosive mixture of nitromethane (CH_3NO_2) and ammonium nitrate (NH_4NO_3) detonated releasing 1.5×10^{27} molecules of H_2O, 1.5×10^{26} molecules of CO_2, and 6.6×10^{26} molecules of N_2, according to the chemical reaction of the corresponding mixture:

$$2CH_3NO_2 + 3NH_4NO_3 \rightarrow 9H_2O + 4N_2 + 2CO_2 \tag{11.1}$$

The experiment included the spectrometric technique, based on a specific photometer, phase coherent radar, and ionosonde for monitoring the background nonperturbed ionospheric plasma, which consists only of simple atomic ions of O^+ (see the next subsection), and the perturbed region caused by such an explosion consisting of more complicated ions, such as N_2^+, NO^+, O_2^+, and H_2O^+, respectively.

During this first stage of the experimental campaign, very important results were found that fully agreed with those observed during the rocket exhaust effects described earlier in this subsection. Thus, the observed data obtained by the combined optical and radio techniques showed that before the chemical release, only atomic ions O^+ ions dominated the ionospheric F-region plasma (this fact will be used in our further discussions on ionization–recombination balance in the nonperturbed ionosphere). Several seconds after the explosion, a huge decrease in the density of O^+ (at least on 3 orders of magnitude) was observed. One hundred and twenty seconds after the explosion, when the rocket was within 30 km of the depletion region, O^+ ions were encountered again with about half the density expected for a nonperturbed ionosphere. During this stage of observation of the depletion region occurring in the F_2-layer, an isotropic diffusive expansion of this hole was found that attained a 30-km radius from the release point location approximately 2 min after the explosion. All these features of the ionospheric O^+ ion density distribution and the cavity spatiotemporal evolution are in reasonable agreement with the experimental observations and theoretical predictions described in Refs. [148–164]. We will return to some of these theoretical frameworks in our further discussions.

A second experimental campaign, called "Logopedo II," that consisted of two high-explosive ionospheric chemical releases occurring on September 11, 1977, was carried out from the same place, Kauai Test Facility, Barking Sands, HI. The similar aim of such an experiment was to study the ionospheric plasma modification by the rocket that produced CO_2, H_2, and H_2O and injected them into the F_2-layer, which consisted of simple atomic O^+ ions. The corresponding explosion detonated at 241 s after the RL, releasing 1.45×10^{27} molecules of H_2O, 2.19×10^{26} molecules of CO_2, and 4.3×10^{26} molecules of N_2 at the altitudinal peak of the F_2-layer, which was indicated as the beginning of the observed "event." The event was about 0.8 km above the rocket trajectory, that is, at an altitude of about 283 km. It was mentioned that the resultant shock wave, coming after the detonation, did not damage the corresponding optical and radio measurements or the rocket devices. We do not describe all the elements of the complex measurements and devices used during this campaign, instead suggesting Ref. [151] to the reader, where the detailed description and the corresponding analysis of the "Logopedo II" results are presented.

We will bring the reader's attention to the main results of such rocket releases that are associated with those observed during the rocket burn and launch caused by the corresponding exhausts. Thus, sounding the nonperturbed background ionosphere before the event at the altitudes of the E-layer of the ionosphere, the existence of electrons (e^-), positive ions (O_2^+), and NO^+ was found, whereas after the event only atomic metallic ions of Na^+, Mg^+, Al^+, Fe^+, and Ca^+ at these altitudes were observed, from which Mg^+ was the dominant species. From 130 km (upper E-layer) to 283 km (the peak of the F_2-layer), the electron (e^-) concentration and the density profile of seven ion species were detected by the corresponding spectrometer after the rocket release (the event), such as N^+, O^+, N_2^+, NO^+, O_2^+, H_3O^+, and H_2O^+. To explain the existence of all positive ion species, a corresponding model of the interactions of the molecules of the explosion products with the molecular and atomic positive ions was prepared, that existed in the undisturbed ionosphere at these altitudinal ranges. The proposed approach, despite its extremely complicated form (consisting of numerous chemical reactions and complicated corresponding schemes and equations), gave a satisfactory explanation to the results of these explosive chemical releases.

The experiments described earlier were carried out either in the middle latitude ionosphere or near the magnetic equator, where coupling between the ionosphere and the magnetosphere (as subsystems of the global EAIM system) are not so evident, as is expected in the polar ionosphere, because many additional electromagnetic and geophysical effects accompanied the rocket burns and launches. Therefore, we bring the reader's attention now on a special Project "Waterhole," which is a chemical release experimental campaign whose major aim was to study perturbations in the *F*-region of the ionosphere by a sounding rocket carrying 88 kg of highly explosive material that after detonation injected an aurora arc into the ionosphere. This campaign, containing the plasma diagnostic instrumentation, was a joint Canadian–US experiment to investigate the "reaction" of the auroral ionosphere on artificially induced plasma depletion. The RL was done from Churchill Research Range (Canada) on April 6, 1980, at 03:33:01 UT. The geometry of the rocket trajectory, its projection down to emission altitude (up to 100 km), the point of the explosive release, and the location and orientation of the aurora arc (contour of peak intensity) at the time of the release are shown in Figure 11.20 (extracted from Ref. [153]). The numbers near trajectory of the rocket flight indicate the altitude of the rocket.

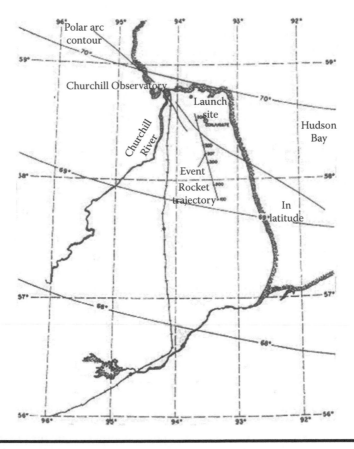

Figure 11.20 **The mapping of special US–Canadian campaign "Waterhole" happening on April 6, 1980. The trajectory of the rocket is shown by a straight line (rearranged from Ref. [153]).**

Besides many other geophysical purposes, the concept of the experiment was to prove artificially made chemical injections or explosions of natural molecules for depleting the background plasma density and to study further for spatiotemporal evolution of the cavity, similar to those observed during "active experiments" with the burn and launch of rockets and SVs. The project "Waterhole" used the same components for the explosion material, as was used in the Logopedo release experiment.

Due to detonation at 88 km, the explosive mixture of nitromethane and ammonium nitrate produced (after complete detonation) 1.5×10^{27} molecules of H_2O, 6.6×10^{26} molecules of N_2, and 3.3×10^{26} molecules of CO_2 at the F-region of the ionosphere, following the chemical reaction described by Equation 11.1. As mentioned earlier, the dominant ion is the atomic positive ion O^+ in the nonperturbed F-region of the ionosphere, which results from the equilibrium between ionization and recombination reactions occurring in the corresponding layer of the ionosphere (see the next subsection). The H_2O and CO_2 molecules, produced from the chemical reaction given in Equation 11.1 and injected into the F-layer, have been exchanged with the ambient ions O^+ to produce molecular ions O_2^+ and H_2O^+. As shown in Ref. [153], the rate of the O^+ exchange is significantly higher than that for ions H_2O^+ and O_2^+. Therefore, a drastic decrease in O^+ density is clearly observed during radio and optical (spectral) observations.

Moreover, because of shorter lifetimes of the molecular ions compared to atomic ion O^+, the total plasma density (plasma is quasineutral, and $N_e \approx N_{H_2O^+} + N_{O_2^+}$) also decreased drastically, that is, an ionospheric cavity (hole) was produced. These two effects are clearly seen from one of the examples of measurement results shown in Figure 11.21 (extracted from Ref. [153]) as the ratio of the density of molecular ions (indicated by XY^+) to O^+ ion density (top panel), and as the relative electron density (relative to its minimum peak inside the hole; bottom panel). The vertical line indicates the time of explosive release at 03:37:14 UT. Thus, from the top panel it follows that the enhancement in ratio inside the hole achieves a factor of more than 10 with simultaneous 2-order depletion of the plasma density magnitude within the hole that expands up to about 54 km from the release explosion location during approximately 2.5 min. The observed expansion is more or less symmetric indicating a quasisymmetrical isotropic diffusion of plasma inside the depletion region (e.g., the hole).

A series of real measurements and the corresponding theoretical models described in Ref. [153] indicate that the reduction in the plasma density inside the cavity, generated at the 300-km altitude of the ionosphere by the release of explosive-produced molecules (mostly water) can achieve more than 1 order of magnitude. Within several minutes, the depletion region can expand to about a 55-km range from the explosive release location point. Other observations

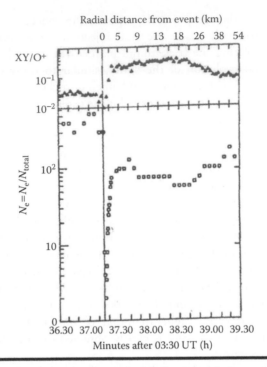

Figure 11.21 *Top view*: **Normalized distribution of the exhaust gas molecules (denoted by XY) to the ambient ion O⁺ before and after the rocket burn (denoted by the vertical line).** *Bottom view*: **Normalized plasma density before and after the rocket burn.**

also show strong interaction of the aurora arc with the ion depletion region via the field-aligned currents, whose density before and after the release (after 130 s) appeared to be the same (of few microampere per square meter) and changes in the optical auroral morphology that was nearly coincident with the previous event of explosive release. After 2–3 min of the event, the measured electron density again returned to the previous level. Finally, we can emphasize, following the conclusions made by the authors of Ref. [153], that in a series of combined optical and radio measurements, a drastic charge exchange of the release molecules with the ambient ions O^+ was clearly shown, which yields not only to its decrease over 2 orders of magnitude but also to a simultaneous increase in the neutral molecule density against the plasma density.

Reference [143], which presents a general review of the past and recent experimental campaigns associated with RL effects, had stated the main feature of these effects: generation of ionospheric holes caused by the rocket exhaust and engine plume, fuel jets, and gaseous streams into ambient nonperturbed

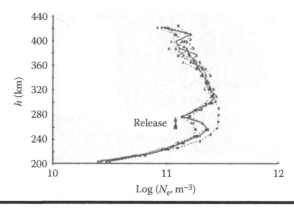

Figure 11.22 **The depletion region temporal dynamics during Spacelab-2 RL in a special campaign. Each profile was obtained for each continuously changed time of the sounding signal record.**

ionospheric plasma. We present additional evidence taken from this special issue in Figure 11.22 (according to Ref. [143]), which clearly shows a steep decrease of total plasma content in the F region of the ionosphere during the rocket release at altitudes from 250 to 300 km.

Ref. [158] mentions that using backscatter radar system, radar echoes from the exhaust plume of the Space Shuttle orbital maneuver engines (OMEs) firing in the ionosphere were found for the first time. Measurements of echo signals scattered from orbital maneuver system (OMS) exhaust were carried out on July 30, 1985, during Spacelab-2 launch over Arecibo, PR. The two engines fired simultaneously for 16 s at an altitude of 317 km, that is, near the F-region peak. The engines' exhaust consisted of N_2O_4 and monomethyl hydrazine (MMH) propellants. Finally, after aeronomic chemical reactions, roughly 33% of the exhaust comprised water vapor (H_2O) creating ice in the plume and moving (with respect to the Shuttle) with speed ~3.07 km s^{-1}. The exhaust left the nozzle with an estimated temperature of ~931 K.

The same features as were predicted theoretically in Refs. [141–143, 146–148,154–164] were found, which are the following:

1. Exhaust from thrusters reacts with neutral atoms of oxygen (O) and ions (O$^+$) at altitudes of the F-region to yield excited gas species with the observed optical emissions.
2. Emission extension and expansion of such an excited molecular gas have been produced by ion–molecular reactions between ambient ions O$^+$ and the OMS exhaust gas which can, finally, achieve a radius of more than 50–60 km.

3. Compression of the ambient plasma electrons to produce higher density of plasma surrounding the OMS.
4. Molecular ions can be rapidly lost via dissociative recombination with ambient electrons after chemical reactions occur between the exhaust molecules and ambient ions O^+ (see the next section).

   Moreover, due to cooling of the ion gas, and ice crystal generation by water vapor, as a main "player" in the corresponding ion–molecular reactions (see further), the ice particles become electrically charged and can, finally, create "dusty plasma" [165–168].
5. Existence of the ice crystals in the exhaust yields the attachment mechanism between them and the ambient plasma electrons in the exhaust plume. This process was supposed to be the exciting mechanism of the "dusty plasma" formation surrounding OMS.
6. Finally, it was found experimentally that charge exchange between ambient ions O^+ and cold exhaust molecules yields low-temperature ion beams, which excite weakly damped ion acoustic waves (the latter feature will be discussed in the last subsection).

All the above-mentioned effects yield creation of a very complicated plasma profile observed at the *F*-region altitudes. Thus, as seen from Figure 11.23a (rearranged from Ref. [158]), before the burnout of OMS engines, the *spike* in the echo signal was decoded evidently. The authors of Ref. [158] related this spike to the reflection from the Shuttle body. At the same time, the lower and the upper peaks were associated with enhancements occurring due to ion–molecular reactions of molecules with the ambient ions O^+. Eleven seconds later, after the burnout of the OMEs, the spike becomes negligible and a deep "hole" (e.g., cavity) with significant decrease of the bottom and upper peaks was observed experimentally (Figure 11.23b; rearranged from Ref. [158]).

To explain the processes associated with enhancements of plasma density surrounding the OMS plume, the process of exhaust plume ion beam production was sketched schematically in Figure 11.24. Here, the chain of processes mentioned above can be described by the following relationship between processes:

$$\left\{ \begin{array}{l} \text{Exhaust} \\ \text{gas} \\ \text{expansion} \end{array} \right\} \leftrightarrow \left\{ \begin{array}{l} \text{Cooling of the} \\ \text{ambient} \\ \text{plasma by} \\ \text{the exhaust gas} \end{array} \right\} \leftrightarrow \left\{ \begin{array}{l} \text{Charge exchange} \\ \text{between exhaust} \\ \text{molecules and} \\ \text{ambient ions} \end{array} \right\} \quad (11.2a)$$

A second stage of exhaust plume dusty plasma generation surrounding OMS and its wake, shown schematically in Figure 11.24, can be described according

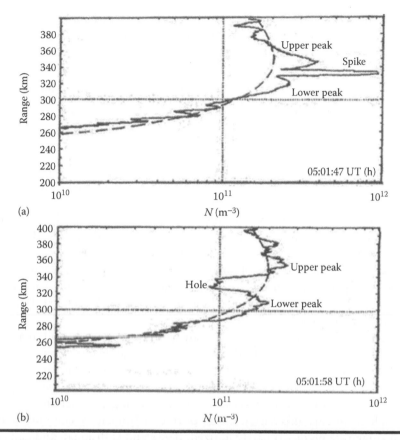

(a)

(b)

Figure 11.23 **Plasma profile (normalized to the electron density 20 s before the start of the Space Shuttle) obtained by ISR: (a) the profile during the rocket burn with the spike bordered by two peaks; (b) the profile after the burn with a central hole between the two peaks.**

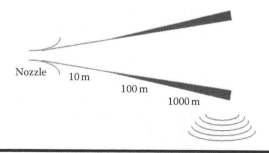

Figure 11.24 **The sketched process of plasma wake creation during plume exhaust molecular gas interaction with the ambient cold ionospheric plasma.**

to Ref. [158] as follows. Due to the cumulative effect of the OME exhaust water vapor mixed with ice crystals, the following chain of the processes can be expected:

$$
\left\{
\begin{array}{l}
\text{Expansion} \\
\text{of exhaust} \\
\text{mixture}
\end{array}
\right\}
\leftrightarrow
\left\{
\begin{array}{l}
\text{Heating of the} \\
\text{ambient plasma}
\end{array}
\right\}
\leftrightarrow
\left\{
\begin{array}{l}
\text{Condensation of the} \\
\text{exhaust dusty} \\
\text{or "dirty" plasma}
\end{array}
\right\}
\quad (11.2b)
$$

As shown in Figure 11.24, during the interaction of exhaust gases with ambient regular plasma, the so-called polycomposite plasma was created, consisting not only of electrons, atoms, and ions of the background but also molecular and atomic ions generated by ion-recombination reactions described earlier. This kind of plasma as "dusty" or "dirty" plasma is mentioned in Refs. [173–178]. We should note that in the corresponding literature [173–178], *dusty* or *dirty* plasma is the name given to plasmas heavily laden with charged dust grains (electrically polarized domains, or cusps) that, together with the background ions and electrons, constitute a new kind of plasma regime (see Ref. [173] and the literature therein).

Consequently, such plasmas initiate a new plasma regime that may include weakly or fully ionized, collisionless, or collisional plasma, which in the scenarios under consideration contains charged and uncharged dust or colloidal gains (namely, ice crystals observed during rocket release experiment [158]) in the broadest sense. Thus, converting this definition to the ionospheric plasma affected by rocket burn and launch, we can outline that such composite plasma is partly ionized, dense and collisional, and can generally be termed as *dusty* or *dirty* plasmas following the definitions made in Ref. [159].

Ref. [159] mentions that modeling was done for altitudinal distribution of the density of the plume exhaust molecules expanded into the surrounding ambient plasma and for ambient electron plasma content using the previous performed theoretical framework [154–156]. Figure 11.25 (rearranged from Ref. [158]) presents the results of numerical computations of the altitudinal distribution of plume release neutrals such as O_2, N_2, and O (left panel), as well as of molecular ions O^+ and electrons e^- (right panel), which are predominant at altitudes beyond 300 km (the rocket release occurred at 317 km).

The increase of N_2 molecules as a product of the aeronomic reactions within the exhaust gas, and deep and slight depletion of the ion and electron components of ambient ionospheric plasma, respectively, is clearly seen around the height of plume gas release.

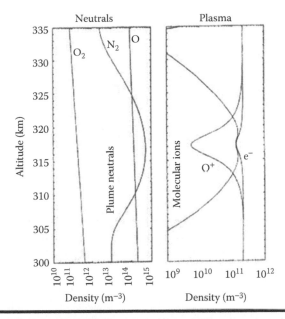

Figure 11.25 Altitudinal distribution of the plume neutrals and the ion and electron components of the ambient ionospheric plasma during the rocket burnout at an altitude of 317 km.

It should be noted that despite a strong depletion region being predicted theoretically in Refs. [140–143,146–164], accounting for the cooling, and then heating process, the effects of ion–molecular exchange reaction occurring in the exhaust plume "dusty" plasma and the complicated plasma profile shown in Figure 11.23a and b, cannot be fully described without accounting for the transport processes prevailing at the altitudes of the *F*-region.

To complete this paragraph, we will present a rocket release campaign called CARE I and CARE II carried out recently and described in Refs. [171,172]. The mutual rocket chemical release and the rocket engine burn occurred during the campaign CARE I on September 19, 2009 at 18:46 h (LT) at an altitude of 280 km, which contained a dusty mixture of AlO_2 molecules with a mass of 111 kg (of the rocket release) and of the exhaust molecular vapor with a mass of 200 kg (produced by the engine burn). The trajectory of the rocket and the horizontal range from the place of RL are shown in Figure 11.26 (extracted from Refs. [171,172]). The radio beacon, as the transmitter operating at dual frequency, 150 and 400 MHz, was used to sound the perturbed ionosphere and to measure the TEC in the ionosphere. A ship located below the place of the

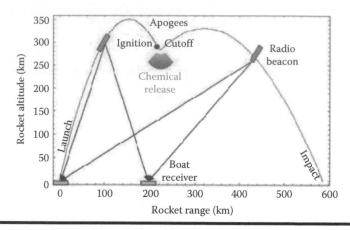

Figure 11.26 Schematically sketched trajectory of the RL in the CARE I campaign with radio beacon, as the receiver and the boat receiver.

rocket release and its engine burn receives the echo signals scattered from the ionosphere (as depicted in Figure 11.26). The measured TEC in vertical directions (denoted as VTEC) before and after the simultaneous rocket release and the engine burn vis-à-vis the altitude is shown in the left panel of Figure 11.27, according to Refs. [171,172].

The derivative of the VTEC vis-à-vis ionospheric altitudes is presented in the right panel of this figure. Both panels clearly show the creation of plasma holes in the perturbed stratified layered ionospheric structures produced by the simultaneous rocket release and the engine exhaust burn.

Additional measurements in the framework of CARE I campaign using UHF-range ISR, located at Millstone Hill observatory (see details in Refs. [171,172]), allow measuring and presenting the integrated TEC profile of plasma density along the radar beam before and after the rocket release and its engine burn (Figure 11.28; extracted from Refs. [171,172]).

It can be clearly seen (and predicted theoretically in Refs. [171,172]) that exhaust molecules can drastically reduce plasma density at the bottom of the x ionospheric F-region due to aeronomic and chemical reactions. Finally, we observe the same features as depicted in Figure 11.27 regarding the beacon radio sounding of the ionosphere perturbed by rocket release and engine burn.

Spectral analysis of the exhaust ions generated by the rocket release with simultaneous radio sounding of the perturbed ionosphere was proposed in Refs. [171,172] using color CCD optical camera. The corresponding spectral technique recovered the major emission species in the release and indicated and identified the electron–ion emission spectral lines. Using spectroscopy with

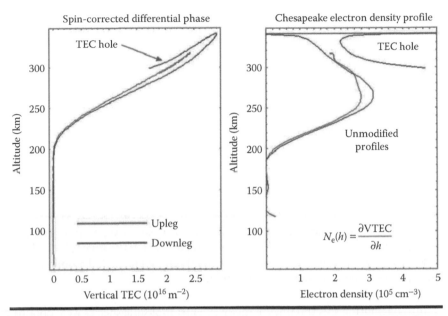

Figure 11.27 TEC and electron density disturbances by the chemical release determined by radio beacon propagation from the CARE I rocket to a ground receiver.

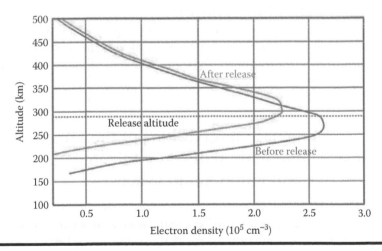

Figure 11.28 Electron density profile around the rocket release point at 280 km of the CARE I rocket campaign obtained by the Millstone Hill ISR on September 19, 2009.

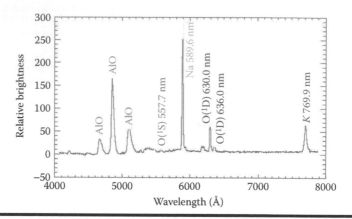

Figure 11.29 **Spectral lines of the molecules AlO and oxygen (O) generated during aeronomic reactions caused by rocket dust and molecular exhaust release and of the red-line emissions (at 769.9 nm) from excited atomic oxygen produced by electron–ion recombination.**

2-min integration spectral lines of dusty exhaust molecule species were obtained (Figure 11.29; extracted from Refs. [171,172]).

As clearly seen, as a result of aeronomic reaction of aluminum (Al), the lines of AlO are formed as one of the species of exhaust molecular gas with oxide (O). Additionally, according to the corresponding chemical reactions of electron–ion recombination, red-line emission (at 769.9 nm) from the excited atomic oxygen was produced (see the left side of Figure 11.29).

Therefore, the complicated plasma evolution registered by the backscatter radar accounts for the ionization–recombination chemical reactions and transport processes such as thermodiffusion, diffusion, and drift of plasma that occur at considerable altitudes.

11.3 Experimental Observations of the Effects Associated with Artificially Injected Clouds and Pump Radio Waves

As mentioned from the outset, the era of rocket, spacecraft, and satellite (also called SVs) burn and launch started simultaneously with active experiments in the near-the-Earth environment (particularly in the ionosphere), using those techniques and methodologies that can be used based on the possibilities of the launched SVs. Moreover, combining the technical possibilities discovered by the SVs with various advanced methods of environmental study using ground-based facilities, complex campaigns have been performed by different countries and groups of countries worldwide.

The other methods of modification of the ionospheric plasma are related to heating of ionospheric plasma by powerful radio waves (called pump waves) from ground-based high-power radar systems. We will briefly present only such experimental campaigns that allow us to understand similar features and processes associated with SV burns, launches, and flights.

11.3.1 Injection of Ion Clouds from Geophysical Rockets

Experiments for injecting various ion beams and clouds in the ionosphere and magnetosphere of the Earth began in the end of the 1950s and continued until the end of the 1990s in the last century (see Refs. [15–22,104–121] and the references therein). The main goal of these experimental campaigns for direct measurements with the help of rockets, spacecrafts, and satellites was to obtain precise information on the geophysical parameters, characteristics, and natural processes occurring in the space near the Earth. We will describe only those that are directly related to the subject matter of our book, that is, describe the effects and peculiarities in the evolution of injected clouds and beams and the plasma structures artificially created by them at the corresponding altitudes of the ionosphere that are found similar to those observed during the rocket burn and launch described in the previous subsection.

Thus, the detailed experiment of rocket injections of beams and clouds based on visual optics and radio sounding observations carried out in space and time domains were studied simultaneously in the United States [15,19,104–110,113–118] and the former USSR [111,112,119–121]. First, we will present a systematic active-space experiment described in Ref. [106], in which the rocket injection of a cloud of barium species was done at the ionospheric altitude of 194 km above the Florida experimental site in 1967. Detailed observations were carried out for 930 s from the moment of barium cloud creation. Analyzing the isoclines of equal optical lightness observed separately from three experimental sites, it was observed that the center of the cloud expands symmetrically and isotropically downward (estimations gave the velocity at 27 m s^{-1}) from 194 to 168 km for 500 s (Figure 11.30; extracted from Ref. [22]) with the plasma concentration inside it at about 10^{16} m^{-3}.

Thereafter, the expansion of the ion cloud becomes asymmetric with deformation of its initial spherical form and generation of fingerlike plasma structures, called strata, elongated along ambient geomagnetic field $\mathbf{B}_0$ lines. During the entire time of observation, the cloud itself moves across $\mathbf{B}_0$ with a velocity of 35 m s^{-1} relative to the neutral wind. The dimensions of the created strata are the length along $\mathbf{B}_0$, $L = 30$ km; the height from the center of the cloud, $h = 10$ km; and the width across $\mathbf{B}_0$, $d = 0.6$ km.

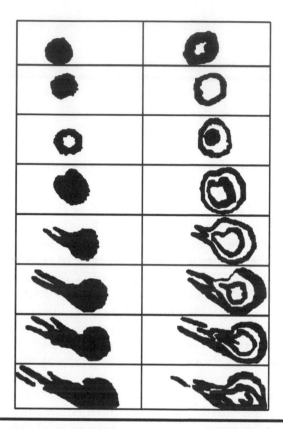

Figure 11.30 The plasma density (left panel) and brightness (right panel) evolution in the space and time domains observed experimentally.

Accounting for these dimensions, the concentration of barium ions inside the strata was estimated to be about $6 \times 10^{12} \, \text{m}^{-3}$, while at the same range of altitude the concentration of background ionospheric plasma does not exceed $\sim 3 \times 10^{12} \, \text{m}^{-3}$, that is, it is lesser. The same result was obtained in another set of experiments of injection of the cloud with neutral molecules of barium, according to Ref. [108] whose result is depicted in Figure 11.31 (extracted from Ref. [22]). Here, dotted line shows the initial position of the injected neutral barium cloud.

It is evident that, during a definite time, after ionization of neutral molecules of barium, the ion cloud is deformed by the floating of its plasma along $\mathbf{B}_0$ with further striation at the strata stretching along $\mathbf{B}_0$ at the back of the cloud.

In conjunction with the above experiments carried out in the United States, the same barium release experiments were carried out in the former USSR from 1975

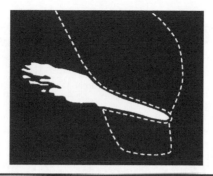

Figure 11.31   Brightness of the ion cloud at the last stage of its evolution compared to initial position of the neutral barium cloud indicated by dashed contour.

(see the references in Refs. [16–18,21,22]). This campaign was called "Spolokh." Following the campaigns, called "Spolokh-1" and "Spolokh-2," precise analysis of observations of evolution of both neutral and ion clouds of barium and the corresponding physical analysis were given in Refs. [107,111,112,120,121]. First, cumulative injections were performed at an altitude of $z - 170$ km close to the vertical direction ($\propto 2°-5°$ to $\mathbf{B}_0$)

Observations were made during $T = 10$–15 min from three sites (in "Spolokh-1") and from two sites (in "Spolokh-2") separated 50–300 km from each other. In Figure 11.32 (extracted from Ref. [22]), the photos of neutral barium and ion clouds obtained from three sites are denoted by 1–3. Observations were absent in some fragments for all three sites. All photos correspond to the following moments of time (from top to bottom, respectively): 20, 50, 150, and 400 s, and had registered the fast symmetrical expansion of the neutral Ba cloud that after 20 s has a radius of about 2 km. After this time, the strata of ions Ba$^+$ stretched along $\mathbf{B}_0$ were created due to ionization of the neutral component of the barium cloud.

After 45 s from that moment, the floating ions of Ba$^+$ in the form of finger-like strata, elongated along $\mathbf{B}_0$, were clearly observed from experimental site 1 and less clearly from site 2.

The following peculiarities: (a) parallel movement of the spherical neutral cloud and the main ion cloud at 32 s time after injection (Figure 11.33a; extracted from Ref. [22]); (b) further, at $t = 100$ s, floating of the ion cloud, generated by the ionization process, from the neutral cloud (Figure 11.33b) were also observed in experimental campaign Spolokh-2. At the later stage of ion cloud evolution (at $t = 15$ min, Figure 11.33c) a tail with a strata structure was clearly observed. By careful optical and spectrometric analysis it is seen that the

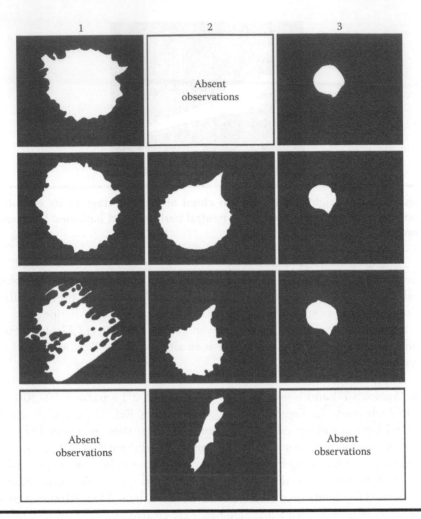

Figure 11.32 Brightness of the barium cloud evolution obtained experimentally from different observation stations, corresponding to the *left, middle,* and *right* panels. Sometimes the data were absent.

estimated concentration in the main ion plasma cloud exceeds the concentration of background ionospheric plasma by more than 1 order.

For 300–400 s after the release, all ions floated from the main plasma structure to the tail structure. Accounting for the homogeneous distribution of ions inside the ion cloud with dimensions $L \approx 20$ km (along $\mathbf{B}_0$), $d \approx 1$ km (across $\mathbf{B}_0$), the average concentration of plasma within the ion cloud was estimated to be about $N_c \approx 10^{13}\,\mathrm{m}^{-3}$ that coincides with the results obtained

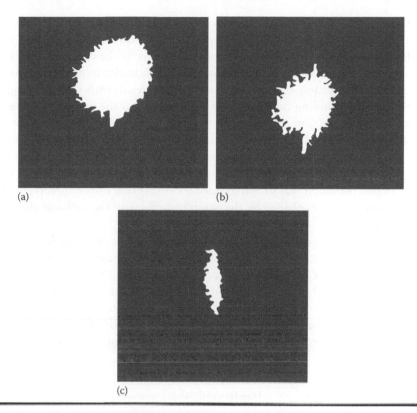

Figure 11.33 Spatiotemporal evolution of the neutral and ion barium clouds obtained in Spolokh-2 experiment. (a) $t = 32$ s, (b) $t = 100$ s, and (c) $t = 15$ min after injection of ion cloud.

using radiolocation measurements during these experiments, which yielded $N_c \approx (6.3 \times 10^{12} - 1.4 \times 10^{13}) \, \text{m}^{-3}$.

This value is about 2 orders greater than the concentration of the background ionospheric plasma at altitudes of 160–170 km. At the same time, estimations made for the tail structure yield $N_c \approx 2 \times 10^{11} \, \text{m}^{-3}$, which is on the same order of concentration of the background ionospheric plasma.

What is interesting to note is that similar results further on evolution of neutral and ionized barium clouds were obtained from optical observations during the entire period until they vanished simultaneously as shown from the experiments carried out in the United States [110] and the former USSR [107]. The following results were obtained. Initially, the neutral cloud spreads isotropically, and the concentration profile of neutral atoms of Ba observed experimentally is close to Gaussian. A certain amount of atoms of neutral Ba decreases due to solar ionization with characteristic time $\tau_i \approx 20$ s. The initial distribution of ions Ba^+ in the ionized component of the plasma cloud was also close to Gaussian with half-width $d_n(\tau_i) \approx 1$ km. The concentration of ions in the center of the cloud was $N_c \approx 6 \times 10^{12} \, m^{-3}$, that is, greater than that of the background plasma. At the initial stage of creation, diffusion spreading of ion clouds with velocity along $\mathbf{B}_0$ with the corresponding coefficient was observed. From the outset, the longitudinal profile of Ba^+ ion concentration was still Gaussian. As for the transverse profile of Ba^+ ion cloud, the essential asymmetry was observed to increase with time.

Because the influence of transverse diffusion on the ion cloud spreading begins only when the moment of time exceeds about 2×10^3 s, that is, the total time of observation, it was shown in Refs. [119–121] that the asymmetry of Ba^+ profile can only be explained by the combined effects of the ambient electrical and magnetic fields and the wind of neutrals.

But what can be assumed about the profile of the ion cloud during the process of its stratification on several strata? It was clearly shown experimentally in Refs. [119–121] that by using precise optical filters after 45 s from the moment of injection, several strata occur floating in the tail in the form of a cloud (Figure 11.34; extracted from Ref. [22]). Here, the temporal dynamic of the brightness profile (i.e., plasma density) is shown in the direction of the floating ions. As can be seen, despite the fact that the "mother" ion cloud remains stable

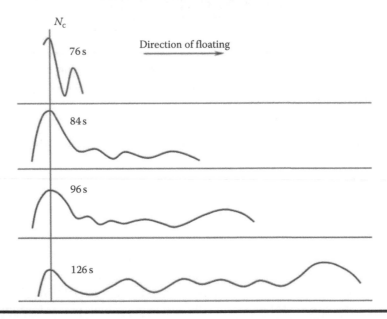

Figure 11.34 Temporal evolution of the main stratum and splitting into secondary strata in the direction of ion drift in ambient electric and magnetic field.

during the time of observation, two to three strata leave the "mother" plasma structure immediately at the beginning of the process of stratification.

Each stratum has asymmetrical brightness profile that is smoother than the sharp profile of the main ion profile. From the commencement of this process, the concentration in each leaving strata exceeds that for the background plasma. Finally, a special campaign comprising four complex experiments described in Refs. [119–121] was carried out using two TV cameras and photographic cameras with high spatial (~50 m) and temporal (~0.2 s) resolution. In these experiments the striation mechanism was observed clearly and the fine filament structure of strata elongated close to the magnetic field lines (with deviations not more than 1°) was recorded in detail. Because in each release similar picture of ion plasma cloud evolution was observed, we present only one of them here.

Thus, in the first stage of the cloud's evolution (within about 30 s of each release) when density of the plasma was about 2 orders of magnitude higher than the background density, the dispersive symmetrically isotropic diffusion stage (Figure 11.35a; extracted from Ref. [22]) occurs. During this time, the cloud began to split into a neutral quasispherical and ionic cloud. About 50–60 s after that, the ion barium cloud began splitting into an ionic cluster and stretched

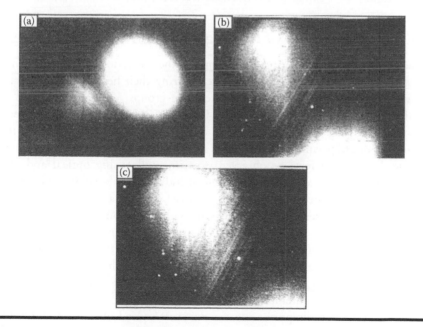

Figure 11.35 (a–c) Evolution of stratified ion plasma grating structure and neutral barium cloud observed in "Spolokh-2" experiment.

along $\mathbf{B}_0$ (Figure 11.35b). In a further process of evolution (from 160 up to 190 s), the striated structure (the grating of the primary strata) was placed into a cluster of $\mathbf{B}_0$ field-aligned plasma filaments (called secondary strata) concentrated within the disturbed region with a 5–6 km size across to $\mathbf{B}_0$ with a 22–25 km size along $\mathbf{B}_0$, respectively (Figure 11.35c).

The transverse scale, $\ell_\perp$, and longitudinal scale, $\ell_\parallel$, of the separate stratum were estimated to be within $\ell_\perp \approx 0.1 - 0.3\,\mathrm{km}$ and $\ell_\parallel \approx 20 - 25\,\mathrm{km}$, that is, the degree of anisotropy is $\ell_\parallel/\ell_\perp \approx 100$. This result is in a good agreement with the estimations obtained in Refs. [111–118], where, for commencement of striation mechanisms, it was noted that the degree of anisotropy should not be less than 100 for the plasma cloud structure.

Thus, we discuss a strong, large-scale plasma structure created in the ionosphere by injection of ion or neutral cloud from a special geophysical rocket. Moreover, it is evident that the ambient magnetic and electric fields, together with the neutral wind occurring at these altitudes, can significantly change the initially symmetrical isotropic expansion of barium ions into the ambient ionospheric plasma and create stratified small-scale fingerlike plasma structures (called *strata*). In other words, we see a similar evolution of ion cloud observed experimentally as also during rocket releases of different kinds of molecular species and rocket exhausts described earlier in the previous subsection.

11.3.2 Observation of Heating-Induced Plasma Structures

The possibility of generating enhancement and depletion of plasma density in the E- and F-regions of the ionosphere during their heating by pump radio waves was separately pointed out by separate groups of researchers from the United States [50–53] and the former USSR [54–62]. We will briefly present some of them that give similar results of the investigation into the disturbed heating-induced plasma structure, in its spatiotemporal evolution. Among many artificially induced features observed experimentally using optical and radio observation, special attention was given to indicate the possibility of creating a region with enhancement (i.e., with increase) of plasma concentration in the D- and lower E-layers, as well as regions with depletion (i.e., with decrease) of plasma density in the ionosphere at altitudes of upper E- and F-layers. The facilities for incoherent scattering of radio waves were used in these experiments for studying large-scale plasma disturbances with the horizontal dimensions $L_h \approx \theta_A h$, where h is the height of heating and θ_A is the width of the diagram of power transmitter antenna.

Thus, systematically precise experiments were carried in Arecibo observatory (Puerto Rico) using high-power (pump) radio wave operating at the frequency $f_{ex} \approx 3.175\,\mathrm{MHz}$ with power inside the powerful beam $P_0 = 300\,\mathrm{kW}$ with gain

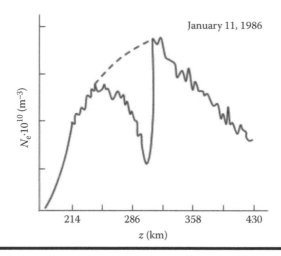

January 11, 1986

z (km)

Figure 11.36 **Deep depletion region generated during artificial heating of the ionospheric plasma at an altitude of about 300 km from ground-based powerful radar.**

$G = 23$ dB. Such a facility allows radiating a radio beam with power density inside it $\sim 50\,\mu\mathrm{W\,m^{-2}}$ at altitudes of perturbed heating region (PHR) location [50–53]. Figure 11.36 (rearranged from Ref. [22]) shows a strong depletion in the profile of TEC along altitude h in the F-layer of the ionosphere (at altitudes of 270–290 km) during nighttime heating process. The corresponding undisturbed profiles of ambient plasma density at the altitudes under consideration are shown by dotted curves.

It is evident that during local heating a depletion region of plasma concentration (i.e., the cavity or "hole") occurs, whose effect was predicted by the self-consistent model of local heating described in Refs. [60–62].

The horizontal dimension of the cavity was compared with the dimension $\sim \theta_A h$, and the cavity was elongated along $\mathbf{B}_0$. Some discrete structures were found inside the cavity, moving insignificantly along the gradient of plasma concentration with effective velocity drift $0 < V_d < 100\ \mathrm{m\,s^{-1}}$.

Diagnostics were carried out by the radio locator working in pulse regime (with $\tau_{\mathrm{pulse}} \approx 52\ \mu\mathrm{s}$ and with the repetition time of 10 μs), which allows obtaining the height profiles of plasma density $N(z)$ with a resolution of 600 m.

When separations with the resonance frequency $f_{\mathrm{ex}} \approx 3.175\ \mathrm{MHz}$ were performed, the specific dynamics of the cavity region was observed. Thus, in Figure 11.37 (extracted from Ref. [22]), the dynamics of the cavity region and its structure are presented after switching off of the heating radar with separations from $f_{\mathrm{ex}} \approx 3.175\ \mathrm{MHz}$ (see curves from the top to the bottom): the top curve corresponds to separation at 1.2 kHz; the lower curve is at 1.7 kHz, the next is

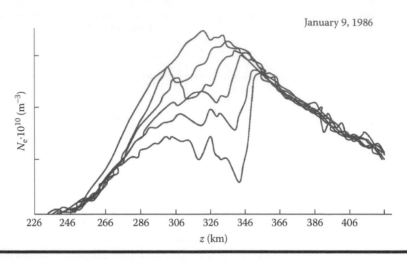

Figure 11.37 Evolution of depletion area and its structure during the heating experiment by using powerful transmitter.

at 2.2 kHz; further, the next is at 2.7 kHz; still further, the next is at 3.2 kHz; and the bottom curve is at 4.2 kHz. This heating regime can be predicted by the model of the so-called adaptive resonance heating of plasma as was predicted theoretically in Refs. [58,59]. As is evident, each minimal value of the plasma density profile within a cavity follows current resonance frequency.

With increase in frequency separation (from 1.2 to 4.2 kHz), the corresponding negative plasma extreme moves downward along the height of the ionosphere.

The cavities presented in Figures 11.36 and 11.37 were only observed during nocturnal time which, according to a theory described in Refs. [58–65], can be related to the absence of competition between the ionization–recombination and the transport processes (diffusion, thermodiffusion, and drift in ambient electrical and magnetic fields) taking place in the upper ionosphere.

At the same time, the altitudinal stratification of the disturbed region (the cavity), as shown in Figure 11.36, was also observed in a set of experiments carried out in the former USSR and discussed in Refs. [54–57]. The distortion of the vertical sounding ionograms during heating, using powerful radio sounder with power $P = 100$–120kW and gain $G \approx 80$, was found in the early morning and late evening hours. Using a special regime of heating-induced irregularity excitation [58,59], the possibility of generation of the small-scale, low-frequency irregularities (also called *plasma turbulences* [54–56]), which can excite and amplify large-scale plasma structures, was shown in these cycles of experiments [54–57].

The results of the experiments discussed above show similar effects in generation of depletion zones (*holes*) in the ionospheric plasma and its further spatiotemporal evolution, as have been observed during rocket burn and launch in the upper ionosphere.

11.4 Similarity between Natural Phenomena and Those Observed in Active-Space Experiments

Many features and processes observed during the formation and further evolution of artificially induced plasma structures can be directly related to natural phenomena occurring in the ionosphere and associated with bubble generation in the equatorial ionosphere, and plasma density inhomogeneities induced by magnetic storms occurring in the northern ionosphere caused by the precipitation of high-energy particles from solar wind and cosmic rain, and coupling between the magnetosphere and the polar and auroral ionosphere.

11.4.1 Observation of Equatorial Plasma Bubbles

As determined in Refs. [33–39,91–94], the bubbles are large-scale plasma depletion regions (*holes*) associated with plasma enhancements called *plumes* that follow the latter in their spatial and time evolution to secure conditions of plasma quasineutrality at each place of the spread of *F*-layer (see also Ref. [22] and references therein). The bubbles and plumes with dimensions across $\mathbf{B}_0$, which change from tens to hundreds of kilometers, are also filled by small- and moderate-scale plasma irregularities with dimensions on the order of tens and hundreds of meters to several kilometers.

More systematically, plume-like and bubble-like spatiotemporal evolution, investigated by ground-based radars located at Jicamarca and Kwajalein, were described and analyzed in Refs. [34–37,90–94], compared to the measured data obtained during satellite observations.

We draw attention now to experiments carried out using Altair backscatter radar located at Kwajalein operating simultaneously at 155.5 and 415 MHz [34,35] combined with measurements carried out by satellite probes [33,37,38]. The ion-density profiles were obtained by satellite probe measurements during the satellite trajectory when it passed the Altair radar in the short-term period of 21:51 to 21:53 h (LT) on August 1978. The authors have identified ion-density variations of interest in terms of four bubbles denoted by letters A to D (top panel), which is shown in Figure 11.38 (extracted from Ref. [33]). The monotonically increase in

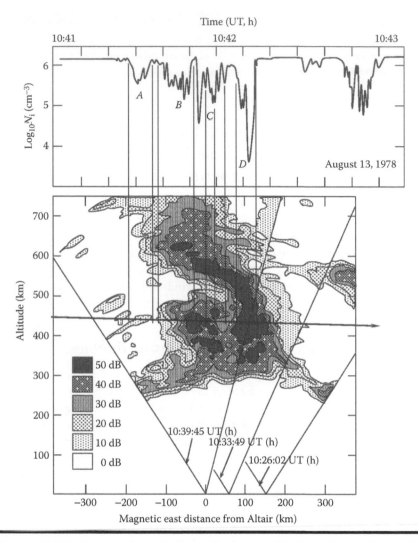

Figure 11.38 Temporal evolution of plasma density along the rocket trajectory, indicated by A and D (top panel) and 2D radio map of echo signals from the corresponding depletion areas (bottom panel), indicated by vertical lines, from A to D.

depth of depletion areas from A to D is evident with bubble D being depleted by more than 2 orders of magnitude from the background plasma.

A special campaign called "Defense Meteorological Satellite Program" (DMSP) whose results were reported in Refs. [93,94] was compared with data obtained during special experiments.

The top panel presents the plasma density variations measured by the satellite during its trajectory, while the bottom panel presents the corresponding radar backscatter radio map. The relationship of bubbles with colocated plumes was determined by the vertical lines between the upper and bottom panels, as shown in Figure 11.38. Contours of equal strength corresponding to the strongest echo-signals (40–50 dB) were associated with plume-like structures observed by radar. Thus, two plume structures are clearly seen at the time of the satellite pass.

The major plume extends upward to an altitude exceeding 700 km and the minor one appears next to the west side of the major plume. This plume pair was associated with large-scale upwelling that has an east–west dimension of approximately 400 km (or even more) as shown in Refs. [33–35]. Bubble A corresponds to a weak patchy backscatter region located to the west of the minor plume, which was interpreted in Ref. [33] by the absence of plasma structures within bubble A. There are more structures with bubbles B and C. Bubble D corresponds to the strongest backscatter region in the neck of the major plume.

More than 15 years later, an experimental campaign carried out in different sites around the magnetic dip of the Equator using the radar system, separately and combined with satellite radio and optical observation of the spread *F* region, was reported regarding observations of the evolution of bubble-like structures and collocated plume-like structures occurring in the equatorial ionosphere, which is given in Refs. [90–94]. Thus, in Ref. [90], a systematic cycle of measurements of plasma density and the structure inside the bubbles and the collocated enhancements of plasma density (plumes) were reported, as well as their altitudinal location, velocity, and direction of drift in the ambient electric and magnetic fields. The corresponding first satellite for these purposes was launched from the People's Republic of China and therefore has the abbreviation ROCSAT-1. This satellite was launched into a 600-km circular orbit with low inclination oriented at an angle of 35°. Equatorial plasma structures were measured over several months in 1999, using an electromagnetic probe operating at 1024 Hz.

A special campaign called DMSP, whose results were reported in Refs. [92,93], was compared with the data obtained during ROCSAT-1 observations that were reported in Refs. [91,94]. We present here only a few characteristic results obtained during these two programs comparing the results obtained at the same time period of observations of the spread of the *F*-region. Thus, Figure 11.39 (extracted from Ref. [91]) shows the northeasterly trajectory of the satellite ROCSAT-1 observed for periods from 13:03 to 13:21 UT and comparing measurements of four DMSP satellites labeled F11, F12, F14, and F15 whose orbits crossed the magnetic equator near 19:15, 20:15, 20:30, and 21:06 magnetic local time (MLT), respectively.

Figure 11.39   Mapping of joint ROCSAT-1 satellite and DMSP satellites (labeled by F11–F15). The trajectory of ROCSAT-1 is indicated by a solid line; trajectories of DMSP satellites are indicated by bold points.

Figure 11.40a (rearranged from Ref. [22]) shows plasma density measured by satellite ROCSAT-1 during the time periods when the satellites F11 and F14 (F12 and F15) intersected ROCSAT-1 trajectory close to 100°E at 12:49 and 13:57 UT (12:49 and 13:30 UT), respectively, plotted as a function of universal time, GLon (geographic longitude), ML at (magnetic latitude), and MLT.

It can be seen that all satellites discovered strong depletion regions with plasma density decreasing by more than 1 order of magnitude compared to the background plasma, that is, from $10^{12}\,\mathrm{m}^{-3}$ to $7\times10^{10}\,\mathrm{m}^{-3}$. At the same time, Figure 11.40b shows the plasma depletion regions during this interval obtained by measurements carried out by DMSP satellites F12 and F15. As is evident, both satellites showed strongly irregular bubble-like structures with decrease of plasma density inside them from $(3-5)\times10^{11}\,\mathrm{m}^{-3}$ to a minimum of $\sim 9\times10^{10}\,\mathrm{m}^{-3}$.

Statistical analysis carried out in Ref. [92] allows for estimating the occurrence of the equatorial plasma bubble structures systematically studied during a whole cycle of satellite observations. Thus, in Figure 11.41 (extracted from Ref. [22]), the occurrence rate of bubbles (%) for March and April 2000 (diamonds) and 2002 (triangles) versus the ML at λ is plotted.

It was found that the satellite ROCSAT-1 while passing the magnetic equator detected more than 90% of bubble-like structures occurring in the equatorial

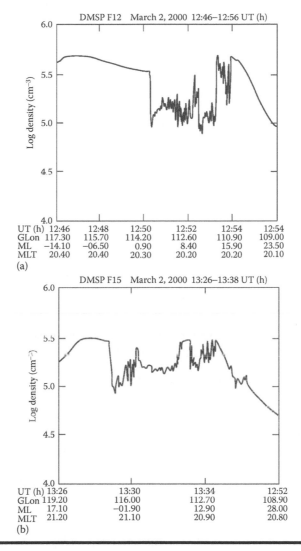

Figure 11.40   Evolution of the depletion zone observed by DMSP satellite F12.

region at attitudes of about 600 km within ±16° region around a dip equator, with a maximum of more than 15% of occurrence of bubbles during the whole cycle of measurements.

Analyzing the results of observations presented in this paragraph obtained by different methods, radio sounding, and optical measurement, we observe a complicated picture of plasma structure evolution every time

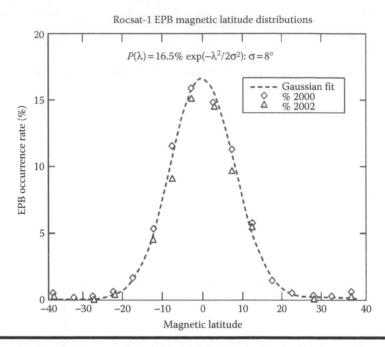

Rocsat-1 EPB magnetic latitude distributions

$$P(\lambda) = 16.5\% \; \exp(-\lambda^2/2\sigma^2): \; \sigma = 8°$$

Figure 11.41 Occurrence of equatorial bubbles (%) obtained experimentally in March and April 2000 (diamonds) and in 2002 (triangles); dashed curve is the Gaussian best-fit to the experimental data.

in the spread F-equatorial region filled by bubbles and plumes, that is, by the enhancement and depleted plasma structures following each other in their spatiotemporal evolution, similar to those cavities and enhancements generated by the rocket exhausts and engine plumes observed in the previous subsections.

11.4.2 Meteor-Induced H_E-Inhomogeneities in Middle-Latitude Ionosphere

The meteor-induced and the natural field-aligned irregularities elongated along the strength lines of the ambient magnetic field $\mathbf{H}_0$ (e.g., H_E-inhomogeneities), which occur in the lower ionosphere at altitudes of 80–130 km, have been observed in the middle-latitude traces in several regions worldwide by separate groups of researchers from the beginning of the 1960s to the beginning of the 1990s [95–103] (see also the references in Ref. [22]) by using traditional radiophysical methods of sounding of the ionosphere, ionosondes, and special radar systems.

Thus, the pioneering systematic experimental campaign was organized in the United States by using a radar system operating at 206 and 106 MHz pulse regimes [95]. Experiments were carried out simultaneously in several states of the south–west United States. It was found that most of the scattered pulse signals recording by the receiver were scattered from heights of the *E*-layer of the ionosphere ($h = 105$–115 km) and were caused by field-aligned plasma irregularities strongly affected by the geomagnetic field, which finally formed an array of plasma strata with width $d \leq \lambda$ (λ is the wavelength) across $\mathbf{B}_0$ and with lengths $L \gg \lambda$ along $\mathbf{B}_0$ [95]. Therefore, the researchers denoted such types of signals as H_E type, and defined the corresponding scattering phenomena as H_E scattering.

The same results, but without any physical explanation of the effects of H_E scattering observed in the *E*-layer of middle-latitude ionosphere were discussed in Ref. [96] based on a cycle of experiments carried out in Japan at frequency $f = 56$ MHz along two traces of 959 km (Kokubunji–Yamagawa) and 935 km (Kokubunji–Hoshika). We note that most of the signals registered at the receiver were of a quasicontinuous type with relaxation time from several minutes to several hours.

Further, other researchers, analyzing scattering of radio waves from plasma irregularities located in the *E*-layer of the middle-latitude ionosphere, have used the same abbreviations to stress the effect of the ambient geomagnetic field on the character of evolution of such anisotropic plasma structures [97–103].

The positive contribution of the first pioneering experiments in our knowledge about meteor-induced and H_E inhomogeneities, located in the *E*-region of the ionosphere, is that they fully differentiated the active zone of specular H_E scattering from zones of meteor activity, allowing a straight comparison of specular H_E-scattering signals with those obtained from pure meteor reflections. At the same time, many peculiarities of signals scattered by field-aligned plasma irregularities were not analyzed in Refs. [95,96], in particular, a "thin" structure of scattered signals, classification of quasicontinuous and burst-type H_E-scattering signals, their nature, and so forth. This was done only after one to two decades, both theoretically and experimentally [98–104]. We will describe, first of all, the results of experiments carried out during the early 1970s in the former USSR at the middle-latitude radio trace with the stationary transmitter and the moving receiver, ranged at a distance of ~1500 km, using a continuous regime of signal radiations operating at 44 and 74 MHz [98,99]. The main aim of the experiments was to investigate the "thin" structure of H_E-scattering signals compared with pure meteor-scatter signals. During these experiments, it was found that most of the several thousand recording signals can be divided into two basic types, according to the form of amplitude envelope, their statistical character, and term of fading:

1. The *burst-like* H_E-scattering signals with decay time of about tens of seconds to several minutes, which are close to those of meteors I–III types [97] and therefore were defined as typical meteor-induced signals;
2. The *quasicontinuous* H_E-scattering signals with decay time of about several minutes to several hours.

As was found during current experimental observations (see also Refs. [99,100]), the existence of continuous H_E-scattering signals was observed over a long period of time that exceeds tens of minutes and continues up to several hours. The existence of such long-term continuous H_E-scattering signals observed at the middle-latitude ionosphere at altitudes of E-region was explained in Refs. [100–103] by the existence of turbulent movements at altitudes from 80 to 120–130 km due to gradients of neutral wind and the generation of two-stream instability in the ionospheric plasma caused by convective ion current distribution. We will return to these phenomena in the next sections, because in our opinion, the same processes can dictate formation of plasma enhancement sporadic irregularities caused by rocket burn and launch.

11.4.3 Natural Plasma Depletions Generated during Magnetic Storms

The same deletion plasma regions were clearly observed at the auroral, subauroral, and even at the middle ionospheric latitudes during the occurrence of magnetic storms. Indeed, much pioneering work and modern investigation into the effects of magnetic storm on plasma disturbance generation caused by perturbations of ambient magnetic field have indicated drastic intensification of plasma disturbances occurring in the ionosphere, from the polar cap to the Equator, during magnetically perturbed periods (see Refs. [44–49,63–70] and the references therein). Thus, Refs. [44,64,66] reported observation of strong plasma structures and current (convection movements of charge particle) flows toward the Equator of the expanded auroral zone during magnetic storms, occurring over a wide period of observations, from 1990 to 2001. It was additionally found in Refs. [45–49,63,67–70] that the dynamics of the storm-time subauroral ionosphere is driven by the ring current–plasmosphere–magnetosphere interactions, as well as by precipitation of high-energy particles entering the polar cap of the ionosphere. All these effects finally result in enhanced flows of the westward convection flow, called subauroral polarization streams (SAPSs), whose region is initially nonregular associated with strong SAPS wave structures and with irregular plasma density troughs. Observations of irregular plasma density troughs occurring during the magnetic storms arising from 1999 to 2001 in the framework of the DMSP by satellites F13 and F14 (the same as were used for investigation of bubble structures at the equatorial ionosphere, see earlier) are summarized in Ref. [70]. During this experimental campaign, depletion plasma zones occurred during magnetic storms and were observed and associated with either SAPS or elevated electron temperature caused by heating of the corresponding regions of the ionosphere, which were in turn caused by the magnetic storm.

Thus, Figure 11.42 (adapted from Refs. [38,44,47]) shows a snapshot of the plasma density profile observations from the DMSP F14 satellite during the

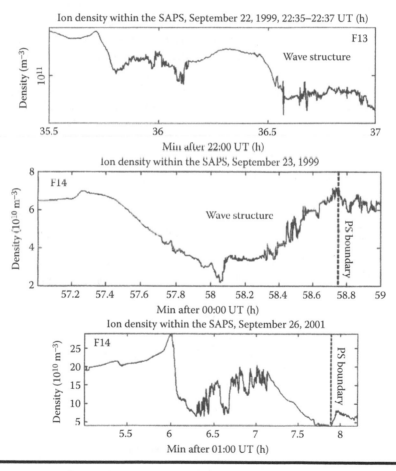

Figure 11.42 **Plasma density fluctuations and depletion zone creation during the magnetic storm observed by different satellites in September 1999 and 2001.**

magnetic storm event near Boston, MA, during September 23, 1999 [44,64]. Figure 11.42 shows the plasma density variation along the satellite track measured by a spherical Langmuir probe sampled at the rate of 24 Hz.

Figure 11.43 (rearranged from Refs. [44,66,70]) shows the corresponding waveform of plasma density deviations, $\delta N/N_0$, obtained by applying 0.1–9.5-Hz bandpass elliptic filter (see details in Ref. [64]). Here, δN is the perturbation of the background ionospheric nondisturbed plasma density (denoted by N_0). The corresponding deviations of plasma densities, shown in Figure 11.43, correspond to those from Figure 11.42, from the top to the bottom.

One can also see that enhanced density irregularities were present well toward the Equator at the boundary of the auroral zone indicated by vertical dashed line (i.e., beyond this zone the subauroral inonosphere is seen to be also perturbed). Note that quite similar SAPS/trough patterns were observed throughout the storm also by the DMSP F13 satellite.

Scanning the corresponding literature [44–49,63–70] and analyzing only the phenomena of ionospheric plasma depletion and deduced plasma cavity

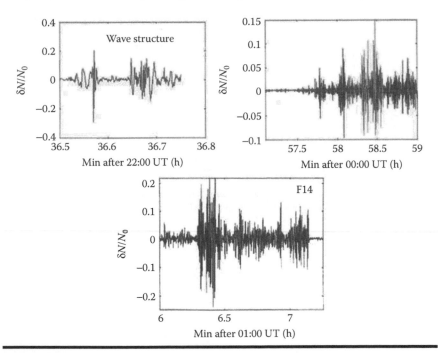

Figure 11.43 Fluctuations of the plasma density (with respect to the ambient plasma N_0) observed by satellites from F12 to F14 at periods of moderate (top panel) and strong (two bottom panels) magnetic storms.

generation during a magnetic storm, as well as its further spatiotemporal dynamics, those data were selected that correspond only to these effects. Thus, Figure 11.43 presents characteristic pictures of depletion zones generated in the subauroral ionosphere in periods of magnetic storms arranged together and observed in Refs. [44,68,70] during DMSP satellite campaigns carried out from 1999 to 2001.

It is evident that all depletion zones can be associated with magnetic storms occurring in the observed periods, and their further spatiotemporal dynamics can be associated with the effects of current wave structures on convective current instability generation in plasma. These plasma instabilities generate irregular plasma structure which affect probing wave amplitude and phase oscillations characterized by the so-called scintillation index, denoted in Refs. [38–44] by S_4, shown at each of the top panels of Figure 11.44 (rearranged from Refs. [44,66,70]). This aspect was considered in detail in Ref. [179], and we will discuss these effects in our further description concerning radio propagation via disturbed ionospheric regions associated with rocket burn and launch effects in Chapter 13.

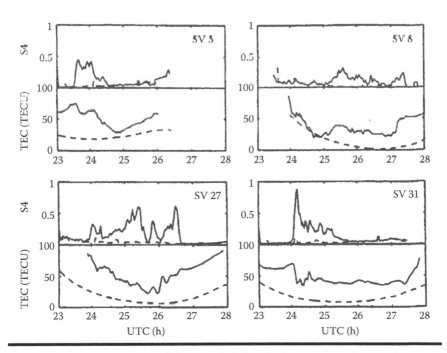

Figure 11.44 Deviations of the scintillation index S4 and the total plasma content (TEC) during the magnetic storm observed at different periods of observations by different satellites.

Completing this chapter, we should note according to recent studies [82–88,71,72,170–178,179] that the geophysical effects of powerful sources of energy radiation of various nature, such as magnetic storms, solar flares and eclipses, and so on, are similar to those observed during the rocket burns and launches. The main goal of the recent experimental and theoretical investigations was the turning out and the indication of similar features in the mutual relations of each subsystem within the whole EΛIM system, as well as their different reactions on the additional geophysical effects. As shown in Refs. [170–176], a solar eclipse is accompanied by ionospheric hole generation (i.e., depletion zones) with the same parameters as observed during rocket burn and launch.

As for the magnetic storms, they are also accompanied by depression of the electron concentration and by wave processes, observed experimentally (see the discussions of the last paragraph). These features are very important for the selection of RL and start from the effects on the ionospheric background caused by other sources of plasma perturbations (see Chapters 8 through 10).

Moreover, as was shown above, peculiarities of the effects caused by pump high-power radiation from ground-based facilities were investigated simultaneously with the experimental investigation of physical processes ranged at distances of 1,000–10,000 km far from the heating ground-based facilities located near the main radiophysical centers worldwide [namely, near Moscow, N. Novgorod (Russia), Tromsø (Norway), Arecibo (Puerto Rico), Fairbanks (Alaska), and so forth].

11.5 Main Results

1. During the burn and launch of rockets and SVs, a new sort of plasma called in the literature as *dusty* or *dirty* plasma is generated at the initial stage of interaction between exhaust gases and liquids (consisting of various composite molecules and atoms) and the ambient plasma (consisting of electrons and atomic or molecular ions, depending on the height of the ionosphere) as a result of the ionization and recombination processes.

2. In such composite dusty plasma, exchange between ions of the exhaust gas or liquid and ions of background plasma occurs yielding complicated chemical reactions of ionization and recombination with ambient electrons and ions.

3. Due to such complicated chemical reactions and owing to the transfer of heat from thermal ions of the exhaust to cool ions and electrons of the ambient plasma, a deep depletion zone (called in the literature as *holes* or *cavities*) and enhanced zones are created in the background ionospheric plasma.

4. The type of plasma perturbations, holes or enhancements, depends on the height of the ionosphere where the rocket burnout, engine jet, or plume exhaust release was done.

5. During further evolution of global plasma disturbances, holes and enhancements, and small- and moderate-scale plasma structures, called irregularities or turbulences, are observed that fill areas of holes and enhancement and create the striated plasma structures within the global zones.

6. The same stratified enhancement and/or depletion zones are evidently observed at equatorial upper ionosphere (called bubbles), at middle-latitude ionosphere affected by meteor trails (called H_E inhomogeneities), and at polar ionosphere affected by magnetic storm, that is, during natural perturbations of the ionosphere.

7. Deep depletion zones with stratified plasma structures within them are clearly seen from artificial modification of the ionosphere by pump electromagnetic waves using high-energy ground-based radars.

8. Strong enhancements of the background ionospheric plasma with its further stratification on plasma structures (called strata) are usually observed during artificial experiments with ion clouds and beam injection or neutral gas or liquids.

9. All artificially or naturally observed modifications of the regular ionospheric plasma show the same or similar features and therefore can be theoretically predicted using unified approaches that were separately performed by many researchers dealing with the study of physical phenomena and processes occurring in each subsystem, perturbed or nonperturbed, of the whole EAIM system.

References

1. Ratcliffe, J. A., *Physics of the Upper Atmosphere*, New York: Academic Press, 1960.
2. Ginzburg, V. L., *Propagation of Electromagnetic Waves in Plasma*, New York: Pergamon Press, 1964.
3. Whitten, R. C. and Poppoff, I. G., *Physics of the Lower Ionosphere*, New York: Prentice Hall, 1965.
4. Ivanov-Kholodnii, G. S. and Nikol'skii, G. M., *The Sun and the Ionosphere*, Moscow: Nauka, 1969 (in Russian).
5. Gershman, B. N., *Dynamics of the Ionospheric Plasma*, Moscow: Nauka, 1974 (in Russian).
6. Gurevich, A. V., *Nonlinear Phenomena in the Ionosphere*, Berlin: Springer-Verlag, 1978.
7. Fel'dshtein, Ya. I., Levitin, A. E., Afonina, R. G., and Belov, B. A., Magnetosphere-ionosphere relations, in *Interplanetary Medium and Magnetosphere*, Moscow: Nauka, 1982, 64–116 (in Russian).

8. Al'pert, Ya. L., *Space Plasma*, Vols. I and II, New York: Cambridge University Press, 1983 and 1990, 305.

9. Volland, H., *Atmospheric Electrodynamics*, Heidelberg: Springer-Verlag, 1984.

10. Gershman, B. N., Erukhimov, L. M., and Yashin, Yu. Ya., *Wave Phenomena in the Ionosphere and Cosmic Plasma*, Moscow: Nauka, 1984 (in Russian).

11. Gel'berg, M. G., *Inhomogeneities of High Latitude Ionosphere*, Novosibirsk: Nauka, 1986 (in Russian).

12. Rees, H., *Physics and Chemistry of the Upper Atmosphere*, Cambridge: Cambridge University Press, 1989.

13. Likhter, Ya. L., Gul'el'mi, A. V., Erukhimov, L. M., and Mikha'lova, G. M., *Wave Diagnostics of the Near-Earth's Plasma*, Moscow: Nauka, 1989 (in Russian).

14. Rawer, K., *Wave Propagation in the Ionosphere*, Dordrecht: Kluwer Academic Publishers, 1989.

15. Grandal, B., Ed., *Artificial Particle Beams in Space Plasma Studies*, New York: Plenum Press, 1982.

16. Filipp, N. D., Oraevskii, V. N., Blaunstein, N. Sh., and Ruzhin, Yu. Ya., *Evolution of Artificial Plasma Irregularities in the Earth's Ionosphere*, Kishinev: Shtiintza, 1986 (in Russian).

17. Mishin, E. V., Ruzhin, Yu. Ya., and Telegin, V. A., *Interaction of Electron Fluxes with the Ionospheric Plasma*, Moscow: Hydrometeorology, 1989 (in Russian).

18. Filipp, N. D., Blaunstein, N. Sh., Erukhimov, L. M., Ivanov, V. A., and Uryadov, V. P., *Modern Methods of Investigation of Dynamic Processes in the Ionosphere*, Kishinev: Shtiintza, 1991 (in Russian).

19. Kohl, H., Ruster, R. R., and Schlegel, K., Eds., *Modern Ionospheric Science*, Berlin: ProduServ GmbH Verlagsservice, 1996.

20. Belikovich, V. V., Benediktov, E. A., Tolmacheva, A. V., and Bakhmeteva, N. V., *Ionospheric Research by Means of Artificial Periodic Irregularities*, Berlin: Copernicus GmbH, 2002.

21. Uryadov, V., Ivanov, V., Plohotniuc, E., et al., *Dynamic Processes in the Ionosphere—Methods of Investigations*, Iasi: TehnoPress, 2006 (in Romanian).

22. Blaunstein, N. and Plohotniuc, E., *Ionosphere and Applied Aspects of Radio Communication and Radar*, Boca Raton, FL: CRC Press, 2008.

23. Singler, S. F., Maple, E., and Bowen, W. A., Evidence for ionospheric currents from rocket experiments near the geomagnetic equator, *J. Geophys. Res.*, 56, 265–281, 1951.

24. Booker, H. G., A theory of scattering by non-isotropic irregularities with application to radar reflection from the Aurora, *J. Atmos. Terr. Phys.*, 8, 204–221, 1956.

25. Brasefield, Ch. J., Some observations of low-level ion clouds, *J. Geophys. Res.*, 64, 141–148, 1959.

26. Cahill, L. J., Investigation of equatorial electrojet by rocket magnetometer, *J. Geophys. Res.*, 64, 489–503, 1959.

27. Van Allen, J. A., McIlwain, C. E., and Ludwig, G. H., Satellite observations of electrons artificially injected into the geomagnetic field, *J. Geophys. Res.*, 64, 877–891, 1959.

28. Newman, P., Optical, electromagnetic, and satellite observations of high-latitude nuclear detonators. Part I, *J. Geophys. Res.*, 64, 923–932, 1959.

29. Peterson, A. M., Optical, electromagnetic, and satellite observations of high-latitude nuclear detonators. Part II, *J. Geophys. Res.*, 64, 933–940, 1959.

30. Frihagen, J. and Jacobson, T., In-situ observations of high-latitude *F*-region irregularities, *J. Atmos. Terr. Phys.*, 33, 519–522, 1971.
31. Titheridge, J. E., The diffraction of satellite signals by isolated ionospheric irregularities, *J. Atmos. Terr. Phys.*, 33, 47–69, 1971.
32. Erukhimov, L. M. and Rizhkov, V. A., Study of focusing ionospheric irregularities by methods of radio astronomy at frequencies of 13–54 MHz, *Geomagn. Aeronom.*, 5, 693–697, 1971.
33. Kelley, M. C., Haerendel, G., Kappler, H., et al., Evidence for a Rayleigh–Taylor type instability and upwelling of depleted density regions during equatorial spread *F*, *Geophys. Res. Lett.*, 3, 448–451, 1976.
34. Tsunoda, R. T., Livingston, R. C., McClure, J. P., and Hanson, W. B., Equatorial plasma bubbles: Vertically elongated wedges from the bottomside *F* layer, *J. Geophys. Res.*, 87, 9171–9180, 1982.
35. Hudson, M. K., Spread *F* bubbles: Nonlinear Rayleigh–Taylor mode in to dimensions, *J. Geophys. Res.*, 83, 3189–3198, 1978.
36. Hanson, W. B. and Bamgboye, D. K., The measured motions inside equatorial plasma bubbles, *J. Geophys. Res.*, 89, 8997–9008, 1984.
37. Zalesak, S. T. and Ossakow, S. L., Nonlinear equatorial spread *F*: Spatially large bubbles resulting from large horizontal scale initial perturbations, *J. Geophys. Res.*, 85, 2131–2141, 1980.
38. Basu, S., McClure, J. P., Hanson, W. B., and Aarons, J., Coordinate study of equatorial scintillations and *in situ* and radar observations of nighttime *F* region irregularities, *J. Geophys. Res.*, 75, 5119–5128, 1980.
39. Woodman, R. F. and La Hoz, C. Radar observations of *F* region equatorial irregularities, *J. Geophys. Res.*, 81, 5447–5458, 1976.
40. Muldrew, D. B. and Vickrey, J. F., High-latitude *F*-region irregularities observed simultaneously with ISIS-1 and Chatanica radar, *J. Geophys. Res.*, 87, 8263–8267, 1982.
41. Hardy, D., Schmidt, L., Gussenhoven, M., et al., Precipitating electron and ion detectors (SSJ/4) for block 5D/Flights 4–10 DMSP satellites: Calibration and data presentation, *Tech. Rep. AFGL-TR-84-0317*, Bedford, MA: Air Force Geophysics Laboratory, Hanscom Air Force Base, 1984.
42. Rich, F. J. and Hairston, M., Large-scale convection patterns observed by DMSP, *J. Geophys. Res.*, 79, 3827–3835, 1994.
43. Farley, D. T., Incoherent scatter radar probing, in *Modern Ionospheric Science*, Kohl, H., Ruster, R., and Schlegel, K., Ed., Katlenburg-Lindau: Copernicus GmbH, 1996, 415–439.
44. Ledvina, B. M., Makela, J. J., and Kitner, P. M., First observations of intense GPS L1 amplitude scintillations at midlatitude, *Geophys. Res. Lett.*, 29, 1659–1662, 2002.
45. Mishin, E., Burke, W. W., and Viggiano, A., Stormtime subauroral density troughs: Ion-molecule kinetic effects, *J. Geophys. Res.*, 109, doi:10.1029/2004JA010438, 2004.
46. Mishin, E., Burke, W., Huang, C., and Rich, F., Electromagnetic wave structures within subauroral polarization streams, *J. Geophys. Res.*, 108, 1309–1315, 2003.
47. Kintner, P. and Ledvina, B., The ionosphere, radio navigation, and global navigation satellite systems, *Adv. Space Res.*, 35, 788–811, 2005.

48. Foster, J. and Burke, W., A new categorization for subauroral electric fields, *EOS Trans. AGU*, 83, 393–401, 2002.

49. Mishin, E. V., Burke, W. J., and Pedersen, T., HF-induced airglow at magnetic zenith: Theoretical considerations, *J. Annal. Geophys.*, 23, 47–53, 2005.

50. *Radio Sci.*, Special issue, 9, 881–1090, 1974.

51. Fejer, J. A., Gonzales, C. A., Ierkis, H. M., et al., Ionospheric modification experiments with the Arecibo heating facility, *J. Atmos. Terr. Phys.*, 47, 1165–1179, 1985.

52. Djuth, F. T., Thide, B., Ierkis, H. M., and Sulzev, M. P., Large-*F*-region electron-temperature enhancements generated by high-power HF radio waves, *Geophys. Rev. Lett.*, 14, 953–956, 1987.

53. Duncan, L. M., Sheerin, J. P., and Behnke, R. A., Observation of ionospheric cavities generated by high-power radio waves, *Phys. Rev. Lett.*, 61, 239–242, 1988.

54. Erukhimov, L. M., Metelev, S. A., Mityakov, N. A., Myasnikov, E. N., and Frolov, V. L., The artificial ionospheric turbulence, *Izv. Vusov. Radiofizika*, 30, 208–225, 1987 (in Russian).

55. Blaunstein, N. Sh., Erukhimov, L. M., Uryadov, V. P., Filipp, N. D., and Tsyganash, I. P., Vertical dependence of relaxation time of artificial small-scale disturbances in the middle-latitude ionosphere, *Geomagn. Aeronom.*, 28, 4, 595–597, 1988.

56. Bochkarev, G. S., Eremenko, V. A., and Cherkashin, Ya. N., Radio wave reflection from quasi-periodical disturbances of the ionospheric plasma, *Adv. Space Res.*, 8, 1, 255–260, 1988.

57. Blaunstein, N. Sh., Boguta, N. I., Erukhimov, L. M., et al., About the dependence of the development and relaxation time of artificial small-scale disturbances in the middle-latitude ionosphere, *Izv. Vuzov. Radiofizika*, 33, 548–551, 1990 (in Russian).

58. Blaunstein, N. Sh., Vas'kov, V. V., and Dimant, Ya. S., Resonant heating of the *F* layer of the ionosphere by a high-power radio wave, *Geomagn. Aeronom.*, 32, 235–238, 1992.

59. Blaunstein, N. Sh. and Bochkarev, G. S., Modeling of the dynamics of periodical artificial disturbances in the upper ionosphere during its thermal heating, *Geomagn. Aeronom.*, 33, 84–91, 1993.

60. Blaunstein, N., Diffusion spreading of middle-latitude heating-induced ionospheric plasma irregularities, *J. Annal. Geophys.*, 13, 617–626, 1995.

61. Blaunstein, N., Changes of the electron concentration profile during local heating of the ionospheric plasma, *J. Atmos. Solar-Terr. Phys.*, 58, 1345–1354, 1996.

62. Blaunstein, N., Evolution of a stratified plasma structure induced by local heating of the ionosphere, *J. Atmos. Solar-Terr. Phys.*, 59, 351–361, 1997.

63. Maynard, N., Burke, W., Basinska, E., et al., Dynamics of the inner magnetosphere near times of substorm onsets, *J. Geophys. Res.*, 101, 7705–7715, 1996.

64. Foster, J. and Rich, F., Prompt midlatitude electric field effects during severe magnetic storms, *J. Geophys. Res.*, 103, 26367–26373, 1998.

65. Burke, W., Rubin, A, Mayanard, M. et al., Ionospheric disturbances observed by DMSP at mid to low latitudes during magnetic storm of June 4–6, 1991, *J. Geophys. Res.*, 105, 18391–19402, 2000.

66. Basu, Su., Basu Sa., Villadares, C. E., et al., Ionospheric effects of major magnetic storms during the international space weather period of September and October 1999: GPS observations, VHF/UHF scintillations and in situ density structures at middle and equatorial latitudes, *J. Geophys. Res.*, 106, 389–399, 2001.

67. Mishin, E. V. and Burke, W. J., Stormtime coupling of the ring current, plasmasphere and topside ionosphere: Electromagnetic and plasma disturbances, *J. Geophys. Res.*, 110, 7209–7216, 2005.
68. Mishin, E. V., Burke, W. J., Basu, Su., et al., Stormtime ionospheric irregularities in SAPS-related troughs: Caused of GPS scintillations at mid latitudes, *AGU Fall Meeting 2003*, Abstract SH52A-07, Colorado, 2003.
69. Pfaff, R., Liebrecht, C., Berthelier, J.-J., et al., DEMETER satellite observations of plasma irregularities in the topside ionosphere at low, middle, and sub-auroral latitudes and their dependence on magnetic storms, *Midlatitude Ionospheric Dynamics and Disturbances*, Kintner, P., Coster, A., Fuller-Rowell, T., et al., Eds., *AGU Geophysical Monograph Series 181*, 297–310, doi:10.1029/181GM26, 2008.
70. Mishin, E. and Blaunstein, N., Irregularities within subauroral polarization stream-related troughs and GPS radio interference at midlatitudes, *Midlatitude Ionospheric Dynamics and Disturbances*, Kintner, P., Coster, A., Fuller-Rowell, T., et al., Eds., *AGU Geophysical Monograph Series 108*, 291–295, doi:10.1029/181GM26, 2008.
71. Garmash, K. P., Kostrov, L. S., Rozumenko, et al., Global ionospheric disturbances caused by a rocket launch against a background of a magnetic storm, *Geomagn. Aeronom.*, 39, 69–75, 1999.
72. Grigorenko, E. I., Lazorenko, S. V., Taran, V. I., Chernogor, L. F., Wave disturbances in the ionosphere accompanying the solar flare and the strongest magnetic storm of September 25, 1998, *Geomagn. Aeronom.*, 43, 718–735, 2003.
73. Gokov, A. M. and Chernogor, L. F., Variations of electron concentration in the midlatitude D-region of the ionosphere during magnetic storm, *Cosmic Sci. Techn.*, 11, 1221, 2005 (in Russian).
74. Grigorenko, E. I., Lysenko, V. N., Taran, V. I., and Chernogor, L. F., Specific features of the ionospheric storm of March 20–23, 2003, *Geomagn. Aeronom.*, 45, 745–757, 2005.
75. Panasenko, S. V. and Chernogor, L. F., Event of the November 7–10, 2004 magnetic storm in the lower ionosphere, *Geomagn. Aeronom.*, 47, 608–620.
76. Chernogor, L. F., Grigorenko, Ye. I., Lysenko, V. N., and Taran, V. I., Dynamic processes in the ionosphere during magnetic storms from the Kharkov incoherent scatter radar observations, *Int. J. Geomagn. Aeronom.*, 7, doi:10.1029/2005GI000125, 2007.
77. Grigorenko, E. I., Lysenko, V. N., Pazyura, S. A., Taran, V. I., and Chernogor, L. F., Ionospheric disturbances during the severe magnetic storm of November 7–10, 2004, *Geomagn. Aeronom.*, 47, 720–738, 2007.
78. Burmaka, V. P. and Chernogor, L. F., Wave activity in the ionosphere during the magnetospheric storm of November 7–10, 2004, *Geomagn. Aeronom.*, 51, 305–320, 2011.
79. Chernogor, L. F., Grigorenko, Ye. I., Lysenko, V. N., Rozumenko, V. T., and Taran, V. I., Ionospheric storms associated with geospace storms as observed with the Kharkiv incoherent scatter radar, *Sun Geosph.*, 3, 81–86, 2008 (in Russian).
80. Grigorenko, E. I., Lysenko, V. N., Taran, V. I., Chernogor, L. F., and Chernyaev, S. V., Dynamic processes in the ionosphere during the strongest magnetic storm of May 30–31, 2003, *Geomagn. Aeronom.*, 45, 758–777, 2005.
81. Domin, I. F., Emel'yanov, L. Ya., Pazura, S. A., Kharitonova, S. V., and Chernogor, L. F., Dynamic processes in the ionosphere during enough moderate magnetic storm of January 20–21, 2010, *Cosmic Sci. Techn.*, 17, 26–40, 2011 (in Russian).

82. Mitra, A. P. and Rowe, J. N., Ionospheric effects of solar flares VI. Changes in *D*-region ion chemistry during solar flares, *J. Atmos. Terr. Phys.*, 34, 795–806, 1972.

83. Gokov, A. M. and Chernogor, L. F., Results of observations of the processes in the lower ionosphere associated with the solar eclipse on August 11, 1999, *Radiophys. Radioastron.*, 5, 348–360, 2000 (in Russian).

84. Kostrov, L. S. and Chernogor, L. F., Results of observations of the processes in the middle ionosphere associated with the solar eclipse on August 11, 1999, *Radiophys. Radioastron.*, 5, 361–370, 2000 (in Russian).

85. Akimov, L. A., Bogovskii, V. K., Grigorenko, E. I., Taran, V. I., and Chernogor, L. F., Atmospheric–ionospheric effects of the solar eclipse of May 31, 2003 in Kharkov, *Geomagn. Aeronom.*, 5, 494–518, 2005.

86. Grigorenko, E. I., Lyashenko, M. V., and Chernogor, L. F., Effects of solar eclipse of March 29, 2006, in the ionosphere and atmosphere, *Geomagn. Aeronom.*, 48, 337–351, 2008.

87. Chernogor, L. F., Precipitation of electrons from the magnetosphere stimulated by the solar eclipse, *Radiophys. Radioastronom.*, 5, 371–375, 2000 (in Russian).

88. Chernogor, L. F., Wave response of the ionosphere to the partial solar eclipse of August 1, 2008, *Geomagn. Aeronom.*, 50, 346–361, 2010.

89. Whalen, J. A., Mapping a bubble at dip equator and anomaly with oblique ionospheric sounding of range spread *F*, *J. Geophys. Res.*, 101, 5185–5194, 1996.

90. Su, S.-Y., Yeh, H. C., and Heelis, R. A., ROCSAT 1 ionospheric plasma and electrodynamics instrument observations of equatorial spread *F*: An early transitional scale result, *J. Geophys. Res.*, 106, 29153–29159, 2001.

91. Burke, W. J., Huang, C. Y., Valladares, C. E., et al., Multipoint observations of equatorial plasma bubbles, *J. Geophys. Res.*, 108, 1–9, 2003.

92. Burke, W. J., Gentile, L. C., Huang, C. Y., Valladares, C. E., and Su, S.-Y., Longitudinal variability of equatorial plasma bubbles observed by DMSP and ROCSAT-1, *J. Geophys. Res.*, 109, 1–12, 2004.

93. Lin, C. S., Immel, T. J., Yeh, H. C., Mende, S. B., and Burch, J. L., Simultaneous observations of equatorial plasma depletion by IMAGE and ROCSAT-1 satellites, *J. Geophys. Res.*, 110, 1–11, 2004.

94. Heritage, J. L., Fay, W. J., and Bowen, E. D., Evidence that meteor trails produce a field-aligned scatter signals at VHF, *J. Geophys. Res.*, 67, 953–961, 1962.

95. Kuriki, I., Further experimental study of scattering propagation caused by the field aligned irregularities in VHF band, *J. Radio Res. Labs* (Japan), 14, 57–77, 1967.

96. Kasheev, B. L., Lebedenetz, V. N., and Lagutin, M. F., *Meteor Phenomena in the Earth's Atomosphere*, Moscow: Nauka, 1967 (in Russian).

97. Wotkins, C. D., James, R., and Nicholson, T. F., Further studies of the effect of the Earth's magnetic field on meteor trails, *J. Atmos. Terr. Phys.*, 33, 1907–1916, 1971.

98. Filipp, N. D. and Blaunstein, N. Sh., Effect of the geomagnetic field on the diffusion of ionospheric inhomogeneities, *Geomagn. Aeronom.*, 18, 423–427, 1978.

99. Filipp, N. D. and Blaunstein, N. Sh., Drift of ionospheric inhomogeneities in the presence of the geomagnetic field, *Izv. Vuzov. Radiofizika*, 21, 1409–1417, 1978 (in Russian).

100. Tanaka, T. and Venkateswaran, S. V., Characteristics of field-aligned *E*-region irregularities over Iioka (36°N), Japan–I, *J. Atmos. Terr. Phys.*, 44, 381–393, 1982.

101. Yamamoto, M., Fukao, S., Woodman, R. F., et al., Mid-latitude *E* region field-aligned irregularities observed with MU radar, *J. Geophys. Res.*, 96, 15943–15949, 1991.

102. Yamamoto, M., Komoda, N., Fukao, S., et al., Spatial structure of the *E* region field-aligned irregularities revealed by the MU radar, *Radio Sci.*, 29, 337–347, 1994.

103. Tsunoda, R. T., Fukao, S., and Yamamoto, M., On the origin of the quasi-periodic radar backscatter from midlatitude sporadic *E*, *Radio Sci.*, 29, 349–365, 1994.

104. Fopple, H., Haerendel, G., Hasera, L., et al., Artificial strontium and barium clouds in the upper atmosphere, *Planet. Space Sci.*, 15, 357–372, 1967.

105. Kaiser, T. R., Pickering, W. H., and Watkins, C. D., Ambipolar diffusion and motion of ion clouds in the Earth's magnetic field, *Planet. Space Sci.*, 17, 519–551, 1968.

106. Rosenberg, N. W., Observations of striation formation in barium ion clouds, *J. Geophys. Res.*, 76, 6856–6864, 1971.

107. Antoshkin, V. S., Zhilinskii, A. P., Petrov, G. G., Rozhanskii, V. A., Tsendin, L. D., et al., About evolution of barium clouds of high density, *Geomagn. Aeronom.*, 19, 1049–1056, 1979.

108. Perkins, F. W., Zabusky, N. J., and Doles, J. H., Deformation and striation of plasma clouds in the ionosphere, 1, *J. Geophys. Res.*, 78, 697–710, 1973.

109. Goldman, S. R., Ossakow, S. L., and Book, D. L., On the nonlinear motion of a small barium cloud in the ionosphere, *J. Geophys. Res.*, 79, 1471–1477, 1974.

110. Baxter, A. J., Lower F-region barium release experiments at sub-auroral location, *Planet. Space Sci.*, 23, 973–983, 1975.

111. Dzubenko, N. L., Zhilinsky, A. P., Zhulin, A., et al., Dynamics of artificial plasma clouds in "Spolokh" experiments: Movement pattern, *Planet. Space Sci.*, 31, 849–858, 1983.

112. Andreeva, L. A., Zhulin, A., Ivchenko, I. S., et al., Dynamics of artificial plasma cloud in "Spolokh" experiments: Clouds deformation, *Planet. Space Sci.*, 32, 1045–1052, 1984.

113. Mitchell, H. G., Fedder, J. A., Huba, J. D., and Zalesak, S. T., Transverse motion of high-speed barium clouds in the ionosphere, *J. Geophys. Res.*, 90, 11091–11103, 1985.

114. Volk, H. J. and Haerendel, G., Striation in ionospherric ion clouds, 1, *J. Geophys. Res.*, 76, 835–844, 1987.

115. Drake, J. F. and Huba, J. D., Dynamics of three dimensional ionospheric plasma clouds, *Phys. Rev. Lett.*, 58, 278–281, 1987.

116. Drake, J. F., Mubbrandon, M., and Huba, J. D., Three-dimensional equilibrium and stability of ionospheric plasma clouds, *Phys. Fluids*, 31, 3412–3424, 1988.

117. Zalesak, S. T., Drake, J. F., and Huba, J. D., Dynamics of three-dimensional ionospheric plasma clouds, *Radio Sci.*, 23, 591–598, 1988.

118. Bernhardt, P. A., Swaetz, W. E., and Kelley, M. C., Sulzer, M. P., and Noble, S. T., Spacelab 2 upper atmospheric modification experiment over Arecibo, 2, Plasma dynamics, *Astro. Lett. Commun.*, 27, 183–198, 1988.

119. Milinevsky, G. P., Romanovsky, Yu. A., Evtushevsky, A. M., Savchenko, V. A., et al., Optical observations in active space experiments in investigation of the Earth's atmosphere and ionosphere, *Cosmic Res.*, 28, 418–429, 1990 (in Russian).

120. Blaunstein, N. Sh., Milinevsky, G. P., Savchenko, V. A., and Mishin, E. V., Formation and development of striated structure during plasma cloud evolution in the Earth's ionosphere, *Planet. Space Sci.*, 41, 453–460, 1993.

121. Blaunstein, N., The character of drift spreading of artificial plasma clouds in the middle-latitude ionosphere, *J. Geophys. Res.*, 101, 2321–2331, 1996.

122. Booker, H. G., A local reduction of *F*-region ionization due to missile transit, *J. Geophys. Res.*, 66, 1073–1079, 1961.

123. Barnes, C., Comment on paper by Henry, G. Booker, A local reduction of *F*-region ionization due to missile transit, *J. Geophys. Res.*, 66, 2580, 1961.

124. Altshuler, S., Moe, M. M., and Molmud, P., The electromagnetics of the rocket exhaust, *Space Tech. Lab. Rept.*, June 15, 1958.

125. *Ionospheric Modifications of Missile Exhaust*, Geophysics Corporation of America (GCA), *Sci. Rep. No. 10* and *GCA Tech. Rep. No 62-15-G*, December 1962.

126. Kellogg, W. W., Pollution of the upper atmosphere by rockets, *Space Sci. Rev.*, 3, 275–316, 1964.

127. Jackson, J. E., Whale, H. A., and Bauer, S. J., Local ionospheric disturbance created by a burning rocket, *J. Geophys. Res.*, 67, 2059–2061, 1962.

128. Golomb, D., Rosenberg, N. W., Weight, J. W., and Barnes, R. A., Formation of an electron-depleted region in the ionosphere by chemical releases, *Space Res.*, 4, 389–398, 1963.

129. MacLeod M. A. and Golomb, D., The evolution of a ionospheric hole, *Upper Atmospheric Phys. Lab.*, *AFCRL Environ. Res. Paper # 101*, April, 1965.

130. Stone, M. L., Bird, L. E., and Balser, M., A Faraday rotation measurement on the ionospheric perturbation produced by a burning rocket, *J. Geophys. Res.*, 69, 971–978, 1964.

131. Wright, J. W., Ionosonde studies of some chemical releases in the ionosphere, *Radio Sci.*, 68D, 189–204, 1964.

132. Felker, J. K. and Roberts, W. T., Ionospheric refraction following rocket transit, *J. Geophys. Res.*, 71, 4692–4694, 1966.

133. Schunk, R. W., On the dispersal of artificially injected gases in the night time atmosphere, *Planet. Space Sci.*, 26, 605–617, 1978.

134. Baum, H. R., The interaction of a transient exhaust plume with a rarefied atmosphere, *J. Fluid Mech.*, 58, 795–804, 1973.

135. Forbes, J. M. and Mendillo, M., Diffusive aspects of ionospheric modification by the release of highly reactive molecules into the *F*-region, *J. Atmos. Terr. Phys.*, 38, 1299–1312, 1976.

136. Pinson, G. T., Apollo/Saturn 5 postflight trajectory—SA-513-Skylab mission, *Tech. Rep. D5 15560 13*, Huntsville, AL: Boeing Co., 1973.

137. Mendillo, M., Hawkins, G. S., and Klobuchar, J. A., An ionospheric total electron content disturbance associated with the launch of NASA's Skylab, *Tech. Rep. 0342*, Bedford, MA: Air Force Cambridge Research Laboratory, 1974.

138. Mendillo, M., Hawkins, G. S., and Klobuchar, J. A., A sudden vanishing of the ionospheric *F* region due to the launch of Skylab, *J. Geophys. Res.*, 80, 2217–2228, 1975.

139. Mendillo, M., Hawkins, G. S., and Klobuchar, J. A., A large-scale hole in the ionosphere caused by the launch of Skylab, *Science*, 187, 343–347, 1975.

140. Mendillo, M. and Forbes, J. M., Artificially created holes in the ionosphere, *J. Geophys. Res.*, 83, 151–162, 1978.

141. Mendillo, M., The effect of rocket launches on the ionosphere, *Adv. Space Res.*, 1, 275–290, 1981.

142. Mendillo, M. and Forbes, J. M., Theory and observation of a dynamically evolving negative ion plasma, *J. Geophys. Res.*, 87, 8273–8285, 1982.

143. Mendillo, M., Ionospheric holes: A review of theory and recent experiments, *Adv. Space Res.*, 8, (1)51–(1)62, 1988.

144. Bernhardt, P. A., The response of the ionosphere to the injection of comically reactive vapor, *Tech. Rep. 17, SU-SEL-76-008*, Stanford, CA: Stanford University, 1976.

145. Pongratz, M. B., Smith, G. M., Sutherland, C. D., and Zinn, J., Lagopedo-two F-region ionospheric depletion experiment, in *Effects of the Ionosphere on Space and Terrestrial Systems*, Goodman, J. M., Ed., Washington, DC: U.S. Government Printing Office, 1978.

146. Anderson, D. N. and Bernhardt, P. A., Modeling the effects of an H_2 gas release on the equatorial ionosphere, *J. Geophys. Res.*, 83, 4777–4790, 1978.

147. Daly, P. W. and Whalen, B. A., Thermal ion results from an experiment to produce an ionospheric hole: Lagopedo Uno, *J. Geophys. Res.*, 84, 6581–6592, 1979.

148. Bernhardt, P. A., High-altitude gas releases: Transition from collisionless flow to diffusive flow in a nonuniform atmosphere, *J. Geophys. Res.*, 84, 4341–4354, 1979.

149. Sjolander, G. W. and Szuszczewicz, E. P., Chemically depleted F_2 ion composition: Measurements and theory, *J. Geophys. Res.*, 84, 4395–4405, 1979.

150. Schunk, R. W., Research note on the dispersal of artificially-injected gases in the night-time atmosphere, *Planet. Space Sci.*, 26, 605–610, 1978.

151. Johnson, C. Y., Sjolander, G. W., Oran, E. S., et al., F region above Kauai: Measurement, model, modification, *J. Geophys. Res.*, 85, 4205–4213, 1980.

152. Zinn, J. and Sutherland, C. D., Effects of rocket exhaust products in the thermosphere and ionosphere, *Space Solar Power Rev.*, 1, 109–117, 1980.

153. Yau, A. W., Whalen, B. A., Creutzberg, F., Pongratz, M. B., and Smith, G., Observations of particle precipitation, electric field, and optical morphology of an artificially perturbed auroral arc: Project waterhole, *J. Geophys. Res.*, 86, 5601–5613, 1981.

154. Bernhardt, P. A., Chemistry and dynamics of SF_6 injections into F region, *J. Geophys. Res.*, 89, 3929–3937, 1984.

155. Bernhardt, P. A., Environmental effects of plasma depletion experiments, *Advance Space Res.*, 2, 129–137, 1982.

156. Bernhardt, P. A., A critical comparison of ionospheric depletion chemicals, *J. Geophys. Res.*, 92, 4617–4628, 1987.

157. Biondi, M. A. and Sipler, D. P., Studies of equatorial 630.0 nm airglow enhancements produced by a chemical release in the F-region, *Planet. Space Sci.*, 32, 1605–1610, 1984.

158. Bernhardt, P. A., Ganguli, G., Kelley, M. C., and Swartz, W. E., Enhanced radar backscatter from space shuttle exhaust in the ionosphere, *J. Geophys. Res.*, 100, 23811–23818, 1995.

159. Bernhardt, P. A., Probing the magnetosphere using chemical releases from CRRES satellite, *Phys. Fluids B*, 4, 2249–2256, 1992.

160. Bernhardt, P. A., Plasma, fluid instabilities in the ionospheric holes, *J. Geophys. Res.*, 87, 7539–7549, 1982.

161. Bernhardt, P. A., Kashiwa, B. A., Tepley, C. A., and Noble, S. T., Spacelab 2 upper atmospheric modification experiment over Arecibo, 1, neutral gas dynamics, *Astrophys. Lett. Communic.*, 27, 169–181, 1988.

162. Guzdar, P., Gondarenko, N., Chaturvedi, P., and Basu, S., Three-dimensional non-linear simulations of the gradient drift instability in the high-latitude ionosphere, *Radio Sci.*, 33, 1901–1912, 1998.

163. Bernhardt, P. A., Huba, J. D., Swartz, W. E., and Kelley, M. C., Incoherent scatter from space shuttle and rocket engine plumes in the ionosphere, *J. Geophys. Res.*, 103, 2239–2251, 1998.

164. Bernhardt, P. A., Huba, J. D., Kudeki, E., et al., The lifetime of a depression in the plasma density over Jicamarca produced by space shuttle exhaust in the ionosphere, *Radio Sci.*, 36, 1209–1220, 2001.

165. Bernhardt, P. A. and Suzler, M. P., Incoherent scatter measurements of ion beam disturbances produced by space shuttle exhaust injections into the ionosphere, *J. Geophys. Res.*, 109, A02 303, 2004.

166. Bernhardt, P. A., Erickson, P. J., Lind, F. D., Foster, J. C., and Reinisch, B., Artificial disturbances of the ionosphere over the millstone hill radar during dedicated burn of the space shuttle OMS engines, *J. Geophys. Res.*, 110, A05 311, 2005.

167. Hester, B. D., Chiu, Y.-H., Winick, J. R., et al., Analysis of space shuttle primary reaction-control engine-exhaust transient, *J. Spacecr. Rockets*, 46, 679–688, 2009.

168. Mendillo, M., Smith, S. M., Coster, A., et al., Man-made space weather, *Space Weather*, 6, doi:10.1029/2008 SW000406, 2008.

169. Kaplan, C. R. and Bernhardt, P. A., The effect of an altitude-dependent background atmosphere on space shuttle engine burn plumes, *J. Spacecr. Rockets*, 47, 700–704, doi:10.2514/1.49339, 2010.

170. Ozeki, M. and Heki, K., Ionospheric holes made by bistatic missiles from North Korea detected with a Japanese dense GPS array, *J. Geophys. Res.*, 115, doi:10.1029/2010 JA015531, 2010.

171. Bernhardt, P. A., Baumgardner, J. L., Bhatt, A. N., et al., Optical emissions observed during the charged aerosol release experiment (CARE I) in the ionosphere, *IEEE Trans. Plasma Sci.*, 39, 2274–2785, 2011.

172. Bernhardt, P. A., Ballenthin, J. O., Baumgardner, J. L., et al., Ground and space-based measurement of rocket engine burns in the ionosphere, *IEEE Trans. Plasma Sci.*, 40, 1267–1286, 2012.

173. Kikuchi, H., *Electrohydrodynamics in Dusty and Dirty Plasmas*, Dordrecht/Boston, MA/London: Kluwer Academic Publishers, 2001.

174. Fu, H. and Scales, W. A., Nonlinear evolution of the ion acoustic instability in artificially created dusty space plasmas, *J. Geophys. Res.*, 116, doi:10.1029/2011 JA016825, 2011.

175. Volland, H., Ed., *Handbook of Atmospheric Electrodynamics*, Vols. I and II, Boca Raton, FL: CRC Press, 1995.

176. Volland, H., Electrodynamic coupling between neutral atmosphere and ionosphere, in *Modern Ionospheric Science*, Kohl, H., Ruster, R. R., and Schlegel, K., Eds., Berlin: ProduServ GmbH Verlagsservice, 1996.

177. Goertz, C. K., Dusty plasmas in the solar system, *Review Geophys.*, 27, 271–292, 1989.

178. Northruo, T. G., Dusty plasmas, *Phys. Scr.*, 45, 475–490, 1992.

179. Blaunstein, N., Pulinets, S., and Cohen, Y., Computation of the main parameters of radio signals in the land-satellite channel during propagation through the perturbed ionosphere, *Geomagn. Aeronom.*, 53, 1–13, 2013.

Chapter 12

Theoretical Aspects of Plasma Structure Creation Associated with Rocket Burn and Launch

12.1 Overview

Accounting for similar features, in this chapter artificially and/or naturally induced in the near-the-Earth space and shown by various experimental campaigns associated with modification of the ionosphere, we will try to give a brief explanation of the observed effects of rocket burn, launch, and flight via currently existing self-consistent theoretical models and frameworks, including those prepared by the authors of this book. This implies introducing the reader to the observed results obtained by various "active experiments" in the ionosphere via vivid and clearly understood theoretical frameworks and phenomenological explanations.

The active rocket and satellite observations described in Chapter 11 have indicated several outstanding features and processes that are associated with the effects of rocket burn, launch, and flight, such as the following:

- Molecular quasisymmetrical (mostly Gaussian) diffusion of the release molecules produced by rocket engine, plume, and/or fuel regularly into the ambient quasineutral ionospheric plasma

■ Additional ionization of the neutral component of the background ionosphere by the expanding molecular cloud produced by the rocket exhaust gases

■ Generation of global depletion zones, called *holes* or *cavities*, covering the upper *E*-region to the upper *F*-region of the ionosphere

■ Generation of large-scale plasma enhancement zones at the *D*-region to the lower *E*-region of the ionosphere

■ Generation of instabilities and plasma waves within the perturbed plasma regions of the ionosphere caused by different ambient factors and sources, such as heating by the exhaust high-temperature gases and plume exhaust, influence of ambient electric, magnetic, and gravitational fields, wind shear, altitudinal gradients of ionospheric plasma profiles, and so forth

■ Creation of small- and moderate-scale plasma irregularities (called plasma *turbulences*) that fill the depletion region during the process of its further temporal and spatial evolution

Therefore, we will further try to briefly explain the main effects and processes mentioned earlier, based on the existing theoretical frameworks, as well as in the authors' own view of the processes occurring in the ionosphere that can be associated to and accompanied with rocket burn and launch described in Chapter 11.

12.2 Evolution of Rocket-Induced Plasma Structures in the Ionosphere

The first pioneering work, in which the author introduced phenomenological views on the process of rocket launch, was presented by Booker [1], in which he suggested that a depletion zone, or *hole*, in the ionosphere could be generated in the wake of a rocket by the exhaust gases rising and spatially expanding into the ambient ionospheric plasma due to their high temperature, which can finally be replaced by air surrounding the wake that has a lower ion content. As given in Ref. [2], another possible explanation for the hole formation was discussed, which explains that the rocket exhaust gases penetrate the near-rocket ambient ionosphere at altitudes of the *F*-layer after coming out of the rocket nozzle and quickly expand by a factor of more than 10^7, until atmospheric pressure equilibrium is reached. This expansion cools the exhaust gases and also reduces the ion density in the ambient plasma from roughly 10^{10} to 10^3 m^{-3}. Because usually the ion density of the *F*-region is above 10^4 m^{-3} during diurnal periods, the expanded exhaust gases create a global hole inside the *F*-layer [3–5]. The density of such a depletion zone (e.g., *cavity*) will be less than 0.1 from that of the surrounding ionospheric plasma of the *F*-region.

The authors of Ref. [6] have also tried to explain a complicated altitudinal profile of ionospheric plasma density (see Figures 11.3, 11.11, and 11.14), occurring from the *E*-layer to the upper *F*-layer, formed by several stages of the rocket burnouts and ignitions of stages. They have suggested that the global hole, observed from 120–130 to 300–320 km, and the plasma enhancements following it (below and above it) are generated by the heating process that can redistribute plasma behind and above the rocket due to rise of high-temperature gases from the exhaust that cools the ambient ionospheric plasma.

However, the suggestions given earlier and mentioned by other researchers [7–10] were made by phenomenological analysis of the observed effects and phenomena caused by rocket burn and launch. First, the simplest physical and mathematical model was proposed in Ref. [9]. This model was based on adiabatic continuum expansion of exhaust gases from the rocket engine or plume (formed in areas behind its trajectory of movement) into the equilibrium steady-state ambient ionospheric plasma. After some estimation of the main parameters taken from the observed experiment, the researchers found that the ionization density in the wake of a rocket, burning and flying through the *F*-region of the ionosphere, can be reduced below the ambient value by a factor of 10–20. At the same time, in the lower ionosphere (around 100 110 km) the ambient plasma density increases as the plume-ionized gas density decreases. Below this altitudinal range, the plume-ionized gas density should be greater that the ambient one, which fully coincides with the experimental data (see Figure 11.5a and b). Unfortunately, the authors ignored diffusive mixing of plume-induced charges with the ambient ionization during the plume gas expansion process, as was suggested by Booker and others [1–3].

In Ref. [11], the authors have tried to simulate the process of formation of huge depletion zones (e.g., *hole*) and its further temporal and spatial evolution. They also found that the main role in the hole creation has effects caused by rocket exhaust gases that stimulate decrease in plasma density in the upper ionosphere (around altitudes of the *F*-region). What is interesting to note is that they first considered the role of shock wave propagation and thermal expansion of rocket exhaust gases in their computer simulation. As we will show further, the authors did not account for actual recombination–ionization reactions, as well as transport processes occurring at altitudes of the *F*-region of the ionosphere, that later, after one or two decades, was done by Mendillo and colleagues [12–17] and by Bernhard and colleagues and other researchers [18–26]; the most adequate reality models will be discussed below.

Thus, Mendillo and colleagues [12–17] have explained the observed phenomena associated with the sudden vanishing of ionospheric plasma and global plasma hole generation at altitudes of the ionospheric *F*-layer by rocket burn and launch using a physical model presented generally and mathematically. In such a model, the researchers have combined the following processes and phenomena to explain the observed features as plasma holes:

■ The aeronomic reactions initiated by the consistent flow of exhaust gas from the rocket engine, plume, or fuel jet
■ The rapid ion–atom and ion–molecule interchange reactions between the ionospheric ion O^+ dominating plasma at altitudes of the *F*-region and the molecules of hydrogen (H_2) and water vapor (H_2O) produced by the exhaust gas plume according to Equation 11.1
■ The diffusion of plume-ionized gas into cool ambient ionospheric plasma
■ The rapid vanishing of the molecular ions during dissociation–recombination reactions in the ambient ionosphere at those altitudes
■ Heat transfer by high-temperature exhaust gas to cool ionospheric plasma

Numerical computations of the TEC, accounting for the above assumptions, supposed by the authors of such a self-consistent theoretical framework evidently showed depletion zone generation at the initial stage of its evolution in a satisfactory agreement with TEC profile measured experimentally in numerous observations mentioned in Chapter 11 (see Figures 11.7 through 11.12 and 11.14). Moreover, theoretical analysis showed that the rate of reaction of exhaust gas molecules with O^+ occurs more rapidly (10 × 10 times) than with the normal ionospheric neutral molecules such as nitrogen (N_2) and exigent (O_2). Further, in Refs. [12–17] the authors analyzed further stages of depletion region evolution and found that the ionized species of the plume plasma can fill the plasma depletion region having a radius of several hundred kilometers due to their higher rate. During its temporal and spatial evolution, this global plasma hole increased from the *E*- to the *F*-region of the ionosphere through exhaust-ionized molecules, and was filled by moderate- and small-scale plasma structures called plasma irregularities or plasma turbulences. Unfortunately, the latter aspects were out in the theoretical analysis described in Refs. [12–17].

It is interesting also to note that during active experiments in the ionosphere related to rocket releases of different kinds of heavy and complicated molecules, consisting of exhaust gas according to Equation 11.1, some research groups have modeled and observed depletion and enhanced region generation in the ionosphere depending on the altitude of the release and the amount of released gas [17–28] numerically and experimentally. We will bring the reader's attention to the researches carried out by Bernhard and colleagues [18–26], because they did not confirm the results obtained in Refs. [12–17] based on chemical reactions of ionization and recombination that occur surrounding the engine fuel or the depleted cloud of plume exhaust. Based also on the actual transport processes accompanying the last stage of plasma structures, enhancements and depletions, evolution and stratification at different altitudes caused by diffusion, drift in crossing ambient electric and magnetic field, and wind movements, they tried to explain the complicated "stratified" plasma profiles observed experimentally (see, e.g., Figures 11.3, 11.11, 11.12, and 11.14). Finally, to describe the effects of stratification of the initial depleted cloud into separate moderate- and small-scale plasma structures, Bernhard [27] analyzed different kinds of plasma instabilities, from gradient drift (GD) and convection-current, to Rayleigh–Taylor, RT (e.g., gravitational), that can be induced by rocket burn and launch effects and can be considered as the main candidate for such plasma irregularities (or plasma turbulences) generation at different latitudes of the ionosphere.

Therefore, based on the previously obtained results, we will analyze each effect mentioned earlier via a glimpse of the actual phenomena occurring at different altitudes of the ionosphere and in different altitudinal regions of the ionosphere, from the equatorial to the auroral, for analyzing the dominant processes happening there. Therefore, in our explanation further on the subject, we will differentiate the effects and processes for different altitudinal regions and different ionospheric latitudes. In our opinion, these will yield better understanding of the peculiarities observed experimentally at different altitudes and in different latitudinal regions of the equatorial, middle, and northern auroral ionosphere.

12.3 Infringement of Ionization–Recombination Balance by Rocket Burn and Launch in the Ionosphere

We discuss the first feature described in the headline above in numerous experimental and theoretical works regarding aeronomical reactions initiated by components of the exhaust gas and the interchange reactions between ions of ambient ionospheric plasma and molecules generated via reactions described by Equation 11.1. Mostly, it was focused on description reactions occurring in the F-region of the ionosphere. We think that this is not a correct approach because the rocket burnouts and ignitions occur not only in the F-region but also in the D- and E-layers of the ionosphere, where, instead of holes, huge enhancement zones of plasma ions are observed experimentally (see, e.g., Figure 11.3). The same enhancements in the lower ionosphere occur during meteor-induced plasma structure evolution (mostly in its initial phase), which are also observed experimentally. Therefore, we think that the processes of ionization and recombination occurring in the three regions of the ionosphere should be differentiated and discussed separately.

12.3.1 Processes in the D-Layer and Lower E-Layer of the Ionosphere

As seen by numerous satellite, rocket, and radar observations, the major ion species comprising the ambient plasma in the D-region of the ionosphere are the NO^+ and O_2^+ ions, which together with electrons (e^-) secure the conditions of its quasineutrality (i.e., $N_{NO^+} + N_{O_2^+} \approx N_e$-) [29,30]. Here, as in Refs. [29,30], we indicated the charge of electrons as $q \equiv e^- < 0$, to differentiate it from the positive atomic ions having the same charge number, that is, $q \equiv Z|e^+| = Z|e^-| > 0, Z \equiv 1$.

In the ambient ionosphere at D-layer altitudes, the effects of ionization and recombination prevail compared to plasma electron and ion transport. Indeed, the characteristic times of plasma transport (diffusion, thermodiffusion, and drift) are quite larger than the characteristic time of ionization–recombination balance establishment. This occurs due to high atmospheric gas density [31,32]. Moreover, at the D-region of the ionosphere, the attachment of the electron to the molecule ($O_2 + O_2 + e \rightarrow O_2^- + O_2$) yields molecular ions and decreases electron concentration as well as the reverse process of the detachment of electrons from the molecular ions and increase of electron concentration plays an important role [31–34].

Next, according to the results obtained in Refs. [12–28] by several groups of researchers, which simultaneously and separately investigated the problem of additional ionization of the ambient ionosphere by high-temperature exhaust gases from the rocket engine or plume, additional ions can be generated in the D-layer and lower E-layer (if the burnout and the ignition were done at altitudes of 50–100 km), such as Na^+, Mg^+, Ca^+ Al^+, Fe^+, N_2^+, NO^+, O_2^+, H_3O^+, H_2O^+, and so forth (depending on the consistency of the engine fuel). Moreover, as mentioned in Refs. [32,33], even considering the formation of plasma structures at altitudes of the lower ionosphere based only on the local chemical ionization–recombination reactions without considering transport processes becomes a very complicated mathematical task. As mentioned in Refs. [33,34], the well-known schemes of the coupling ionization–recombination chemical reactions, containing several tens of components and maybe some hundreds of reactions, have not yet been performed and are therefore complicated to analyze.

As promised to the reader from the beginning, we perform a simple scheme that is important for explaining the mechanism of the creation of exhaust and ambient ions accounting for the following assumption.

■ First, performing this scheme, we suggested that in the D-layer of the ionosphere the chemical reactions comprise the ionization (with intensity I) of neutral molecules and the creation of positive ions and electrons, indicated by M^+ and N_e, respectively, produced by both solar radiation (Q_S) and high-temperature exhaust gas streams (Q_E), that is, $I = Q_S + Q_E$.

■ Second, negative ions in plasma are produced by the attachment of various neutral molecules with ambient plasma electrons. Because we have simple atomic molecules and complicated two-atom or three-atom molecules (see, e.g., Equation 11.1), we will denote them as M_1^-, M_2^-, and M_3^-, respectively. The corresponding rates of their generation are denoted, as in Refs. [33, 34], by β_1, β_2, and β_3, respectively. At the same time, accounting for the processes of detachment of electrons from the corresponding negative ions, we denoted the corresponding rates by γ_1, γ_2, and γ_3, respectively.

All steps are clearly seen from the proposed scheme shown in Figure 12.1. The process of recombination of electrons with positive ambient ions ($N_e \leftrightarrow M^+$) occurring with the rate of α_e, and the negative ions M_1^-, M_2^-, and M_3^-, occurring with the rates α_1, α_2, and α_3 introduced in this scheme, we completed a coupling between ionization and recombination reactions for "electron–ion" gas (e.g., plasma) and neutral gas equilibrium established in the D-region of the ionosphere.

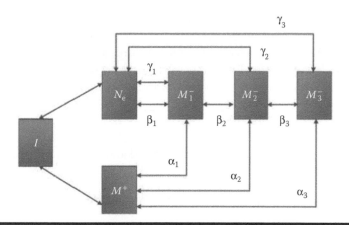

Figure 12.1 **Schematically presented chemical reactions occurring in the lower ionosphere described by a system (12.1).**

The performed scheme yields, in the same manner as was done in Refs. [32,33] for the regular nonperturbed ionosphere, the following differential equation for each charge species of plasma, accounting for both the ambient plasma ions and the exhaust-ionized gas from plume or engine fuel stream:

$$\frac{dN_e}{dt} = I - \alpha_e N_e M^+ - \beta_1 N_e + \gamma_1 M_1^- + \gamma_2 M_2^- + \gamma_3 M_3^-, \qquad (12.1a)$$

$$\frac{dM^+}{dt} = I - \alpha_e N_e M^+ - \alpha_1 M_1^- M^+ - \alpha_2 M_2^- M^+ - \alpha_3 M_3^- M^+, \qquad (12.1b)$$

$$\frac{dM_1^-}{dt} = \beta_1 N_e - \beta_2 M_1^- - \gamma_1 M_1^- - \alpha_1 M_1^- M^+, \qquad (12.1c)$$

$$\frac{dM_2^-}{dt} = \beta_2 M_1^- - \beta_3 M_2^- - \gamma_2 M_2^- - \alpha_2 M_2^- M^+, \qquad (12.1d)$$

$$\frac{dM_3^-}{dt} = \beta_3 M_2^- - \gamma_3 M_3^- - \alpha_3 M_3^- M^+. \qquad (12.1e)$$

The equations in this coupled system of chemical reactions for the concentrations of charged particles consisting finally of ambient and exhaust-induced ions depend on each other because the species are satisfied by the conditions of quasineutrality of the "disturbed-by-rocket" plasma, that is,

$$M^+ \approx N_e + \sum_{k=1}^{3} M_k^-. \qquad (12.2)$$

This condition allows us to eliminate one of the equations of the system (Equation 12.1a–e) and describe the concentration of the corresponding plasma component by the others. However, to solve this general system is not an easy task initially. Therefore, to have the correct physical explanation of the phenomena observed experimentally, we will present an adequate reaction scheme with some "effective" negative molecular ion $M_{eff}^- = M_1^- + M_2^- + M_3^-$, determined by the effective attachment coefficient $\gamma_{eff} = \left(\gamma_1 M_1^- + \gamma_2 M_2^- + \gamma_3 M_3^- \right) / M^-$. In this case, the last three equations in (12.1) can be reduced and transformed to one, that is,

$$\frac{dM_{eff}^-}{dt} = \beta_{eff} N_e - \gamma_{eff} M_{eff}^- - \alpha_i M_{eff}^- M^+. \qquad (12.3)$$

This yields to the following circumstance: the total number of negative ions are replaced by a certain effective ion, whose production rate is equal to the attachment rate, β_{eff}, of electrons to such an "effective" molecule and the detachment from this "effective" ion characterized by the effective coefficient γ_{eff}. The mentioned transformations yield to the deduced system of three equations only, that is,

$$\frac{dN_e}{dt} = I - \alpha_e N_e M^+ - \beta_{eff} N_e + \gamma_{eff} M_{eff}^-, \tag{12.4a}$$

$$\frac{dM^+}{dt} = I - \alpha_e N_e M^+ - \alpha_i M_{eff}^- M^+, \tag{12.4b}$$

$$\frac{dM_{eff}^-}{dt} = \beta_{eff} N_e - \gamma_{eff} M_{eff}^- - \alpha_i M_{eff}^- M^+. \tag{12.4c}$$

It is interesting to note that such a system has been used for a long time to investigate ionization–recombination reactions in the nonperturbed *D*-layer [35] under equilibrium steady-state regime. At the same time, analysis of this simplified system will allow us to explain the existence of enhancing plasma region in the lower ionosphere obtained experimentally (see further).

12.3.2 Processes in the Upper E-Layer and the F-Layer of the Ionosphere

Now we will consider a case in which rocket burnout and ignition occur at the upper *E*-layer and the *F*-layer of the ionosphere. The main ions of the ambient plasma at altitudes of the *E*-layer are NO^+ and O_2^+, whereas in the *F*-region it is O^+. We also account for the main ion species produced by chemical reactions inside the exhaust gas and presented earlier, and denote all of them by M^+. Numerous experiments and numerical estimations of the ionization–recombination balance disturbance in this region of the ionosphere, carried out in Refs. [36–41], have shown that at the *E*-layer and beyond (up to the F_2-layer), the reactions of the attachment and detachment are not actual, and hence can be ignored further in our discussions. At the same time, the researchers should account for both chemical reactions of ionization and recombination and transport processes occurring from 90–100 to 160–180 km. In this case, the main equations that describe the ionization–recombination balance in this region of the ionosphere, accounting for the transport process, can be written for both electron and ion components of plasma as follows:

$$\frac{\partial}{\partial t}\frac{dN_e}{dt} + \nabla \cdot \mathbf{j}_e = I - \alpha_e N_e M^+, \tag{12.5a}$$

$$\frac{\partial}{\partial t}\frac{dM^+}{dt} + \nabla \cdot \mathbf{j}_i = I - \alpha_e N_e M^+. \tag{12.5b}$$

Here, I describes the ionization process that produces both the ambient plasma and exhaust ion gas ions ($I = Q_S + Q_E$), and $\mathbf{j}_e$ and $\mathbf{j}_i$ are the current densities of electrons and ions in plasma.

Equations 12.5a and 12.5b describe the balance of charged particle number in the ionosphere, accounting for their transport by diffusion, thermodiffusion, and drift processes, defined by $\mathbf{j}_e$ and $\mathbf{j}_i$, whose expression we do not present here due to their complexity; instead we suggest the books denoted by Refs. [35,41] to the reader. As also shown in Refs. [29,30,35–40], at the E- and F-regions of the ionosphere, the *dissociative recombination* is the main chemical reaction that describes attachment of the molecular ion M^+ with electron e^-, which is accompanied by the dissociation of the molecular ion on two excited neutral atoms A_1^* and A_2^*:

$$M^+ + e^- \rightarrow A_1^* + A_2^*. \tag{12.6}$$

For example, for the molecular ion of oxygen O_2^+ or the oxide of nitrogen NO^+, we have, respectively, a reaction of the following type:

$$O_2^+ + e^- \rightarrow O^* + O^*, NO^+ + e^- \rightarrow N^* + O^*. \tag{12.7}$$

Therefore, for the dissociative recombination, the amount of electrons attached with ions in the volume of 1 cm³ for 1 s can be described in Equation 12.5 by the effective recombination coefficient:

$$\alpha = \alpha_1 n_{NO^+} + \alpha_2 n_{O_2^+}, \tag{12.8}$$

where:

$n_{NO^+} = N_{NO^+}/N_e$ and $n_{O_2^+} = N_{O_2^+}/N_e$ are the relative (to electron component of the quasineutral plasma) concentrations of ions NO^+ and O_2^+, respectively

α_1 and α_2 are the reaction coefficients of dissipative recombination of these ions

It should also be noted that apart from dissociative recombination, *radiative recombination* also exists in the ionosphere, related to the attachment of the electron by the atomic ion and radiation of the photon. This process prevails for the D-region of the ionosphere and is described earlier. The following estimations, carried out in Refs. [35–40], have shown that this type of recombination can be ignored at the E-layer and beyond.

Now the interactions between the main ions of plasma and neutral molecules, such as N_2 and O_2, are determined by the following reactions:

$$N_2 + O_2^+ \rightarrow NO^+ + N, O_2 + O^+ \rightarrow O_2^+ + O. \tag{12.9}$$

Such reactions are characterized by the coefficients of ion–molecule interactions, k_1 and k_2, respectively, which are fully described in Refs. [12–17].

Moreover, in the E-layer the coefficients of dissociative recombination described by Equation 12.7 are $\alpha_1 = 10^{-1}$ and $\alpha_2 = 2 \times 10^{-1}\,\mathrm{m^3 s^{-1}}$, while those described by Equation 12.9, presented below, are $\beta_1 \approx \beta_2 \approx 10^{-7}\,\mathrm{m^3 s^{-1}}$. At the same time, in the F-layer of the ionosphere $\alpha_1 = (3\text{–}5) \times 10^{-2}$ and $\alpha_1 = (2\text{–}5) \times 10^{-2}\,\mathrm{m^3 s^{-1}}$, while for dissociative reactions, described by Equation 12.9, they are: $\beta_1 = 3 \times 10^{-6}$ and $\beta_2 = 2 \times 10^{-6}\,\mathrm{m^3 s^{-1}}$.

Similar reactions with ambient ion O^+ were analyzed in Refs. [12–23] for other components of the exhaust plume gas or engine fuel jet, such as H_2, H_2O, and more complicated molecules. As a result, molecular ions can be generated at altitudes of the upper ionosphere that have never been observed in the nonperturbed ionospheric plasma at considerable heights.

These estimations show that the most important chemical ion–electron or ion–neutral reactions occurring in the high-temperature exhaust gas cloud produced by the rocket engine or plume at altitudes beyond the D-layer and lower E-layer, are similar to those described by Equations 12.7 and 12.9 for the ambient nonperturbed ionospheric plasma species, which are described by the coefficients of *dissociative* recombination, α_1 and α_2.

More general coefficient of recombination that account for the altitudes of the nonperturbed regular ionosphere, below and above the critical range of 150–160 km, where the process of plasma transportation (diffusion, thermodiffusion, and drift) prevail over chemical ionization and recombination processes [41] are given in Refs. [42–45]. We now present this coefficient that accounts for both the ambient regular processes and those caused by the exhaust gas molecules produced by the rocket engine, plume, or fuel jet, that is

$$R = \frac{N_e \left(\sum_{k=1}^{K} \beta_k N_{M_k} \right)}{\left[1 + \sum_{k=1}^{K} \left(\beta_k N_{M_k} \middle/ \alpha_k N_e \right) \right]}. \tag{12.10}$$

This formula reduces to that which is obtained in Refs. [42–45] for the regular nonperturbed ionosphere (namely, not perturbed by the rocket releases), which fully covers scenarios that occur in the lower ionosphere and are described by Equation 12.9. So, the proposed expression given in Equation 12.10 covers chemical reactions occurring at altitudes below and above the critical range of 150–160 km, where the process of plasma transfer prevails over chemical processes beyond this range [32,39,41].

At the *F*-layer, as was shown by numerous researchers [12–23], the parameter β_k that describes the rate of ion–neutral collisions inside the exhaust gas, is changed at the range of magnitude $\beta_k \in (1.0 \times 10^{-6}, 3.0 \times 10^{-5})\text{m}^3\text{s}^{-1}$, whereas the coefficient of ion–electron recombination, $\alpha_k \in (2.0 \times 10^{-2}, 3.0 \times 10^{-1})\text{m}^3\text{s}^{-1}$, that is, much more higher compared to the first one. Therefore, as it was found that the processes of recombination are faster (by 2–3 orders of time) than the processes of ion production by an exhaust gas of molecules generated according to aeronomical and chemical reactions inside the exhaust gas, such as Equation 11.1, depending on the initially consistent fuel taken for the rocket engine.

It was also found in Refs. [31,35,41] that the coefficients of recombination described by Equation 12.8 strongly depend on the temperature (in Kelvin) and decrease as follows:

$$\alpha_1 \approx 5 \times 10^{-7} \left(\frac{300}{T_e}\right)^{1.2}, \quad \alpha_2 \approx 2.2 \times 10^{-7} \left(\frac{300}{T_e}\right)^{0.7}, \tag{12.11}$$

which fully coincides with the analogous dependences obtained in Refs. [17–19,22]. The above formulas and reactions allow us to explain complicated evolution of the plasma cavities (*holes*), with enhanced and depletion zones, following each other. We only note that all the processes observed further depend on where and at what altitude the rocket engine ignition, shutoff, or burnout takes place.

12.4 Plasma Enhancements and Depletions Induced by Rocket Burn and Launch in the Ionosphere

12.4.1 D-Layer of the Ionosphere

First, we should note that at altitudes of *D*-layer to lower *E*-layer, the process of *radiative* recombination, $\alpha_R(h)$, which is predominant at altitudes from 50 to 90–100 km, can be described simply as $\alpha_R(h) \sim \alpha_R N_e^2(h)$ with the rough constant rate of $\alpha_R \approx 10^{-6}\text{m}^3\text{s}^{-1}$ [30,31,33,35], accounting for the conditions of plasma quasineutrality as shown by Equation 12.2. Further, as shown in Refs. [31,32] by analyzing the chemical reactions caused by neutral molecules and molecular ions, the rates of ionization, attachment, and detachment are much higher than the rate of radiative recombination at the lower ionosphere altitudes.

Indeed, the analysis of Equation 12.4 made in Refs. [31,32,35] showed that the processes of attachment and detachment occur in the *D*-layer within the timescale of a 0.01–1 s, whereas for the process of recombination the timescale is measured in minutes. Therefore, the difference of about 2 orders of magnitude between the characteristic timescales of these two different processes allows us to consider them as practically independent. Moreover, we can suppose that despite the fast process associated with the increase of electrons attachment rate, leading to a decrease in their density, the slow process associated with the decrease of the radiative recombination rate will finally increase the density of electrons. Finally, in the lower ionosphere at altitudes of 60–100 km, the enhancements in plasma profile occur in the lower ionosphere. The additional ions of the exhaust gas can also increase the total plasma content at the lower ionosphere.

Therefore, the effect of increase of ion and electron component of the disturbed plasma and generation of exhaust gas-induced plasma enhancement region below 100–120 km should be expected during rocket burn and launch in the *D*-layer. This effect was clearly shown by the corresponding observations of rocket burn and launch (see Figure 11.3). Additional measurements of the scattered signal from the lower ionosphere during the burnout of the rocket engine occurs at altitudes of about 73–75 km have shown that for a few seconds (which correspond to estimations of the high rate of chemical ionization reaction presented above for the *D*-region) the exhaust ion gas due to interaction with the dense atmospheric gas consisting of neutral molecules dominated at these altitudes and has the necessary potential to generate enhancement ions.

A similar scattering effect observed in the lower ionosphere was found during the active plasma heating experiment discussed earlier. For instance, it was found experimentally in Ref. [32] that a scattering effect (Figure 12.2) was

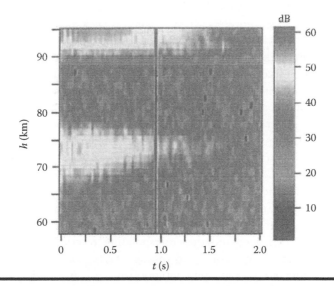

Figure 12.2 **The strength of the scattered radio signals (dB) from the heating-induced plasma enhancements located at the *D*-layer and lower *E*-layer of the ionosphere.**

evident from the plasma enhancement structures in the *D*- and lower *E*-regions at altitudes of 60–80 and 90–100 km while heating of the lower ionosphere.

Figure 12.2 shows the altitudinal distribution of the amplitude of scattering signal from plasma enhancement regions versus the time of observation. It is clearly seen that small-scale enhancements of plasma density (i.e., plasma irregularities or turbulences) fill the lower ionospheric regions, causing strong scattering effects of the probing signals sent by ground-based radar systems (see details in Ref. [32]). We will discuss in the next section about the mechanism of plasma stratification on several plasma eddies (i.e., turbulent plasma structures) and the source of such turbulent structure generation.

12.4.2 Upper E-*Layer and* F-*Layer of the Ionosphere*

As mentioned earlier, the range of altitudes from 100 to 200 km is a more complicated region where the chemical ionization–recombination reactions between plasma species are sufficiently strong and the transport process (diffusion, thermodiffusion, and drift) is not so weak (with respect to the *D*-layer and lower *E*-layer). Therefore, they both should be taken into account. Moreover, all the theoretical models presented in Refs. [12–34] mostly accounted for the effects occurring at the *F*-region of the ionosphere, that is, at the altitudes ranged beyond 250 km, where the transport processes are predominant and the effects of chemical reactions are insufficient [32,35,39,41]. Only in a few of these researches, the authors took into account the transport processes accompanying exhaust engine and plume gases expansion into the ambient ionospheric plasma. As mentioned in the self-consistent theoretical frameworks [42–46], analyzing the initial and further stages of evolution of heating-induced plasma structures at the *E*-layer of the ionosphere ranged from 100 to 200 km, the processes of diffusion and thermodiffusion should be taken into account together with processes of ionization and recombination.

In the case under consideration, we have a more complicated picture of evolution of the heating-induced plasma structures, because it should be taken into account not only as a reaction of ambient plasma ions with the plasma electrons, existing in the regular ionosphere (which was done in Refs. [47–51]), but also as the exhaust ions generated during expansion of the neutral exhaust engine fuel or plume gas (according to, let us say, the reactions in Equation 11.1).

Therefore, in our brief explanation of the problem, we should modify the recombination coefficient and the corresponding reactions, described earlier by formulas given as Equations 12.8 through 12.10, for presentation of some numerical results of our own computations. Thus, accounting for the reactions with exhaust gas described by Equation 11.1 and rearranging the chemical reactions given as Equations 12.1 through 12.5, now for the corresponding ions CO_2^+ and H_2O^+, instead of ambient ions, NO^+ and O_2^+, prevailing in the regular ionosphere at altitudes from 100 to 200 km [32,35,39,41], we instead get Equation 12.10:

$$R = \frac{N_e(\beta_1 N_{H_2O} + \beta_2 N_{CO_2})}{\left(1 + \dfrac{\beta_1 N_{H_2O}}{\alpha_1 N_e} + \dfrac{\beta_2 N_{CO_2}}{\alpha_2 N_e}\right)}, \tag{12.12}$$

where:

α₁ and α₂ are described by Equation 12.8, and their estimations as well as estimations of β₁ and β₂, are presented above

Here, again we take into account the quasineutrality of the ionospheric plasma, that is, $N_e \approx N_{H_2O^+} + N_{NO^+} + N_{O_2^+} + N_{CO_2^+} \equiv N$ (N is quasineutral plasma concentration). Then, the densities of the molecular ions can be found by the local balance equations of the ions:

$$N_{O_2^+} = \frac{N_e}{\left(1 + \dfrac{\beta_1 N_{H_2O}}{\alpha_1 N_e} + \dfrac{\beta_2 N_{CO_2}}{\alpha_2 N_e}\right)}, \tag{12.13a}$$

$$N_{H_2O^+} = N_{O_2^+} \frac{\beta_1 N_{H_2O}}{\alpha_1 N_e}, \tag{12.13b}$$

$$N_{CO_2^+} = N_{O_2^+} \frac{\beta_2 N_{CO_2}}{\alpha_2 N_e}. \tag{12.13c}$$

Equations 12.12 and 12.13 fully determine the balance of ionization and recombination in the ionosphere at altitudes beyond the *D*-layer when exhaust ion gas occurs during burnout or ignition of the engine fuel or plume. We should note that if there are other ion species, instead of CO_2^+, H_2O^+ will be generated, and the corresponding formulas given as Equations 12.12 and 12.13 should be written in the same manner for the corresponding species. The same procedure should be followed for the *F*-layer ranged from 250 to 350 km, where, instead of the molecular ion O_2^+ (or NO^+), the atomic ion O^+ is predominant [12–25], and in Equations 12.13a through 12.13c, its concentration should be introduced in the product with the corresponding fraction.

Now we estimate "local heating" effect on the ambient plasma by the high-temperature exhaust plume or engine gas. As shown in Chapter 3, according to computations carried out in Refs. [52–53], the super-heavy and heavy rocket engine each has power $\sim 10^{11}$–10^{10} W, respectively. According to computations made in Refs. [52–54], only an insignificant part ($\sim 0.1\%$–1.0%) of the total power of the reactive fuel jet can be transferred to the ambient plasma. Thus, we found that the effective power of the "artificial heater" can change in a wide range from about 10–100 to 100–1000 MW.

If so, we can present the "local heater" at the same manner, as was done in Ref. [45], via the relative measure Q/N_e [where $Q(z_0)$ is the power of the external local heater located at an altitude of z_0, $N_e \equiv N$ is the "composite" plasma concentration], that describes the temperature rate of the "heating" process in Kelvin per second ($K\,s^{-1}$), that is,

$$Q/N_e(K\,s^{-1}) \approx \frac{(6.5 \times 10^{10}) \times P_a(100\ \text{MW})}{L(\text{km})h_0^2(100\ \text{km})N_e(m^{-12})},\qquad(12.14)$$

where:
L is the scale of the depletion or enhancement region
h_0 is the height of the source of local heating

A self-matched system of hydrodynamic equations of diffusion–thermodiffusion, combined with chemical ionization and recombination reactions, describing a nonstationary, one-dimensional (along the height z of the ionosphere) problem of the first stage of evolution of plasma structure elongated along an ambient magnetic field $\mathbf{B}_0$ (i.e., along the height of the ionosphere), can be presented, accounting for the corresponding changes made above, with respect to what was done in Ref. [45], as follows:

$$\frac{\partial N_I}{\partial t} = \frac{\partial}{\partial z}\left[D_{aN\|}(z)\frac{\partial N_I}{\partial z}\right] + \frac{\partial}{\partial z}\left[D_{eT\|}(z)\frac{\partial T_e}{\partial z} + D_{iT\|}(z)\frac{\partial T_i}{\partial z}\right]$$

$$+\frac{2}{3}\frac{Q}{N_e}-R,\qquad(12.15a)$$

$$\frac{\partial T_e}{\partial t} = \frac{\partial}{\partial z}\left[\frac{\kappa_{e\|}(z)}{N_e}\frac{\partial N_e}{\partial z}\right] + \delta_{ei}v_{ei}(T_e-T_i)$$

$$-\delta_{em}v_{em}(T_e-T_m) + \frac{2}{3}\frac{Q}{N_e},\qquad(12.15b)$$

$$\frac{\partial T_i}{\partial t} = \frac{\partial}{\partial z}\left[\frac{\kappa_{i\|}(z)}{N_I}\frac{\partial N_I}{\partial z}\right] + \delta_{ei}v_{im}(T_i-T_m) + \frac{2}{3}\frac{Q}{N_e}.\qquad(12.15c)$$

Here, $N_I = N_{H_2O^+} + N_{NO^+} + N_{O_2^+} + N_{CO_2^+}$ is the concentration of the mixture of ions in "dusted" plasma; the components of tensors of diffusion, thermodiffusion, and conductivity, which in the case of nonisothermal plasma $(T_e \neq T_i)$ can be written in the following forms:

$$D_{aN\parallel} \approx \frac{D_{ii} + \gamma D_{ii\parallel}}{1 + \gamma}; D_{Te\parallel} \approx \frac{2 + p(1+\gamma)}{(1+p)(1+\gamma)} \frac{N_e}{m\nu_{em}}; \qquad (12.16a)$$

$$D_{Ti\parallel} = \frac{2 + p(1+\gamma)}{(1+p)(1+\gamma)} \frac{N_I}{\bar{M}_i \nu_{im}}; D_{ee\parallel} \approx \frac{T_e[1 + \gamma p(1 + T_i/T_e)]}{m\nu_{em}(1+p)}; \qquad (12.16b)$$

$$D_{ii\parallel} = \frac{T_e[p + (T_i/T_e)(1+p)]}{\bar{M}_i \nu_{im}(1+p)}; \kappa_e = \frac{N_e T_e}{m\nu_{em}(1+p)}; \qquad (12.16c)$$

$$\kappa_I = \frac{N_I T_i}{\bar{M}_i \nu_{im}(1+p)}; \qquad (12.16d)$$

where:
the parameters $D_{aN\parallel}$, $D_{T\alpha\parallel}$, $D_{\alpha\alpha\parallel}$, and κ_α are the longitudinal components of coefficients of ambipolar diffusion, thermodiffusion, mutual diffusion, and thermoconductivity for electrons ($\alpha = e$) and ions ($\alpha - i$), respectively, relative to the ambient magnetic field $\mathbf{B}_0$, that for simplicity we directed along the height axis
p is the coefficient of ionization usually used in the ionized plasma [39–43]

Here, as before, we take a mean mass of the ion content of the "dusted" plasma. All other parameters of the ionospheric plasma are as defined in Chapter 1.

In our numerical experiment on virtual heating, we suppose that the "heating source" was located at the fixed altitude $z_0 = 130$ km, that is, model a situation where the exhaust gas plume (or engine fuel) expanded in the cool ionosphere at altitudes of *E*-region. Figure 12.3 presents the result of numerical simulations for the case of the "local heater" located at an altitude of $z_0 = 130$ km, that is, within the *E*-region of the ionosphere, with low heating power of ~50 MW.

The curve "1" corresponds to the moment of "switching off" of the heating process (after 20 s of "heating"). The curve denoted by "2" in Figure 12.3 corresponds to the observation time of 10 s after "switching on" of the heating source. As is clearly seen, when the heating process is not so strong, the chemical ion-exchange reactions prevail, yielding an increase of amount of exhaust ions in the dusty plasma with a not essential increase of total (electrons plus monatomic and polyatomic ions) concentration of the perturbed plasma (only on 3%–5%).

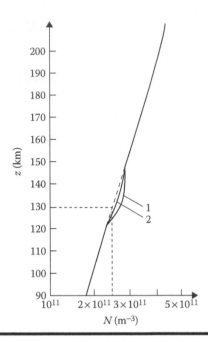

Figure 12.3 Plasma density profile and the enhanced zone creation during local heating occurred at an altitude of 130 km. Twenty seconds after the heating process the profile is denoted by 1; 2 corresponds to 10 s after switching off the heater. Nonperturbed ionospheric profile is shown by dashed curve.

The same local heating, occurring at an altitude of 290 km, that is, within the *F*-region of the ionosphere, where the transport processes are predominant, as was expected and as is clearly seen from Figure 12.4, after 20 s of "local heating" of a deep depletion region (*hole* or *cavity*) elongated at the altitudinal range of 50–60 km (with respect to background plasma density presented by the dashed curve) can be created.

Now, if we model the real experiment shown in Figure 11.5 or 11.7, that is, locate the virtual "heating source" at different altitudes from 170 to 230 km, thereby modeling the situation when the rocket releases occur at different altitudes during its flight (Figure 12.5).

The first release was modeled, as a "local heater," with a power of ~100 MW at an altitude of 170 km (denoted by 1), the second heating process was started after 2 s (at an altitude of 175 km) with a power of ~50 MW (denoted by 2), then after 10 s, with a power of 25 MW at an altitude of 190 km (denoted by 3). The fourth heater "switching on" (denoted by 4) was done after 18 s at an altitude of about 232 km with a power of ~10 MW, and finally, the fifth release heating process was after 25 s with the same power of 10 MW (denoted by 5).

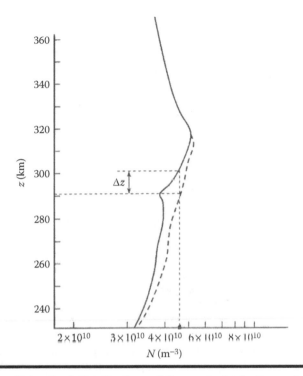

Figure 12.4 Altitudinal distribution of the perturbed plasma density (solid curve) in the local heating of the ambient ionospheric plasma (dashed curve) by the exhaust gases during the rocket release made at the height of ~290 km.

This numerical experiment, presented in Figure 12.5, simulates moving vehicle with five stages of ignition and engine fuel exhaust or plume exhaust. The dotted curve represents the background plasma profile, and the vertical dotted curve indicates the disturbed plasma density profiles at an altitude of z_{0i}, $i = 1, ..., 5$, of the "heating" source location.

The results of the same numerical experiment with movements of the virtual local heater that moves from the altitude of 200 km (the first exhaust gas ignition) to 240 km (the second stage of ignition) are shown in Figure 12.6.

The "heating" process was of about 10 s with a delay of 10 s between the first and the second stage of ignition. The numbers I and II indicate the profile of perturbed plasma during heating, and that "observed" 20 s after switching off of the heating process, respectively. Dashed curve represent the nonperturbed profile of the background ionospheric plasma.

It is evident that using the proposed self-consistent model described earlier and accounting for both transport processes and chemical reactions occurring

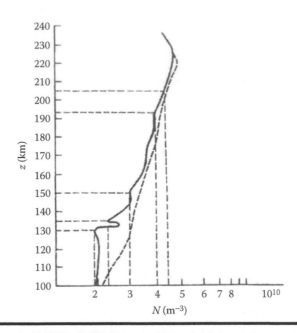

Figure 12.5 **Altitudinal distribution of the perturbed plasma density (solid curve) in the local heating of the ambient ionospheric plasma (dashed curve) by the exhaust gases during the moving rocket ignitions made at the heights of ~130, 135, 150, 195, and 205 km.**

in the "dirty" plasma (consisting of not only ambient, but also exhaust gas ions), we can explain the complicated plasma density profiles, consisting of depletion and enhanced zones following each other (as observed with naturally induced bubbles and plumes) over the wide range of altitudes—from the upper *E*-layer to the upper *F*-layer (comparing with experimentally observed profiles shown in Figures 11.3, 11.5, 11.7, and 11.11).

The presented examples of the numerical experiment show that depending on the height of the ionosphere, where the engine-exhaust or plume-exhaust heating will take place, power was expanded by "dusty" or "dirty" plasma into the ambient cool ionospheric plasma, the corresponding enhanced or depletion global plasma structures can be generated at the initial stage of their evolution. As shown further, at the later stages of global plasma structure evolution, the corresponding instabilities are generated (depending on the ionospheric altitudes and regions) causing, finally, creation of small-scale plasma turbulences or irregularities filling the disturbed plasma regions. Finally, a stratified plasma structure can be clearly observed experimentally. This will be the subject of the next subsections.

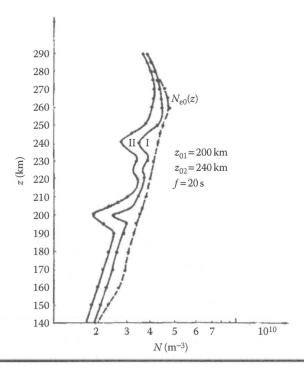

Figure 12.6 Altitudinal distribution of the perturbed plasma density (solid curve) in the local heating of the ambient ionospheric plasma (dashed curve) during the moving rocket releases made at the heights of ~200 and 240 km. Number I indicates the profile during heating process of 10 s, number II indicates the plasma profile 20 s after "switching off" the local heater.

12.5 Generation of Plasma Instabilities in the Ionosphere by Rocket Burn and Launch

As is well known, different kinds of waves can be generated in plasma [55,56] (see also Chapter 3, where this aspect is fully described regarding rocket-induced effects and phenomena). Now, we will discuss the possibility of generating instabilities in plasma as the future source of plasma irregularities (or plasma turbulences). More precisely, the possibility of generating three kinds of plasma instabilities during ion gas release in the upper ionosphere (from 300 to 400 km) was analyzed by Bernhardt in Refs. [25,27] at the further stages of the "dirty" plasma evolution consisting of exhaust gas ions and ambient ionospheric plasma ions, which, according to the proposed concept of creating small- and moderate-scale plasma irregularities, filled the depletion zones generated at the initial stage of evolution of the "dirty" plasma.

We will discuss the possibility of generating plasma irregularities or turbulences in the next subsection. Now, we will try to understand what types of plasma instabilities excited by the rocket releases in ionosphere can be actually generated. The problem is that in Ref. [27] the author describes the rocket releases occurring at altitudes beyond 300 km, that is, in the upper *F*-region. Interpreting the obtained results briefly, we will try to explain the problem spreading the range of interests. We suppose that the rocket releases, currently made and those that will be in future (let us call them "virtual"), are observed at the lower, middle, and upper ionosphere, from the equator to auroral and polar ionospheric zones. To predict such future rocket burn and launch effects, the reader should be introduced to a brief description of the types of instabilities responsible for the creation of plasma irregular (turbulent) structures that fill the enhanced and depletion zones at the last stage of their evolution.

12.5.1 Plasma Instability Generation in the Upper *F-Layer* Depletion Zones

It was found in Ref. [27] that instabilities can be triggered or quenched by chemical releases, during which dusty gases, such as water vapor, molecules of hydrogen, nitrogen, oxygen, carbon oxide, and so forth, react with the ions of atomic oxide, prevailing in the *F*-region of the ionosphere and produce molecular ions in the ambient plasma. Due to the process of recombination, these ions are quickly combined with electrons (see reactions in the previous sections). The enhancing plasma region neutralization at these altitudes leads to plasma "hole" generation and fluid instabilities. After the injected gases dissipate, chemical quenching becomes unimportant, and at this stage the depletion zones becomes unstable to fluid instabilities. In Ref. [27], Bernhardt suggested that such plasma irregularities (turbulences) can be produced by three types of instabilities, the *wind-driven gradient-drift instability* (GDI), the *current-convective instability* (CCI), and the *gravitational Rayleigh–Taylor instability* (RTI).

Let us briefly describe the Bernhardt assumptions following Ref. [27]. Thus, according to schematic illustrations introduced in Ref. [27], presented in Figure 12.7 (top panel) rearranged according to notations given in this chapter, the GDI can be driven by the neutral wind vector $\mathbf{U}_m$ oriented normal to the vector of the ambient magnetic field $\mathbf{B}_0$. Here, the dashed circular contour represents the neutral chemical release located at the *F*-region peak, whereas the solid contours represent the electron concentration depletions (*holes*).

The vertical sinusoidal line denotes the locations where GDI may be excited. Neglecting the effects of field-aligned currents in the process of diffusion of ambient plasma electrons and molecular ions created due to exchange ionization and recombination processes occurring at the *F*-region of the ionosphere, the corresponding instability growth rate γ was found. Without entering into description of complicated

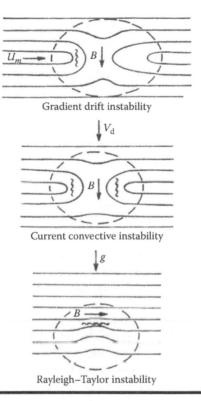

Gradient drift instability

Current convective instability

Rayleigh–Taylor instability

Figure 12.7   Geometrical sketch of three types of plasma instabilities.

mathematics presented in Ref. [27], it should be outlined that the parameter $\gamma = -i\omega$ fully depends on the direction and magnitude of the neutral wind velocity, as well as on the gradient of difference between the electron and molecular ion concentration [27]. It was also found in Ref. [27] that the GDI growth is maximum on the steep, upwind side of the hole, as sketched in Figure 12.7 (top panel).

As for the CCI generation process, which is schematically shown in Figure 12.7 (middle panel), rearranged from Ref. [27] by using our notations, it is driven by an electron drift velocity V_d, whose vector is elongated along $\mathbf{B_0}$; all other notations are the same as in Figure 12.7 (top panel). In Ref. [27], it was stated that this kind of instability can be produced by reactive gas releases and fully depends on the magnitude of the drift velocity of streaming electrons, and therefore, is located on a ring encircling the magnetic field lines through the center of the hole, as sketched in Figure 12.7 (middle panel). Here, we should note that in Ref. [27], some very important nonlinear effects were not taken into account. Below, we will briefly discuss the additional features that play an important role in the linear analysis carried out in Ref. [27].

Finally, the RTI for rocket releases below the *F*-region peak of the ionosphere can be driven by gravitation forces, determined by the acceleration vector **g** directed across the ambient magnetic field lines, as is shown schematically in Figure 12.7 (bottom panel), rearranged from Ref. [27] using our notations. As shown in Ref. [27], the RTI growth achieves maximum in regions where the plasma gradient is opposite to the vector **g**.

Generally speaking, in Ref. [27] it was found for all three kinds of instabilities that were in the process of filling the plasma holes by irregularities depends on two factors, that is

1. If the diffusive recovery rate (that in Ref. [27] was assumed to be equal to the chemical loss rate) is much larger than the instability growth rate, the ionospheric cavity (*hole*) will be filled by plasma due to its diffusion before the instability can grow.
2. If the instability growth rate is larger than the time of diffusion process, the hole becomes unstable for sufficiently long time for field-aligned plasma irregularities (turbulences) to be formed.

To enter into a detailed description of these types of instabilities and their importance in different altitudinal ranges and for different latitudes of the ionosphere, we should introduce the reader to their definitions and present our understanding of the problem, following Refs. [35,39,41] where all kinds of ionospheric instabilities were analyzed and discussed, and the same style of explanation were given as we presented earlier.

12.5.2 Plasma Instabilities Generated in the D-Layer by Rocket Release

Let us consider that the rocket release was observed at the *D*-layer or lower *E*-layer of the ionosphere. As mentioned in Chapter 1, at altitudes of 50–100 km ionospheric plasma is weakly magnetized (i.e., $\omega_H \Omega_H \ll \nu_e \nu_i$, $\nu_e = \nu_{ei} + \nu_{em}$, $\nu_i = \nu_{ei} + \nu_{im}$; all parameters are as defined in Chapter 1). Further, the transport processes (diffusion, thermodiffusion, and drift) are weak with respect to ionization–recombination reactions (see discussions presented earlier). In this case, most candidates of plasma instabilities and plasma wave generation at the *D*-layer are the so-called recombination instabilities.

12.5.2.1 Recombination Instabilities

We start with recombination instability, which creates small-scale plasma disturbances in the lower ionosphere [57,58]. As shown earlier, such instability is driven due to heating of plasma electrons by high-temperature exhaust molecular ions hidden in the ambient plasma during the rocket release. Thus, with increase of plasma temperature, the rate of recombination of plasma-charged particles decreases (according to Equations 12.8 and 12.10) within the areas of heating of the lower ionosphere (usually the *D*- and lower *E*-regions). Moreover, at the lower ionosphere, changes of the coefficient of recombination presented by Equation 12.8 (or more generally by Equation 12.10), relate to the coefficient of attachment of electrons to molecules and atoms, $\beta(T_e)$, and creation of negative ions [32]:

$$\beta(T_e) = \beta_0(T_{e0})\frac{T_{e0}}{T_e}\exp\left(\frac{700}{T_{e0}} - \frac{700}{T_e}\right), \qquad (12.17)$$

which also decreases with increase of temperature $T_e = T_{e0} + \delta T_e$, where:

T_{e0} is the nonperturbed temperature of the ambient ionospheric plasma

δT_e is the perturbed temperature caused by high-temperature exhaust molecular ions expanded into the ambient plasma

We will now present the plasma wave frequency in complex form by introducing the increment of instability growth $\omega = \omega_{Re} + i\gamma, \gamma \equiv \gamma_R$. Then, following the solutions of the dispersion equation of plasma wave propagating with vector **k** and with a complex wave frequency $\omega = \omega_{Re} + i\gamma, \gamma \equiv \gamma_R$, obtained in Refs. [57,58], we can present the increment of recombination instability γ_R of the disturbed ionospheric plasma in a physically clear form. Thus, we simplified the dispersion equations, accounting for a weak degree of ionization, magnetization, and quasineutrality ($N_e \approx N_i$) of the ionospheric plasma. In such a quasi-isotropic plasma with mobility of electrons and ions, $\mu_e = e/m_e(\nu_{em} + \nu_{ei})$ and $\mu_i = Ze/M_i(\nu_{im} + \nu_{ci})$, we get the following:

$$\gamma_R = -\frac{3}{2}\frac{dR'}{dt}\frac{\mu_e(\mathbf{E}_0 \times \mathbf{U}_0)}{\delta_{em}(\omega_H/\nu_{em})(\mu_e + \mu_i)} - R' - \frac{D_{i0}|\mathbf{B}_0|}{(\Omega_H/\nu_{im})}\frac{\mu_i\mu_e}{(\mu_e + \mu_i)}, \qquad (12.18)$$

where:

$\mathbf{U}_0 = \mathbf{V}_{i0} - \mathbf{V}_{e0}$ is the difference between the nonperturbed velocity of plasma ions and electrons

$D_{i0} = C_s^2/\nu_i$ is the coefficient of nondisturbed ion diffusion

$\nu_i = \nu_{im} + \nu_{ei}$ is the total frequency of ion–neutral and ion–electron collisions

$C_s^2 = (T_e + T_i)/M_i$ is the velocity of ion sound waves

δ_{em} is the part of heating energy transferred from light high-mobile electrons (with charge e) to molecules and heavy ions (with charge Ze, $Z \gg 1$ is the charge number), and

other parameters of plasma are as described in Chapter 1

Here, the parameter $R' = 2\alpha_{\text{eff}}(T_e)N_0$ is the coefficient associated with the recombination process at ionospheric altitudes, where the interactions of molecular ions are predominant with respect to those of atomic ions, that is, straightly corresponds to any situation that occurs in the rocket releases at lower ionosphere, and $\alpha_{\text{eff}}(T_e)$ is the effective recombination rate described by Equations 12.8 and 12.11.

As shown in Refs. [32,58,59], at altitudes of the *D*- and lower *E*-layers of the ionosphere, the wind of neutral molecules and atoms orients them along the geomagnetic field $\mathbf{B}_0$ due to interactions with plasma ions, and causes additional polarization of plasma disturbances. In the sites of depletions (i.e., with the negative deviations of concentration, $\delta N < 0$), the polarization field increases the temperature of plasma. Finally, by the action of thermodiffusion forces plasma is pushed out from the depletion areas [58,59].

It should be noted that in Refs. [32,58,59], thermal instabilities were described by using pure hydrodynamic approach. This approach is valid only for description of excitation of recombination instability at *mid-latitude D*-layer.

As for *equatorial* or *high-latitude* (polar or auroral) *D*-layer, the ionospheric plasma is usually nonequilibrium and nonstationary. In this case, short-length plasma waves can be generated, having length less than or on the same order of the free path (or gyroradius, if plasma is magnetized) of plasma ions. In such scenarios, the hydrodynamic description is not valid for analyzing such plasma disturbances, and we need to use kinetic equations for distribution functions of plasma electrons and ions (the same statement was mentioned in Chapter 1 based on the phenomenological approach). The kinetic approach was used by many authors regarding the analysis of plasma instabilities in the *D*- and *E*-layers of the ionosphere (see, e.g., Refs. [60–62]). Later, we briefly discuss such instabilities for which the hydrodynamic approach is not valid.

12.5.2.2 Kinetic Instability

As was shown in Refs. [60–62], when the drift velocities of plasma exceed the absolute value of the thermal velocities of plasma ions, that is, $V_d > |\mathbf{v}|_{Ti}$, reaction of interactions between the ions and neutrals leads to anisotropy of velocity of ions, and their distribution function becomes anisotropic, that is,

$$f_i(\mathbf{v}) = \frac{1}{\left(2\pi|\mathbf{v}|_{Tm}^2\right)^{3/2}} I_0\left(\frac{|\mathbf{v}|_\perp V_d}{|\mathbf{v}|_{Tm}^2}\right)\exp\left(-\frac{|\mathbf{v}|_\parallel^2 + |\mathbf{v}|_\perp^2 + V_d^2}{|\mathbf{v}|_{Tm}^2}\right), \quad (12.19)$$

where:

$|\mathbf{v}|_\parallel$ and $|\mathbf{v}|_\perp$ are the components of the full velocity $\mathbf{v}$ along and across the ambient magnetic field $\mathbf{B}_0$

We took into account the following approximate relation here, introduced in Ref. [61], $T_\perp/T_\parallel \approx 1 + V_d^2/2\,|\mathbf{v}|_{Tm}^2$. In Equation 12.19, $I_0(\bullet)$ is the Bessel function of the zeroth order. As shown in Refs. [159,160], the temperature anisotropy leads to the generation of the so-called the *Rozenbluth* plasma instability (defined by names of the corresponding researchers [61]). The condition of creation of such a thermal instability is given in Ref. [61]:

$$\int \frac{\partial f_i(\mathbf{v})}{\partial \mathbf{v}_\perp} \frac{d\mathbf{v}_\perp}{\mathbf{v}_\perp} > 0. \tag{12.20}$$

Introducing Equation 12.19 into Equation 12.20 gives

$$\left[\frac{V_d^2}{|\mathbf{v}|_{Tm}^2} - 2\right] I_0\left(\frac{V_d^2}{4|\mathbf{v}|_{Tm}^2}\right) > \frac{V_d^2}{|\mathbf{v}|_{Tm}^2} I_2\left(\frac{V_d^2}{|\mathbf{v}|_{Tm}^2}\right), \tag{12.21}$$

where $I_2(\bullet)$ is the Bessel function of the second order. From the constraint Equation 12.21, it follows that $V_d/v_{Tm} > 1.8$, that is, the instability can be generated for drift velocities of $V_d \geq 1000$ ms^{-1} in the ionospheric plasma. These values coincide with those obtained by experimental estimations presented in Chapter 3.

This electrostatic instability has the frequency ω_{CI} and the increment γ_{CI} on the order of low-hybrid frequency of ions Ω_{Hi}, that is, $\omega_{CI} \approx \gamma_{CI} \propto \Omega_{Hi}$, and the wave number of $k \propto \omega_{CI}/V_d$, that is, the wave length of $\lambda \sim 0.1$ m. Here, the ratio $k_\perp/k_\parallel \approx 1/100$, that is, the high-anisotropic plasma short-length plasma waves can be created due to the excitation of Post-Rozenbluth plasma thermal instabilities. The corresponding small-scale plasma irregularities (turbulences), generated by these types of plasma instabilities within the enhanced or depletion zones, will be discussed in the next subsection.

12.5.3 Plasma Instabilities Generated in the Upper E-Layer to the Upper F-Layer by Rocket Release

We now suppose that the rocket releases were observed at altitudes from 150 to 300 km and greater. At altitudes of upper *E*-layer to upper *F*-layer (more precisely, F_2-layer), as mentioned in Chapter 1, plasma is partly ionized and magnetized, that is, $\nu_{ei} > \nu_{em}$, $\nu_{ei}\nu_{em} < \omega_H\Omega_H$, that is, it is anisotropic. Moreover, the transport processes prevailed, and the ionization–recombination processes can be neglected. In this case, as shown in the literature, there are several types of candidates for plasma irregularity generation in the upper ionosphere that can be called *current-induced instabilities*, including the *current-convective*, the *GD* (or $\mathbf{E}_0 \times \mathbf{B}_0$), the *two-stream*, and the RTIs.

We do not enter deeply into this subject because it is fully presented in the literature (see Ref. [41] and references therein). We will describe these types of instabilities via the prism of real and "virtual" rocket releases that were observed and may be observed in future, bringing the reader's attention to the physical conditions of their generation, from the equator to the polar ionospheric latitudes, without entering into detailed mathematical description of the process.

12.5.3.1 Current-Convective Instability

Plasma instability and the corresponding plasma waves can be driven in the upper ionosphere by convective longitudinal currents (along $\mathbf{B}_0$) which occur on the "map" of sharp gradients of the plasma density [63].

The increment of such instability can be simplified and written as follows:

$$\gamma_{CC} \approx \frac{j_{0\|} \cdot \nabla(\ln N_0)}{(\omega_H/\nu_{em})} \frac{k_\perp}{k_\|}. \tag{12.22}$$

This increment depends strongly on the intensity of the longitudinal current and on the gradient of plasma density. Thus, we can estimate this increment for the middle-latitude ionospheric plasma, following Refs. [60,63], taking for the longitudinal current density the value of $j_{0\|} \approx 3 \times 10^{-6}\,\mathrm{Am^{-2}}$, for the plasma density of $N_0 = 10^{11}\,\mathrm{m^{-3}}$, for the parameter of anisotropy $k_\perp/k_\| = \sqrt{\omega_H \Omega_H/\nu_{em} \nu_{im}} > 10$. Finally, we will get the increment of about $\gamma_{cc} = 10^{-3}\,\mathrm{s^{-1}}$. As mentioned in Ref. [64], in the turbulent ionospheric plasma at high (*polar* or *auroral*) latitudes, it is possible to generate plasma disturbances with smaller anisotropy than at the *middle* latitudes.

At the same time, in Ref. [65] it was shown that the longitudinal current with density $j_{0\|} > j_{0\|kr} \approx 3 \times 10^{-6}\,\mathrm{Am^{-2}}$ generates ion cyclotron waves (ICWs) in the upper ionosphere. The possibility of generation of ICW at altitudes of maximum of *F*-layer and below (up to 150–200 km) was investigated in Refs. [66,67]. Following Ref. [67], we obtain the increment of excitation of such plasma waves:

$$\gamma_{CC} = \frac{m_e(\nu_{em} + \nu_{ei})}{2M_i} \frac{k^2}{k_\|^2}\left(\frac{k_\| V_{eo\|}}{\omega_0} - 1\right). \tag{12.23}$$

From Equation 12.23, it follows that the increment of instability of ionospheric plasma is positive, that is, $\gamma > 0$, and the phase velocity of ICWs, $V_{ph} = \omega_0/k_\|$, along the ambient magnetic field $\mathbf{B}_0$, is smaller than the electron velocity $V_{eo\|} = -j_{0\|}/eN_0$. It was also found that the ICW frequency $\omega_{CC}^2 \gg \omega_{Hi}^2$ is much larger than the gyrofrequency of ions.

12.5.3.2 Gradient-Drift Instability

In the actual ionosphere, for strongly elongated plasma structures where the longitudinal (along $\mathbf{B}_0$) scale of plasma structure exceeds the scale of neutral atmosphere H_m height (see definitions in Chapter 1), altitudinal dependence of the background plasma density $N_0(z)$ yields to generation of the so-called GDIs (see, e.g., Refs. [68,69]). Moreover, as shown in Chapter 1, if the first-order parameter of plasma, as density, is a function of altitude, the second-order (collision frequencies) and the third-order (transport coefficients) parameters of the ionospheric plasma are also functions of the ionospheric altitude z.

As given earlier, we need to differentiate the effects of such instability generation at altitudes of E-region from those occurring at altitudes of the F-region of the ionosphere. It was shown both theoretically and experimentally that in the E-layer of the ionosphere, GDI occurs when the plasma density disturbances δN are converted to the direction of the ∇N_0 by the perturbation drift with velocity $\delta \mathbf{V} = \delta \mathbf{E} \times \mathbf{B}_0 / |\mathbf{B}_0|^2$. The corresponding scheme of GDI (or $\mathbf{E} \times \mathbf{B}$ instability) is presented in Figure 12.8a.

As for the *middle-latitude* or *equatorial* E-layer, for the plasma disturbances with dimensions that exceed the standard height H_m along $\mathbf{B}_0$, that is, for $l_{\parallel} > H_m \approx 10-50\,\text{km}$ (i.e., for $k_{\parallel} \equiv l_{\parallel}^{-1} > H_m^{-1} \approx 10^{-1} - 2 \times 10^{-2}\,\text{km}^{-1}$), and for $l_{\perp} \geq 10\,\text{km}$ (i.e., for $k_{\perp} < 10^{-1}\,\text{km}^{-1}$) across $\mathbf{B}_0$, we get: $l_{\parallel}/l_{\perp} \equiv k_{\perp}/k_{\parallel} << 10-100$. In such a situation, the increment of GDI excitation can be obtained, following Refs. [68] and rearranging it according to the notations introduced in this chapter:

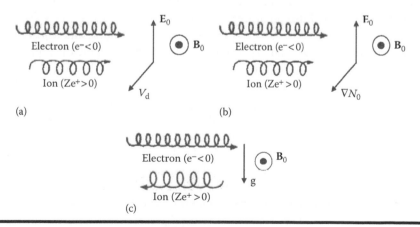

(a)

(b)

(c)

Figure 12.8 Schematic presentation of the (a) gradient-drift (or $\mathrm{E} \times \mathrm{B}$ instability), (b) Farley–Buneman instability, and (c) Rayleigh–Taylor instability.

$$\gamma_{GD} = \frac{\Omega_H \nu_{em}}{\omega_H \nu_m H_m} \frac{\mathbf{k} \mathbf{U}_0}{k_\parallel} + \frac{D_i \nabla(\ln N_0)}{(\omega_H/\nu_{em}) H_m} \frac{k_\perp}{k_\parallel} + \frac{|\nabla(\ln N_0) \cdot \mathbf{U}_0|}{(\omega_H/\nu_{em})} \frac{k_\perp^2}{k_\parallel^2} -$$

$$D_{i0} k_\parallel^2 \left(1 - \frac{T_e}{T_e + T_i} \frac{1}{H_m k_\parallel} - \frac{T_e}{T_e + T_i} \frac{\Omega_H \nu_{em}}{\omega_H \nu_m} \frac{1}{H_m^2 k_\parallel^2} \right),$$

$$(12.24)$$

where:

$\mathbf{U}_0 = \mathbf{V}_{i0} - \mathbf{V}_{e0}$, $D_{i0} = C_s^2/\nu_i$ is the coefficient of nondisturbed ion diffusion, as earlier

$C_s^2 = (T_e + T_i)/M_i$ is the velocity of ion sound waves

D_i is the coefficient of diffusion of plasma ions

Other parameters of plasma are described in Chapter 1.

The last term in Equation 12.24 describes a decay (due to diffusion spread) of plasma waves with arbitrary direction of longitudinal wave vector component $k_\parallel$. At the same time, from the expression given in Equation 12.24, it follows that for plasma waves propagating downward ($k_\parallel < 0$) in the ionosphere, the altitudinal dependence of the background ionospheric plasma, $N_0(z)$, increases the growth rate of plasma waves and decreases their attenuation due to the diffusion process ($\sim D_i$). For plasma waves propagating upward ($k_\parallel > 0$), the effect of altitudinal dependence of parameters of plasma is the opposite: their dependence on the height decreases the velocity of growth of plasma waves and increases their attenuation due to diffusion process. This occurs because of changing of phase velocity of waves, $V_{ph} \approx V_{i0} = V_d + U_0$, with altitude z. At the lower altitudes of the ionosphere, the phase velocity is greater and plasma waves grow faster, and their diffusion relaxation is slower. From Equation 12.24 it can also be seen that dissipation of waves due to the diffusion process is minimal when $k_\parallel \to H_m^{-1}$ ($l_\parallel \to H_m$).

In Ref. [68], the above nonlocal approximation continued, and it was shown that many features observed experimentally can find their physical explanation only in the framework of full nonlocal nonlinear equations. Because the task of evaluating nonlinear equations nowadays is a complicated issue, in Ref. [68] it was proposed that the following approach gives a good explanation of the main features of the large-scale experimental observations. It is based on the introduction into the formulation of the long-wavelength problem of the additional electron mobility and diffusion caused by the short-wavelength turbulence occurring in the plasma background. Introducing such a procedure, the corresponding increment of GDI excitation and growth were derived for the wavelengths on the order of ~1 km, that is, for the wave numbers on the order of $k_0 = (\sigma_H/\sigma_P) L^{-1}$:

$$\gamma = \frac{\mu_{iH} - \mu_{eH}}{(\mu_{iP} - \mu_{eP})(1 + \varsigma)} \frac{\mu_{eH} E_z}{L \left[1 + (k_0/k_\parallel)^2 \right]} - 2\alpha N_0. \qquad (12.25)$$

Here, $\sigma_H = eN_0(\mu_{iH} - \mu_{eH})$ and $\sigma_P = eN_0(\mu_{iP} - \mu_{eP})$ are the Hall (across $\mathbf{B}_0$) and Pedersen (along $\mathbf{E}_0$), $\varsigma = (\nu_{em}\nu_m/\omega_H\Omega_H)^{-1}$, $L = [\nabla N(z)]^{-1}$ is the scale length of the equilibrium plasma density gradient, and α is the coefficient of the radiative recombination (see previous section). Because the quadratic term in Equation 12.25 is presented in the denominator of Equation 12.25, the increment of plasma modes growth decreases as the magnitude of $k_{\parallel}$ becomes comparable to or smaller than k_0. Finally, at even longer wavelengths, all the modes will eventually be stabilized due to the effects of radiative recombination. Additional analysis carried out in Ref. [68] showed that the dispersive nature of the kilometer-wavelength plasma modes depends critically on the value of k_0, that is, on the scale length L, and on the ratio of the Hall and Pedersen conductivities.

In Ref. [69], for a description of GDI-induced plasma waves occurring at *high-latitude* (polar or auroral) ionosphere, quasilocal approach was used. Following the same notations as given earlier, and taking $D_{0i} = C_s^2/(\nu_{im} + \nu_{ei})$, we finally get

$$\gamma_{GDI} = \frac{k_0\mathbf{E}_0 \cdot [\nabla(\ln N_0) \times \mathbf{k}_\perp]_{\mathbf{B}_0}}{k_\perp^2} - \frac{C_s^2 k_\perp^2}{\Omega_H}, \qquad (12.26)$$

where:

$[\nabla(\ln N_0) \times \mathbf{k}_\perp]_{\mathbf{B}_0}$ is the vector product projection in the direction of the vector of ambient magnetic field $\mathbf{B}_0$

Equation 12.26 was obtained by the assumption that in the upper high-latitude ionosphere $k_\perp < 10^{-1}\,\mathrm{m}^{-1}$ and the second (negative) term in Equation 12.25 is sufficiently small.

Precise analysis carried out in Refs. [68,69] has shown that the GDI can be taken into account as the real candidate of small- and moderate-scale plasma disturbance generation that can fill the enhanced and depletion regions created in the upper ionosphere at the initial stage of the rocket releases, described earlier.

12.5.3.3 Two-Stream Instability

Despite the fact that in Ref. [27] this kind of instability was not taken into account, we will present it as a candidate of short- and moderate-scale plasma disturbance generation at the middle and outer ionosphere, from equatorial to polar latitudes [70–77]. The physical concept of how such instabilities are generated in the ionosphere is based on the existence of high relative velocities of electrons and ions, that is, strong currents (even in the homogeneous plasma) caused by strong ambient electric field $\mathbf{E}_0$ and/or winds of neutral molecules (or atoms), $\mathbf{U}_0$. In the ionospheric plasma under a strong ambient electric field $E_0 \geq 30\text{–}50\,\mathrm{mV\,m}^{-1}$, two-stream instability is mostly created in high-latitude ionosphere, perturbed middle-latitude ionosphere, and equatorial ionosphere [70–77]. Such instability is the main source of plasma disturbance generation in the ionosphere, and we analyze it more consistently compared to other kinds of ionospheric instabilities. As a partial case of CCIs occurring in the ionospheric plasma, two-stream instability for crossing electric E_0 and magnetic B_0 fields can be schematically presented as shown in Figure 12.8b.

The two-stream instabilities are usually called the Farley–Buneman (FB) instabilities because Buneman first analyzed conditions of formation of current instability in the collisionless adiabatic plasma in the absence of an ambient magnetic field [71], and then Farley obtained criteria of formation of such current instabilities at the upper *E*-region in the equatorial and polar iono-sphere [70–72], assuming the existence of magnetized but isothermal plasma $(T_e \approx T_i = T)$. According to their criteria, in the isothermal plasma $(T_e \approx T_i = T)$, the appearance of two-stream instability (defined also as *ion-acoustic wave* [70–72]) occurs, if the relative electron stream (compared with the ion stream) exceeds the ion-sound velocity $C_s = \sqrt{T/M_i}$, and in the nonisothermal plasma $(T_e \neq T_i)$, it exceeds the magnitude of the thermal velocity of ions $|\mathbf{v}|_{T_i}$ [70–72].

Both authors, Buneman and Farley, analyzed instability of low-frequency electromagnetic waves perturbations $(\omega \ll \nu_{im} < \nu_{em})$, for the case of magnetized electrons $(\omega_H \gg \nu_{em})$ and weakly magnetized ions $(\Omega_H \ll \nu_{im})$, that is, for conditions of the *E*- and *F*-layers of the ionosphere (see definitions in Chapter 1).

In the FB-instability analysis, we denote the frequency of such unstable waves as $\omega_0 \equiv \omega_{FB}$, and the corresponding increment (or growth of their rate) as $\gamma \equiv \gamma_{FB}$, assuming, following [70–72], that the excited plasma waves strongly stretched along $\mathbf{B}_0$, that is, $k_\perp/k_\parallel > \sqrt{\omega_H \Omega_H / \nu_{em} \nu_{im}} \approx 10–100$. In this case, following Refs. [70–72] and using our notations, we get

$$\gamma_{FB} = \frac{1}{\nu_{im}(1 + \omega_H \Omega_H / \nu_{em} \nu_{im})} \left[\omega_{FB}^2 - k^2 C_s^2 \right], \qquad (12.27)$$

where:

$\mathbf{V}_d$ is the drift velocity of plasma electrons

All other parameters are defined and introduced earlier as also in Chapter 1.

In Refs. [60,73,74], Equation 12.27, yielding from the standard FB model, was modified for the case of *high-latitude* ionosphere ranging from the *E*-layer to the *F*-layer, from which we get the increment of two-stream instability:

$$\gamma_{FB} = \frac{1}{\nu_{im}(1 + \omega_H \Omega_H / \nu_{em} \nu_{im})} \left[\omega_{FB}^2 \left(1 - \frac{6k^2 v_{Ti}^2}{v_{im}^2} \right) - k^2 C_s^2 \right], \qquad (12.28)$$

where:

v_{Ti} is the magnitude of thermal velocity of plasma ions

It is important to point out that as shown by many researchers "dealing" with FB instability (see, e.g., Refs. [70–77] and references therein), the hydrodynamic and kinetic descriptions of this kind of instability give the same result only for $0 < k < k_{\text{max}}$, where

$$k_{\text{max}} = \frac{v_{\text{im}}}{v_{\text{Ti}} |\mathbf{V}_d|} \left[\frac{1}{12} \left(V_d^2 - C_s^2 \left[1 + \frac{1}{(\omega_H \Omega_H / v_{\text{em}} v_{\text{im}})^2} \right] \right) \right]^{1/2}, \quad (12.29)$$

where:

$|\mathbf{V}_d|$ is the magnitude of the drift velocity of electron component of plasma

With the increase of the ambient electric field E_0, the drift velocity increases and becomes predominant (compared with v_{Ti}), and k_{max} also increases, expanding the applicability of the hydrodynamic approach.

As for conditions of *middle-latitude* ionosphere at altitudes of the D- and E-layers of the ionosphere, it was shown in Ref. [41] that small-anisotropic meteor-induced plasma waves can be explained by the generation of FB instabilities when observed experimentally. For such meteor-induced plasma waves with relatively small degree of anisotropy ($k_\perp / k_\parallel \ll \sqrt{(\omega_H \Omega_H / v_{\text{em}} v_{\text{im}})} \approx 10$) we can obtain that [41]

$$\omega_{\text{FB}} = \frac{\mathbf{k} \cdot \mathbf{V}_d}{\dfrac{(1 + \omega_H \Omega_H / v_{\text{em}} v_{\text{im}})}{\omega_H \Omega_H / v_{\text{em}} v_{\text{im}}} + \dfrac{\omega_H v_{\text{im}} \cdot k_\parallel^2}{\Omega_H v_{\text{em}} \cdot k_\perp^2}}, \quad (12.30)$$

which is close (for $k_\parallel^2 / k_\perp^2 \ll 1$) to that obtained in Refs. [70–72]:

$$\omega_{\text{FB}} = \frac{(\omega_H \Omega_H / v_{\text{em}} v_{\text{im}})}{(1 + \omega_H \Omega_H / v_{\text{em}} v_{\text{im}})} \mathbf{k} \cdot \mathbf{V}_d, \quad (12.31)$$

that is, with the decrease of anisotropy of plasma disturbances, the frequency ω_0 and the phase velocity $V_{\text{ph}} = \omega_0 / k$ of plasma waves are decreased correspondingly according to the standard models [70–72]. For condition $V_d < C_s$, the two-stream instability is not generated in the homogeneous plasma.

12.5.3.4 Rayleigh–Taylor Instability

We note from the beginning that the definition of RTI given in Ref. [27] is not exact. Such instability can be generated in the existence of sharp gradients of concentration of ionospheric plasma on the order of $\kappa \equiv \nabla(\ln N_0) = 10^{-3}\,\text{m}^{-1}$

combined with the existence of an ambient electric or gravity field (see Ref. [41] and the corresponding references therein). The corresponding plasma waves were observed both in the polar and the equatorial ionosphere at the *F*-region, called *spread* F-*layer* [78,79–84]. Schematic presentation of the gravitation field drift caused by RTI is shown in Figure 12.8c.

Taking into account the angle of inclination of the geomagnetic field denoted by *I*, we can present the increment of RTI by the following expression [78,79]:

$$\gamma_{RT} = \frac{|\mathbf{g}|\,\kappa}{v_{im}} \cos I. \tag{12.32}$$

Estimations presented in Ref. [79] state that $\gamma_{RT} \approx (7-8) \times 10^{-3}\,\mathrm{s}^{-1}$. At altitudes higher than the maximum of *F*-layer (which correspond to the rocket releases described earlier), RTI causes growth of plasma waves at altitudes of 500–700 km, where $v_{im} \approx 10^{-2}\,\mathrm{s}^{-1}$ [41,78,79].

Other researchers [82,83] investigating the process of RT-type instability generation in the *equatorial* ionosphere, found possibilities to produce large- and global-scale depletion zones called *bubbles* in the *F*-region of the equatorial ionosphere (see Figure 11.28), which finally form the equatorial spread *F*-layer observed experimentally.

Completing this section, we outline that all mentioned instabilities, depending on the height of their generation, from the *D*- to *F*-layers, and on the latitude, from the equator to the polar cap, contribute more or less to the cumulative effect of plasma perturbations creation. To show this, we will give some evident example. In the real ionospheric active experiments described in this chapter, in the case of the inhomogeneous ionosphere and the existence of gradients of plasma density, $\kappa = \nabla(\ln N_0)$, orthogonal to ambient magnetic field $\mathbf{B}_0$, and parallel to ambient large-scale electric field $\mathbf{E}_0$, plasma waves can grow due to the joint effect of GDI, FBI, and RTI.

Following the approaches proposed in Refs. [41,75,78], we can obtain the increment of such complex GDI–FBI–RTI instability, as follows:

$$\gamma_{\Sigma} = \frac{1}{v_{im}(1 + \omega_H \Omega_H/v_{em} v_{im})}\left[\begin{array}{l} \omega_0^2 + \dfrac{\omega_H v_{im}}{v_{em}(1+p)}\dfrac{|\kappa \cdot \mathbf{E}_0|}{|\mathbf{B}_0|} \\[2ex] + \dfrac{|\mathbf{g}|\,\kappa}{v_{im}}\cos I - C_s^2 k^2 \end{array}\right], \tag{12.33}$$

where:

$p = \nu_{ei}/\nu_{em}$ is the parameter of ionization introduced in Chapter 1

For the value of drift velocity of $V_d \approx 200\,\mathrm{ms}^{-1}$, in the case of nonisothermal plasma $C_s = \left[(T_e + T_i)/M_i\right]^{1/2} \approx 350\,\mathrm{ms}^{-1}$, and for the wave number of $k \approx 1\,\mathrm{m}^{-1}$ [63,78], and taking into account the parameters of plasma at the considered altitudes of $\nu_{im} = 2 \times 10^3\,\mathrm{s}^{-1}, \nu_{em}(1 + p) \approx 1.6 \times 10^5\,\mathrm{s}^{-1}, \Omega_H = 2 \times 10^2\,\mathrm{s}^{-1}, \omega_H = 10^7\,\mathrm{s}^{-1}$ (see definitions in Chapter 1), plasma instability can be developed if $\kappa > 3.5 \times 10^{-3}\,\mathrm{m}^{-1}$. The horizontal gradients of such plasma waves were observed in the polar ionosphere [85].

12.6 Generation of Plasma Turbulences in the Ionosphere by Rocket Burn and Launch

Brief theoretical analysis carried out in the preceding sections and the obtained results there allow us to state that the instability of ionospheric plasma is a major source of generation of a wide spectrum of plasma disturbances, from small-scale to large-scale structures, observed during active experiments and the rocket releases occurring from the D-layer to the upper F-layer of the ionosphere, from the Equator to the polar cap. These are naturally induced (by meteors, magnetic storm, bubbles) and artificially induced (by pump radio wave, ion beam injection, and rocket exhaust release), which vary over the wide ranges of scales, plasma densities, orientations with respect to ambient magnetic and electric field, and can be explained by introducing some specific instabilities and waves generated in the ionospheric plasma, accounting for their "cumulative" effect.

In other words, it is difficult to find some universal instability that is responsible for the creation of plasma turbulences/irregularities in the ionosphere associated with rocket releases and rocket burn and launch. Each of these instabilities may play an important role in creation and further evolution of plasma structures, depending on the height, latitude, and ambient conditions of ionospheric plasma. These aspects we will discuss later in this section. We will start, as in the previous sections, with analysis of the situation in the lower ionosphere, that is, at altitudes of D-layer and lower E-layer.

12.6.1 Evolution of Plasma Irregularities/Turbulences in the Lower Ionosphere Associated with Rocket Release and Burn

Theoretical frameworks regarding the rocket releases as was mentioned earlier, usually have the described situation occurring at the upper ionosphere ranging from 250 to 300 km, where most releases were practically performed (see Refs. [11–25]). As the reader now understands, the processes observed there are totally different to those occurring at the lower ionosphere ranged from 50 to 100 km. Numerous experiments and theoretical researches carried out at the lower ionosphere (see Refs. [32,36] and bibliography there) showed that the horizontal wind of neutrals (molecules and atoms) that creates the motion of plasma, as polytrophic component liquid, spreading through the neutral gaseous medium, is not responsible for generation of small-scale plasma disturbances, called *turbulences* or *irregularities*. Most important and efficient in such structure generation is the vertical wind, having complicated profile along the ionospheric altitude [32]. Its behavior is completely random (e.g., stochastic) and, therefore, performs random chaotic motions of charged particles, electrons, and molecular ions, due to interactions of plasma component with neutral atoms and molecules.

Therefore, at considerable altitudes, turbulent plasma motion should be accounted for as an additional factor for the strong ionization–recombination process described earlier in the previous section.

Now, we raise a question: what does *turbulence* mean? According to traditional Kolmogorov–Obukhov concept of turbulence, any large-scale perturbation of the atmospheric gas transfers its energy to smaller scale through random chaotic motions caused by ambient factors (wind, pressure, humidity) and some additional steady-state breaking factors [47,48].

Further, we will base our notion not only on instabilities caused by chemical reactions and heating process, as was mentioned in Section 12.3, following the scheme presented in Figure 12.1, but also try to briefly introduce the effects of plasma turbulence, via its own stabilization factor in further evolution of large-scale plasma enhancements and small-scale turbulent structures, filling these global perturbed ionospheric zones observed experimentally during the rocket release at the lower ionosphere (see Figures 11.3, 11.14, and 12.2).

12.6.1.1 Role of Chemical Reactions and Heating-Induced Instabilities in Plasma Structure Creation

As mentioned earlier, chemical reactions as attachment and recombination processes as well as heating processes can generate the local plasma instabilities surrounding the rocket engine or plume, working as a "local heater" in the vicinity of the rocket and within its wake. If so, we can suppose that the same instabilities can generate small-scale plasma turbulences (called "eddies" [48]) locally presented in the vicinity of the moving rocket or in the wake of *composite* plasma, consisting the polytrophic atomic and molecular ions, the electrons, and the neutral atoms and molecules. According to reactions given in Equations 12.1 through 12.4 described in the previous section, following Figure 12.1 we can now approximately estimate the concentration of such small-scale plasma perturbation obeying a turbulent motion of background and dirty plasma at this stage of discussions. If we, as mentioned earlier, consider only one *effective* polytrophic ion of the exhaust gas (because it is impossible to obtain a complete solution of the whole system given in Equation 12.1 illustrated by Figure 12.1, as an average of other ions created by ion–electron chemical reactions described by a system given in Equations 12.3 and 12.4, we can estimate, following Ref. [32], the plasma density disturbance generated by a fast process of attachment, δN_A, and the plasma density disturbance generated by a slow process of recombination, δN_R. Indeed, for the first process a time scale can be estimated as follows:

$$\tau_A = (\beta_{eff} + \gamma_{eff})^{-1}. \tag{12.34a}$$

Regarding the second process, a time scale corresponds to

$$\tau_R = (\alpha_{eff} N_{e0})^{-1}. \tag{12.34b}$$

As shown from experimental measurements, the first time scale related to the processes of attachment and detachment does not exceed a few seconds, whereas the second term that corresponds to the process of recombination can reach several minutes [49]. If so, we suppose following Ref. [49], that these processes can be considered as separate and independent, because their time scales differ on 2 orders of magnitude [49]. If so, we can, after straight derivations, find the plasma disturbances generated at the lower ionosphere by the two instabilities, attachment and recombination, respectively:

$$\delta N_A \equiv N - N_{e0} = -\frac{\Delta\beta_{eff}\chi}{\beta_{eff}(1 + \chi)} N_{e0}, \tag{12.35a}$$

$$\delta N_R \approx \frac{N_{e0}}{2}\left[\frac{\Delta\alpha_{eff}}{\alpha_{eff}} + \frac{\Delta\beta_{eff}\chi}{\beta_{eff}(1 + \chi)}\frac{(\alpha_e - \alpha_i)}{\alpha_{eff}}\right]. \tag{12.35b}$$

Here, $\Delta\alpha_{eff}$ and $\Delta\beta_{eff}$ are the changes of the coefficient of recombination and the attachment caused by rocket exhausts compared to those for the nonperturbed ionosphere, $\chi = \sum_{k=1}^{K} M_k^-/K \approx \beta_{eff}/\gamma_{eff}$ is the average mass of the effective negative ion; all other parameters and characteristics were defined after formulas represented by Equations 12.3 and 12.4.

It is seen that the magnitudes of both plasma disturbances are roughly equal to each other, but opposite in sign. This means that the fast process related to the increase of the electron attachment rate yields the decrease of the composite plasma density within the area filled by δN_A. At the same time, the slow process caused by the decrease of the recombination process (due to increase of temperature of dirty plasma surrounding the rocket, see Section 12.3), finally increases the plasma density within the area filled by δN_R. This result completely matched the result obtained in Ref. [50]. The approximate solution of the system given in Equation 12.35 can roughly explain the effects of plasma enhancement observed at the lower ionosphere by the existence of the recombination process as being slower with respect to the process of attachment and detachment, that is, the process of polytrophic ion exchange process with those consisting in the ambient ionospheric plasma. At the same time, many experiments described in Ref. [32] are in contradiction with the results of the model presented in Ref. [49] using only one molecular ion in consideration. This is because in Ref. [49], the authors used only one sort of ion, the light one. We used in the system given in Equations 12.3 and 12.4 some effective ion with average mass and charge, combining both heavy and light atomic and molecular ions that according to the reactions given in Equation 12.1 should be within the composite (e.g., dirty) ionospheric plasma.

To improve the model proposed in Ref. [49], a heavy ion was introduced as an additional component of the composite plasma in Ref. [51]. We do not enter into many complicated derivations but only mention that this can only partly improve results, because by having a larger relaxation time a heavy ion can significantly decrease the time of recombination with ambient plasma electrons, that is, increasing the relaxation time of plasma disturbances generated by the chemical instabilities. The authors have found some contradiction here, because, the maintenance of the equilibrium between light and heavy ions in the process of their exchange reduces the electron density in such a composite plasma, decreasing the total electronic content of the ionospheric plasma. Also found by the researchers was the existence of turbulent vertical (along the height of the ionosphere) motions of the neutral gas at lower altitudes, that is, existence of the atmospheric turbulence can inhibit the formation of plasma irregular structures from long-live heavy ions.

As shown in Figure 12.1, extracted from Ref. [32], in regular nonperturbed ionospheric plasma, consisting of electrons, one type of light ions and neutral molecules (atoms), a local heating of plasma, as was shown earlier generates chemical (attachment and recombination) and heating instabilities, yielding decrease of the lifetime of plasma disturbances occurring at lower ionospheric altitudes. As for the composite plasma, containing both exhaust heavy and light ions, the process of plasma instability generation, and then creation of plasma perturbed structures, is more complicated. As predicted from Ref. [51] the lifetime of plasma disturbances arising in the two-ion composite plasma was about several minutes, whereas it is shown by numerous experiments described in Chapter 11 as the lifetime of plasma enhancements observed there is too long—up to several tens of minutes. An atmospheric turbulence, as was shown in Ref. [32], can additionally decrease the lifetime of such positive plasma structures. So, again we find a contradiction between the theoretical predictions and experimental observations. To avoid such contradiction, we propose the following theoretical frameworks fully performed in Refs. [86,87], about the existence of a new kind of turbulence called helical, which can accelerate generation of short-length plasma waves in the composite (called in Ref. [26] *dust* and *dirty*) plasma that can transfer their energy to the long-length plasma waves. Let us briefly consider this process to prove long-time evolution of plasma disturbances, from small- to large-scale observed experimentally in the lower ionosphere.

12.6.1.2 Role of Turbulence in Plasma Disturbance Generation and Damping

To introduce the reader to the problem in more detail, we define again the environmental conditions that occur surrounding the rocket after the exhausts spread into the ambient nonperturbed plasma. *A priori*, we now consider turbulent background plasma as stationary and homogeneous, but not time-invariant (i.e., according to Refs. [86,87], plasma is *helical*). The type of plasma within the wake of the rocket is composite consisting of polytrophic atomic and molecular ions and electrons, as discussed in Section 12.2. Such composite (dusty or dirty) plasma is affected by the ambient electric and magnetic field and the turbulent vertical motion of the neutral molecular and atomic gas. Due to the tendency of such composite plasma to quasineutrality of an inner polarization (also called ambipolar [41]) electric field between the ion and electron components occurs. In such a situation, small-scale turbulences are not generated from large-scale turbulences as follows from the Kolmogorov–Obukhov traditional concept (see above). In the noninvariant (helical) turbulent plasma, which is now in *non-Kolmogorov* regime [87], small-scale plasma turbulences (e.g., irregularities) become helical also and can generate additional turbulent pulsations (i.e., plasma fluctuations), as a result of their interactions with large-scale plasma perturbations occurring earlier in the plasma background surrounding the moving space vehicle. The proposed concept of helical turbulence occurring in the dense collision composite plasma is due to changes of inner and outer fields, gradients of density, pressure, temperature, and other parameters of plasma, which allows us to introduce helical turbulence and its effect on plasma disturbance creation and further evolution in later descriptions. Thus, in the helical turbulent plasma, as was shown in Refs. [86,87], the amplification of plasma perturbation seed eddies (i.e., short-wave incoherent structures [48]) yields the consequent appearance of large-scale coherent turbulent structures, as we illustrate with the help of a simple scheme presented in Figure 12.9, following a general description of plasma wave–wave

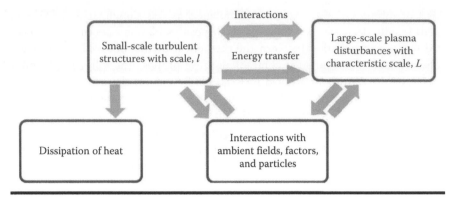

Figure 12.9 Schematic representation of the generation of various-scale plasma disturbances by turbulence during the rocket release.

interactions and energy exchange described in Ref. [41]. As follows from this scheme, interactions between small- (short-length) and large-scale (long-length) plasma structures (e.g., waves) can generate the process of amplification of large-scale structures by energy transfer (or *energy pumping* [87]) from small- to large-scale plasma turbulences. This process can essentially increase the lifetime of small- to large-scale plasma irregularities observed in the lower ionosphere during rocket burn and launch.

Indeed, as seen from Figure 12.2, in simple plasma affected by turbulent motions after local heating, the heating-induced plasma irregularities are short-lived structures (not exceeding several seconds), whereas in rocket release experiments actually the corresponding lifetime of plasma irregularities observed at lower altitudes can exceed several minutes and more (see Chapter 11). Therefore, to overcome such a contradiction we later consider the evolution of composite plasma accompanied by the helical turbulent mechanism of movement during which exchange of energy between the small- and large-scale structures allows explanation of their long-term evolution. In such an approach, the parameters and characteristics of plasma, such as the velocities of charge particles, their concentration and charge, as well as ambient electric and magnetic field, become random and turbulent [26,86,87], that is, for the quasineutral plasma, $N_{0e} \approx N_{0i} + N_{0I}$, we have the electron and ion velocity of turbulent motion, total plasma concentration, perturbed ambient electrical, and magnetic fields, respectively:

$$\mathbf{V} = \mathbf{V}_0 + \mathbf{V}^t + \delta\mathbf{V}, \langle \mathbf{V}^t \rangle = \langle \delta\mathbf{V} \rangle = 0, \tag{12.36a}$$

$$\tilde{N} = N_0 + N^t + \delta N, \langle N^t \rangle = N_0, \langle \delta N \rangle = 0, \tag{12.36b}$$

$$\mathbf{E} = \mathbf{E}_0 + \mathbf{E}^t + \delta\mathbf{E}, \langle\mathbf{E}^t\rangle = \mathbf{E}_0, \langle\delta\mathbf{E}\rangle = 0, \tag{12.36c}$$

$$\mathbf{B} = \mathbf{B}_0 + \mathbf{B}^t + \delta\mathbf{B}, \langle\mathbf{B}^t\rangle = \mathbf{B}_0, \langle\delta\mathbf{B}\rangle = 0. \tag{12.36d}$$

Here, the angle brackets $\langle\rangle$ and superscript t refer to the ensemble average of turbulent pulsations regarding the above characteristics, δX are the corresponding random fluctuations of these characteristics. In such a nonlinear regime of plasma turbulence instability generation, the main equations of plasma dynamics become very complicated for each component of the composite plasma (in our further explanation we use this terminology instead of "dusty" or "dirty" plasma, introduced in Ref. [26]).

Recalling the subject matter of our book and what we promised our readers from the beginning, we do not enter into discussions of these complicated equations, presenting and analyzing only the growth rate of nonlinear turbulent plasma instability (i.e., increment of turbulent instability factor) or the damping rate (i.e., decrement of turbulent stabilization factor) through the parameters of the characteristics of the turbulence, following Refs. [48,87] for the case of the vertical axis (that is along the ionospheric altitude), directed along vector $\mathbf{k}$ of turbulent small-scale plasma waves:

$$\gamma^t = -\frac{1}{2}\left[\left(\langle\mathbf{V}^t \cdot \nabla \times \mathbf{V}^t\rangle\tau - 2\eta\langle\mathbf{V}^t \cdot \nabla \times \mathbf{B}^t\rangle\tau\right)|\mathbf{k}|\right]$$
$$- E_T\tau + \frac{1}{2}\eta\langle\mathbf{V}^t \cdot \nabla \times \mathbf{B}^t\rangle\tau. \tag{12.37}$$

From Equation 12.37, it can be easily found that the largest value of the increment (decrement), γ^t_{max} for the wave number of the short-length waves k_0:

$$k_0 = \frac{\left|\langle\mathbf{V}^t \cdot \nabla \times \mathbf{V}^t\rangle\tau - 2\eta\langle\mathbf{V}^t \cdot \nabla \times \mathbf{B}^t\rangle\tau\right|}{4E_T\tau}, \tag{12.38a}$$

$$\gamma^t_{max} \equiv \gamma(k_0) = \frac{\left(\langle\mathbf{V}^t \cdot \nabla \times \mathbf{V}^t\rangle\tau - 2\eta\langle\mathbf{V}^t \cdot \nabla \times \mathbf{B}^t\rangle\tau\right)^2}{16E_T\tau}$$
$$+ \frac{1}{2}\eta\langle\mathbf{V}^t \cdot \nabla \times \mathbf{B}^t\rangle\tau. \tag{12.38b}$$

Here, $\eta = (\sigma\mu)^{-1}$ is the magnetic diffusivity, E_T and τ are the characteristic energy and lifetime of the small-scale helical turbulences. From Equation 12.38, one can easily find the characteristic spatiotemporal scales of the large-scale plasma structures, L and T,

$$L \sim k_0^{-1} = \frac{4E_T\tau}{\left|\langle\mathbf{V}^t \cdot \nabla \times \mathbf{V}^t\rangle\tau - 2\eta\langle\mathbf{V}^t \cdot \nabla \times \mathbf{B}^t\rangle\tau\right|}, \qquad (12.39)$$

$$T \sim [\gamma(k_0)]^{-1} = \left\{ \frac{\left(\langle\mathbf{V}^t \cdot \nabla \times \mathbf{V}^t\rangle\tau - 2\eta\langle\mathbf{V}^t \cdot \nabla \times \mathbf{B}^t\rangle\tau\right)^2}{16E_T\tau} + \frac{1}{2}\eta\langle\mathbf{V}^t \cdot \nabla \times \mathbf{B}^t\rangle\tau \right\}^{-1}, \qquad (12.40)$$

and of the small-scale turbulences, l and τ,

$$l \sim E_T^{1/2}\tau. \qquad (12.41)$$

$$\tau \sim \frac{l}{E_T^{1/2}}. \qquad (12.42)$$

From a system given in Equations 12.39 through 12.42, the relation between small- and large-scale turbulent temporal and spatial plasma structures can be written as follows:

$$\frac{\tau}{T} = \left(\frac{l}{L}\right)^2 \left(1 + \frac{\eta\langle\mathbf{V}^t \cdot \nabla \times \mathbf{B}^t\rangle\tau}{2E_T\tau}L^2\right). \qquad (12.43)$$

This relation recovers the interactions between small- and large-scale spatiotemporal plasma structures for polytrophic turbulent plasma, which models a composite plasma consisting of atomic and molecular ions of the exhaust gas or liquid, and electrons and ions of ambient ionospheric plasma. A schematically sketched relation between such structures, shown in Figure 12.9, together with formulas presented earlier, allow us to estimate the time of relaxation of small-scale plasma turbulences filling large-scale enhancement plasma zones.

Thus, accounting for the temporal and spatial scales of large-scale structures and the spatial scale of small-scale disturbances that can be estimated from experimental observations, described in Chapter 11 and in the previous chapters, we get T to be of several hours, L of several tens of kilometers (along altitude), and l of several hundred meters. If so, the following simplified formula, $\tau \sim (l/L)^2 T$, allows us to estimate roughly that τ is several tens of minutes, that is, on the same order of magnitude as was observed experimentally at the lower ionosphere from 50 to 120 km (see Chapter 11).

At the same time, we should note that generally a turbulent process leads to the saturation of instability growth due to turbulent diffusion (called Bohm's diffusion), whose rate exceeds the rate of traditional laminar diffusion of plasma. It can be caused by the same wave–wave interaction and by the inverse (with respect to that shown in Figure 12.9) transform of energy from the long-scale to the short-scale wave modes, as well as by wave–particle interactions. A decrease of increment of instability, that is, stabilization of plasma perturbations in the lower ionosphere, occurs in the case when additional fluctuations of electric and magnetic fields, plasma densities, and velocities (according to a system given by Equation 12.36) lead to decrease of plasma density gradient $\kappa = \nabla(\ln N_0)$ and relaxation of perturbed plasma concentration to its steady-state (equilibrium) configuration.

Thus, as was shown in Refs. [88,89] at altitudes of lower ionosphere, where the plasma is weakly magnetized, that is, in the system given by Equation 12.36, one can follow Equation 12.36d for it, a quasiequilibrium level of waves in plasma can be found from the saturation condition, combining the regular non perturbed ambient plasma, $\kappa = \nabla(\ln N_0)$ with the gradient of average density of turbulent plasma, $\kappa^t = N_0^{-1}\nabla N^t$, which yields the following:

$$\kappa + \kappa^t = 0. \tag{12.44}$$

Equation 12.44 states that the gradients of plasma density in a perturbed turbulent state equals that in the equilibrium nonperturbed state. It was found for the lower E-layer conditions that the relative amplitude of plasma disturbances $\delta N/N_0$ fully depends on velocity $\mathbf{V}_{0i}$ for movement of induced plasma ions, their rate of diffusion D_i, and mobility μ_{iP} in the direction of the ambient electric field (called Pedersen mobility, see earlier), that is,

$$\delta N \propto \frac{\mu_{iP}k^2}{\left(i(\omega - \mathbf{k}\cdot\mathbf{V}_{0i}) - D_i k^2\right)}. \tag{12.45}$$

Despite the fact that in Ref. [90], the author dealt with two-liquid plasma dynamic (consisting only of electrons and single-charge atomic ions), we can roughly adapt the results to the composite plasma, consisting of electrons and mono- and polytrophic atomic and molecular ions. Following Ref. [90], the corresponding quasiequilibrium level of plasma turbulence via a nonlinear increment of plasma instability, which in steady-state regime equals $\gamma_{NL} = 0$, can be presented as follows:

$$\gamma_{QL} \equiv \gamma_L - D_{0i}k^2 - \alpha_{eff}N_0 - \frac{\varepsilon_0 \omega_{iP}^2 k^2}{M_i v_{im} \omega_{iH}^2 N_0}\left\langle |\delta E|^2 \right\rangle \left(1 + \frac{\Omega_H^2}{v_{im}^2}\right), \quad (12.46)$$

where:

ε_0 is the permittivity of the ambient plasma

ω_{iP} and ω_{iH} are the Pedersen and Hall angular frequencies of waves propagating in the direction along $\mathbf{E}_0$ and across $\mathbf{B}_0$, respectively

D_{0i} is the coefficient of isotropic ion diffusion

Other parameters are as described earlier as also in Chapter 1.

The linear increment of weakly perturbed turbulent plasma is presented in the following form [88,89]:

$$\gamma_L \propto -4\mu_{eH} \frac{k_\perp}{\tilde{\chi}} \frac{k_\parallel^2}{\delta a} \left|\frac{\delta N}{N_0}\right|^2. \quad (12.47)$$

Here, $\tilde{\chi} = N_0^{-1}\left[\partial \tilde{N}(t)/\partial y\right]$ is the one-dimensional (1D) gradient of average density of turbulent plasma, the y-axis is the vertical axis, which for simplicity of the problem is directed along the magnetic field $\mathbf{B}_0$, subscripts $\parallel$ and $\perp$ determine components of wave vectors along and perpendicular to $\mathbf{B}_0$. Accounting for conditions of weakly turbulent plasma (linear perturbation regime), from Equation 12.47 in Refs. [88,89] it was found that $k = 2\pi/l \approx 1\,\mathrm{m}^{-1}$, ($l \approx 6$ m), $\chi = 10^{-2}\,\mathrm{m}^{-1}$, $k_\parallel \approx k_\perp$, and $\mu_{iP}/\mu_{iH} \approx 10^{-1}$, and for condition given in Equation 12.44 (when the gradients of plasma density in turbulent and in equilibrium states are equal), that $|\delta N/N_0| \approx 10^{-2}$, that is, of about 1%.

The corresponding estimations of plasma disturbances associated with the nonlinear instability of turbulent plasma defined by Equation 12.46 according to Ref. [90] were estimated as $|\delta N/N_0| \approx 3 \times 10^{-2}$, that is, about 3%. Thus, in the nonlinear regime we can expect stronger plasma disturbances with respect to those observed in weak turbulent plasma. All these estimations, of course, are approximate because they account for a purely turbulent regime occurring separately without accounting for processes of chemical reactions and local heating of the exhaust jet during the rocket releases. Therefore, we need to emphasize that, despite the fact that all estimations of plasma disturbances caused either by chemical reactions and local heating or by turbulent movements of short- and long-wave plasma structures "deal" only with two-liquid plasma consisting of electrons (e^-) and single-charge ions ($i = e^+$) and the results obtained earlier can be adapted to the polytrophic electron–ion plasma, consisting of heavier atomic and molecular ions, as shown in our previous discussions, the heavy ions are long-living particles (regarding recombination instability) and have longer time of relaxation and damping (regarding turbulent plasma effects). Therefore, we guess that our approach to take some effective ion with the "average" mass and charge and introduce effective recombination and ionization coefficients in all chemical reactions yields correct results on the prediction and estimation of temporal and spatial dimensions of plasma perturbations generated in rocket burn and launch. Such an approach gives qualitatively correct results coinciding with observed experimental data obtained in the lower ionosphere, as well as with the corresponding "active experiments" related to rocket releases.

12.6.2 Evolution of Plasma Irregularities in Middle and Upper Ionosphere Associated with Rocket Releases

Starting the analysis of the process of small- and moderate-scale plasma disturbance generation at altitudes of the upper E- to the upper F-region, we state that at altitudes above 150 km ambient nonperturbed ionospheric plasma is magnetized, partly ionized, and fully determined by transport processes, which prevail in the recombination processes. They are fully negligible for attachment and detachment reactions. At the same time, due to altitudinal gradient of plasma density and its turbulent character caused by the rocket releases and the corresponding exhaust gases and liquid, the situation can dramatically change compared to that observed during active experiments described in Chapter 11. Unfortunately, all authors mentioned previously [12–27] dealt with one species of the exhaust gases, water, nitrogen, oxygen, and so forth in their models and numerical simulations. They all understood that it is very complicated to derive full system, from Equations 12.1 through 12.5, describing chemical reactions together with dynamics and evolution (diffusion, drift, and thermodiffusion) of each component of polytrophic composite ion–electron plasma.

Therefore, as in our previous explanation of the effects occurring at lower ionosphere, we briefly discuss the matter regarding the middle and upper ionosphere, by introducing some effective ion as was done in Sections 12.2 and 12.3, and analyze the corresponding sources that excite the short- and the long-length plasma disturbances that can be associated with rocket burn and launch.

The same qualitative analysis as shown earlier, without entering into complicated derivations of the corresponding equations and integrals, can be made following Refs. [91–93], to explain nonlinear growth rate of plasma instabilities and damping (caused by saturation mechanisms), as well as the evolution of the plasma perturbed structures in space and time domains, observed experimentally and described in the previous chapters. Despite the use of the same quasilinear two-fluid theory by the researchers as given in Refs. [88–90], it can be adapted for explanation of nonlinear generation of different kinds of CCIs, including two-stream (FB) and GD, occurring in the composite plasma created after the process of rocket gas releases of polytrophic molecules and atoms. This can be done because this self-consistent approach accounts for the vector of plasma gradient κ, which in the general case elongates along and across the vertical axis (i.e., along and across the ambient magnetic field $\mathbf{B_0}$), which is correct for middle-latitude and high-latitude ionosphere. For the lower latitudes, namely, the equator, the gravitational RT instability should be involved in the whole process of nonlinear process of plasma disturbance evolution.

We present the increment of nonlinear growth $\gamma_{\mathbf{k}}^{\mathrm{NL}}$ (or decrement of nonlinear damping) following Refs. [91–93] through the linear increment $\gamma_{\mathbf{k}}^{L}$ (decrement) using our unified notation introduced in this chapter and by introducing turbulent motions of the composite plasma consisting of some effective ion, as was done in Sections 12.2 through 12.4:

$$
\begin{aligned}
\gamma_{\mathbf{k}}^{\mathrm{NL}} &= \frac{\Delta\omega_{\mathbf{k}}^2}{\nu_{\mathrm{im}}(1 + q_{\mathrm{H}}Q_{\mathrm{H}})} + \frac{q_{\mathrm{H}}\mathbf{k} \cdot \kappa\bar{\omega}_{\mathbf{k}}}{k^2(1 + q_{\mathrm{H}}Q_{\mathrm{H}})} + \frac{(\bar{\omega}_{\mathbf{k}}^2 - k_{\perp}^2 C_s^2)}{\nu_{\mathrm{im}}(1 + q_{\mathrm{H}}Q_{\mathrm{H}})} \\
&= \frac{\left[\bar{\omega}_{\mathbf{k}}^2 + \Delta\omega_{\mathbf{k}}^2 - (\omega_{\mathbf{k}}^L)^2\right]}{\nu_{\mathrm{im}}(1 + q_{\mathrm{H}}Q_{\mathrm{H}})} + \frac{q_{\mathrm{H}}\mathbf{k} \cdot \kappa(\bar{\omega}_{\mathbf{k}} - \omega_{\mathbf{k}}^L)}{k^2(1 + q_{\mathrm{H}}Q_{\mathrm{H}})} + \gamma_{\mathbf{k}}^L.
\end{aligned}
\tag{12.48}
$$

Here, we rewrite all parameters and notations, following Ref. [41], where $q_{\mathrm{H}} = \omega_{\mathrm{H}}/\nu_{\mathrm{em}}$, $Q_{\mathrm{H}} = \Omega_{\mathrm{H}}/\nu_{\mathrm{im}}$, and $\bar{\omega}_{\mathbf{k}} \equiv \omega_{\mathbf{k}}^t$ are the angular frequencies of the turbulent plasma modes in the $\mathbf{k}$-domain, $\Delta\omega_{\mathbf{k}}$ is the shift of plasma frequency (due to Doppler effects) in the $\mathbf{k}$-domain, caused by the laminar drift of plasma disturbances in the ambient electrical and magnetic fields; all other parameters of plasma are the same as above. The angular frequencies of linear plasma waves (in the $\mathbf{k}$-domain) [91–93] is as follows:

$$
\omega_{\mathbf{k}}^L = \frac{q_{\mathrm{H}}Q_{\mathrm{H}}}{(1 + q_{\mathrm{H}}Q_{\mathrm{H}})}\mathbf{k} \cdot \tilde{\mathbf{V}}_{\mathrm{d}}.
\tag{12.49a}
$$

The linear increment (decrement) of linear mode growth (damping) [91–93] is as follows:

$$
\gamma_{\mathbf{k}}^L = \frac{q_{\mathrm{H}}Q_{\mathrm{H}}}{(1 + q_{\mathrm{H}}Q_{\mathrm{H}})}\left\{\frac{\left[(\omega_{\mathbf{k}}^L)^2 - k_{\perp}^2 C_s^2\right]}{\nu_{\mathrm{im}} q_{\mathrm{H}}Q_{\mathrm{H}}} + \frac{\mathbf{k} \cdot \kappa\omega_{\mathbf{k}}^L}{Q_{\mathrm{H}}k^2}\right\}.
\tag{12.49b}
$$

The drift velocity is presented for the turbulent composite plasma, containing electrons and some effective heavy ion M^+, in more general form, as is given in Refs. [94,95] for the laminar regular plasma, that is,

$$\mathbf{V}_d^{M^+} = \mathbf{U}_{m\|} + \mu_\|^{M^+} \mathbf{E} + \mu_\perp^{M^+} \left(\mathbf{E} + \frac{\mathbf{U}_m \times \mathbf{B}_0}{cB_0} \right)$$

$$+ \mu_\Lambda^{M^+} \frac{\mathbf{E}_0 \times \mathbf{B}_0}{B_0} + \frac{\mathbf{U}_{m\perp}}{1 + Q_H^2}. \tag{12.50}$$

Here, the movement of the exhaust ion–electron composite plasma is determined by faster unipolar diffusion coefficients ($D_{e\|}$ along $\mathbf{B}_0$ and $D_{i\perp}$ across $\mathbf{B}_0$), by drift in $\mathbf{E}_0 \times \mathbf{B}_0$ direction and by components of velocity of neutral wind, $\mathbf{U}_{m\|}$ along $\mathbf{B}_0$ and $\mathbf{U}_{m\perp}$ normal to $\mathbf{B}_0$; $\mu_\|^{M^+}$, $\mu_\perp^{M^+}$, and $\mu_\Lambda^{M^+}$ are the mobility of ions along $\mathbf{B}_0$, normal to $\mathbf{B}_0$ (and along $\mathbf{E}_0$—Pedersen), and across $\mathbf{B}_0$ (Hall), respectively.

As for the upper *E-* and *F-*regions of the ionosphere, whose plasma is magnetized and ionized, that is, $q_H Q_H \gg 1$ and $\nu_{ei} \gg \nu_{em} \gg \nu_{im}$, the same estimation of the quasiequilibrium level of plasma turbulence determined by Equation 12.44 shows that plasma disturbances associated with the effects of turbulent stabilization regime, will not exceed $|\delta N/N_0| \approx (2.5\text{–}3.0) \times 10^{-2}$, that is, 2.5%–3.0%. These estimations made for middle and upper ionosphere are in good agreement with those made in Refs. [88,89] and presented earlier.

We now present the effect of plasma instability growth and the conditions of their stabilization for plasma clouds created during the rocket releases, following the results obtained in Refs. [96–98] for analyzing the drift processes of plasma clouds in the ionosphere. Let us suppose from the outset that the concentration of ions within the perturbed plasma cloud, N_M^+, is large compared to the concentration of background ionospheric plasma N_0, that is, $N_M^+/N_0 \gg 1$. It was shown that the stabilization effect can be observed at the initial stage of plasma cloud evolution due to the creation of the shear of the azimuthally directed drift velocity created in the background plasma by the azimuthally directed currents. Their stabilizing effect is observed if the azimuthally directed drift velocity exceeds its threshold value [96–98]:

$$V_{th} \sim V_A[A/(A + 2)], \tag{12.51}$$

where:

$A = \delta N(0,0,0)/N_0$ is the relative initial plasma density perturbation within the plasma exhaust cloud created surrounding the moving vehicle (in the wake and in the vicinity of the rocket)

$V_A \sim c^{-1}\mathbf{E}_{0\perp} \times \mathbf{B}_0$ is the azimuthally directed drift velocity

c is the speed of light

$\mathbf{E}_{0\perp}$ is the ambient transverse (to $\mathbf{B}_0$) electric field component

It was also shown that the shear of the azimuthally directed drift velocity decreases with an increase of the plasma structure dimensions along $\mathbf{B}_0$. According to Refs. [96–98], the criterion of plasma drift stabilization can be presented (using notations introduced in this chapter) as

$$\frac{T_e(1 + T_i/T_e)}{eE_{0\perp}} > \left(\frac{\ell_{\parallel}}{\ell_{\perp}}\right)^2 \frac{(1 - V_{th})}{q_H Q_H},\tag{12.52}$$

where:

   $\ell_{\parallel}$ and $\ell_{\perp}$ are the dimensions of plasma structures that fill the cloud zone along and normal to $\mathbf{B}_0$

As seen from Equation 12.52, with increase in $E_{0\perp}$, that is, with the increase of drift velocity V_A, the threshold of stabilization increases. Hence, the absence of stabilizing effects can be obtained accompanied by growth of drift instability in the strong ambient electric field. At the same time, with increasing parameters of plasma density perturbations A within the initial plasma structure ($A > 10$) the factor $A/(A + 2) \approx 1$ and, as seen from Equation 12.52 for $E_{0\perp}$ = constant, the threshold of plasma drift stabilization becomes smaller than that in the inverse case $A < 10$, where $A/(A + 2) < 1$. Thus, with increasing plasma perturbations inside the plasma cloud, the effect of plasma drift stabilization is strongly manifested and a smooth spreading of plasma structure in the process of its evolution is observed (see experiments with barium cloud release described in Chapter 11).

As also found in Refs. [98–100], the effect of stabilization of $\mathbf{E}_0 \times \mathbf{B}_0$ instability strongly depends on the criterion of "stretching" of plasma irregularities that fill the perturbed region and the amplitude of initial plasma instability and its transversal dimension to $\mathbf{B}_0$. It was found that the maximal growth rate which is needed for the development of the "stretching" mechanism, is on the order [99]

$$\gamma_{max} \sim \frac{cE_{0\perp}}{B_0 \ell_{\perp}}.\tag{12.53}$$

At the same time, as shown in Refs. [96–98], the background plasma is an important "player" during the evolution of the artificially induced plasma cloud, consisting of heavy ions, such as barium, strontium, and so forth. The electrons of the ambient plasma move along $\mathbf{B}_0$ and the ions of the ambient plasma move across $\mathbf{B}_0$ trying to compensate rising and exciting plasma structures, creating the so-called depletion zones. As a result, a "short circuit" current is generated in the background plasma to conserve the quasineutrality of plasma inside the globally perturbed area. The reader can find in Ref. [100] a corresponding theoretical background of such phenomena.

Following Refs. [96–98], the plasma instability growth can be developed if the growing perturbation of the ambient electric field δE is not cancelled by longitudinal electron currents due to the "short circuit" effect in the background plasma, that is, when the time of the "mean" electron's run along $\mathbf{B}_0$, $\tau_\| = \ell_\|/V_\|$, exceeds the inverse increment of plasma instability, $\tilde{\tau}_\| \sim \gamma_{max}^{-1}$, that is, $\tau_\| > \tilde{\tau}_\|$. Because $\delta E_\|/\delta E_\perp \sim \ell_\perp/\ell_\|$, we have $\tau_\| \approx m_e(\nu_{em} + \nu_{ei})\ell_\|^2/e\ell_\perp\delta E_\perp$, and the "stretching" criterion can be formulated as follows [98]:

$$\left(\frac{\ell_\|}{\ell_\perp}\right) \geq \left(\frac{\omega_H}{\nu_{em} + \nu_{ei}}\frac{\delta E_\perp}{E_{0\perp}}\right)^{-1/2}. \qquad (12.54)$$

Using estimations made in Refs. [96–98] for $\delta E_\perp/E_{0\perp} \approx 0.1$, we finally obtain the "stretching criterion" for any plasma structure which completely depends on the plasma conditions at the altitude of its generation

$$\chi \equiv \left(\frac{\ell_\|}{\ell_\perp}\right) > \left(0.1\frac{\omega_H}{\nu_{em} + \nu_{ei}}\right)^{-1/2}. \qquad (12.55)$$

From Equation 12.55 accounting for plasma parameters estimated in Ref. [98], we get the corresponding criterion of plasma structure stratification on strata: $\ell_\|/\ell_\perp \geq 3 \times 10^1$. At the same time, the striation mechanism observed in Ref. [96] for every anisotropic plasma structures is fully confirmed by the analysis presented earlier and by estimations obtained from Equations 12.53 through 12.55. Consequently, following the above estimations it can be found that with a decrease of the anisotropy of the initial plasma perturbed structure, that is, for $\ell_\|/\ell_\perp < \times 10^1$, the threshold of drift stabilization becomes smaller.

Next, because at altitudes of $h > 150$ km, ions are magnetized and $(1 + Q_H^2) \gg 1$, they do not move in the direction of $\mathbf{U}_{m\perp}$ but only in the $\mathbf{E}_0 \times \mathbf{B}_0$ direction. Here, in the linear case, described in Refs. [98–100], due to the dispersion mechanism of plasma cloud spreading, three plasmoids are created, the first of which is associated with the *electronic* and spreads with slower velocity across $\mathbf{B}_0$, the second is *ionic* and spreads faster along $\mathbf{B}_0$, whereas the third spreads in the $\mathbf{E}_0 \times \mathbf{B}_0$ direction. Using the results of the theoretical prediction described earlier, it is clearly explained in Ref. [98] that the asymmetry of the density profile of ions Ba^+ across $\mathbf{B}_0$ is obtained experimentally in Refs. [96,98] and shown in Figures 11.30 through 11.34. Such dispersive spreading leads to essential increasing of the barium cloud, which was also observed experimentally [30–36]. Moreover, a sharp decrease of the amount of ions in the barium cloud after $t \geq 400$ s can obviously be described by the influence of dispersive drift mechanism in the process of plasma ordinary diffusion at the quasilinear and linear stages of evolution of the ion cloud. Now, taking from a theoretical framework proposed in Refs. [96–98], a critical time $t_{kr} = d_n(\tau_0)/b_{i\perp}E_\perp^*$ of commencing the drift process, as well as an information on the initial plasma cloud transverse dimension $d_n(\tau_0)$, one can estimate the magnitude of the transverse electric field $E_\perp^*$, $E_\perp^* \approx 2$–$6\,\mathrm{mV\,m^{-1}}$, which is in good agreement with estimations made during experimental optical observations [96–99].

12.7 Main Results

1. At the *initial stage* of the rocket release, the exhaust gases and liquids created by the rocket engine jet or plume expand within the ambient regular ionosphere, generate composite polytrophic electron–ion plasma surrounding the rocket and within its wake by ionization–recombination chemical reactions. Accompanied by additional heating of ambient plasma via high-temperature exhaust gas, both these phenomena finally create large-scale enhancements at the lower ionosphere at altitudes of D-layer and lower E-layer of the ionosphere and global depletion zones (*holes* or *cavities*) at the middle and outer ionosphere at altitudes of upper E- and F-regions. The time of this initial stage does not exceed roughly several tens of seconds.

2. Then, during several tens of minutes, the created global positive (enhancement) or negative (cavity) plasma structures move in the lower ionosphere for several minutes with the velocity of horizontal laminar wind $\mathbf{U}_m$ of neutrals combined with vertical turbulent wind $\mathbf{U}^t$ of plasma turbulence. For the middle and outer ionosphere, the dense plasma cloud moves with the velocity of neutral wind $\mathbf{U}_m$, damping due to unipolar and quasineutral diffusion (as in the case of $\mathbf{B}_0 = 0$ [41]), and only part of the ions float in the crossing $\mathbf{E}_0 \times \mathbf{B}_0$ fields, fully coinciding with the vision of the drift spreading following from a classical drift theory described in Ref. [41] and from observation data obtained experimentally in Chapter 11.

3. At the *later stage* of its evolution ($t > 10$ min), the density of the plasma within the composite ion–electron plasma structure (cloud) becomes smaller due to ambipolar diffusion together with wind and drift, turbulent and laminar, creating a nonlinear regime, when $N_M^+/N_0 > 1$ or $N_M^+/N_0 \geq 1$. Here, the effects of short-circuit currents through the depletion regions and the effects of the background plasma become predominant. Drift spreading of ions occurs faster than in the case of ambipolar diffusion and most plasma is floating in a $\mathbf{E}_0 \times \mathbf{B}_0$ direction. The picture of spreading becomes more complicated, because in this case and as been shown in Refs. [96–98],

the shape, form, and orientation of large-scale plasma structure (cloud) in space changes nonuniformly. In this case, disturbance $\delta\mathbf{E}$ of the ambient electric field $\mathbf{E}_0$ is sufficiently small and currents of particles caused by this disturbance mostly flow through the background plasma.

4. At the *final stage* of its evolution, the ion cloud transforms into small-scale plasma structures—strata (i.e., short-length modes) with density lesser than that for background ionospheric plasma. During this stage, due to various CCIs described above, creation of stratified structure filling the depletion or enhanced zones is observed, where each plasmoid is stretched along geomagnetic field lines.

5. To explain this phenomenon, various physical mechanisms were analyzed and discussed earlier, and as given in Refs. [88–93], three possible types of instabilities, two-stream (FB), gravitational (RT), and GD, were analyzed and it was found that the more probabilistic source of striation of the global plasma structure on strata structure is GDI discussed earlier in detail. It was found that the increment of GDI growth (as a limit of striation of plasma plasmoids into grating structure) can be estimated for stretched along $\mathbf{B}_0$ irregularities as $\gamma \approx V_d/\ell_\perp$, where V_d can be estimated from Equation 12.50 and $\ell_\perp$ is the transverse (to $\mathbf{B}_0$) scale of primary strata (plasmoid).

6. Using the above formulas and results obtained in Refs. [96–98], we can predict the process of plasma instability generation, the small-scale plasma turbulences (irregularities) creation, during the evolution of a weakly ionized ($A < 10$), strongly anisotropic ($\ell_\parallel/\ell_\perp < 3 \times 10^1$) plasma structure drifting in a strong ambient electric field ($E_{0\perp} > 5\,\mathrm{mV\,m^{-1}}$), as well as the perturbed plasma stabilization in the case of strongly ionized ($A > 10$) weakly anisotropic ($\ell_\parallel/\ell_\perp < 3k \times 10^1$) plasma structure evolution in a weak ambient electric field ($E_{0\perp} \leq 1\text{–}5\,\mathrm{mV\,m^{-1}}$), along with the process of GDI generation.

7. The results presented in this chapter depend on many parameters of plasma irregular (turbulent) structures artificially induced by the rocket releases and can be predicted strictly using modern radio physical and optical methods described in Chapter 11 and the previous chapters.

References

1. Booker, H. G., A local reduction of *F*-region ionization due to missile transit, *J. Geophys. Res.*, 66, 1073–1079, 1961.
2. Barnes, C., Comment on paper by Henry G. Booker, A local reduction of *F*-region ionization due to missile transit, *J. Geophys. Res.*, 66, 2580, 1961.
3. Altshuler, S., Moe, M. M., and Molmud, P., The electromagnetics of the rocket exhaust, *Space Tech. Lab. Rep.*, June 15, 1958.

4. *Ionospheric Modifications of Missile Exhaust*, Geophysics Corp. of America (GCA), *Sci. Rep. No. 10* and *GCA Tech. Rep. No 62-15-G*, December 1962.

5. Kellogg, W. W., Pollution of the upper atmosphere by rockets, *Space Sci. Rev.*, 3, 275–316, 1964.

6. Jackson, J. E., Whale, H. A., and Bauer, S. J., Local ionospheric disturbance created by a burning rocket, *J. Geophys. Res.*, 67, 2059–2061, 1962.

7. Golomb, D., Rosenberg, N. W., Weight, J. W., and Barnes, R. A., Formation of an electron-depleted region in the ionosphere by chemical releases, *Space Res.*, 4, 389–398, 1963.

8. Malcolm M. A. and Golomb, D., The evolution of a ionospheric hole, *Upper Atmospheric Phys. Lab.*, *AFCRL Environ. Res. Paper # 101*, April, 1965.

9. Stone, M. L., Bird, L. E., and Balser, M., A Faraday rotation measurement on the ionospheric perturbation produced by a burning rocket, *J. Geophys. Res.*, 69, 971–978, 1964.

10. Wright, J. W., Ionosonde studies of some chemical releases in the ionosphere, *Radio Sci.*, 68D, 189–204, 1964.

11. Felker, J. K. and Roberts, W. T., Ionospheric rarefraction following rocket transit, *J. Geophys. Res.*, 71, 4692–4694, 1966.

12. Mendillo, M., Hawkins, G. S., and Klobuchar, J. A., A sudden vanishing of the ionospheric F region due to the launch of Skylab, *J. Geophys. Res.*, 80, 2217–2228, 1975.

13. Mendillo, M., Hawkins, G. S., and Klobuchar, J. A., A large-scale hole in the ionosphere caused by the launch of Skylab, *Science*, 187, 343–347, 1975.

14. Mendillo, M. and Forbes, J. M., Artificially created holes in the ionosphere, *J. Geophys. Res.*, 83, 151–162, 1978.

15. Mendillo, M., The effect of rocket launches on the ionosphere, *Adv. Space Res.*, 1, 275–290, 1981.

16. Mendillo, M. and Forbes, J. M., Theory and observation of a dynamically evolving negative ion plasma, *J. Geophys. Res.*, 87, A10, 1982, 8273–8285.

17. Mendillo, M., Ionospheric holes: A review of theory and recent experiments, *Adv. Space Res.*, 8, (1)51–(1)62, 1988.

18. Anderson, D. N. and Bernhardt, P. A., Modeling the effects of an H_2 gas release on the equatorial ionosphere, *J. Geophys. Res.*, 83, 4777–4790, 1978.

19. Bernhardt, P. A., High-altitude gas releases: Transition from collisionless flow to diffusive flow in a nonuniform atmosphere, *J. Geophys. Res.*, 84, 4341–4354, 1979.

20. Johnson, C. Y., Sjolander, G. W., Oran, E. S., et al., F region above Kauai: Measurement, model, modification, *J. Geophys. Res.*, 85, 4205–4213, 1980.

21. Bernhardt, P. A., Chemistry and dynamics of SF_6 injections into F region, *J. Geophys. Res.*, 89, 3929–3937, 1984.

22. Bernhardt, P. A., Environmental effects of plasma depletion experiments, *Adv. Space Res.*, 2, 129–137, 1982.

23. Bernhardt, P. A., A critical comparison of ionospheric depletion chemicals, *J. Geophys. Res.*, 92, 4617–4628, 1987.

24. Biondi, M. A. and Sipler, D. P., Studies of equatorial 630.0 nm airglow enhancements produced by a chemical release in the F-region, *Planet. Space Sci.*, 32, 1605–1610, 1984.

25. Bernhardt, P. A., Ganguli, G., Kelley, M. C., and Swartz, W. E., Enhanced radar backscatter from space shuttle exhaust in the ionosphere, *J. Geophys. Res.*, 100, 23811–23818, 1995.

26. Kikuchi, H., *Electrohydrodynamics in Dusty and Dirty Plasmas*, Dordrecht: Kluwer Academic Publishers, 2001.

27. Bernhardt, P. A., Plasma, fluid instabilities in the ionospheric holes, *J. Geophys. Res.*, 87, 7539–7549, 1982.

28. Guzdar, P., Gondarenko, N., Chaturvedi, P., and Basu, S., Three-dimensional non-linear simulations of the gradient drift instability in the high-latitude ionosphere, *Radio Sci.*, 33, 1901–1912, 1998.

29. Volland, H., Ed., *Handbook of Atmospheric Electrodynamics*, Vols. I and II, Boca Raton, FL: CRC Press, 1995.

30. Volland, H., Electrodynamic coupling between neutral atmosphere and iono-sphere, in *Modern Ionospheric Science*, Kohl, H., Ruster, R. R., and Schlegel, K., Eds., Berlin: ProduServ GmbH Verlagsservice, 1996.

31. Volland, H., *Atmospheric Electrodynamics*, Heidelberg: Springer-Verlag, 1984.

32. Belikovich, V. V., Benediktov, E. A., Tolmacheva, A. V., and Bakhmet'eva, N. V., *Ionospheric Research by Means of Artificial Periodic Irregularities*, Berlin: Copernicus GmbH, 2002.

33. Mitra, A. P. and Rowe, J. N., Ionospheric effects solar flares VI. Changes in *D*-region ion chemistry during solar flares, *J. Atmos. Terr. Phys.*, 34, 795–806, 1972.

34. Wisemberg, J. and Kockarts, G., Negative ion chemistry in the terrestrial *D* region and signal flow graph theory, *J. Geophys. Res.*, 85, 4642–4652, 1980.

35. Al'pert, Ya. L., *Space Plasma*, Vol. I and II, New York: Cambridge University Press, 1983 and 1990, 305.

36. Whitten, R. C. and Poppoff, I. G., *Physics of the Lower Ionosphere*, New York: Prentice Hall, 1965.

37. Ivanov-Kholodnii, G. S. and Nikol'sky, G. M., *The Sun and the Ionosphere*, Moscow: Nauka, 1969 (in Russian).

38. Gershman, B. N., *Dynamics of the Ionospheric Plasma*, Moscow: Nauka, 1974 (in Russian).

39. Gurevich, A. V., *Nonlinear Phenomena in the Ionosphere*, Berlin: Springer-Verlag, 1978.

40. Uryadiv, V., Ivanov, V., Plohotniuc, E., et al., *Dynamic Processes in the Ionosphere— Methods of Investigations*, Iashi: TehnoPress, 2006 (in Romanian).

41. Blaunstein, N. and Plohotniuc, E., *Ionosphere and Applied Aspects of Radio Communication and Radar*, Boca Raton, FL: CRC Press, 2008.

42. Blaunshtein, N. Sh., Vas'kov, V. V., and Dimant, Ya. S., Resonant heating of the *F* layer of the ionosphere by a high-power radio wave, *Geomagn. Aeronom.*, 32, 235–238, 1992.

43. Blaunshtein, N. Sh. and Bochkarev, G. S., Modeling of the dynamics of periodical artificial disturbances in the upper ionosphere during its thermal heating, *Geomagn. Aeronom.*, 33, 84–91, 1993.

44. Blaunstein, N., Diffusion spreading of middle-latitude heating-induced iono-spheric plasma irregularities, *J. Annal. Geophysic.*, 13, 617–626, 1995.

45. Blaunstein, N., Changes of the electron concentration profile during local heating of the ionospheric plasma, *Atmos. Solar-Terr. Phys.*, 58, 1345–1354, 1996.

46. Blaunstein, N., Evolution of a stratified plasma structure induced by local heating of the ionosphere, *J. Atmos. Solar-Terr. Phys.*, 59, 351–361, 1997.

47. Kadomtsev, B., *Plasma Turbulence*, New York: Academic Press, 1965.

48. Blaunstein, N., Arnon, S., Silberman, A., and Kopeika, N., *Applied Aspects of Optical Communication and LIDAR*, Boca Raton, FL: CRC Press, 2010.

49. Belikovich, V. V. and Razin, V., Formation of artificial periodic inhomogeneities in the *D*-region of the ionosphere with attachment and recombination processes taken into consideration, *Izv. Vuzov. Radiofizika*, 29, 251–256, 1986 (in Russian).

50. Tomko, A. A., Ferraro, A. G., Lee, H. S., and Mitra, A. P., A theoretical model of *D*-region ion chemistry, modification during high-power wave heating, *J. Atmos. Terr. Phys.*, 42, 273–285, 1980.

51. Belikovich, V. V. and Benediktov, E. A., About short-time variations of the plasma parameters in the lower part of the *D*-region ionosphere, *Geomagn. Aeronom.*, 26, 680–682, 1986.

52. Grandal, B., Ed., *Artificial Particle Beams in Space Plasma Studies*, New York: Plenum Press, 1982.

53. Garmash, K. P., Gokov, A. M., Kostrov, L. S., et al., Radiophysical investigations and modeling of processes in the ionosphere perturbed by sources of different nature, *Radiophys. Electron.*, 405, 157–177, 1998; 407, 3–22, 1999 (in Russian).

54. Garmash, K. P., Rozumenko, V. T., Tyrnov, O. F., Zimbal, A. M., and Chernogor, L. F., Radiophysical investigations of processes in the near Earth plasma perturbed by high-energy sources, *Successes in Modern Radioelectronics*, 7, 3–18, 1999; 8, 3–19, 1999 (in Russian).

55. Chernogor, L. F. and Rozumenko, V. T., Wave processes, global- and large-scale disturbances in the near Earth plasma, in *Proc. Int. Conf. Astronomy in Ukraine-2000 and Perspective, Kinematics and Physics of Sky Bodies*, Annex K, 514–516, 2000.

56. Fejer, B. G., Natural ionospheric plasma waves, in *Modern Ionospheric Science*, Kohl, H., Ruster, R., and Schlegel, K., Eds., Berlin: Luderitz and Bauer, 1996, 216–273.

57. D' Angelo, N., Recombination instability, *Phys. Fluids*, 10, 719–723, 1967.

58. Rumiantsev, S. A. and Smirnov, V. S., About origin of recombination instability in the high-latitude ionosphere, *Geomagn. Aeronom.*, 20, 1107–1109, 1980.

59. Erukhimov, L. M., Kagan, L. M., and Savina, O. N., About heating mechanism of formation of small-scale plasma inhomogeneities at the heights of *E*-layer, *Izv. Vuz. Radiofizika*, 24, 1032–1034, 1983.

60. Gershman, B. N., Ignat'ev, Yu. A., and Kamenetskaya, G. H., *Mechanisms of Formation of Ionospheric Sporadic E_s-Layer at Various Latitudes*, Moscow: Science, 1976 (in Russian).

61. Post, R. F. and Rosenbluth, M. N., Electrostatis instability in finite mirror confined plasmas, *Phys. Fluids*, 9, 730–749, 1966.

62. Ott, E. and Farley, D. T., Microstabilities and production of short wavelength irregularities in the auroral *F*-region, *J. Geophys. Res.*, 80, 4599–4602, 1975.

63. Ossakow, S. L. and Chaturvedi, P. K., Current convective instability in the diffuse aurora, *Geophys. Res. Lett.*, 6, 332–334, 1973.

64. Gel'berg, M. G., Formation of small-scale inhomogeneities above the maximum of *F*-layer of the auroral ionosphere, *Geomagn. Aeronom.*, 19, 629–632, 1979.

65. Kindel, S. M. and Kennel, C. F., Topside current instabilities, *J. Geophys. Res.*, 76, 3055–3078, 1971.

66. Chaturvedi, P. K. and Kaw, P. K., Current driven ion-cyclotron waves on collisional plasma, *Plasma Phys.*, 17, 447–452, 1975.

67. Chaturvedi, P. K., Collisional ion-cyclotron waves in the auroral ionosphere, *J. Geophys. Res.*, 81, 6169–6171, 1976.

68. Ronchi, C., Similon, P. L., and Sudan, R. N., A nonlocal linear theory of the gradient drift instability in the equatorial electrojet, *J. Geophys. Res.*, 94, 1317–1326, 1989.

69. Gel'berg, M. G., Gradient-drift instability of ionospheric plasma, in *Propagation of Radio Waves in the Polar Ionosphere*, Apatiti: KFAN USSR, 1977, 3–37 (in Russian).

70. Farley, D. T., The plasma instability resulting in field-aligned irregularities in the ionosphere, *J. Geophys. Res.*, 68, 6083–6097, 1963.

71. Buneman, O., Excitation of field-aligned sound waves by electron streams, *Phys. Rev. Lett.*, 10, 285–287, 1963.

72. Farley, D. T., Theory of equatorial electrojet plasma waves: New developments and current status, *J. Atmos. Terr. Phys*, 47, 729–744, 1985.

73. Wang, T. N. C. and Tsunoda, R. T., On a crossed field two-stream plasma instability in the auroral plasma, *J. Geophys. Res.*, 80, 2172–2182, 1975.

74. Fel'dshtein, A. Ya., About influence of heating of particles at the Farley–Buneman instability in the auroral ionosphere, *Geomagn. Aeronom.*, 20, 333–334, 1980.

75. Dimant, Y. S. and Sudan, R. N., Kinetic theory of low-frequency cross-field instability in a weakly ionized plasma, I, *Phys. Plasmas*, 2, 1157–1168, 1995; ibid II, 1169–1181.

76. Dimant, Y. S. and Sudan, R. N., Kinetic theory of the Farley–Buneman instability in the E region of the ionosphere, *J. Geophys. Res.*, 100, 14605–14623, 1995.

77. Schmidt, M. J. and Gary, S. P., Density gradients and Farley–Buneman instability, *J. Geophys. Res.*, 78, 8261–8265, 1974.

78. Gel'berg, M. G., *Inhomogeneities of High Latitude Ionosphere*, Novosibirsk: Nauka, 1986 (in Russian).

79. Gershman, B. N., About conditions of formation of Rayleigh–Taylor instability in the region *F* of the ionosphere, in *Ionospheric Irregularities*, Yakutsk: Academy of Science of USSR, 1981, 3–15 (in Russian).

80. Huang, C.-S. and Kelley, M. C., Nonlinear evolution of equatorial spread *F*, 1. On the role of plasma instabilities and spatial resonance associated with gravity wave seeding, *J. Geophys. Res.*, 101, 283–292, 1996.

81. Kelley, M. C., Haerendel, G., Kappler, H., et al., Evidence for a Rayleigh–Taylor type instability and upwelling of depleted density regions during equatorial spread *F*, *Geophys. Res. Lett.*, 3, 448–451, 1976.

82. Ossakow, S. L., Spread-*F* theories—A review, *J. Atmos. Terr. Phys.*, 43, 437–452, 1981.

83. Migliuolo, S., Nonlocal dynamics in the collisional Rayleigh–Taylor instability: Applications to the equatorial spread *F*, *J. Geophys. Res.*, 101, 10975–10984, 1996.

84. Huang, C.-S. and Kelley, M. C., Nonlinear evolution of equatorial spread *F*, 2. Gravity waves seeding of the Rayleigh–Taylor instability, *J. Geophys. Res.*, 101, 293–302, 1996.

85. Unwin, R. S., The evening diffuse radio aurora, field-aligned currents and particle precipitation, *Planet. Space Sci.*, 28, 547–557, 1980.
86. Branover, H., Eidelman, A., Golbraikh, E., and Moiseev, S. S., *Turbulence and Structures: Chaos, Fluctuations, and Helical Self-Organization in Nature and the Laboratoty*, New York: Academic Press, 1999, 270.
87. Moiseev, S. S., *The Helical Mechanism of Generation of Large-Scale Structures in Continuous Media*, Singapore: World Scientific, p. 451.
88. Sato, T., Nonlinear theory of the cross-field instability explosive mode coupling, *Phys. Fluids*, 14, 2426–2435, 1971.
89. Sato, T., Nonlinear stabilization and nonrandom behavior of macro-instabilities in plasmas. II. Numerical verification, *Phys. Fluids*, 17, 3162–3169, 1974.
90. Kamenetskaya, G. H., About excitation of longitudinal waves by current of the equatorial current stream, *Geomagn. Aeronom.*, 9, 351–353, 1969.
91. Hamza, A. M. and St-Maurice, J.-P., A turbulent theoretical framework for the study of current-driven E region irregularities at high latitudes: Basic derivation and application to gradient free situations, *J. Geophys. Res.*, 98, 11587–11599, 1993.
92. St-Maurice, J.-P. and Hamza, A. M., A new nonlinear approach to the theory of E region irregularities, *J. Geophys. Res.*, 106, 1751–1759, 2001.
93. Hamza, A. M. and Imamura, H., On the excitation of large aspect angle Farley–Buneman echoes via three-wave coupling: A dynamical system model, *J. Geophys. Res.*, 106, 24745–24754, 2001.
94. Sudan, R. N. and Keskinen, M. J., Theory of strong turbulent two-dimensional convection of low-pressure plasma, *Phys. Fluids*, 22, 2305–2314, 1979.
95. Sudan, R. N., Unified theory of type I and II irregularities in the equatorial electrojet, *J. Geophys. Res.*, 88, 4853–4860, 1983.
96. Blaunstein, N., Tsedilina, E. E., Mishin, E. V., and Mirzoeva, L. I., Drift spreading and stratification of inhomogeneities in the ionosphere in presence of an electric field, *Geomagn. Aeronom.*, 30, 656–661, 1990.
97. Blaunstein, N., Milinevsky, G. P., Savchenko, V. A., and Mishin, E. V., Formation and development of striated structure during plasma cloud evolution, *Planet. Space Sci.*, 41, 453–460, 1993.
98. Blaunstein, N., The character of drift spreading of artificial plasma clouds in the middle-latitude ionosphere, *J. Geophys. Res.*, 101, 2321–2331, 1996.
99. Drake, J. F., Mubbrandon, M., and Huba, J. D., Three-dimensional equilibrium and stability of ionospheric plasma clouds, *Phys. Fluids*, 31, 3412–3424, 1988.
100. Rozhansky, V. A. and Tsendin, L. D., *Transport Phenomena in Partly Ionized Plasma*, Philadelphia, PA: Taylor and Francis, 2001.

Chapter 13

Influence of Rocket Burn and Launch on Ionospheric Radio Communication

13.1 Overview

The reaction of the ionosphere on radio propagation is a very important issue in radio communication between terrestrial antennas and space vehicles (SVs) or satellites. The problem of wave propagation and scattering in the ionosphere has become increasingly important in recent decades because the near-the-Earth environment, including the atmosphere, ionosphere, and magnetosphere, all play a significant role in determining the efficiency and quality of the land–rocket, land–satellite, or satellite–satellite communication channel.

Thus, an increasing demand by mobile–satellite networks designed to provide a global radio coverage using constellations of low- and medium-Earth-orbit satellites is observed. Such systems form regions called mega-cells (see definitions in Refs. [1–3]) that consist of a group of cochannel cells, and clusters of spot beams from each satellite, which move rapidly across the Earth's surface. In these scenarios only local environmental features such as ionospheric, atmospheric, and terrestrial, which are very close to the desired radio path, contribute significantly to the propagation process. This is because the same propagation effects, such as multiray reflection, diffraction, and scattering of radio waves, occur in over-the-Earth communication links [4–11].

However, unlike land communication channels, prediction of ionospheric communication channels tends to be highly statistical in nature, because coverage across very wide areas must be taken into consideration, while still accounting for large variations due to local environmental features. The reader can find more detailed information in Refs. [1–3].

Additional features in ionospheric plasma are observed during burn, launch, and flight of rocket and SV (see Chapter 12) caused by numerous plasma instabilities generated by exhausts, rocket plume, and gas jet, finally creating global and large-scale plasma structures surrounding the moving SV and in its wake, which are filled by moderate- and small-scale plasma disturbances during further evolution. These features can significantly destroy ionospheric communication links, causing multiplicative noises yielded by fast and slow fading of radio signals. The sources of such kinds of fading can be diffraction, attenuation, scattering, and reflection phenomena caused by the whole spectra of plasma irregularities (turbulences) generated around the moving SV.

As mentioned in Chapter 1 (and thereafter in the chapters that follow), the above effects arising in the vicinity of the moving SV depend on its velocity V_0 with respect to the velocity V_k of the external exhaust particle fluxes injected by the rocket engine or plume that impinge upon it. However, as shown in Ref. [12], the nature of the plasma disturbances in the vicinity of the moving SV is not just defined by the ratio V_0/V_k. The ratio of the linear dimensions of the rocket to the electronic and the multi-ionic Debye lengths, or to the Larmor radii of the electrons and the multi-ions is also important (all definitions have been introduced briefly in Chapter 1; see also Refs. [7,12]). Therefore, as given in Chapter 1, the rockets were differentiated by large, moderate, and small scale, depending on whether the dimensions were larger, at the same level or smaller than the plasma characteristic lengths mentioned earlier.

There are other problems when we try to understand the wave processes and resonances taking place not only in the near-the-Earth background plasma but also those effects that are related to plasma disturbances created in the multi-component exhaust plasma surrounding the moving rocket. To understand these phenomena, as mentioned in Chapter 12, the reader needs to go deeper into complicated mathematical derivations and we, as authors, should use the corresponding complicated mathematical tool. Because this aspect is out of the scope of this book, in Section 13.2 we will briefly introduce the reader to the types of frequency resonances and their branches that can occur in the perturbed plasma. Despite the fact that this subject is fully presented in the literature for two-component plasma (electrons and protons), in this section we give an explanation of the situation occurring in the multicomponent ion–electron plasma that, as was shown in Chapters 11 and 12, can be generated by the interaction of the rocket's exhaust gas and liquid with ambient ionospheric plasma. In Section 13.3, types of plasma waves are briefly described, which can be observed propagating through the ionospheric plasma. They are ultralow frequency (ULF)—from 1 to 100 Hz, extremely low frequency (ELF)—from 100 Hz to 3 kHz, very low frequency (VLF)—from 3 to 30 kHz, low frequency (LF)—from 30 to 300 kHz, medium frequency (MF)—from 300 kHz to 3 MHz, high frequency (HF)—from 3 to 30 MHz, very high frequency (VHF)—from 30 to 300 MHz, and so forth. In Section 13.4, the guiding effects of the ionospheric radio propagation in ELF/VLF/LF ranges are considered. Section 13.4 deals with radio wave amplitude attenuation and fading for various frequency bands, where absorption of ULF/VLF radio waves and MF/HF radio waves is discussed separately. Section 13.5 deals with the capture of radio waves of MH/HF ranges into Earth–ionosphere waveguide. In Section 13.6, the effects of global and large-scale ionospheric structures on radio propagation of waves at MF/HF/VHF range are presented. Here, the effects of moderate- and small-scale irregularities (turbulences), which fill global and large-scale plasma cavities and enhancements, on MF/VHF bands radio wave propagation are described. Section 13.7 deals with MF/VHF range radio wave amplitude attenuation and fading, as well as its phase deviations occurring in the perturbed ionospheric regions filled by large-, moderate-, and small-scale plasma irregularities. Section 13.8 presents information on Doppler frequency shifting and its spread for radio waves of various frequency bands, caused by turbulent motion and drift of the perturbed plasma. In Section 13.10, depolarization of radio signals of various frequency bands are discussed. Here, we combined the two approaches proposed in the literature to find changes of the polarization ellipse shape and its spatial displacement at the depolarization angles, as well as the depolarization loss caused by plasma disturbances of various origins in ambient regular ionosphere. Finally, in Section 13.11, a summary is presented on the main effects caused by the global-, large-, moderate-, and small-scale plasma perturbations, which can be supposed as the main candidates for probing radio signal attenuation and fading, occurring in the proximity of the moving rocket or SV, as well as inside the turbulent wake around and behind the SV.

13.2 Spectral Branches, Resonances, and Waves in the Ionospheric Plasma

As mentioned earlier, different physical approaches and mathematical models should be introduced for explanation of the subject, depending on the types of plasma waves, spectra, and resonances occurring in the lower, middle, and outer ionospheric plasma.

13.2.1 Key Definitions and Characteristics

First, we present some fundamental definitions of plasma particle characteristics, which are important for further description of the subject. Thus, the conditions of plasma waves and instability excitation, their nature, and the plasma-oscillation spectra all vary essentially, depending on when and whether the plasma is weakly magnetized and ionized (in the D-layer and the lower E-layer) or partly magnetized and ionized (in the upper E-layer to the upper F-layer). In other words, it is important to understand when the average energy of the ambient magnetic field ($\sim B_0^2/2$) exceeds the cumulative kinetic energy of the plasma particles $\sim N(T_e + T_{ieff})$ and plasma density is sufficient to account for the collisions between charged particles with respect to their interactions with neutral molecules (atoms). If we now introduce the Alfven velocity $V_A = c(\Omega_H/\Omega_0) = c/n_A$, the above statement reduces to (in our further explanation we use notations that were introduced in Chapter 12):

$$\left(V_A/V_{eff}\right)^2 \gg 1 \text{ or } \approx 1, \text{ or } \ll 1, \tag{13.1}$$

where:

$V_{eff} \approx \sqrt{(T_e + T_{ieff})/M_{ieff}}$ is the effective velocity of the ions in the multicomponent plasma (as given in Ref. [7])

Here, several kinds of light and heavy ions with effective mass $M_{ieff} = K^{-1}\sum_{k=1}^{K} M_i^k$ and effective temperature $T_{ieff} = K^{-1}\sum_{k=1}^{K} T_i^k$ are expressed in energy values (in Joule), $\Omega_H = ZeB_0/M_{ieff} = 2\pi f_{Hi}$ is the Larmor (or gyro-) frequency of ions, $\Omega_0 = \sqrt{N(Ze)^2/\varepsilon_0 M_{ieff}} = 2\pi f_{0i}$ is the Langmuir (or own) frequency of the "effective" ion in plasma, ε_0 is the permittivity of the ambient ionospheric plasma.

This means that the plasma is weakly magnetized, if $f_{He} \leq \nu_{em}$ and $f_{Hi} \ll \nu_{im}$, and strongly magnetized, if $f_{He} \gg \nu_{em}$ and $f_{Hi} \geq \nu_{im}$. Weakly ionized plasma satisfies the following constraints: $\nu_{ei} \ll \nu_{im} \ll \nu_{em}$, whereas strongly ionized plasma satisfies converse conditions: $\nu_{ei} > \nu_{em} \gg \nu_{im}$.

Here, we also suggest that plasma is quasineutral, that is, $N_e \approx \sum_{k=1}^{K} N_i^k = N$. As for electron component of ionospheric plasma, excitation of its Langmuir oscillations is defined by the frequency $\omega_{0e} = \sqrt{N_e^2/\varepsilon_0 m_e} = 2\pi f_{0e}$; all other parameters were introduced in previous chapters. Using these definitions, we can now introduce the reader to the resonances occurring in ionospheric plasma, types of plasma waves passing through the ionosphere, and the corresponding frequency spectra.

13.2.2 Spectral Branches and Resonances in Two-Component Cold Plasma

To present all types of plasma waves, their spectra, and possible resonances of probing radio waves passing the perturbed ionosphere, we need to analyze their spatiotemporal (or frequency) dispersion, that is, of warm, collision, multicomponent ion–electron plasma, for which the corresponding electromagnetic, hydrodynamic, and kinetic equations are very complicated for presentation in our explanation, because they are out of scope of this book. On the other hand, as promised to the reader in Chapter 12, in order to give phenomenological knowledge and preliminary understanding of the aspects of radio propagation in the perturbed ionosphere, we use a ray theory of plane plasma wave propagation in the magnetically active anisotropic plasma consisting of, as was done in Chapter 12, one kind of the "average" ions and electrons. For a general description of the problem, we take the reader to the excellent books, where this subject is described more in detail for one-, two-, and multicomponent plasmas [4,7–9,13], respectively.

As mentioned in Refs. [4–12], the key parameter (for understanding the problem of wave generation in plasma and its resonances) is the refractive index n that determines the phase velocity $v_{ph} - cn$ of the waves and their group velocity $v_{gr} - c/[\partial(n\omega)/\partial\omega]$. Generally speaking, this parameter is complex, consisting of its real part, the index n, the imaginary part, and the parameter of attenuation χ, that is, $\tilde{n} = n - i\chi$.

Variations of the refractive index in the frequency domain describe dispersion of plane electromagnetic waves in the plasma under consideration. For analyzing the refractive index of multicomponent collisional magnetized warm plasma, first we describe two-component plasma, consisting of effective ions and electrons. However, as mentioned in Chapter 12, even in two-component "electron–ion" warm, magnetized, collision plasma the corresponding formulas for the refractive index and for the parameter of attenuation are extremely cumbersome. Further, the problem becomes increasingly complicated as the number of ions in such composite plasma increases. We start with the most simple case of two-component, electron (e^-) and proton (e^+) collisionless cold plasma, for which we ignore the influence of the ions in the charged particle interactions.

13.2.2.1 Two-Component Collisionless Cold Plasma

For the considered scenario the ion component and how much batches of ions are present in the plasma are not important.

For such a plasma, we can present the real and imaginary parts of the complex refractive index as follows [4,7]:

$$n = \sqrt{\frac{\varepsilon}{2} + \sqrt{\frac{\varepsilon^2}{4} + \left(\frac{\sigma}{f}\right)^2}}, \qquad (13.2a)$$

$$\chi = \sqrt{-\frac{\varepsilon}{2} + \sqrt{\frac{\varepsilon^2}{4} + \left(\frac{\sigma}{f}\right)^2}}. \qquad (13.2b)$$

Here, σ is the conductivity and ε is the permittivity of plasma.

In such plasma, there are three resonant branches that are related to the quasilongitudinal waves in the two-component plasma, to which three main frequencies of plasma wave oscillations correspond directed along the wave vector **k**: $f_1 = \omega_1/2\pi$, $f_2 = \omega_2/2\pi$, and $f_3 = \omega_3/2\pi$ as shown in Figure 13.1 rearranged according to Ref. [7], using the notations introduced in Chapter 12.

The first branch defined by $f_1(\theta)$ is called *high frequency*, which is changed in limits: (1) the lower limit is $f_1 = f_0 \equiv \omega_0/2\pi$, for $\theta = 0$; and (2) the upper limit is $f_1 = \left(f_0^2 + f_{He}^2\right)^{1/2}$, $f_{He} = \omega_H/2\pi$, for $\theta \to \pi/2$. The latter frequency is usually called the *upper-hybrid frequency* and is denoted by f_{UFR}. It defines

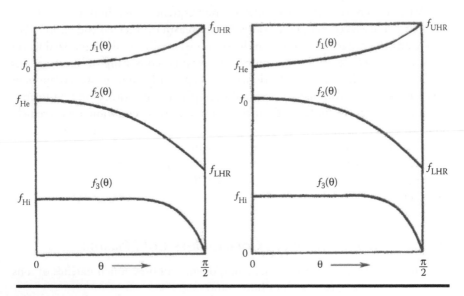

Figure 13.1 Resonance branches of two-component plasma.

HF resonances in plasma [4,7–9] and can be presented for a collisionless cold plasma as [4,7–9]:

$$f_{UHF} \approx \sqrt{\left(f_0^2 + f_{He}^2\right) + \left(f_{0i}^2 - \frac{f_{He}f_{Hi}}{1 + f_{He}^2 / f_{0e}^2}\right)}. \tag{13.3}$$

The second resonance branch determined by the frequency $f_2 = \omega_2/2\pi$ is called a *low frequency*. Its upper limit is f_H, if $f_H > f_0$ (Figure 13.1, left panel), or f_0, if $f_H < f_0$ (Figure 13.1, right panel). However, for $\theta \to \pi/2$, f_2 limits to zero, which is not correct physically and shows that the ion under consideration is required to be introduced. Accounting for the influence of the ion component of plasma, it was found that the lower limit for $\theta \to \pi/2$ is called the *lower-hybrid resonance frequency* and is denoted by f_{LHR}. It can be presented as given in Refs. [4,7–9], which is as follows:

$$f_{LHF} \approx \sqrt{\left(f_{He} \cdot f_{Hi}\right)\frac{1 + f_{Hi}^2/f_{0i}^2}{1 + f_{He}^2/f_{0e}^2}} \approx \sqrt{\frac{\left(f_{He} \cdot f_{Hi}\right)}{1 + f_{He}^2/f_{0e}^2}}. \tag{13.4}$$

The third branch of resonance frequencies, defined by $f_3 = \omega_3/2\pi$, is called the ELF or VLF branch, which also takes into account the interactions between electrons and ions [4,7]. Its upper limit is $f_3 = f_{Hi} = \Omega_H/2\pi$ and low limit is (Figure 13.1), which is as follows:

$$f_3 \approx \frac{f_{Hi}\cos\theta}{\sqrt{\left(f_{0i}^2 + f_{Hi}^2\right)/f_{0e}^2}} \to 0 \text{ for } \theta \to \frac{\pi}{2}. \tag{13.5}$$

Hence, low-hybrid frequency lies at the boundary of LF branch of plasma resonances and ELF/VHF branches of plasma resonances, and is changed in limits from f_{Hi} to zero.

13.2.2.2 Two-Component Collisional Cold Plasma

Introducing now collisions between charged particles, electrons, and ions, and overcoming the problem of multiple ion components, we introduce, as in Chapter 12, the "effective ion" and consider the two-component "electron–ion" collision plasma containing one type of *effective* ion with charge $q_i = Ze^+$. If so, the ratio of charged particle masses, electron, and one kind of ion (m_e/M_i) is changed on $m_e/M_{ieff} = m_e/\sum_{k=1}^{K}(\alpha_k/M_{ik})$, where the relative ion densities, $\alpha_k = N_{ik}/N_e$, were introduced to satisfy condition of quasineutrality of a multicomponent plasma: $\sum_{k=1}^{K} \alpha_k = 1$ (or $\sum_{k=1}^{K} N_{ik} = N_e \equiv N$).

In this more general case, rearranging different complicated presentations of the square of n dependence on frequency f, usually described in the literature [4,7–9], we simplified this presentation and show now the rearranged dispersion diagram in Figure 13.2. Here a square of the refractive index, $n^2(f)$, versus frequency f is plotted for propagation conditions nearly parallel to the vector $\mathbf{B}_0$ of ambient magnetic field (called quasilongitudinal propagation; left panel) and nearly across to the vector $\mathbf{B}_0$ (called quasitransversal propagation, right panel).

As clearly seen, for $n < 0$, the plasma waves are damped fast and cannot transport energy—they are of no use. To explain all notations, we briefly introduce the reader to the main definitions and parameters indicated in Figure 13.2. Thus, at frequencies f_0, f_1, and f_2 that correspond to condition $n = 0$, the cutoff frequencies are called and conditions of total reflection of the waves in plasma are defined. At $n^2 \to \infty$ the resonance frequencies are gyro-frequencies

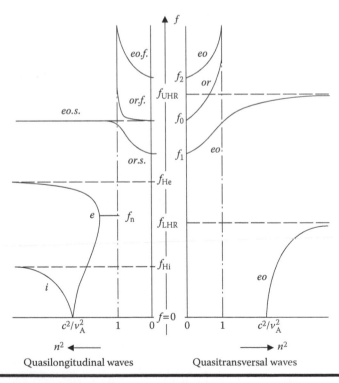

Figure 13.2 Refractive index of two types of waves, longitudinal and transversal, vs their frequency, and their branches in the spectral diagram.

(or Larmor) of electrons (f_{He}) and ions (f_{Hi}), respectively, and the lower and upper hybrid frequencies (f_{LHR} and f_{UHR}).

At these resonance frequencies, the corresponding waves can be excited within the ionosphere. At $f\to0$, the refractive index limits to the value n_A, defined earlier as an Alfven refractive index determined by the Alfven velocity V_A.

There are two kinds of waves for longitudinal propagation: ion and electron whistler (or acoustic) waves. The first is indicated by "i" and the second—by "e." The frequency of the first wave (also called *i-wave* [14]) lies between zero frequency and f_{Hi}, whereas the frequency of the second wave (also called the *e-wave* [14]) is limited by f_{Hi} (lower bounded) and by f_{He} (upper bounded). These waves are circular polarized: the *i*-wave rotates counterclockwise and the *e*-wave rotates clockwise, if one looks in the direction of ambient magnetic field. We will discuss their spectral properties and resonances later.

In the case of transversal propagation, the two waves can be created in the ionosphere due to double refraction caused by magnetic properties of ionospheric plasma. Finally, this effect yields two different waves having different refraction indexes, n_1 and n_2, and propagating with different phase velocities, $v_{ph1} = c/n_1$ and $v_{ph2} = c/n_2$. In literature, these waves are called the *ordinary* and *extraordinary* waves. In Figure 13.2 (right panel), we denoted them by *or* and *eo*, respectively, when they travel transverse to ambient magnetic field. At the same time, when they travel along the ambient magnetic field, we also differentiate them on "fast" (denoted in the left panel of Figure 13.2 by *or.f.* and *eo.f.*, respectively) and "slow" (denoted in the left panel of Figure 13.2 by *or.s.* and *eo.s.*, respectively).

As is seen from Figure 13.2 (right panel), for $n^2 < 1$, that is, for waves with frequencies f satisfied by the constraint $f_{0e} > f$, the refraction index become imaginary and in this frequency domain the wave cannot propagate—a strong and fast damping occurs at the thin layer (called *skin layer*) within the plasma. When $f = f_{0e} \equiv f_0$ (Figure 13.2), $n_1 = 0$, the radiated frequency of the ordinary wave equals the plasma frequency (e.g., the Langmuir frequency of electrons). So, we see that, if the ordinary wave propagates in the ionosphere in regions with increasing plasma concentration along directions of increase of the density, N, we get $f = f_{0e} \equiv f_0$, and the full reflection of the fast ordinary wave occurs. The same phenomenon occurs for $f = f_1$, when full reflection is observed for slow ordinary wave.

As for the extraordinary wave, for $f > f_{Hi}$ slow extraordinary wave propagates, whereas for $f > f_{He}$ there are two roots of the refractive index that satisfy the condition $n_2 = 0$, when

$$f_1 = \frac{f_{0e}}{\sqrt{1 + f_{He}/f_{0e}}}, \qquad (13.6a)$$

for the slow extraordinary waves and

$$f_2 = \frac{f_{0e}}{\sqrt{1 - f_{He}/f_{0e}}},$$ (13.6b)

for the fast extraordinary waves where full reflection of these waves occurs. All these scenarios are described by illustrations in Figure 13.2.

Both waves, *ordinary* and *extraordinary* are polarized elliptically and their fields have *counterclockwise* (e.g., left-hand) rotation (coincides the direction of ion rotation in the magnetic field) and *clockwise* (e.g., right-hand) rotation (coincides the direction of electron rotation in the magnetic field) looking toward the direction of ambient magnetic field, respectively. In points of full reflection, these two waves become linearly polarized, normal to the direction of each other. In the vicinity of the region of f_0, f_1, and f_3, it is possible to excite both extraordinary and ordinary quasilongitudinal waves (see Figure 13.2, left panel).

In the *collisional plasma*, when collisions between charged particles, electrons, and ions, cannot be ignored, both components of the incident wave weakly penetrate beyond the regions, where $f_{0e} < f$ or $f_{0e} < f \pm f_{He}$. At these boundaries reflection of such waves occurs practically. Accounting for collisions that can be defined by ν_{eff}, we can present the conditions of full reflection of both kinds of waves as:

$$n^2 - \chi^2 = 0.$$ (13.7)

It is important to note that to characterize the effects of refraction for both types of waves, the key parameter, such as a critical number of collisions (in units s^{-1}, the same as a frequency) can be introduced according to Refs. [4,7–9]:

$$\nu_{kr} = \pi f_{He} \frac{\sin^2 \theta}{\cos \theta} \left(s^{-1} \right).$$ (13.8)

As was found in Refs. [4,7–9], that, if $\nu_{eff} = \nu_{kr}$ and $f = f_{0e}$, the refractive indexes of the ordinary and extraordinary waves are equal (i.e., $n_1 = n_2$) and the character of polarization and attenuation are also the same (i.e., $\chi_1 = \chi_2$ and $\eta_1 = \eta_2$). Here, $\eta_{1,2} = E_{\Lambda 1,2}/E_{\perp 1,2}$ are the parameters of polarization of two waves with the components of the wave's electric field directed across and perpendicular to k (whose direction is usually denoted by $\parallel$). In this case, only one type of wave propagates in the ionosphere; the double ray refraction is totally absent.

In the case of quasilongitudinal propagation $(\theta \to 0)$ for both types of waves, ordinary and extraordinary, if $v_{kr} \ll \sqrt{(f^2 - f_{0e}^2)^2 / f^2 + v_{eff}^2}$, we get, following Refs. [4,7–9], but using our notations:

$$\left. (n - i\chi)^2 \right|_{1,2} = 1 - \frac{f_{0e}^2}{f\left[(f - iv_{eff}) \pm f_{He} \cos\theta \right]}. \tag{13.9}$$

Here, both waves are polarized circularly in opposite directions of rotations (see earlier). For $v_{kr} \gg \sqrt{(f^2 - f_{0e}^2)^2 / f^2 + v_{eff}^2}$, the conditions of propagation in plasma are close to quasitransversal $(\theta \to \pi/2)$ and we get for two types of waves the following expressions rearranged from Refs. [4,7], using our notations:

$$\left. (n - i\chi)^2 \right|_1 = 1 - \frac{f_{0e}^2}{\left[\left(f^2 - iv_{eff}f \right) + \left(f^2 - f_{0e}^2 - iv_{eff}f \right)\cot^2\theta \right]}, \tag{13.10a}$$

$$\left. (n - i\chi)^2 \right|_2 = 1 - \frac{f_{0e}^2}{\left[\left(f^2 - iv_{eff}f \right) - \frac{\left(f^2 \cdot f_{He}^2 \right)\sin^2\theta}{\left(f^2 - f_{0e}^2 - iv_{eff}f \right)} \right]}. \tag{13.10b}$$

Here, both waves are nearly linearly polarized in the two cross-perpendicular directions. In the general case of wave propagation, $0 < \theta < \pi/2$, when $v_{eff} \neq 0$, the waves are elliptically polarized, that is, changed only with changes of the ratio $\mu = f_{0e}/f$.

We should also note that for arbitrary angle from $0 < \theta < \to \pi/2$ in the frequency range $f \ll f_{Hi}^{(1)}$, the refractive indexes for the ionic and electronic waves equal:

$$n_1 = \frac{n_A}{\cos\theta}, \ n_2 = n_A \tag{13.11}$$

and

$$f_1 = \frac{kV_A}{2\pi}\cos\theta, \ f_2 = \frac{kV_A}{2\pi}. \tag{13.12}$$

Here, as above, n_A is the Alfven refractive index, $k = 2\pi f / v_{ph}$ is the magnitude of the wave vector $\mathbf{k}$, V_A is the Alfven's wave velocity, which are simply related as $n_A = c/V_A = f_{0i}/f_{Hi}$.

General dependences of $n_i^2(f)$ and $f(k)$ for the whole spectra of frequencies, from ELF to HF range, with the resonance regions $f_1(\theta)$, $f_2(\theta)$, and $f_3(\theta)$, shown in Figure 13.1, are presented in Figure 13.3a and b, respectively,

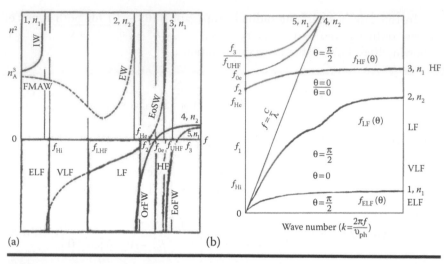

Figure 13.3 **(a) Refractive index of different types of waves propagating in two-component cold plasma vs their radiated frequency. (b) Different types of waves generated in plasma vs the wave number k and their branches in the spectral diagram.**

for the cold magnetized collisionless two-component plasma, containing electrons and one kind of ion, whose mass, velocity, and temperature were defined earlier. The corresponding dispersion equations for such plasma are sufficiently cumbersome, for which the low-hybrid frequency can be presented approximately as follows:

$$f_{LHF} = \sqrt{\frac{m_e}{M_{eff}} \frac{f_{He}^2 f_{0e}^2}{\left(f_{He}^2 + f_{0e}^2\right)}}. \tag{13.13}$$

Finally, we present in Figure 13.3a and b the results obtained in Refs. [4,7–9] for five branches, denoted by digits, from 1 to 5, which we rearranged by using our notations.

First, we saw that propagation of two types of ELF waves is possible: one of them is ionic or slow magnetically acoustic wave (for $0 \le f \le f_{Hi}$), which for $f \ll f_{Hi}$ is transferred into Alfven wave, and for $f \to f_{Hi}$, it transfers into cyclotron wave, called *ionic whistlers* (IW) (see Figures 13.3a, curve 1).

Second, depending on the angle θ between $\mathbf{k}$ and $\mathbf{B}_0$, the resonance frequencies of this ionic whistler branch are changed from $f = 0$ to $f = f_{Hi}$, where $n_1(f_{Hi}) \to \infty$.

For $f > f_{Hi}$, the refractive index becomes imaginary—the ionic whistlers cannot propagate and finally vanish (see Figures 13.2 and 13.3a). At the same time, the other ELF wave for $f < f_{Hi}$ arises, called the fast magnetoacoustic wave, which for $0 \leftarrow f \ll f_{Hi}$ sometimes is called modified Alfven wave. After $f = f_{LHF}$ this electronic wave vanishes $[n_2(f_{LHF}) \to \infty]$ if $\theta = \pi/2$ (called the lower hybrid resonance or LHR), or when $f = f_{He}$ for $\theta = 0$ (called the electronic gyroresonance) (see Figure 13.3b). The range of frequencies $f_{LHF} \leq f \leq f_{He}$ corresponds to the resonance branch $f_2(\theta)$ depicted in Figure 13.1. These waves correspond to the LF range and they are usually called the *electronic whistlers* (see Figure 13.3a).

Third, resonances are absent at the range $f_{Hi} \leq f \leq f_{LHF}$ for cold plasma, but for nonisothermic plasma, where $T_e > T_{ieff}$, excitation of the three branches of longitudinal *ion-acoustic waves* at the range of $0 \leq f \leq \sqrt{f_{Hi}^2 + f_{0i}^2}$ or at $f \leq \sqrt{(kV_s)^2 + f_{Hi}^2}$ is possible, whereas in Chapter 12, $V_s = \sqrt{(T_e + T_i)/M_i}$ is the velocity of the nonisothermal sound.

Fourth, $f > f_{He}$ corresponds to the range of middle and high frequencies (MF and HF), Here, two extraordinary waves—slow and fast (see also Figure 13.3a), and one ordinary fast wave can be excited. In cold plasma, as was shown in Figure 13.1, only one resonance branch $f_1(\theta)$ arises that corresponds to a range $f_{0e} \leq f \leq f_{UHF}$, if $f_{0e} > f_{He}$, or this resonance range corresponds to $f_{He} \leq f \leq f_{UHF}$, if $f_{0e} < f_{He}$ (Figures 13.2 and 13.3a). For $\theta = 0$ this branch is limited to the *plasma resonance*, and for $\theta = \pi/2$—to the *upper hybrid resonance* (as is clearly seen from Figures 13.2 and 13.3). As also follows from the presented illustrations that HF waves in cold plasma do not have resonances in the range of $f_{He} \leq f \leq f_{0e}$. Moreover, HF waves cannot propagate in the range of $f_{He} \leq f \leq f_2$, because here $n_2^2 < 0$; f_2 is the frequency where $n_2^2 = 0$ for slow extraordinary wave (EoSW, Figure 13.3a, curve 3). At the range of $f_{0e} \leq f \leq f_3$, the fast extraordinary wave (EoFW) also cannot propagate ($n_2^2 < 0$), where f_3 is the frequency in which $n_1^2 = 0$ for fast extraordinary wave (Figure 13.3a, curve 5).

Frequencies f_2 and f_3, where $n_2^2 = 0$ and $n_1^2 = 0$, respectively, are known as the *cutoff frequencies*, that can be defined as [4,7,9]:

$$f_2 = -\frac{f_{He}}{2} + \left(f_{0e}^2 + \frac{f_{He}^2}{4} \right)^{1/2} , \quad f_3 = \frac{f_{He}}{2} + \left(f_{0e}^2 + \frac{f_{He}^2}{4} \right)^{1/2} . \quad (13.14)$$

Thus, as follows from the above formulas and from the illustrations in Figure 13.3a and b, the refractive index of a collisional cold plasma containing just one kind of ion goes to zero only in the range of MF waves (for $f_{He} < f \leq f_2$) and only in the range of HF waves (for $f_{0e} \leq f \leq f_3$).

13.3 Wave Types in Nonisothermal Multicomponent Ionospheric Plasma Perturbed by Rocket Engine and Plume Exhaust Gases

As mentioned in Chapter 12 and from the beginning of this chapter, the situation with burnout and release of exhaust gases and rocket plume and jet can dramatically change the ambient cold ionospheric plasma containing electrons and a maximum of two kinds of ions. In this scenario, we should now consider multicomponent "electron–ion" collision warm plasma and show which types of waves can be excited in such a composite plasma, in addition to those observed from illustrations presented in Figures 13.1 through 13.3. Finally, we expect and compare these additional plasma features with those observed experimentally and described in Chapter 3. Because such multicomponent plasmas yield many new features, we will analyze them separately.

13.3.1 Multicomponent Isothermal Collisional Plasma

We consider the spectral properties and the corresponding behavior of plasma refractive index containing several kinds of ions and electrons. Such a multicomponent plasma can be generated in conditions occurring in the actual ionosphere during rocket burn and launch. It is related to the corresponding exhaust gases and "dirty plasma" interaction with ambient cold ionospheric plasma (see Chapter 12). As mentioned in the literature [4–13], such multicomponent plasma plays an essential role in the propagation of ULF/ELF waves ($0 \leq f \sim f_{\mathrm{Hi}}$), VLF waves ($f_{\mathrm{Hi}} < f < f_{\mathrm{LHR}}$), including the beginning of the range of LF waves, where $f_{\mathrm{Hi}} \ll f < f_{\mathrm{He}}$.

As shown by numerous investigations, the presence of several kinds of ions changes expressions of the refractive index and the attenuation parameter only in the range of frequencies comparable with the ion gyro-frequencies $f_{\mathrm{Hi}}^{(k)}$, where k determines the amount of different ions in the ionospheric plasma. In other words, the dispersive properties of plasma can be dramatically changed only for waves with frequencies $f^2 \ll f_{\mathrm{He}}^2$, that is, in the ULF/ELF and VHF ranges, including part of LF waves with frequencies located in the vicinity of LH. Because general formulas are very complicated, considering the limit cases $\theta = 0$ and $\theta = \pi/2$ allows us to understand, without loss of generality, the main properties of these waves and their resonances arising in such multicomponent plasma.

When $\theta = 0$ is the refractive index for the ordinary wave (called in ULF/ELF/VLF frequency range—*ionic wave* or *slow magnetoacoustic wave* [4,7–9]) it can be written as follows:

$$n_1^2 = 1 - \frac{f_{0e}^2}{f \cdot f_{He}} + \frac{\left(f_{0i}^{(1)}\right)^2}{f\left(f_{Hi}^{(1)} - f\right)} + \frac{\left(f_{0i}^{(2)}\right)^2}{f\left(f_{Hi}^{(2)} - f\right)} + \dots \qquad (13.15a)$$

For the same case of $\theta = 0$, the refractive index for the extraordinary wave (called in ULF/ELF/VLF frequency range—*electronic wave* or *fast magnetoacoustic wave* [4,7–9]) can be written as:

$$n_2^2 = 1 - \frac{f_{0e}^2}{f \cdot f_{He}} - \frac{\left(f_{0i}^{(1)}\right)^2}{f\left(f_{Hi}^{(1)} - f\right)} - \frac{\left(f_{0i}^{(2)}\right)^2}{f\left(f_{Hi}^{(2)} - f\right)} - \dots \qquad (13.15b)$$

Let us assume, for example, that plasma contains three kinds of ions—light, medium, and heavy—for which the following constraints are valid: $f_{Hi}^{(1)} > f_{Hi}^{(2)} > f_{Hi}^{(3)}$ for the gyro-frequencies (or Larmor), $f_{0i}^{(1)} > f_{0i}^{(2)} > f_{0i}^{(3)}$ for the Langmuir frequencies, and $f_{LHF}^{(1)} > f_{LHF}^{(2)} > f_{LHF}^{(3)}$ for low-hybrid resonance frequencies of the three kinds of ions present in plasma that can be defined by Equation 13.4 for each kind of ion.

In this case, essentially changes in the f_{LHF}, defining *LHRs*, with respect to that described by Equation 13.4, should be expected and a more complicated formula for the three kinds of ions describe this frequency branch now:

$$f_{LHF} = f_{He}\sqrt{\frac{\left(f_{0i}^{(1)}\right)^2 + \left(f_{0i}^{(2)}\right)^2 + \left(f_{0i}^{(3)}\right)^2}{f_{0e}^2 + f_{He}^2}}. \qquad (13.16)$$

Moreover, the so-called ion–ion hybrid frequencies can be risen from a complicated dispersal equation of the following form, quadratic to f^2 [7]:

$$\left(f^2 - f_{LHF12}^2\right)\left(f^2 - f_{LHF23}^2\right)$$

$$\equiv f^4 - f^2\left(f_{LHF12}^2 + f_{LHF23}^2\right) + f_{LHF12}^2 \cdot f_{LHF23}^2 = 0, \qquad (13.17)$$

where:

$$f_{LHF12}^2 + f_{LHF23}^2 = \frac{\left(f_{0i}^{(1)}\right)^2\left[\left(f_{Hi}^{(2)}\right)^2 + \left(f_{Hi}^{(3)}\right)^2\right] + \left(f_{0i}^{(2)}\right)^2\left[\left(f_{Hi}^{(1)}\right)^2 + \left(f_{Hi}^{(3)}\right)^2\right] + \left(f_{0i}^{(3)}\right)^2\left[\left(f_{Hi}^{(1)}\right)^2 + \left(f_{Hi}^{(2)}\right)^2\right]}{\left(f_{0i}^{(1)}\right)^2 + \left(f_{0i}^{(2)}\right)^2 + \left(f_{0i}^{(3)}\right)^2} \qquad (13.18a)$$

$$f_{\text{LHF12}}^2 \cdot f_{\text{LHF23}}^2 = \frac{\left(f_{0i}^{(1)}\right)^2 \left(f_{Hi}^{(2)}\right)^2 \left(f_{Hi}^{(3)}\right)^2 + \left(f_{0i}^{(2)}\right)^2 \left(f_{Hi}^{(1)}\right)^2 \left(f_{Hi}^{(3)}\right)^2 + \left(f_{0i}^{(3)}\right)^2 \left(f_{Hi}^{(1)}\right)^2 \left(f_{Hi}^{(2)}\right)^2}{\left(f_{0i}^{(1)}\right)^2 + \left(f_{0i}^{(2)}\right)^2 + \left(f_{0i}^{(3)}\right)^2}. \tag{13.18b}$$

In many cases, namely, for the conditions of outer ionosphere, Equations 13.18a and b can be simplified and presented in the simple symmetrical form:

$$f_{\text{LHF12}}^2 = \frac{\left(f_{0i}^{(1)}\right)^2 \left(f_{Hi}^{(2)}\right)^2 + \left(f_{0i}^{(2)}\right)^2 \left(f_{Hi}^{(1)}\right)^2 + \left(f_{0i}^{(3)}\right)^2 \left(f_{Hi}^{(1)}\right)^2}{\left(f_{0i}^{(1)}\right)^2 + \left(f_{0i}^{(2)}\right)^2 + \left(f_{0i}^{(3)}\right)^2}, \tag{13.19a}$$

$$f_{\text{LHF23}}^2 = \frac{\left(f_{0i}^{(1)}\right)^2 \left(f_{Hi}^{(2)}\right)^2 + \left(f_{0i}^{(2)}\right)^2 \left(f_{Hi}^{(1)}\right)^2 + \left(f_{0i}^{(3)}\right)^2 \left(f_{Hi}^{(1)}\right)^2}{\left(f_{0i}^{(1)}\right)^2 + \left(f_{0i}^{(2)}\right)^2 + \left(f_{0i}^{(3)}\right)^2}. \tag{13.19b}$$

For three-component plasmas, the two values of the cut-off frequency in the refractive index frequency dependence should be added at the ELF range together with three zero points at the MF and HF ranges, as shown in Figure 13.3a. This peculiarity is clearly seen from plots presented in Figure 13.4, in which for the case of three-component plasmas, accounting for collisions between the charged particles, electrons, and ions, the dependences of refraction indexes n_1^2 and n_2^2 are presented schematically for ELF to VHF ranges (left panel—$\theta = 0$; right panel—$\theta = \pi/2$) rearranged from Refs. [7,8]. For the case of $\theta = \pi/2$, as was shown in Refs. [4,7–9], the refractive index of the ionic (ordinary) wave is $n_1 < 0$.

As clearly seen from Figure 13.4 (right panel), the refractive index for electronic (extraordinary) wave $n_2 > 0$ only at the discrete ranges of frequency band:

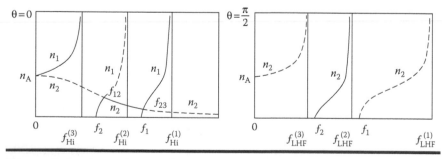

Figure 13.4 Dependence of refractive index on resonance frequencies of propagating waves in three-component cold plasma.

$0 \le f \le f_{\text{LHF}}^{(3)}$, $f_2 \le f \le f_{\text{LHF}}^{(2)}$, $f_1 \le f \le f_{\text{LHF}}^{(1)}$, where f_1 and f_2 have been defined earlier and presented in Figure 13.1.

13.3.2 Nonisothermal Warm Plasma

In addition to that discussed in Section 13.2, in the real nonisothermal warm ionospheric plasma for the whole frequency range of $f > f_{0i}$, excitation of longitudinal *electron-acoustic* waves propagation up to two upper limits is possible: $f > f_{0e}$ and $f > f_{\text{UHF}}$. Thus, for $T_e > T_i$, the whole range of frequencies can be filled by additional plasma resonances, which leads to the existence of additional HF waves, namely, *electron* and *ion cyclotron waves*, and so forth. Let us briefly discuss such peculiarities of warm collisional nonisothermal plasma.

In warm plasma generated by high-temperature plume or engine exhaust gases, which expand into the cold ambient ionospheric plasma, heating of the ambient electrons and ions takes place. As shown in Refs. [6,7,10], accounting for spatial dispersion of the plasma due to particle movements, because ions are much more inertial (with respect to electrons, in $\sim M_i / m_e$ times) the effect of the motion of ions can be completely neglected and just the motion of plasma electrons is accounted for. Despite this fact, mathematical equations describing a spatial dispersion (via wave vector $\mathbf{k}$), that is, describing relations between the refractive index $n^2(f, \mathbf{k})$ and the frequency $\omega(\mathbf{k})$, are difficult to use.

It is interesting to note that this procedure can be simplified for the wide spectra of HF ($f \ge f_{0e}$), gyrotropic or Langmuir ($f \ge f_{\text{He}}$), and low frequency (LF, $f_{\text{LHF}} \le f \le f_{\text{He}}$) waves. At the same time, as was shown in Refs. [4,7,8], a new branch of $n_3^2(f)$ can be arisen that describes the properties of potential *longitudinal* HF waves. As was shown in Ref. [7], for $f \to f_{\text{He}}$, this additional refractive index (to well-known $n_1^2(f)$ and $n_2^2(f)$ presented in Figures 13.2 through 13.4) becomes negative, that is,

$$n_3^2(f) \approx -\frac{\left(f^2 - f_{\text{He}}^2\right)}{2\beta_e^2 f_{\text{He}}^4 \cos^2 \theta} < 0. \tag{13.20}$$

This means that at the electron gyro-frequency a longitudinal HF wave is highly damped and, therefore, cannot propagate through plasma. However, in the vicinity of the second-harmonic gyro-resonance, $f = 2f_{\text{He}}$, in a sufficiently wide range of frequencies $n_3^2(f) > 0$, the wave does not disappear. In this case, it can be found that [7]:

$$n_3^2(f) \approx -\frac{\left(f^2 - 4f_{\text{He}}^2\right)\left[3f^2 - f_{0e}^2(4 - \cos^2 \theta)\right]}{12\beta_e^2 f^2 f_{0e}^2 \sin^4 \theta} \tag{13.21}$$

Here, $\beta_e^2 = (V_e / c)^2$, V_e is the thermal velocity of electrons in the warm plasma, c is the light speed; all other parameters were defined earlier.

For $\theta \to 0$, we get $n_3^2(f) \to \infty$. This could not confuse us, because a magnetic field does not act upon electrons moving in the direction of the vector $\mathbf{B}_0$. Thus, for $\theta = 0$, the magnetic field does not influence the branch describing propagation of the longitudinal resonant waves. The same results were obtained in Refs. [4–9] for isotropic nonmagnetized plasma (i.e., when $\mathbf{B}_0 = 0$). In the latter case, the longitudinal *Langmuir* waves mentioned earlier can propagate through plasma with slight damping when their phase velocity V_{ph} exceeds the thermal electron velocity $V_e = \sqrt{2T_e / m_e}$ ($T_e \gg T_i$), or, equivalently, when their wavelength λ is much greater than the electron Debye radius $D_e = \sqrt{\varepsilon_0 T_e / N_e e^2} = V_e / 2\sqrt{2}\pi f_{0e}$. As earlier, here the temperatures $T_{e,i}$ are measured in energy units (Joule).

For $\theta = \pi/2$, for frequencies quite far from the usual resonance frequencies of collisionless cold plasma, f_{0e} and f_{He}, for which $(f^2 - f_{0e}^2)/f^2 \gg \beta_e^2$ and $(f^2 - 4f_{He}^2)/f^2 \gg \beta_e^2$, the refractive index that defines the *HF resonance* branches for plasma waves can be found in the following manner:

$$n_3^2(f) \approx -\frac{\left(f^2 - f_{UHF}^2\right)\left(f^2 - 4f_{He}^2\right)}{1.5\beta_e^2 f^2 f_{0e}^2}. \tag{13.22}$$

Now we raise a question: What kind of situation occurs in such a multicomponent warm plasma where ELF, VLF, and beginning of LF waves, that is, according to the above definitions, lie at the ranges of $0 < f \le f_{Hi}$, $f_{Hi} \le f \le f_{LHF}$, and $f_{LHF} \le f < f_{He}$, respectively? As follows from our previous discussions, in this wide LF spectral range, one of the branches of the refractive index n_1, as in the case of cold plasma, will be described as ELF *ion* waves: Alfven, hydromagnetic, and ion whistlers (see Figures 13.2 through 13.4). For the second branch of the refractive index n_2, it describes ELF, VHF, and LF *electron* waves: modified Alfven, fast magnetoacoustic, and electron whistlers. In nonresonance regions, the expressions of these refractive indexes are not changed much from those corresponding to the HF-range. They both account for the thermal motion of charged particles, electrons (and lesser ions). Therefore, we will not present them here.

Concluding this section we outline that accounting for the multicomponent, magnetized, nonisothermal warm plasma, a whole spectra of HF and slow-frequency wave excitation can be expected during creation of such "dirty" plasma by the rocket burn and launch, namely, ion-acoustic, electron-acoustic (fast and slow), magnetoacoustic, electron and ion cyclotron waves, and so forth. Some of these waves are the participants in the processes that accompany the rocket burn and launch. These waves were recorded during special rocket experiments described in Chapters 4 through 10, as evidence of the theoretical explanations carried out in this section.

Moreover, as is shown in Figure 11.29 (see Chapter 11), the existence of several light to heavy species in the exhaust gaseous cloud produced by engine burn can generate several types of waves and, therefore, are the most likely candidates for additional molecule generation due to aeronomic and chemical reactions between the exhaust gas species and the neutral and ion components of the background ionosphere, whose emission lines can be registered spectrally by using a special spectroscopic technique. As shown in Figure 11.29, this technique of using CCD optical camera yields the registration of such ingredients and molecules inside the exhaust cloud streaming from the rocket engine as happened in the registration of the molecule AlO after interaction with aluminum (Al) [one of the main species of the exhaust release] with background oxygen (O) during aeronomic reactions occurring between the exhaust gas and the neutral molecular component of the ionospheric plasma at considerable altitudes (see Chapter 11 and the corresponding references therein).

13.4 Ionospheric Guiding Effects for ELF/VLF/LF-Range Radio Wave Propagation

It is very interesting to note that ULF-band waves with frequencies lying at the range of $0 < f < 100$ Hz cannot propagate in the Earth–ionosphere propagation channel due to their extremely long wavelengths. At the same time, waves that coincide with the ELF/VLF/LF range are the most likely candidates for Earth–ionosphere waveguide propagation. Let us briefly describe this phenomenon and analyze the effects that can bring plasma global enhancements or cavities (*holes*) in such long wave propagation all over the world.

The waves under consideration with wavelengths varying from a few kilometers to hundreds of kilometers can reflect and attenuate at the ground surface as well as in the ionosphere at altitudes of about 90 km (lower *E*-layer) during nighttime and close to 70 km of the ionospheric *D*-layer during daytime conditions. Moreover, if the processes of reflection and attenuation on the ground depend on frequency, distance, and the features of the terrain [14–16], in the case of the ionosphere they additionally depend in a complicated manner on diurnal and nocturnal time periods, seasons, latitude, and conditions of the ambient electrical and magnetic fields [4–16].

For ELF waves (with a range of 100 Hz $< f <$ 3 kHz), it is impossible to use description ray theory and the corresponding ray-tracing numerical technique for their propagation effects. For them only waveguide mode propagation

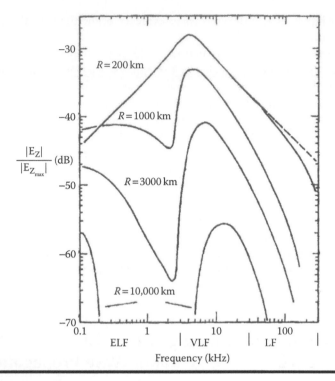

Figure 13.5 Guiding effects of the ionosphere on ELF-, VLF-, and LF-wave propagation.

theory is appropriate. To show this peculiarity of ELF waves, we present spectral amplitude of the vertical electric field of the steric waveform from an ideal return stroke (e.g., the "natural source" of ELF waves) in Figure 13.5, recording during special experiments described in Ref. [14]. This signal is affected by the Earth–ionosphere waveguide, characterized by its transfer function T during its propagation within the waveguide. Curves for various distances from the "natural source" are presented versus radiated frequency (kilohertz). It is clearly seen for a distance of 200 km that only the ground wave is important (when $T = 1$) [14].

Its attenuation on the ground in the whole spectra of frequencies is marked by the solid upper curve, whereas the dashed line corresponds to a perfectly conducting smooth terrain. For ELF waves, the zero mode begins to dominate for a window around 100 Hz, as seen from Figure 13.5. The VLF and LF waves propagating within the Earth–ionosphere waveguide can be described both by mode and ray theory. At distances less than 100 km (depending on radiated frequency) ray theory is appropriate. Thus, the ground-reflected ray and the ray

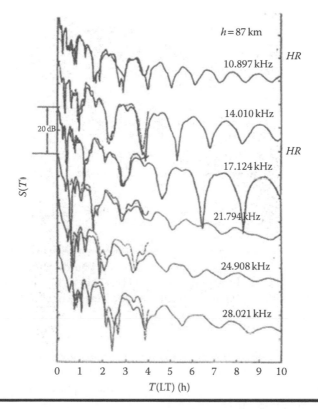

Figure 13.6 **Waveguide mode propagation due to reflection from the ground surface and the ionospheric layer of the lower ionosphere at $h = 87$ km.**

reflected from the lower ionosphere (*D*-layer or *E*-layer), after the first hopping, interfere with each other.

Figure 13.6, rearranged from Ref. [8], shows an interference pattern of the total ray amplitude obtained experimentally along the trace Hawaii–San Diego (USA) versus distance from the source for different radiated frequencies. The reflection height was approximately $h = 87$ km. A deep interference occurs between ground-reflected and ionosphere-reflected modes. For higher waveguide modes, the interference pattern indicates a sinusoidal-like height profile with the field minima and maxima at specific altitudes.

A similar interference pattern created by the ground-reflected and the first-hop waveguide mode was observed experimentally by using a 15-kHz transmitter [8], which is shown in Figure 13.7 (rearranged from Ref. [14]). Here, a plot of the signal amplitude (solid curve) and phase (dashed curve) is presented versus the distance between the transmitter and the receiver. The transmitter was a

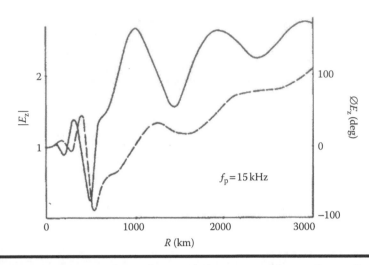

Figure 13.7 Distance dependence of the waveguide modes radiated by electrical dipole at a frequency of 15 kHz. The right vertical axis indicates variations of the signal amplitude, the left axis shows signal phase variations.

vertical electric dipole radiating at 15 GHz. The topside of such a waveguide is about 70–80 km, corresponding to altitudes of the daytime *D*-layer, where the plasma concentration N (electrons and ions; plasma is quasineutral [10]) is about 10^8 and 10^9 m^{-3}, respectively.

At nighttime ionosphere, the topside of the waveguide varies from about 85 to 95 km, which corresponds to the same values of plasma concentration. A deep minimum of the signal amplitude at a distance of about 500 km between both terminals is evident. It can be explained as the result of vector summation of the last interference minimum of the ground wave mode and the first sky (i.e., ionospheric) wave mode. The complicated oscillation form of this minimum can be explained as the result of interference between the first and the second ionospheric wave modes. Moreover, at this point of minimum, the two kinds of the reflected wave modes, ground and sky, have approximately the same amplitudes and are out of phase.

As also seen from the illustrations plotted in Figures 13.5 and 13.6, the multihop waves reflected from the ionosphere as topside of the waveguide become essential beyond the distance of about 1000 km. In the case of VLF/LF-range waves, the attenuation of multihop modes increases with mode number. At the same time, mode number one (principal waveguide mode) is weakly attenuated and dominates at larger distances of up to 1000 km and more.

To introduce an appropriate theory of wave mode propagation, we note that for ELF/VLF-range waves their wavelength λ is on the same order of the vertical height of the Earth–ionosphere waveguide, H. If so, one can use mode theory similar to that described in microwave propagation within an electric waveguide, that is, describe wave propagation by the waveguide modes. This we will present later using the approach described and analyzed in Ref. [3].

13.5 Capturing MF/HF-Range Radio Waves into Waveguide Created in Layered, Perturbed Ionosphere

We should note, unfortunately, that the model of the Earth–ionosphere waveguide with the ideal smooth "walls," described in the previous section, is satisfied only for ELF/VLF/LF-range waves, because all "roughness features" located at the ground surface or irregular plasma structures existing in the ionosphere, have dimensions that are much less than the wavelengths of the waves under consideration. The regular plasma layers of the ionosphere lead, in principle, to guiding of the MF- and the HF-radio waves extremely long distances around the Earth [6,17]. In other words, the layered quasiregular ionosphere works for such waves as the waveguide where radio waves (called *waveguide modes*) reflect from the upper and lower boundaries of the ionospheric layers and therefore can be channeled due to multiple reflections at long distances. In the literature, this effect was defined as *long-range* ionospheric radio propagation [6,17–21].

On the other hand, global or large-scale irregular plasma structures, created during the rocket exhaust gases and plasma clouds injections or during heating of the ionosphere (called *cavities*, or *holes*, and enhancements), or caused by other natural phenomena, described in Chapter 12, can be expected as candidates for channeling of MF and HF waves between the ionospheric D- and E-layer, E- and F-layer, that can be considered as layered global plasma-guiding structures.

Thus, as mentioned in Chapter 12, global and large-scale plasma structures with moderate- and small-scale plasma irregularities filled inside, can work as natural or artificially induced *lenses*, capturing the MF/HF-range waves into the ionospheric waveguide and channeling their energy to long distances.

It was shown both theoretically and experimentally that to obtain the guiding effect for probing radio waves propagating inside such plasma-layered waveguides, it is necessary for the guiding modes to satisfy the following constraints [6,17]:

$$\omega \cdot \sec\beta_c < \omega_{e0}(Z_1),$$

$$\omega \cdot \sec\beta_c < \omega_{e0}(Z_2), \qquad (13.23)$$

where, as earlier:

ω is the angular frequency of probing radio wave

ω_e is the frequency of background plasma oscillations defined above

β_c is the critical angle (see its meaning below)

Z_1 and Z_2 are the lower and upper boundaries of the plasma waveguide shown in Figure 13.8

Constraints (13.23) state the possibility of the waveguide modes to reflect periodically from bottom-side to top-side "walls" during mode propagation inside the plasma-layered waveguide. These waveguide modes can propagate to long distances with sufficiently low attenuation of about 5—10 dB for each 1000 km [6,17].

If we now assume that the regular layered ionosphere is spherically symmetric containing layers with horizontal dimensions (with respect to $\mathbf{B}_0$, that is, oriented vertically to the ground surface) larger than the wavelength, the radio wave transmitted under the angle α_0 from the ground transmitter with frequencies satisfied constraints (13.23), cannot achieve the critical height Z_c located in

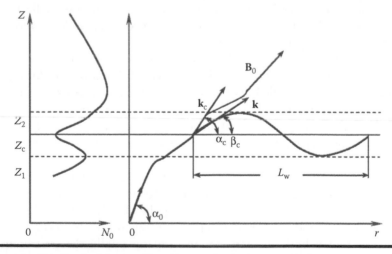

Figure 13.8 The height profile of ionospheric density (left panel) and geometry of radio wave capture in the layered, perturbed ionosphere.

the middle axis of the waveguide under angles β_c. Capturing of probing waves energy in the ionospheric waveguide can only be achieved if rays arriving at the waveguide under angles $\alpha_0 > \beta_c$ are oriented to achieve the angle α_c inside the waveguide according to constraints (13.23), as shown in Figure 13.8. This additional displacement of the angle α from α_0 to α_c can be achieved due to moderate-scale or small-scale irregularities filling the global structures, holes, or enhancements which cause scattering effects of the probing radio waves. Here, the angle between the passing ray and the corresponding axis is denoted as α_c (Figure 13.8).

We note now that the corresponding theory is correct only if spatial gradients of parameters of plasma waveguide are sufficiently smooth, that is, when

$$\frac{d(\ln \xi)}{dx} \cdot \frac{\Lambda_\xi}{2\pi} \ll 1, \tag{13.24}$$

where:

ξ is the parameter of the plasma waveguide

Λ_ξ is the scale of spatial oscillations of wave modes inside the layered waveguide between Z_1 and Z_2 (see Figure 13.8)

Because Λ_ξ is generally larger than 100 km in the channel between the E-layer and the F-layer; the characteristic scale of the horizontal gradients inside the plasma waveguide, L_w (see Figure 13.8), can exceed 500 km [6,17], that is, they are much larger than the wavelengths of the modes under consideration.

Following Ref. [6,17], we now assume that radio waves arriving and entering into the ionosphere under the angle α cross the waveguide along the path between and (see Figure 13.8). The virtual middle axis of the waveguide is placed at height Z_c. The ray trajectory inside the waveguide is characterized by the critical angle $\beta_c(Z_c)$. Then, the boundaries of the capturing wave mode are defined by the following condition [6,17]:

$$\varepsilon(z) = \varepsilon(Z_c)\cos^2\beta_c, \ Z_1 < Z_c < Z_2. \tag{13.25}$$

Scattering of radio waves at the small-scale plasma irregularities, that fill the holes or enhancements, leads to the turning of reflected waves relative to vector k_c. All waves, for which the angle of scattering is larger than the difference, $\alpha_c - \beta_c$, will be captured inside the ionospheric waveguide. The angle β_c can be determined from the following relationship [6,17]:

$$\varepsilon(Z_1) = \varepsilon(Z_c)\cos^2\beta_c, \tag{13.26}$$

which yields for small β_c to

$$\beta_c = \left[\frac{\omega_{e0}^2(Z_1) - \omega_{e0}^2(Z_c)}{\omega^2} \right]. \tag{13.27}$$

Following Refs. [6,17], after straightforward derivations made in Ref. [12], we can obtain the total coefficient of capturing of radio waves as a sum of waveguide modes, each of which will now propagate along the specific guiding trajectory characterized by the angle β_c:

$$\Gamma_\Sigma(\alpha) = - \int_{-\beta_c}^{\beta_c} \cos\beta_c' \, d\beta_c' \int_0^{2\pi} \Gamma_0(\alpha, \beta_c', \phi) D(\phi, \alpha') d\phi, \tag{13.28}$$

where:

ϕ is the angle in the azimuth (e.g., horizontal) plane measured from the x-axis

$$D(\phi, \alpha') = \begin{cases} 1 & -\beta_c < \alpha' < \beta_c \\ 0 & |\alpha'| > \beta_c \end{cases}. \tag{13.29}$$

Equation 13.28 determines the coefficient of capturing of radio waves into a layered ionospheric channel with frequencies less than the maximum usable frequencies (MUF), that is, the LF/MF/HF-range waves. From Equation 13.28, it follows that the possibility of channeling radio waves in the ionospheric layered waveguide is determined by the intensity of small-scale plasma irregularities.

Thus, the rocket's engine ignites or plume exhaust releases with a great probability that can be suggested as the most likely candidate of a source, which artificially generates such plasma irregularities in the ionosphere with outer dimensions $\ell < \omega/c$. This source artificially excites the corresponding plasma instabilities and, as final product, generates small-scale turbulences filling the global holes or enhancements, evidently observed experimentally during rocket burn and launch described in Chapter 12.

Moreover, according to research reflected from Refs. [6,17–22], the optimal capturing effect of radio waves with frequencies of 1–30 MHz occurs in the case when, at the ionospheric altitudes of the waveguide, plasma irregularities exist with horizontal dimensions (to geomagnetic field $\mathbf{B}_0$) $\ell_\| = 100$–500 m. Because the relative plasma disturbances of plasma irregularities, obtained experimentally and described in Chapter 12 are on the order $\delta N/N_0 \sim 10^{-3}$–$10^{-2}$, the capture of probing MF/HF waves (for f lies from 1 to 30 MHz) into the ionospheric waveguide can be resolved for the coefficient $\Gamma_\Sigma(\alpha)$ of 10^{-4}–10^{-2} [6,17–23].

13.6 Propagation of HF/VHF/UHF-Range Radio Waves in Perturbed Ionosphere

For more than 50 years, the problem of propagation of radio waves of MF to VHF bands, from moderate- to high-range frequencies (i.e., with wavelength from tens of meters to a few kilometers) through the ionosphere were intensively investigated by radiophysical societies. It was found that such MF- to VHF-range waves in the regular nonperturbed ionosphere propagate along the plane of a great circle (see Refs. [6–8,17]).

13.6.1 Effects of Global and Large-Scale Irregular Structures on MF/HF/VHF-Range Radio Wave Propagation

Existence of global or large-scale irregular structures, as *holes* (*cavities*) or enhancements, having horizontal scale L, larger than the first Fresnel zone, $d_F \approx \sqrt{\lambda R}$, where λ is the wavelength and R is the distance from the ground facility to the inhomogeneous area of the ionosphere containing such irregular structures, the MF/HF/VHF-range waves change their direction from the plane of great circle. Moreover, additional investigations carried out during the recent decades have shown that even weakly ionized large-scale irregular plasma structures can increase the thickness of the reflecting layer deviate the radio wave trajectory due to horizontal and vertical changes of the height of the reflected layer. As follows from Refs. [6,17], radio wave reflection can occur only for frequencies lesser than MUF, f_{MUF}. For $f > f_{MUF}$, radio waves freely penetrate the ionosphere due to refraction effects, and later become "transparent" for such radio waves. In other words, the thickness of the ionospheric layers becomes transparent for waves with $f > f_{MUF}$. However, this effect strongly depended on the angle of incidence θ (or on the grazing angle, $\alpha = 90° - \theta$) of the radio waves at the ionospheric layer filled by large-scale holes or enhancements. Finally, it was shown that the reflected signals create the interference structure of reflected waves due to the oscillatory character of their strength, resulting in additional irregular modulation of the radio wave amplitude at the receiver. To understand the effects of large-scale irregular structures in the layered spherically symmetric ionosphere, the extended models of ray tracing, analytical and numerical (compared to the well-known Hamilton equations), were created [24–31], in order to examine the influence of losses on ray paths for HF/VHF propagation through the E- and F-layers of the natural regular ionosphere, taking into account their real plasma density profiles and the corresponding collision frequencies (see definitions in Chapter 1).

Here, we draw the reader's attention to the behavior of radio waves under consideration, propagating in the perturbed ionosphere in conditions of strong changes of the plasma density height profiles during the heating process that could be caused by rocket releases, engine ignitions, and plume exhausts, described in Chapter 11. Despite the fact that for lossless and smooth plasma height profiles the problem is understood quite well [24–31], new effects were found [32] for irregular profiles generated in the case of local heating, such as the following:

- Significant enhancements [or depletions (holes), depending on the height of heating, see Chapter 12] of plasma density
- The corresponding irregular plasma structures in its altitudinal profile, facilitating reflections of probing radio waves passing such disturbed regions at frequencies from hundreds of kilohertz to tens of megahertz, even up to hundreds of megahertz
- Changes of the ray trajectories (creating one-hop and more radio paths)
- Changes in the field strength of probing radio waves, passing the perturbed ionospheric layers.

To show this for scenarios related to rocket burn and launch, we will use those profiles that were obtained in Chapter 12 for specific situations in local heating presented in Figures 12.4 through 12.7. Because the redistribution of plasma density and changes in its profile during local heating, described in Chapter 12, lead to increasingly complex dispersion relations between plasma characteristics and parameters, and to very complicated wave equations describing the ray-tracing effects, it was proposed in Ref. [32] that a numerical model and the corresponding ray equation formalism based on the extended Hamilton equations by introducing the dispersion relation $F(\mathbf{k}, \omega, \mathbf{r}, t) = 0$ in the complex space for considering whether the propagation vector $\mathbf{k} = (k_x, k_y, k_z)$ and the angular frequency ω may be complex [29–31]:

$$\mathbf{v} = \frac{d\mathbf{r}}{dt} = -\frac{\partial F/\partial \mathbf{k}}{\partial F/\partial \omega},$$

$$\frac{d\mathbf{k}}{dt} = \frac{\partial F/\partial \mathbf{r}}{\partial F/\partial \omega} + i\boldsymbol{\beta}, \qquad (13.30)$$

$$\frac{d\omega}{dt} = \frac{\partial F/\partial t}{\partial F/\partial \omega} + i\mathbf{v} \cdot \boldsymbol{\beta}.$$

Here, $\mathbf{r} = (r_x, r_y, r_z)$ is the local vector in the space domain, t is the time which also serves to define a parameter along the ray trajectory, and $\boldsymbol{\beta}$ is the vector that guarantees a real group velocity ($\mathbf{v}$) along the ray trajectory defined as

$$\beta = -\left[\mathrm{Re}\left(\frac{\partial \mathbf{v}}{\partial \mathbf{k}} + \frac{\partial \mathbf{v}}{\partial \omega} \cdot \mathbf{v} \right) \right]^{-1} \cdot \mathrm{Im} \left[\begin{array}{c} \dfrac{\partial \mathbf{v}}{\partial \mathbf{k}} \cdot \dfrac{\partial F / \partial \mathbf{r}}{\partial F / \partial \omega} - \dfrac{\partial \mathbf{v}}{\partial \omega} \cdot \dfrac{\partial F / \partial t}{\partial F / \partial \omega} \\ + \dfrac{\partial \mathbf{v}}{\partial \mathbf{r}} \cdot \mathbf{v} + \dfrac{\partial \mathbf{v}}{\partial t} \end{array} \right]. \tag{13.31}$$

The Earth's magnetic field in the proposed model was approximated by a localized dipole field with the dipole located near the center of the Earth. In these conditions, the Appelton–Hartree (sometimes called Appelton–Lassen [24–27]) dispersion relation, obtained for the cold, nonisothermal, collisional, magnetized ionosphere, that is, for the actual ionospheric conditions, occur in the *E*- and the *F*-regions of the ionosphere. It can be presented as follows [29–31]:

$$F = k^2 - \frac{\omega^2}{c^2}\left[1 - \frac{X}{(1 - iZ) - \dfrac{Y_{\parallel}^2}{2(1 - X - iZ)}} \pm \sqrt{\frac{Y_{\parallel}^4}{4(1 - X - iZ)^2} + Y_{\perp}^2} \right] = 0, \tag{13.32}$$

where:

$$X = \omega_{0e}^2 / \omega^2, \quad Z = \nu_e / \omega, \quad Y_{\parallel} = |\mathbf{Y}|\cos\psi, \quad Y_{\perp} = |\mathbf{Y}|\sin\psi \tag{13.33}$$

are the longitudinal and transverse components of the normalized geomagnetic field vector **Y** with respect to the direction of **k**, and $|\mathbf{Y}|$ is the magnitude of vector **Y**.

As shown in the previous sections, the electron plasma frequency denoted by ω_{0e} can be defined as $\omega_{0e}^2 = e^2 N_{0e} / (m_e \varepsilon_0)$, the gyro-frequency of electrons, ω_{eH}, was defined in Chapter 1, m_e and e are the mass and the charge of the electron component of plasma, respectively. Here, $\nu_e = \nu_{em} + \nu_{ei}$ is the collision frequency related to losses, ε_0 the permittivity of plasma, $N_{0i} \approx N_{0e} \equiv N_0$ the background plasma density (plasma is quasineutral), and ψ *is* the angle between vectors **k** and **Y** (Figure 13.9; according to Refs. [31,32]).

The derivatives in the ray, Equations 13.30 and 13.31, accounting for Equations 13.32 and 13.33, are quite complicated for analytically solving for the real model of ionospheric plasma, whose parameters vary with height and the conditions of the rocket release experiments (see Chapters 11 and 12). The corresponding data for numerical computation of the problem were obtained from the corresponding solutions of the heating problem. The variations of ray paths, according to frequency, grazing angle α, collision frequency and plasma density profile, before, during, and after the heating process, were examined for the *E*- and *F*-layers of the ionosphere.

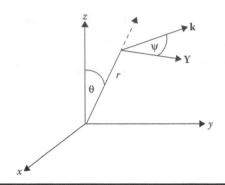

Figure 13.9   Geometry of the problem of radio wave propagation through the ionosphere.

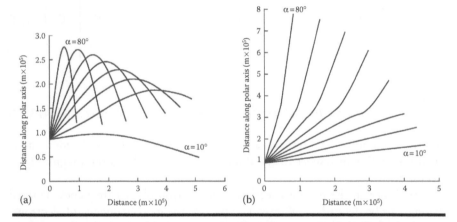

(a)

(b)

Figure 13.10   (a) Reflection of rays from the regular ionosphere at $f = 10$ MHz. (b) Reflection of rays from the regular ionosphere at $f = 30$ MHz.

First, we present some computations of the ray trajectories for the nonperturbed regular middle-latitude ionosphere. The results are shown in Figure 13.10a and b for the ordinary mode of probing waves for $f = 10$ and 30 MHz) versus grazing angles $\alpha(\alpha = 90° - \theta, \theta = 10°, 20°, \ldots, 80°)$.

The figures show that the frequency of probing radio waves passing through the ionosphere is at the same level or higher than the MUF of the investigated ionospheric layers for the ordinary mode, that is, for $f < 20$ MHz [31,32]. As seen from the illustrations in the quasiregular nondisturbed ionosphere for probing waves of 10 MHz, that is, less than 20 MHz, and for small grazing angles $\alpha = 10° - 30°$, only the one-hop propagation is observed due to reflection from the ionospheric quasiregular plasma layers. With increasing frequencies of the

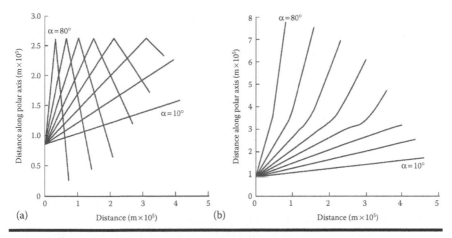

Figure 13.11 (a) Trajectories of the reflected and refracted waves from the perturbed ionosphere for various grazing angles $\alpha \in [10°, 80°]$ and for $f = 10$ MHz. (b) Trajectories of the reflected and refracted waves from the perturbed ionosphere for various grazing angles $\alpha \in [10°, 80°]$ and for $f = 30$ MHz.

probing waves (from 10 to 30 MHz), the possibility of obtaining propagation channels with higher grazing angles (up to $\alpha = 50°–60°$) increases (Figure 13.10). This effect is recognized in commonly used propagation ionospheric channels for frequencies of probing waves that are close to the MUF for each layer of the ionospheric.

Now we will consider the perturbed ionosphere caused by local heating of plasma by the hot exhaust gases of the rocket engine or plume. Thus, taking into account a virtual experiment with the local heating of plasma described in detail in Chapter 12 (see Figure 12.4), the same ray trajectories derived according to the above equations can be obtained. Results are shown in Figure 13.11a and b for the same variations of grazing angles of probing radio waves with an angular step $\Delta\alpha = 10°$, and for different frequencies $f = 10$ (a) and 30 MHz (b), respectively.

Following from Figure 13.11b, by using local heating of the ionospheric *F*-layer, a long-distance propagation channel can be identified due to reflections from disturbed ionospheric regions for higher frequency bands, up to the upper HF range for a sufficiently wide range of grazing angles $10° < \alpha < 30°$.

As clearly seen from illustrations in Figure 13.11a and b, by using a modified profile of the ionospheric plasma during its heating larger ranges of one-hop propagation can be obtained for probing waves with frequencies close to MUF ($f < 20$ MHz) of the undisturbed ionosphere for a wide range of grazing angles α.

Further, the existence of a heat-induced ionospheric region at a wide range of ionospheric altitudes allows using the HF probing waves for one-hop propagation at long distances due to reflections from the disturbed ionospheric sporadic layers. Thus, for frequencies 10 MHz $< f <$ 30 MHz of probing waves, efficient one-hop propagation for distances more than 500 km and for a wide range of grazing angles $10° < \alpha < 40°–50°$ can be achieved (Figure 13.11a).

It is evident from illustrations in Figure 13.11b that by using local heating of the ionospheric *F*-layer, a long-distance propagation channel can be identified due to reflections from disturbed ionospheric regions for higher frequency bands, up to the upper HF range for a sufficiently wide range of grazing angles $10° < \alpha < 30°$.

We also present the effects of the complicated heating of the ionosphere, from the *E*- to the *F*-layer of the ionosphere, described in Chapter 12. Thus, this heating regime is close to the heating effects observed during several steps of the rocket burnout and ignition of the rocket engines or plume exhaust releases, illustrated by Figure 12.5 (see Chapter 12). The results of the computations are presented in Figure 13.12a and b for various grazing angles of probing waves and the frequencies from the VHF range.

It is evident that with the increase of probing wave frequency for a significantly wide range of grazing angles, $\alpha < 30°$, the ray trajectories of one-hop propagation are greater and can reach 1000 km and more. The example

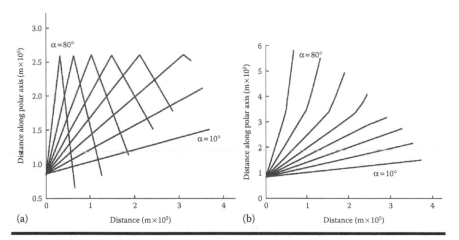

(a)   (b)

Figure 13.12 **(a) Trajectories of the reflected and refracted waves from the perturbed ionosphere for various grazing angles** $\alpha \in [10°, 80°]$ **and for** $f = 100$ **MHz. (b) Trajectories of the reflected and refracted waves from the perturbed ionosphere for various grazing angles** $\alpha \in [10°, 80°]$ **and** $f = 300$ **MHz.**

shown earlier reflects the fact that the trajectory of radio waves through the ionosphere itself depends on the absorption and the plasma concentration gradients involved.

It appears that for the strong plasma depletions (or cavities) or enhancements, generated in the disturbed ionosphere due to the process of forcing out plasma during a relatively short heating period, the presence of the Earth's magnetic field does not have any appreciable effect on the radio wave trajectories. A more important parameter for radio propagation of HF/VHF ranges at extremely long distances, caused by reflections from disturbed ionospheric regions, is the grazing angle and operational frequency of the ground-based facilities. Thus, for different kinds of heating processes, local (see Figure 12.4) or global irregular (see Figure 12.5), it has been found that with an increase of grazing angles for the same operational frequency of probing waves or with an increase of operational frequency for the same grazing angles, the effect of channeling of one-hop modes in the Earth–ionosphere waveguide channel becomes increasingly unrealistic.

13.6.2 Effects of Moderate- and Small-Scale Irregularities Filled Perturbed Ionospheric Regions on HF/VHF/UHF-Range Radio Wave Propagation

We define moderate-scale irregularities, following Refs. [6,10,17,33–35], as those whose dimension, $\ell_\perp$ transverse to $\mathbf{B}_0$, is smaller than the radius of the first Fresnel zone, that is, $\ell_\perp \leq \sqrt{\lambda R}$ (see definitions of the corresponding parameters in the previous paragraph). Usually, $0.1\,\mathrm{km} < \ell_\perp < 10\,\mathrm{km}$. Hence for small-scale plasma irregularities, their dimensions do not exceed several tens of meters.

The difference between these irregularities compared to the large-scale ones, is that they not only change the structure of the signal reflected from the ionosphere during vertical and oblique sounding of the ionospheric layers filled by such kinds of irregularities, but also cause changes in the phase and amplitude of radio signals passing through the disturbed ionospheric regions, namely, those created by the rocket burn and launch of orbital and geostationary satellites. To understand these phenomena, we further present some analytical estimations following Refs. [10,6,17,33,34].

Fluctuations of signal amplitude A are caused by the so-called *ray-refraction effect* based on geometric-optics approximation, where the corresponding deviations of the normalized wave amplitude $\Delta A/A$ or intensity $\Delta I/I$ (because $\Delta I/I \approx 2\Delta A/A$, for $\Delta A/A \ll 1$ [6,17]) are determined by changes of the ray tube in the transverse to ray direction $\mathbf{r}(x, y)$ (i.e., to vector $\mathbf{k}_0 \| \mathbf{R}$). These changes can be defined by the value of $2(\partial\theta_R / \partial r_\perp)R$. Here θ_R is the angle of refraction,

R the distance between the transmitter and receiver antennas introduced above, and $k_0 = 2\pi / \lambda$. We can define θ_R according to Refs. [6,10,17] as:

$$\theta_R = k_0^{-1}(\partial \Delta \Phi / \partial r) \approx 2\Delta \Phi / k_0 l_\perp. \tag{13.34}$$

Here, $\Delta \Phi$ is the phase deviation of the wave caused by irregularities filling the ionospheric layer, and $l_\perp$ the characteristic scale of irregularities in the plane orthogonal to the trajectory of ray propagation. Then, the mean-square value of wave intensity I fluctuations, $F_I = \langle (\Delta I)^2 \rangle / \langle (I)^2 \rangle$, approximately equals [10,6,17]:

$$F_I \approx q^2 (\Delta \Phi_0)^2, \quad (\Delta \Phi_0)^2 = \langle (\Delta \Phi)^2 \rangle. \tag{13.35}$$

The wave parameter q defines the ratio of the area of the first Fresnel zone, $\sim \lambda R$, and of the area filled by an irregularity, that is, $q = \lambda R / S_I = 4R / k_0 l_\perp^2 \ll 1$. From Equation 13.35 it follows that for $q^2 (\Delta \Phi_0)^2 \ll 1$ (i.e., for $l_\perp^2 > \lambda R$), fluctuations of the recorded signal amplitude and phase are sufficiently small. Now it is clear that the large-scale irregularities with $l_\perp > \sqrt{\lambda R}$, affected by the influence of "Fresnel filtering" [10,17], cannot change the conditions of radio propagation in HF/VHF/UHF bands. Only moderate- and small-scale irregularities, for which $l_\perp < \sqrt{\lambda R}$, should be taken into account because they affect large fluctuations of amplitude and phase of radio signals transmitted via the ionospheric channels.

The value $(\Delta \Phi_0)^2$ shows that the radio signal phase changes can be easily estimated, if we assume now that after passing only one single irregularity it equals $(\Delta \Phi_{00})^2 = k_0^2 \langle (\delta \varepsilon)^2 \rangle l_R$, where $\delta \varepsilon$ is the fluctuation of plasma permittivity due to the existence of moderate-scale irregularities, and l_R is the scale of irregularity along the wave. If we also assume now that the layer with thickness ΔR is filled by N independent irregularities (the so-called, the *striated structures*, see Chapters 11 and 12), then we get

$$(\Delta \Phi_0)^2 = (\Delta \Phi_{00})^2 N = k_0^2 \langle (\delta \varepsilon)^2 \rangle l_R \Delta R. \tag{13.36}$$

Taking into account that for $\omega \gg \omega_{0e}$ the parameter $(\delta \varepsilon) = \tilde{N}(\omega_{0e}^2 / \omega^2)$ [6,17], we finally get

$$(\Delta \Phi_0)^2 = \langle (\tilde{N})^2 \rangle \frac{\omega_{0e}^2}{\omega^2} l_R \Delta R. \tag{13.37}$$

The mean-square of plasma density fluctuations $\langle (\tilde{N} = \delta N / N_0)^2 \rangle$ relates to the spectrum of these fluctuations $U_N(K)$ in the K-domain ($K \sim l^{-1}$) as [6,10,17]

$$\left\langle (\tilde{N})^2 \right\rangle = \int_{K_1}^{K_2} U_N(K')\mathrm{d}K', \tag{13.38}$$

where:

the wave numbers, K_1 and K_2, correspond to maximum $l_{max} = 2\pi/K_1$ and minimum $l_{min} = 2\pi/K_2$ scales of plasma irregularities, respectively

If now, according to Refs. [10,17], we suggest the existence of a three-dimensional (3D) polynomial spectrum $U_N(K) \sim K^{-p}$, which yields that $\left\langle (\tilde{N})^2 \right\rangle \sim K^{-p+3}$ and $(\Delta\Phi_0)^2 \sim K^{-p+2}$, we can easily find the relative mean-square signal intensity deviations versus K for various power density spectral (PDS) parameters p:

$$F_I = \left\langle (\Delta I)^2 \right\rangle / \left\langle (I)^2 \right\rangle \sim K^{-p+6}. \tag{13.39}$$

From these estimations, it follows the phase fluctuations of probing radio waves, passing a layer with moderate-scale or small-scale irregularities, are sufficiently strong and the parameter of their degree of perturbations corresponds to the constraint $p > 2$. These fluctuations increase with increase of the scale of irregularities along the radio path, $l_R = 2\pi/K$. It was also found [36–39] that for two-dimensional (2D) spectrum $U_N(K)$ of plasma density fluctuations, $\left\langle (\tilde{N})^2 \right\rangle \sim K^{-p+2}$ and $(\Delta\Phi_0)^2 \propto K^{-p+1}$ can be obtained, which finally yields

$$F_I \sim K^{-p+5}. \tag{13.40}$$

For smaller irregularities, $l \ll \sqrt{\lambda R}$ for weak fluctuations of signal phase, $(\Delta\Phi_0)^2 \ll 1$, we get for fluctuations of signal intensity $F_I \approx 2(\Delta\Phi_0)^2$ and for fluctuations of signal amplitude $F_A \approx (1/2)(\Delta\Phi_0)^2$. Here, spectral characteristic of amplitude fluctuations for the so-called plane ionospheric screen problem is simply related to $U_N(K)$ [41–43]:

$$F_A(\Omega = K_x V_x) \sim K_x^{-p}, \tag{13.41}$$

that is, the amplitude fluctuations of the signal are characterized by the power of 1D spectrum. Here, V_x is the approximate horizontal velocity of the moving SV. The above-mentioned allows investigating the 3D form of the spectra of moderate-scale irregularities, $F_A(\Omega)$, depending on the angle Ω between the vector of the geomagnetic field $\mathbf{B}_0$ and the vector of satellite velocity, as well as on the angle between $\mathbf{B}_0$ and the visible ray at the satellite [33,34].

Thus, simultaneous study of spectra $F_A(\Omega)$ and $F_\Phi(\Omega)$ during sounding of the ionosphere allows for obtaining information about the spectrum of plasma density fluctuations $U_N(K)$ in the broad range of the wave numbers, as well as over whole altitudinal interval where these plasma disturbances are localized in the perturbed ionosphere [10,6,17].

Another method of investigation of moderate-scale and small-scale irregularities of various origins was described in detail in Refs. [10,34,35]. Using this method together with satellite direct radio sounding of the ionosphere, the information on spectrum, $U_N(K_\perp)$, of artificially induced turbulences (AIT) can be found. It was shown in Refs. [10,17], for moderate kilometer-scale irregularities the power of spectrum changes from $p = 3$ to $p = 4$, whereas for irregularities with scales of hundreds of meters, the power decreases slightly, that is, $p = 1.5–2$.

As for *small-scale* irregularities with scales of tens of meters or less, this was shown in Refs. [17,33,34], $p = 2–3$. About the effects of these kinds of ionospheric irregularities we shall discuss later.

To investigate the effects of moderate-scale and small-scale irregularities of various origin, a plane-screen (i.e., 1D) model was analyzed in Refs. [40–43]. In Refs. [36–39], a curved-screen (i.e., 2D) ionospheric model was prepared, which then was expanded for spherically symmetric 3D model of the ionosphere that can be applicable for various scenarios occurring in the artificially or naturally perturbed ionosphere [10,33,34]. In the later works, to investigate the signal intensity deviations in the perturbed ionosphere and the corresponding scintillation index evaluations, the theoretical self-consistent framework was presented, according to which, filling by moderate-scale irregularities with scales of $l \approx (1–10)d_F$ in the strongly perturbed ionospheric global structures, with the outer scales of $L_0 = (10–100)d_F$, that is, for $p = 4$ and $p = 3$, the scintillation index, respectively, is:

$$\sigma_I^2 = \frac{8\sqrt{2}}{3\sqrt{\pi}L_0^2} d_F^2 \left\langle (\Delta\Phi)^2 \right\rangle, \tag{13.42}$$

$$\sigma_I^2 = \frac{\pi}{2L_0^2} d_F^2 \left\langle (\Delta\Phi)^2 \right\rangle. \tag{13.43}$$

As for weak or local plasma perturbations of scales $L_0 = (10 – 100)d_F$ containing small-scale irregularities (i.e., for $p = 2$) with the scales $l < 10^{-2} d_F$, the following expression was obtained:

$$\sigma_I^2 = \frac{2\sqrt{2}}{\sqrt{\pi}L_0} d_F \left\langle (\Delta\Phi)^2 \right\rangle. \tag{13.44}$$

Figure 13.13a and b present the root-mean-square (RMS) of the scintillation index, $\sigma_I \equiv \sqrt{\sigma_I^2}$, calculated for weak $\left(\sqrt{\langle(\Delta\Phi)^2\rangle} = 0.1 \text{ rad}\right)$ and moderate

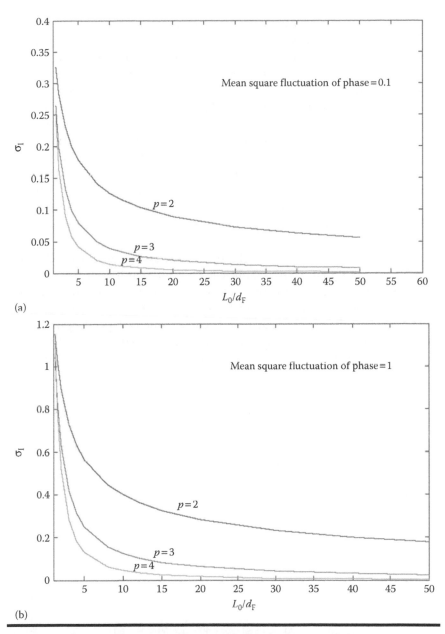

Figure 13.13 (a) The RMS of the scintillation index, σ_I, vs the outer scale L_0 for $1.5d_F \le L_0 \le 50d_F$, $\sqrt{\langle(\Delta\Phi)^2\rangle} = 0.1\,\text{rad}$. (b) The RMS of the scintillation index, σ_I, vs the outer scale L_0 for $1.5d_F \le L_0 \le 50d_F$ for $\sqrt{\langle(\Delta\Phi)^2\rangle} = 1\,\text{rad}$.

$(\sqrt{\langle(\Delta\Phi)^2\rangle} = 1 \text{ rad})$ signal phase fluctuations versus normalized outer scale (to d_F) of the perturbed ionospheric region (hole or enhancement) for various PDS parameters $p = 2, 3, 4$ and different scales of ionospheric irregularities.

It is clearly seen that for $p = 2$ the scintillation index with increase of phase fluctuations limits to the unit, whereas for more larger PDS index ($p > 2$), σ_I exceeds the unit within the range of $0 < L_0/d_F < 1$, which can explain the focusing properties of the ionospheric layer consisting of various irregularities and strong variations of signal phase after passing the perturbed ionosphere.

In Figure 13.14, the RMS of the signal phase fluctuations, $\sqrt{\langle(\Delta\Phi)^2\rangle}$, (in radian), is presented versus the zenith angle of a satellite for various frequencies from 1 MHz (HF range) to 10 GHz (UHF band) and for $\delta N/N_0 = 0.1$ (for the total plasma content of $N_0 = 10^{11}$ m^{-3} taken for the *F*-layer of the ionosphere that corresponds to the virtual local heating experiment presented in Figure 12.4).

As clearly seen from Figure 13.14, the frequency dependence of RMS fluctuations of the signal phase is sufficient only for zenith angles greater than $60°-65°$ for strong perturbations of the background ionospheric plasma, that is, for $\delta N/N_0$ exceeding several percentages. With increase of the radiated frequency, the phase fluctuations become strongly dependent on $\delta N/N_0$. Thus, as was shown in a detailed analysis carried out in Refs. [10,33,34], for the frequency band from 1 to 10 GHz, usually used in satellite–land

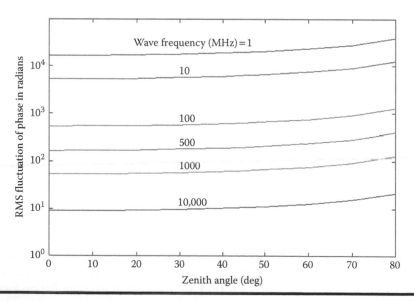

Figure 13.14 RMS of phase deviations (in radians) vs zenith angle for the radiated wave frequencies from $f = 1$ to 10,000 MHz; ionization of plasma is 10%.

communications, for zenith angles of 50°–60°, the RMS of phase fluctuations increases from 1 to 5 rad ($\delta N/N_0 = 0.1\%$) to 20 rad ($\delta N/N_0 = 1\%$) and 100 rad ($\delta N/N_0 = 10\%$). According to presented numbers of estimations, we should note that in our computations we used a wide spectrum of frequencies, covering HF and VHF range usually used in land–ionosphere–land communication links, as well as VHF and UHF range usually used in satellite–land communication links.

13.7 Attenuation of HF/VHF-Range Radio Waves in Perturbed Ionosphere

Usually, in the special literature regarding absorption of radio waves in the ionosphere [3–8,10,17], the cumulative intensity integral absorption along the radio-path is usually investigated. We will follow this approach and present below this parameter versus grazing angles and radiated frequency of probing radio waves of HF/VHF range. Thus, to estimate losses due to absorption, that is, the part of energy of radio signal, which is absorbed in the perturbed region of the ionosphere, from D to F, we propose here, following Refs. [10,6,17], to use a special value of absorption in decibels (dB):

$$A_\omega = 10\log\frac{I_0}{I} \approx \frac{4.3}{c}\int\frac{\omega_{pe}^2(\nu_{em}+\nu_{ei}+\nu_{ee})}{[\omega^2+(\nu_{em}+\nu_{ei}+\nu_{ee})^2]}ds, \text{ [dB]}, \qquad (13.45)$$

where:
   ν_{em}, ν_{ei}, and ν_{ee} are the frequencies of electron–neutral, electron–ion, and electron–electron collisions, respectively

The parameter of absorption, described by Equation 13.45 can usually be estimated by measuring radiometric absorption at the fixed frequencies and by knowledge of frequency dependence of coefficient of absorption χ introduced in the previous sections.

Accounting now in Equation 13.45 for the fact that the length of the radio path, s, and the height of the ionosphere, h, are related as $h = s{\cdot}\sin\alpha$, where α is the grazing angle of the ground-based antenna with respect to ground surface, we can estimate the total absorption of radio signal along the radio path. In the process of numerical computations, the corresponding perturbed profiles of plasma content, taken from Figures 12.4 and 12.5 (see Chapter 12) were normalized on the nondisturbed plasma density $N_0(h)$ presented in Figure 13.15 according to Refs. [3,5] for the mid-latitude regular daytime and nocturnal

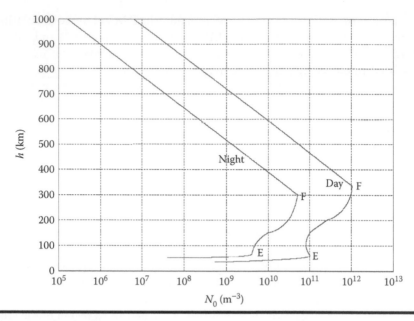

Figure 13.15 Profile of $N_0(h)$ during moderate solar activity.

ionosphere, and computations of the integral given in Equation 13.45 were carried out for $h_{min} = 50$ km and $h_{max} = 500$ km.

In Figures 13.16a and b and 13.17a and b, the parameter of integral absorption of the wave, passing the perturbed region of the ionosphere, is presented in dB versus the carrier frequency of the probing radio signal sent through the locally perturbed ionosphere under different grazing angles, respectively.

The corresponding profile of disturbed plasma density $N(h)$ is taken from data presented in Figures 12.4 [for all figures indicated by (a)] and 12.5 [for all figures indicated by (b)] (see Chapter 12). In other words, we consider later two scenarios described in Chapter 12: a local heating (described by the virtual heating experiment presented in Figure 12.4) and a global irregular heating (described by Figure 12.5), which are shown in Figure 13.16a and 13.17a, and in Figure 13.16b to 13.17b, respectively.

As follows from the above illustrations, with increase of the power of the "heater," from weak and local [Figures 13.16a and 13.17a] to strong and global [Figures 13.16b and 13.17b], absorption of radio signals increases roughly three times at frequencies from 10 to 500 MHz, while at 500–600 MHz, absorption does not depend on the frequency but only on the grazing angle.

At the same time, with increase of the grazing angle, the tendency of decrease of signal absorption for weak storms compared with strong storms is the same,

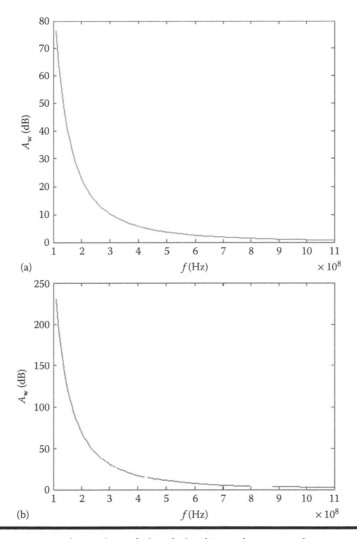

(a)

(b)

Figure 13.16 **(a) Absorption of signal (in dB) vs frequency for $\alpha \approx 0.175$ rad ($\alpha \approx 10°$) for a virtual heating experiment presented in Figure 12.4 (see Chapter 12). (b) Absorption of signal (in dB) vs frequency for $\alpha \approx 0.175$ rad ($\alpha \approx 10°$) for a virtual heating experiment, presented in Figure 12.5 (see Chapter 12).**

twice or three times. Results of computations presented earlier allow us to conclude that heating of the ionosphere, local or global, caused by hot exhaust gases from the rocket engine or plume, can significantly perturb ionospheric plasma up to 500 MHz, that is, for waves at the HF and VHF ranges, causing sufficient attenuation of radio signals up to 500 MHz.

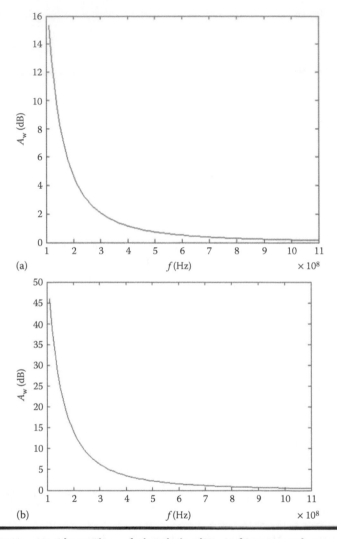

(a)

(b)

Figure 13.17 **(a) Absorption of signal (in dB) vs frequency for $\alpha \approx 1.05$ rad ($\alpha \approx 60°$) for a virtual heating experiment presented in Figure 12.4 (see Chapter 12). (b) Absorption of signal (in dB) vs frequency for $\alpha \approx 1.05$ rad ($\alpha \approx 60°$) a virtual heating experiment, presented in Figure 12.5 (see Chapter 12).**

This effect depends on the grazing angle and can be insignificant for grazing angles beyond $\alpha \geq 20°$. Thus, in HF- and VHF-range wave propagation via the perturbed ionosphere, with increase of the radiated frequency of radio signal for the same grazing angle or for increase of grazing angle for the same frequency, the tendency of significant decrease of signal energy absorption is evident.

At the same time, we note that for local perturbations made by rocket burn and launch, for radar and radio communication applications, instead of cumulative absorption defined by Equation 13.45, another characteristics of signal fading was proposed, called the *point-to-point* amplitude and phase deviations of the signal along the radio path [33,34]. These two characteristics allow us to understand how the perturbed parameters of the ionospheric plasma cause attenuation of radio wave amplitude and change of its phase.

To understand these radio physical effects, we will consider an arbitrary monochromatic radio wave propagating along the axis z [33,34]:

$$U(z,t) = A(z,t) \cdot \exp\left[j(\omega t - kz) \right].$$ (13.46)

In this case, the wave number k can be described through the coefficient of wave amplitude attenuation, α, and the coefficient determining changes of wave phase in real time, β, that is,

$$k = \alpha + j\beta.$$ (13.47)

Unlike the usual description of parameters α and β, we will present them through collision frequencies of plasma particles using elements of kinetic theory (see Refs. [4,6,7]). For numerical computations, we introduce the additional notations and parameters to describe α and β, that is, $X = \omega_e^2/\omega^2$, $Z = v/\omega$, and ω_e is the background plasma (e.g., electron, because plasma is quasineutral) frequency in the perturbed ionosphere [3–8]. In these notations, after straightforward derivations, we get [33,34]:

$$k^2 = \frac{\omega^2}{c^2}\left[1 - \frac{X}{1 - iZ} \right] = \frac{\omega^2}{c^2}\left[1 - \frac{X(1 + iZ)}{1 + Z^2} \right] = \frac{\omega^2}{c^2}\left[1 - \frac{X + iZX)}{1 + Z^2} \right]$$ (13.48a)

or

$$k^2 = \frac{\omega^2}{c^2}\left[1 - \frac{X}{1 + Z^2} - i\frac{ZX}{1 + Z^2} \right].$$ (13.48b)

By introducing now normalized parameters (to the wave number in free space k_0), $\tilde{\alpha} = \dfrac{\alpha}{\alpha_0}$ and $\tilde{\beta} = \dfrac{\beta}{\beta_0}$, we finally get

$$\tilde{\alpha} = \frac{1 + \left(\dfrac{v}{\omega}\right)^2 - \dfrac{\omega_e^2}{\omega^2}}{1 + \left(\dfrac{v}{\omega}\right)^2} = \frac{\omega^2 + v^2 - \omega_e^2}{\omega^2 + v^2},$$ (13.49a)

$$\tilde{\beta} = \frac{ZX}{1+Z^2} = \frac{\left(\dfrac{\nu}{\omega}\right)\dfrac{\omega_e^2}{\omega^2}}{1+\left(\dfrac{\nu}{\omega}\right)^2} = \frac{\nu\omega_e^2}{\omega(\omega^2+\nu^2)}, \tag{13.49b}$$

where:

$\nu = \nu_{em} + \nu_{ei} + \nu_{ee} = \nu_{em}(1 + p + p')$

$p = \nu_{ei}/\nu_{em}$

$p' = \nu_{ee}/\nu_{em}$

α_0 and β_0 are the corresponding radio wave parameters, obtained for nondisturbed background ionospheric plasma

Deviations of radio signal parameters, α and β, in the perturbed ionospheric region, were found by using parameters of the disturbed ionosphere, that is, variations in the plasma frequency ω_e/ω_{e0}, plasma concentration N/N_0 ($N = N_0 + \delta N$, $\delta N < N_0$), and plasma conductivity σ/σ_0, where ω_{e0}, N_0, and σ_0 are the background nondisturbed ionospheric parameters described in Refs. [3–6] in CGSE unit system as:

$$\omega_{e0} = \left(4\pi N_0 e^2/m_e\right)^{1/2}, \tag{13.50a}$$

$$\varepsilon_0 = 1 - \left\{\omega_0^2 / \left[\omega^2 + (\nu_{em} + \nu_{ei})^2\right]\right\}, \tag{13.50b}$$

$$\sigma_0 = \left\{\omega_0^2(\nu_{em} + \nu_{ei}) / 4\pi\left[\omega^2 + (\nu_{em} + \nu_{ei})^2\right]\right\}. \tag{13.50c}$$

Here, as in previous chapters, e and m_e are the charge and mass of plasma electron; other parameters are as defined earlier.

To present not only effects of HF/VHF-range waves, but also beyond these frequency-band waves, namely, UHF-range waves, we show in Figure 13.18a and b deviations of the normalized radio signal parameters, α and β (see definitions above). The computation results are shown in Figure 13.18a (for normalized attenuation) and 13.18b (for normalized phase velocity) versus altitude of the ionosphere for different frequencies of probing waves varying from 50 MHz to 2 GHz (i.e., covering HF/VHF/UHF bandwidths). Thus, we use those frequencies which are actual both in land–ionosphere–land or ionosphere–ionosphere propagation ($f < 500$–600 MHz) and those, which are usually used in satellite radio monitoring of the ionosphere ($f > 900$ MHz).

As seen from Figure 13.18a, with the increase of frequency of the probing wave, the effect of attenuation of the wave energy becomes weaker, and

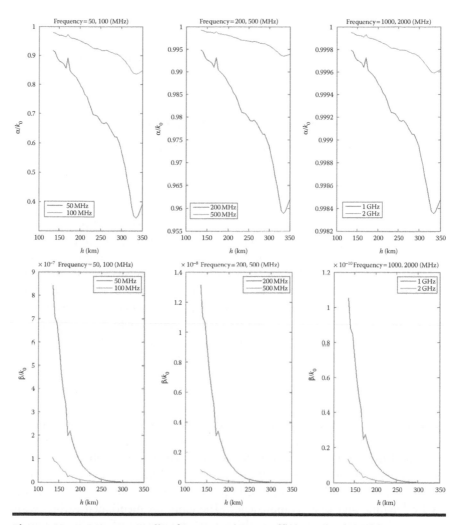

Figure 13.18 **(a) Normalized attenuation coefficient $\tilde{\alpha}$ of probing wave vs ionospheric altitudes for frequencies varied from 50 MHz to 2 GHz–the top panel. (b) Normalized phase velocity $\tilde{\beta}$ of probing wave vs ionospheric altitudes for frequencies varied from 50 MHz to 2 GHz–the bottom panel.**

the probing radio wave propagates in the same manner as in the nondisturbed ionosphere. This effect depends strongly on which altitudes the propagation process is observed. Thus, with increase of ionospheric altitude, attenuation affects stronger for all frequencies under consideration.

This effect can be easily explained. When the frequency increases (or the wavelength decreases compared to the dimensions of plasma irregularities), the effects of scattering from small-scale irregularities that filled the global or large-scale perturbations (such as holes, cavities, and enhancements) become weaker and, instead of defocusing effect with strong fading, we observe focusing effect with weak fading [6,17].

13.8 Doppler Frequency Shift and HF/VHF/ UHF-Range Radio Wave Spread

As was mentioned in Chapters 11 and 12, the complicated aerodynamic, chemical, and electro-hydrodynamic processes associated with the rocket engine and plume hot exhaust gases interaction with the background cold ionospheric plasma, finally creates a "dirty" multicomponent (ion–electron) plasma, whose movements accompanied by turbulent motion of small-scale to large-scale irregularities (plasma modes with various wavelengths) created surrounding the moving SV and in the wake behind it. These complicated processes can interrupt any communication link between SVs, rockets or satellites, and the ground-based stations by introducing additional shift of Doppler frequency and, what is more important, the spread of DS in the frequency domain, which together can be an additional source of frequency- and time-selective fading [3] occurring in the land-to-land or satellite-to-satellite radio communication links, mostly in HF/ VHF/UHF ranges.

Let us briefly illuminate the problem of Doppler frequency shift and DSs spread (i.e., broadening). Such complicated processes of plasma instabilities generation that finally create small- and moderate-scale plasma disturbances (called *waves* or *modes* [10]), allow us to mention again that to enter deeper into the problems, we need to deal with the so-called nonlinear effects of plasma instabilities generation. These effects are totally different compared to those effects described in Chapter 12 based on linear two-fluid theory based on both hydrodynamic and kinetic approaches. Therefore, as mentioned earlier, without entering into complicated mathematical description of nonlinear fourth-order equations of the turbulent plasma dynamics, we take the reader to the Refs. [10,44–56], where all aspects of plasma waves generation and the corresponding types of plasma instabilities creation are fully presented. In other words, we will present phenomenological analysis of the possible processes accompanying Doppler frequency shifting and the DS width spread based on the corresponding literature, as well as on our vision of the problem.

Thus, in Refs. [44–47], the self-consistent theoretical analysis of the nonlinear broadening of the DSs was analyzed via the 2D spectral fluctuations of the density of plasma disturbances (i.e., modes or waves), caused by the turbulent motions of the perturbed background plasma [44,45]

$$I_{\mathbf{k}}(\mathbf{k},t) = \left(\frac{L_0}{2\pi}\right)^2 \left\langle \left|\frac{\delta N(\mathbf{k},t)}{N_0}\right|^2 \right\rangle, \tag{13.51}$$

where:

$\mathbf{k}$ is the plasma density wave vector

$\delta N(\mathbf{k},t)$ are the perturbations of plasma density

N_0 is the plasma density of the nonperturbed background ionosphere

L_0 is the outer scale of the initial plasma turbulence (i.e., irregularity)

The above spectral function after averaging over the angles θ between the vector $\mathbf{k}$ and the drift velocity $\mathbf{V}_d$ of plasma can be derived and presented in the form of the angle-average temporal power spectra:

$$I_{\mathbf{k}} \equiv I(\mathbf{k}) = \frac{1}{2\pi} \int_0^{2\pi} I(\mathbf{k},0)d\theta. \tag{13.52}$$

For the analysis of the nonlinear broadening of the Doppler frequency, the frequency spectra $I_{\mathbf{k}}(\mathbf{k},f)$ were derived in Ref. [47] by using Fourier transform of the spectra $I_{\mathbf{k}}(\mathbf{k},t)$ from Equation 13.51 at the time interval Δt of about 20 s after achievement of the quasiequilibrium state (called the *steady-state regime* [10]). In Figure 13.19a–c, the frequency spectra $I_{\mathbf{k}}(\mathbf{k},\omega)$ are shown for the turbulent small-scale turbulence (i.e., wave mode) with scale (i.e., wavelength) of $\Lambda = 9.2$ m and for the angles of $\theta = \pi/6, \pi/4, \pi/2$. In Figure 13.19a and b, the experimental shapes of $I_{\mathbf{k}}(\mathbf{k},\omega)$ are also presented by dashed curves. The obtained illustrations show that the nonlinear interaction of plasma turbulent modes that the authors of Ref. [47] associated with the gradient-drift instability (GDI) generation (see definition in Chapter 12). This interaction yields the nonlinear dependence between ω and $\mathbf{k}$, that is, nonlinear effects of plasma dispersion [10]. This means that each wave vector $\mathbf{k}$ in such mode interactions corresponds in the frequency spectra to not only one Doppler frequency $f_{\mathbf{k}}$, but to a wide band of frequencies of width $\Delta f_{\mathbf{k}}$.

As shown in Ref. [10] following Refs. [44,45], the value of Δf generally depends on the orientation of the turbulent modes' vector $\mathbf{k}$ and achieves maximum magnitude for $\mathbf{k}$ orthogonal to drift velocity $\mathbf{V}_d$ (i.e., for $\theta = \pi/2$), as is evidently seen from Figure 13.19c.

The average width of the frequency spectra can be found as follows [44,45,47]:

$$\langle \Delta f \rangle \equiv \frac{\langle \Delta \omega \rangle}{2\pi} = \frac{1}{2\pi} \int_0^{2\pi} \Delta f(\theta)d\theta. \tag{13.53}$$

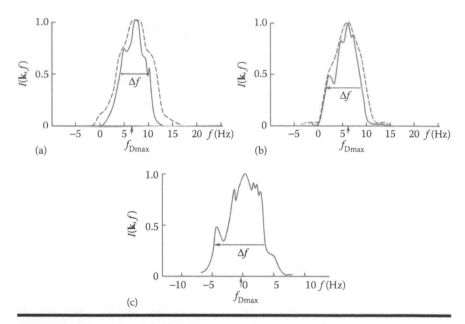

Figure 13.19 **(a) Spectral intensity of plasma density fluctuations vs Doppler shift frequency. The dashed curves represent the result of experimental observations. (b) Spectral intensity of plasma density fluctuations vs Doppler shift frequency. The dashed curves represent the result of moderate plasma turbulences. (c) Spectral intensity of plasma density fluctuations vs Doppler shift frequency. The dashed curves represent the result of strong turbulent plasma.**

Estimations made in Refs. [46,47] show that based on experimental observations of the dependence of the average bandwidth spreading of Doppler frequency, $\langle \Delta f \rangle$, versus the drift velocity $\mathbf{V_d}$ of plasma electrons (or ions, plasma is quasineutral) have shown that the average bandwidth $\langle \Delta f \rangle$ increases nonlinearly $\sim V_d^2$ with increase of the threshold level of the velocity of turbulent plasma movements. The data were obtained experimentally in Refs. [46,47] for different wave-mode scales. They also predict such a nonlinear ($\sim V_d^n$, $n = 2$–2.4) dependence of the DS width spread versus $\mathbf{V_d}$.

We note, following Ref. [10], that the possibility of nonlinear transformation of energy from the small-scale irregularities (turbulences), which fill the global perturbed regions (holes or enhancements), to the large-scale plasma structures, and vice versa, observed experimentally and described in Chapters 11 and 12, were analyzed in Refs. [44–47] based on the two-fluid theory (electron and ion) of plasma, as a liquid, dynamic. For example, it was predicted in Refs. [46,47] that there was a possibility of nonlinear transformation of energy from linearly excited large-scale modes into the linear damping small-scale modes. Moreover, it was shown that the steady-state 2D plasma density fluctuation spectrum $I_k(\mathbf{k},t)$ defined by Equation 13.51 or 13.52 is isotropic and proportional to $|\mathbf{k}|^{-p}$, where p lies between 3 and 4 (for moderate and strong turbulent plasma, respectively), and equals $p = 2$ for weakly turbulent plasma.

The results obtained in Refs. [44–47], are in a full agreement with those obtained in Refs. [33,34] and described above, in which the spectra of intensity and phase fluctuation of the probing radio wave, passing the perturbed turbulent ionospheric region, were analyzed.

Moreover, based on the self-consistent theory [44,45], the broadening of the plasma power spectra in the nonlinear regime of strong turbulent plasma motion, accounting for the corresponding short wave–long wave interactions (in such manner the authors called interaction between the small-scale and large-scale plasma irregularities), was compared with those obtained by using the linear theory. This theory can be correctly used only for description of the process of damping of weak plasma irregularities (turbulences) to the steady state regime (or equilibrium state) [10].

Thus, in Figure 13.20, the numerically obtained angle-averaged spectral width $\langle \Delta f \rangle = \langle \Delta \omega \rangle / 2\pi$, defined by Equation 13.53 is shown versus k/k_0 (the upper digits along the horizontal axis) and the corresponding wavelength λ [in meter] (the bottom digits along the horizontal axis) for the three plasma drift velocities $V_d = 75, 100, 125$ m s^{-1}, rearranged from Refs. [46,47] by using our notations. Here, dots, circles, and triangles show $\langle \Delta f \rangle$ as a result of the model experiment; the solid lines represent simulation results of the nonlinear growth rate of the corresponding GDIs (see definitions in Chapter 12). The dotted curve corresponds to the linear growth rate $\gamma_k = \gamma_k^l$ [10].

It can be seen from the nonlinear theory that by introducing the broadening effects of the spectra I_k in the nonlinear turbulent regime, gives the best fit to direct results of simulation of the DS spread width, on the order $\langle \Delta f \rangle = 1 - 4$ Hz (solid lines) with respect to linear damping rate γ_k^l (dotted curve) following the linear theory that gives much less values, from 0.1 to 1 Hz.

However, all theoretical frameworks mentioned above used the two-fluid plasma theory based on hydrodynamic approach, whereas in Refs. [48,49] the consistent kinetic theory was proposed for the description of the two-stream instability [called the Farley–Buneman instability (FBI), see definition in Chapter 12] evolution in the E-layer of the ionosphere. The authors found that FBI growth and unstable turbulent modes generation are very sensitive to signs of flow angle θ introduced in Refs. [48,49]: $\cos\theta \sim \mathbf{k} V_d$; as above, θ is the angle between the $\mathbf{k}$-vector and the drift velocity $\mathbf{V}_d$. The corresponding instability growth and unstable short-wavelength turbulent modes generation were found at the negative angles θ ($\sim\mathbf{k}\mathbf{E}_0$) and their growth stabilization—at the positive

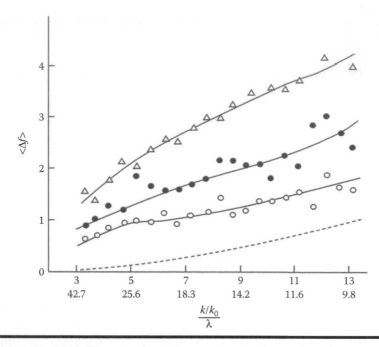

Figure 13.20 Doppler spread width vs normalized wave number (top numbers) and wavelength (bottom numbers).

angles θ. The second mechanism is associated with thermal perturbations of the plasma conductivity due to the monotonically increasing dependence of the electron–neutral collision frequency on temperature, the effect that plays an essential role in the FBI evolution during the local heating of plasma by various natural or artificial sources [50,51].

It was found both theoretically and experimentally that at all latitudes of the ionosphere the GDI and FBI can be considered as the main candidates of creation of plasma irregularities (turbulences) in the perturbed ionospheric regions, from the *E*- to the upper *F*-layer. We should note that a more general self-consistent two-fluid theory performed in Refs. [52–56] was performed to explain nonlinear evolution of both kinds of main plasma instabilities, FBI and GDI. Therefore, the same qualitative analysis, as was discussed above, can be done, following Refs. [52–57], to explain nonlinear growth rate, spreading of frequency width of the power spectra of unstable small-scale modes and the nonlinear DS, observed experimentally [58–63].

Without loss of generality we will present the qualitative analysis for our specific case of generation of small-scale turbulent plasma structures due to the effects associated with the rocket burn and launch occurring in the perturbed ionosphere described in Chapters 11 and 12. First, we will start with weakly perturbed ionosphere, for which a linear theory can be easily adapted. As shown in Refs. [52–56], in such a plasma, the Doppler frequency spectral width is close to zero, that is, $\langle \Delta f_{\mathbf{k}} \rangle \to 0$, and the mean Doppler shift frequency equals that obtained from the linear theory, that is,

$$\langle f_{\mathbf{k}} \rangle \equiv f_k^L \approx C_s / l_{\perp}, \tag{13.54}$$

where:

$C_s = \sqrt{\overline{T_i} / \overline{M_i}}$ is the velocity of the ion-acoustic waves

$l_{\perp}$ is the small-scale turbulence inner scale across ambient magnetic field $\mathbf{B}_0$

$\overline{T_i}$ and $\overline{M_i}$ are the temperature and mass of the average ion in the multi-ion composite plasma generated surrounding the moving SV and in its wake (see analysis presented in Chapters 11 and 12)

The Doppler shift frequency according to quasilinear theory [10], can be presented in the following manner:

$$f_k^L = \frac{V_d}{\lambda \left(1 + \omega_{II} / (\nu_{cm} + \nu_{ci}) \right)}. \tag{13.55}$$

The above scenario is presented in Figure 13.21a. It is evident that in the unperturbed weakly turbulent ionospheric plasma, as was expected from classical theory [10], the width of the plasma density fluctuations is sufficiently narrow (limits to the δ-function shape), that is, $\langle \Delta f_{\mathbf{k}} \rangle \ll \langle f_{\mathbf{k}} \rangle$. In such a scenario, owing to the absence of the outer sources of perturbation, the created unstable plasma modes attenuate fast due to coupling between modes and their interactions with plasma-charged particles, electrons, and ions.

Now, if we consider existence of the artificial source of "local heating," as a rocket's exhausts expanded into the ambient multicomponent cold plasma, the secondary small-scale plasma turbulences can quickly extract energy from the source and this process leads to a fast quasilinear (QL) growth of plasma instabilities. In this case, instead of Equation 13.54, we get [10]:

$$\langle f_{\mathbf{k}} \rangle + \langle \Delta f_{\mathbf{k}} \rangle \approx C_s^2 / l_{\perp}^2. \tag{13.56}$$

In this case, for all turbulent modes the nonlinear phase velocity can be expressed as:

$$V_{ph}^{QL} \equiv \langle \Delta f_{\mathbf{k}} \rangle l_{\perp} = \frac{C_s}{\sqrt{1 + \langle \Delta f_{\mathbf{k}} \rangle^2 / f_{\mathbf{k}}^2}}, \tag{13.57}$$

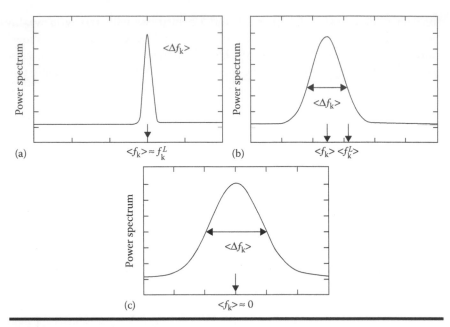

Figure 13.21 **(a) Power spectrum vs Doppler spectral width for nonperturbed or weakly turbulent plasma. (b) Power spectrum vs Doppler spectral width for moderate plasma turbulences generated by ambient sources of various origin. (c) Power spectrum vs Doppler spectral width for strong turbulent plasma perturbed by a strong ambient sources of various origin.**

which becomes smaller than the ion-acoustic wave velocity in plasma with the threshold of $V_{ph}^{QL} \approx 0.7 \cdot C_s$. This is shown in Figure 13.21b.

It is evident that $\langle f_{\mathbf{k}} \rangle \neq f_{\mathbf{k}}^L$, and a broadening of DS is observed with the average width $\langle \Delta f_{\mathbf{k}} \rangle$ that can exceed the average DS $\langle f_{\mathbf{k}} \rangle$.

Finally, we consider a case in which strong turbulent conditions occur with the $\langle \Delta f_{\mathbf{k}} \rangle$ sufficiently broad (as seen from Figure 13.21c).

In this case due to chaotic movements of plasma, when its turbulent velocity exceeds the deterministic velocity, the mean DS frequency limits to zero, that is, $\langle f_{\mathbf{k}} \rangle \to 0$. In such a scenario, as was shown in Ref. [57], due to wave-coupling process, the small-scale wave modes receive periodically energy from the large-scale wave modes, obtaining the stable amplitudes during a very short time. This effect was observed during the rocket releases experiments described in Chapter 11.

This effect was also observed during the rocket release experiments described in Chapter 11 (see Figure 11.17c–d), where for a strong turbulence, generated by the hot exhaust gases injected in the background cool ionospheric plasma, a huge broadening of the Doppler power spectra is vividly observed during the 60-s period of turbulence formation and evolution.

In Ref. [10], the threshold of the DSs width spread for a nonlinear increment of instability growth was obtained for $\langle f_{\mathbf{k}} \rangle \equiv f_{\mathbf{k}}^{L}$ in the following manner:

$$\left(f_{\mathbf{k}}^{NL} \right)^{2} = \langle f_{\mathbf{k}} \rangle^{2} + \left(\Delta f_{\mathbf{k}}^{L} \right)^{2}, \tag{13.58}$$

where the DSs width spreading is given for a linear process of turbulent modes growth as follows:

$$\left\langle \Delta f_{\mathbf{k}}^{L} \right\rangle = -\nu_{\text{im}} \left(1 + \omega_{\text{H}} \Omega_{\text{H}} / \nu_{\text{em}} \nu_{\text{im}} \right) \gamma_{k}^{L}. \tag{13.59}$$

Here, the increment of plasma wave growth according to linear theory can be presented as follows [10]:

$$\gamma_{k}^{L} = \frac{\varsigma}{1+\varsigma} \left[\frac{\omega_{\text{H}} V_{\text{d}} \cos\theta}{L_{0} \nu_{\text{em}} (1+\varsigma)} - \frac{k^{2} C_{\text{s}}^{2}}{\nu_{\text{im}}} \right] \tag{13.60}$$

where:

$\varsigma = \nu_{\text{em}} \nu_{\text{im}} / \omega_{\text{H}} \Omega_{\text{H}}$

all other parameters of plasma have been defined earlier and in Chapter 12

The obtained results are consistent with observations of DS width deviations of the turbulent moving of plasma disturbances with velocities from 350 to 600 m s^{-1} (see Refs. [58–63]). Moreover, they allow us to conclude that the high-temperature (with respect to the background cold ionospheric plasma) multicomponent plasma that is involved in the turbulent motions occurring around and behind the moving SVs can be said to be an outer source of plasma instability generation and small- to large-scale plasma turbulence. The later wave modes, due to their coupling and interactions, yield the existence of Doppler shift frequency and the broadening of the DSs width from dozens of Hertz to a few Hertz, depending on the degree of nonlinear processes accompanied such turbulent plasma structures generation. The estimated magnitudes of the above parameters are in a good agreement with those obtained during experimental observations of rocket burn and launch described across Chapters 1 to 10 (see also Refs. [63–68]).

13.9 Depolarization of HF/VHF/UHF-Band Radio Waves

As mentioned in Chapter 12 and the beginning of this chapter, two kinds of waves propagate in the ionosphere, ordinary and extraordinary, that can be linearly, circularly, or elliptically polarized with rotations of the vector of

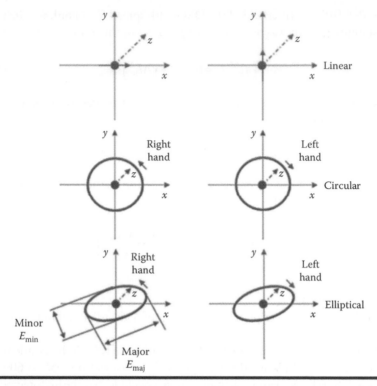

Figure 13.22   Three types of electromagnetic wave polarization: linear, circular, and elliptical.

electric field in opposite directions, one clockwise and the other counterclockwise (or positive and negative), depending on the conditions and processes occurring in the ionosphere, regular or perturbed (see Refs. [6,8] and references therein).

Before briefly delving into the subject, let us determine the main parameters and characteristics of the polarization as a phenomenon. For the simplification of the problem, we consider propagation of the plane wave. The alignment of the electric field vector E of a plane wave relative to the direction of propagation **k** and the magnetic component of the wave, **B**, defines the *polarization* of the wave [1–3]. There are several types of wave polarizations, and these are schematically shown in Figure 13.22, according to Ref. [3].

- *Linear polarization*: If the vector **E** is transverse to the direction of wave propagation **k** and to the vector **B**, the wave is determined as TE wave or *vertically* polarized. Conversely, when the vector of the magnetic field

B is transverse to **k**, the wave is determined as TM wave or *horizontally* polarized. Both of these waves are linearly polarized, as the electric field vector **E** has a single direction along the entire wave vector **k**.

– *Circular polarization*: If two-plane linearly polarized waves of equal amplitude and orthogonal polarization (vertical and horizontal) are combined having a 90° phase difference between them, the resulting wave will be a *circular polarized* (CP) wave, in which the motion of the electric field vector will describe a circle around the propagation vector. The field vector will rotate by 360° for every wavelength traveled. As mentioned in Refs. [1–3], circularly polarized waves are most commonly used in the land–satellite and satellite–satellite communications, as they can be generated and received using antennas which are oriented in any direction around their axis without loss of power [1–3]. They may be generated as either *right-hand circular polarized* (RHC, or clockwise) or *left-hand circular polarized* (LHC or counterclockwise), depending on the direction of vector **E** rotation.

– *Elliptical polarization*: In the most general case, the components of two waves could be of unequal amplitude, or their phase difference differs by 90°. The combination result is an elliptically polarized wave, where vector **E** still rotates at the same rate as for circular polarized wave, but varies in amplitude with time. In the case of elliptical polarization, the axial ratio, $AR = E_{maj}/E_{min}$, is usually introduced [1–3]. AR is defined to be positive for left-hand polarization and negative for right-hand polarization.

Now we define the effect of depolarization and the corresponding losses following Refs. [1–3]. According to IEEE standardization, the polarization state of an antenna is defined as the polarization state of the wave transmitted by the antenna. It is characterized by a sense of rotation and spatial orientation of the ellipse, if it is elliptically polarized. If the receiving antenna has a polarization that is different from that of the incident wave, a *polarization mismatch* will occur. A polarization mismatch causes the receiving antenna to extract less power from the incident wave. The *polarization loss factor* (PLF) is used as a measure of the degree of polarization mismatch. It is defined as the square power of the cosine angle between the polarization states of the antenna in its transmitting mode and the incoming wave.

$$PLF = |\cos\gamma|^2, \qquad (13.61)$$

where:
  γ is the angle of depolarization that will be defined later (Figure 13.23)

Generally speaking, an antenna is designed for a desired polarization. The component of the electric field in the direction of the desired polarization is

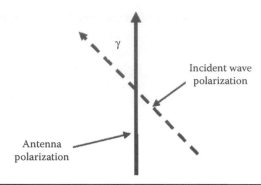

Figure 13.23 Definition of the angle of depolarization γ.

called the *copolar component*, whereas the undesired polarization, usually taken in orthogonal direction to the desired one, is known as *cross-polar component*. The latter can be due to a change of polarization characteristics during the propagation or scattering of waves that is known as *polarization* or *Faraday rotation* [3,8,10].

Because elliptical polarization is often used in ionospheric wireless communication links designs, we will briefly present the main characteristics of the polarized ellipse versus the vectors of electromagnetic field components, following Refs. [8,69,70] and analyze the depolarization loss effects and the changes of the angle of depolarization when the wave passes a perturbed ionospheric region produced by the local heating caused by rocket burn and launch, based on examples described in Chapters 11 and 12.

If we now rewrite the desired equations from Ref. [8], by changing the notations introduced in them in Chapter 12, we can express the depolarization effects via the parameters of the elliptical polarization, introduced in Refs. [8,69,70]. Thus, following Ref. [8], we introduce the angle θ between the ambient magnetic field vector $\mathbf{B}_0$ and the vector of the wave refraction $\mathbf{n}$ caused by influence of the perturbed ionosphere on wave propagation, and will define it as:

$$\cos\theta = \frac{\mathbf{n}\cdot\mathbf{B}_0}{|\mathbf{n}|\cdot|\mathbf{B}_0|}. \tag{13.62}$$

If we will denote the direction along the $\mathbf{B}_0$ (along the altitude defined by the z-axis) by $\|$, and normal to $\mathbf{B}_0$ (x-axis) by $\perp$, and across $\mathbf{B}_0$ (y-axis) by Λ, we get the axes ratio of the polarization ellipse in the wave front (i.e., in the (x,y)-plane):

$$AR = j\frac{\zeta}{\left(1\pm\sqrt{1+\zeta^2}\right)}, \tag{13.63}$$

where:

"+" corresponds to the right-hand rotation of the ellipse (called the *positive* [8])

"−" corresponds to the left-hand rotation of the ellipse (called the *negative* [8])

The parameter ς can be defined via the parameters of the multicomponent electron–ion collisional plasma introduced in Chapter 12: $X = f_{0e}^2/f^2$, $Y = (f_i^{(k)}/f)^2$, $Z = (\nu_{em} + \nu_{ei} + \nu_{im})/2\pi f$, f is the probing wave frequency, $f_i^{(k)} = \omega_i^{(k)}/2\pi = Z_i^{(k)}e/2\pi \cdot M_i^{(k)}$ is the frequency of plasma ion of the sort "k"; other parameters are defined 2 previous sections. In such notations, we get:

$$\zeta = \frac{2\cos\theta}{\sin^2\theta}\frac{(1 + jZ - X)}{Y}. \tag{13.64}$$

Using such notations, we can finally obtain the ratio of the radio wave intensity distribution in the (x,y)-plane [69,70]:

$$AR = \frac{E_{maj}}{E_{min}} = \sqrt{\frac{\sigma_\perp^2}{\sigma_\Lambda^2}}, \tag{13.65}$$

where, combining with results obtained in Ref. [8], we finally get:

$$\sqrt{\frac{\sigma_\perp^2}{\sigma_\Lambda^2}} = \frac{j\zeta}{1 \pm \sqrt{1 + \zeta^2}}. \tag{13.66}$$

The angle of depolarization in such notations equals:

$$\gamma = \left|\sin^{-1}\sqrt{\frac{\sigma_\perp^2}{\sigma_\perp^2 + \sigma_\Lambda^2}}\right|. \tag{13.67}$$

The straightforward analytical computations, based on the theoretical framework described in Chapter 12 for the multicomponent plasma and on the algorithm of derivation of the parameters of the ellipse [8,69,70], allows us to estimate the degree of depolarization γ and the depolarization losses PLF for some specific cases occurring during active experiments in the ionosphere similar to the rockets burn and launch.

Thus, for the virtual local heating experiment, presented in Figure 12.4 (according to Figure 11.5), we get: $\gamma = 6° - 8°$ and PLF can achieve ~5 dB (for 1 MHz) to ~10 dB (for 100 MHz) for the elevation angle θ of the radar (with respect to ground surface) of ~30°. These two parameters decrease roughly in one degree and in 1.3 dB, respectively, with increase of θ each 10°, achieving $\gamma = 1° - 2°$ and PLF ~1–2 dB for $\theta = 90°$ (i.e., for vertical sounding of the ionosphere. In more complicated case of global heating shown in Figure 12.5 (according to Figure 11.3), we get for $\theta = 30°$: $\gamma = 9°$ and PLF ~9 dB ($f = 1$ MHz) to $\gamma = 13°$ and PLF ~15 dB ($f = 100$ MHz), which decreases for each 10°, achieving several degrees (for the angle of depolarization) and several decibels (for PLF). The magnitudes of the estimated parameters varies in the wide range of values, depending on what kind of the "average" ion we took into account in our computations, the light, moderate, or heavy, but finally cannot exceed values $\gamma = 15° - 20°$ and PLF ~20–25 dB for frequencies ranging from 1 to 100 MHz, the waveband useful for tracking the rockets in the ionosphere. As for higher frequencies, from several hundred megahertz to several gigahertzes, as was expected, the depolarization angle and PLF do not exceed few dozens of degree and even one decibel, respectively. These estimations are in a good agreement with the results obtained earlier for other propagation parameters of VHF/UHF-range waves.

Therefore, we can suppose that the global-scale holes (or cavities) and enhancements that can be produced by the rocket's exhaust gases at different altitudes of the ionosphere (see Chapter 11) can change parameters of the polarization ellipse and yields the significant depolarization loss and the essential angle of depolarization occurrence of the probing LF/HF-range radio waves passing the perturbed regions of the ionosphere associated with the rocket burn and launch.

13.10 Main Results

1. The phenomenological analysis of the processes accompanying rocket burn and launch, which may be taken as one of the main candidate of the "sources" of distortion of probing radio wave parameters in the land–land or land–satellite communication links, shows that all effects depend, first of all, on the dimensions of plasma irregularities produced in the perturbed regions of the ionosphere, on the wavelengths of the probing waves, and on the inner conditions of the rocket burn and launch (i.e., on the rocket release conditions).

2. For the ELF range to the VLF range probing radio waves (i.e., with lengths exceeding tens to hundreds of kilometers), the effects of channeling of probing waves in the Earth–ionosphere waveguide channel is actual.

3. For LF range, MF range, even to HF range probing radio waves passing the global perturbed regions filled by the large-scale irregularities, new effects can be observed such as:

    – Changing of the probing waves direction from the plane of great circle;
    – Increase the thickness of the reflecting layer, that is, change of the height of the ionospheric layered waveguide;

- Deviation of the radio wave trajectory due to horizontal and vertical changes of the height of the reflected layer;
- Focusing effect, if the large-scale structures (holes or enhancements) are not filled by the moderate-scale and small-scale irregularities (turbulences);
- Defocusing effect, if the large-scale structures (holes or enhancements) are filled by the moderate- and small-scale irregularities (turbulences);
- Capturing of probing waves into the ionospheric layered waveguides.

4. For HF-range to ultrahigh frequency (UHF) range probing waves an essential attenuation of the wave amplitude and deviation of the wave phase occur when it passes the perturbed global or large-scale region filled by the moderate- or small-scale irregularities.

5. For HF range to UHF range probing waves, the scattering effects from the moderate- and small-scale irregularities (turbulences) becomes predominate yielding the signal intensity fluctuations, strong deviations of the wave phase, shifting of the Doppler frequency, and essential broadening of DS width (depending on the degree of instability of turbulent plasma structures filled the perturbed ionospheric region. The corresponding broadening of the DS width can achieve from a few Hertz to several (even tens) of Hertz with DS exceeding tens of Hertz, which are in a good agreement with rocket burn and launch observations [63–68].

6. As for depolarization effects, as was shown experimentally and proved above according to theoretical framework introduced in Refs. [8,69,70], the turbulent plasma structure-filled perturbed regions can significantly change the initial shape and the spatial orientation of the polarization ellipse (which yields the increase of the depolarization angle), as well as can significantly increase the depolarization losses of the probing radio waves passing the perturbed (ionospheric region).

References

1. Saunders, S. R., *Antennas and Propagation for Wireless Communication Systems*, New York: John Wiley & Sons, 1999.
2. Evans, B. G., *Satellite Communication Systems*, London: IEE, 1999.
3. Blaunstein, N. and Christodoulou, Ch., *Radio Propagation and Adaptive Antennas for Wireless Communication Links: Terrestrial, Atmospheric and Ionospheric*, New Jersey: Wiley Interscience, 2007.
4. Ginzburg, V. L., *Propagation of Electromagnetic Waves in Plasma*, New York: Pergamon Press, 1964.
5. Gurevich, A. V. and Shvartzburg, A. B., *Nonlinear Theory of Radiowave Propagation in the Ionosphere*, Moscow: Science, 1973 (in Russian).

6. Gurevich, A. V. and Tsedilina, E. E., *Extremely-Long-Range Propagation of Short Waves*, Moscow: Science, 1979.

7. Al'pert, Ya. L., *Space Plasma*, Vols. I and II, New York: Cambridge University Press, 1983 and 1990.

8. Rawer, K., *Wave Propagation in the Ionosphere*, Dordrecht: Kluwer Academic Publishers, 1989.

9. Kohl, H., Ruster, R. R., and Schlegel, K. (Eds.), *Modern Ionospheric Science*, Berlin: ProduServ GmbH Verlagsservice, 1996.

10. Blaunstein, N. and Plohotniuc, E., *Ionosphere and Applied Aspects of Radio Communication and Radar*, Boca Raton, FL: CRC Press, 2008.

11. Volland, H., *Atmospheric Electrodynamics*, Heidelberg: Springer-Verlag, 1984.

12. Alpert, Ya. L., Gurevich, A. V., and Pitaevsky, L. P., *Artificial Satellites in the Rarefied Plasma*, Moscow: Science, 1964 (in Russian).

13. Kikuchi, H., *Electrohydrodynamics in Dusty and Dirty Plasmas*, Dordrecht: Kluwer Academic Publishers, 2001.

14. Volland, H., Electrodynamic coupling between natural atmosphere and ionosphere, in *Modern Ionospheric Science*. Kohl, H., Ruster, R., and Schlegel, K., Eds., Katlenburg-Lindau, Max Planck Institute fur Awronomie, 1996, 102–135.

15. Richmond, A. D., Ionospheric electrodynamics, in *Handbook of Atmospheric Electrodynamics*, Vol. II, Volland H., Ed., Boca Raton, FL: CRC Press, 1995, 249–290.

16. Yakovlev, O. I., Yakubov, V. P., Uryadov, V. P., and Pavel'ev, A. G., *Propagation of Radiowaves*, Moscow: Lenand, 2009 (in Russian).

17. Gel'berg, M. G., *Inhomogeneities of High-Latitude Ionosphere*, Novosibirsk: Science, USSR, 1986 (in Russian).

18. Krasnushkin, P. E., *Method of Normal Waves to the Problem of Long Radio Communication*, Moscow: Moscow University Publisher, 1947 (in Russian).

19. Borisov, N. D. and Gurevich, A. V., To the theory of short radio waves in the horizontal-inhomogeneous ionosphere, *Izv. Vuzov. Radiofizika*, 19, 1275–1284, 1976 (in Russian).

20. Erukhimov, L. M., Matugin, S. N., and Uryadov, V. P., To question of radiowave propagation in the ionospheric waveguide channel, *Izv. Vuzov. Radiofizika*, 18, 1297–1304, 1975 (in Russian).

21. Gurevich, A. V., Erukhimov, L. M., Kim, V. Yu., et al., Influence of scattering on capture of radio waves in the ionospheric channel, *Izv. Vuzov. Radiofizika*, 18, 1305–1316, 1975 (in Russian).

22. Erukhimov, L. M. and Trahtengertz, V. Yu., About some effects of scattering of radio waves in the ionosphere, *Geomagn. Aeronom.*, 9, 834–841, 1969.

23. Mit'akov, N. A., Rapoport, V. O., Trahtengertz, V. Yu., About scattering of an ordinary wave near the point of reflection at the small-scale irregularities, *Izv. Vuzov. Radiofizika*, 18, 1273–1278, 1975.

24. Suchy, K., The velocity of wave packet in an anisotropic absorbing media, *J. Plasma Phys.*, 8, 33–51, 1966.

25. Suchy, K., Ray tracing in an anisotropic absorbing medium, *J. Plasma Phys.*, 8, 53–56, 1966.

26. Suchy, K., The propagation of wave packet in inhomogeneous anisotropic media with moderate absorption, *Proc. IEEE*, 62, 1571–1577, 1974.

27. Bennett, J. A., Complex rays for radio waves in the absorbing ionosphere, *Proc. IEEE*, 62, 1577–1585, 1974.
28. Bennett, J. A., Non-uniqueness of "real ray" equation when the ray direction is complex, *Proc. IEEE*, 65, 1599–1601, 1977.
29. Censor, D., Ray propagation and self-focusing in non-linear absorbing media, *Phys. Rev.*, A18, 2614–2617, 1978.
30. Dyson, P. L. and Bennett, J. A., Exact ray path calculations using realistic ionosphere, *IEEE Proc. H*, 139, 407–413, 1992.
31. Sonnenschein, E., Censor, D., Rutkevich, I., and Bennett, J. A., Ray tracing in an absorptive collisional and anisotropic ionosphere, *J. Atmos. Solar-Terr. Phys.*, 59, 2101–2110, 1997.
32. Sonnenschein, E., Blaunstein, N., and Censor, D., HF ray propagation in the presence of resonance heated ionospheric plasmas, *J. Atmos. Solar-Terr. Phys.*, 60, 1605–1623, 1998.
33. Blaunstein, N., Modeling of radio propagation in the land-satellite link through the stormtime ionosphere, in *Proc. Int. Conf. on Mathematical and Informational Technologies*, Kopaonik, Serbia; Budva, Montenegro, August 27–September 5, 71–76, 2009.
34. Blaunstein, N., Pulinets, S., and Cohen, Y., Computation of the main parameters of radio signals in the land-satellite channels during propagation through the perturbed ionosphere, *Geomagn. Aeronom.*, 53, 1–13, 2013.
35. Uryadov, V., Ivanov, V., Plohotniuc, E., et al., *Dynamic Processes in Ionosphere and Methods of Investigations*, Iasi: Technopress, 2006 (in Romanian).
36. Erukhimov, L. M. and Rizhkov, V. A., Study of focusing ionospheric irregularities by methods of radio-astronomy at frequencies of 13–54 MHz, *Geomagn. Aeronom.*, 5, 693–697, 1971.
37. Alimov, A. A. and Erukhimov, L. M., About distribution of fluctuations of the shortwave signals, *Izv. Vuzov. Radiofizika*, 16, 1540–1551, 1973.
38. Rino, C. L. and Fremouw, E. J., The angle dependence of single scattered wavefields, *J. Atmos. Terrestr. Phys.*, 39, 859–868, 1977.
39. Ga'lit, T. A., Gusev, V. D., Erukhimov, L. M., and Shpiro, P. I., About the spectrum of phase fluctuations during sounding of the ionosphere, *Izv. Vuzov. Radiofizika*, 26, 795–801, 1983.
40. Booker, H. G., A theory of scattering by non-isotropic irregularities with applications to radar reflection from the Aurora, *J. Atmos. Terr. Phys.*, 8, 204–221, 1956.
41. Booker, H. G., Application of refractive scintillation theory to radio transmission through the ionosphere and the solar wind and to reflection from a rough ocean, *J. Atmos. Terr. Phys.*, 43, 1215–1233, 1981.
42. Booker, H. G. and Majidi, A. G., Theory of refractive scattering in scintillation phenomena, *J. Atmos. Terr. Phys.*, 43, 1199–1214, 1981.
43. Crain, C. M., Booker, H. G., and Fergusson, S. A., Use of refractive scattering to explain SHF scintillations, *Radio Sci.*, 14, 125–134, 1974.
44. Sudan, R. N., Akinrimisi, J., and Farley, D. T., Generation of small-scale irregularities in the equatorial electrojet, *J. Geophys. Res.*, 78, 240–248, 1973.
45. Sudan, R. N. and Keskinen, M. J., Theory of strong turbulent two-dimensional convection of low-preasure plasma, *Phys. Fluids*, 22, 2305–2314, 1979.

46. Ferch, R. L. and Sudan, R. N., Numerical simulation of type II gradient drift irregularities in the equatorial electrojet, *J. Geophys. Res.*, 82, 211–219, 1977.

47. Keskinen, M. J., Sudan, R. N., and Ferch, R. L., Temporal and spatial power spectrum studies of numerical simulations of type II gradient drift irregularities in the equatorial electrojet, *J. Geophys. Res.*, 84, 1419–1430, 1979.

48. Dimant, Y. S. and Sudan, R. N., Kinetic theory of low-frequency cross-field instability in a weakly ionized plasma, I, *Phys. Plasmas*, 2, 1157–1168, 1995; ibid. II, 1169–1181.

49. Dimant, Y. S. and Sudan, R. N., Kinetic theory of the Farley-Buneman instability in the E region of the ionosphere, *J. Geophys. Res.*, 100, 14605–14623, 1995.

50. Gurevich, A. V. and Karashtin, A. N., Small-scale thermal diffusion instability in the lower ionosphere, *Geomagn. Aeronomom.*, 24, 733–741, 1984.

51. Gurevich, A. V., Borisov, N. D., and Zybin, K. P., Ionospheric turbulence induced in the lower part of the *E* region by the turbulence of the neutral atmosphere, *J. Geophys. Res.*, 102, 379–388, 1997.

52. Hamza, A. M. and St-Maurice, J.-P., A turbulent theoretical framework for the study of current-driven E region irregularities at high latitudes: Basic derivation and application to gradient free situations, *J. Geophys. Res.*, 98, 11587–11599, 1993.

53. Hamza, A. M. and St-Maurice, J.-P., Self-consistent fully turbulent theory of auroral *E* region irregularities, *J. Geophys. Res.*, 98, 11601–11613, 1993.

54. Hamza, A. M. and St-Maurice, J.-P., Large aspect angles in auroral *E* region echoes: A self-consistent turbulent fluid theory, *J. Geophys. Res.*, 100, 5723–5732, 1995.

55. St-Maurice, J.-P. and Hamza, A. M., A new nonlinear approach to the theory of *E* region irregularities, *J. Geophys. Res.*, 106, 1751–1759, 2001.

56. Hamza, A. M. and Imamura, H., On the excitation of large aspect angle Farley-Buneman echoes via three-wave coupling: A dynamical system model, *J. Geophys. Res.*, 106, 24745–24754, 2001.

57. Blaunstein, N., Theoretical aspects of wave propagation in random media based on quanty and statistical field theory, *J. Prog. Electromagn. Res.*, 47, 135–191, 2004.

58. St-Maurice, J.-P., Prikryl, P., Danskin, D. W., et al., On origin of narrow non-acoustic coherent radar spectra in the high-latitude *E*-region, *J. Geophys. Res.*, 99, 6447–6474, 1994.

59. Haldopis, C. A., Nielsen, E., and Ierkis, H. M., STARE Doppler spectral studies of westward electrojet radar aurora, *Planet. Space Sci.*, 32, 1291–1300, 1984.

60. Hall, G. and Moorcroft, D. R., Doppler spectra of the UHF diffuse radio aurora, *J. Geophys. Res.*, 93, 7425–7440, 1988.

61. Schlegel, K., Turunen, T., and Moorcroft, D. R., Auroral radar measurements at 16-cm wavelength with high range and time resolution, *J. Geophys. Res.*, 95, 19001–19009, 1990.

62. Del Pozo, C. F., Foster, J. C., and St.-Maurice, J.-P., Dual-mode *E* region plasma wave observations from Millstone Hill, *J. Geophys. Res.*, 98, 6013–6032, 1993.

63. Foster, J. C., Tetenbaum, D., Del Pozo, C. F., et al., Aspect angle variations in intensity, phase velocity, and altitude for high-latitude 34-cm *E*-region irregularities, *J. Geophys. Res.*, 97, 8601–8617, 1992.

64. Garmash, K. P., Rozumenko, V. T., Tyrnov, O. F., Tsymbal, A. M., and Chernogor, L. F., Radio physical investigations of processes in the near-the-Earth

plasma disturbed by the high-energy sources, *Foreign Radioelectronics: Success in Modern Radioelectronics*, 8, 3–19, 1999 (in Russian).

65. Kostrov, L. S., Rozumenko, V. T., and Chernogor, L. F., Doppler radio sounding of perturbations in the middle ionosphere accompanied burns and flights of cosmic vehicles, *Radiophys. Radioastron.*, 4, 227–246, 1999 (in Russian).

66. Chernogor, L. F. and Rozumenko, V. T., Wave processes, global- and large-scale disturbances in the near Earth plasma, in *Proc. Int. Conf. Astronomy in Ukraine-2000 and Perspective, Kinematics and Physics of Sky Bodies*, Annex K, 514–516, September, 2000, Kiev: Academy of Science of Ukraine.

67. Burmaka, V. P., Kostrov, L. S., and Chernogor, L. F., Statistical characteristics of signals of Doppler HF radar during sounding the middle ionosphere perturbed by rockets' launchings and solar terminator, *Radiophys. Radioastron.*, 8, 143–162, 2003 (in Russian).

68. Burmaka, V. P., Panasenko, S. V., and Chernogor, L. F., Modern methods of spectral analysis of quasi-periodical processes in geo-space, *Success in Modern Radioelectronics*, 3–24, 2007 (in Russian).

69. Daniels, D. J., Gunton, D. J., and Scott, H. F., Introduction in sub-surface radar, *IEE Proc. F*, 135 (4), 1988, 278–321.

70. Kanare'kin, D. B., Pavlov, N. F., and Potekhin, V. A., *Polarization of Radiolocation Signals*, Moscow: Soviet Radio, 1966 (in Russian).

ECOLOGICAL PROBLEMS IN NEAR-THE-EARTH SPACE ACTIVITY

V

ECOLOGICAL PROBLEMS IN NEAR-THE-EARTH SPACE ACTIVITY

Chapter 14

Ecological Aspects of Rocket Burns and Launches

14.1 Overview

Cosmonautics underwent a complicated, but very glorious, epoch for more than the past half century. With the help of space vehicles (SVs), people of the Earth studied all planets of the solar system, their satellites, asteroids, and comets. Pioneer 10, Voyager 1, and Voyager 2 SVs started about 40 years ago with messages from the global populace. During the time past, they were far away from our planet, at distances exceeding the radius of the solar system by several times. What is amazing is that radio communication with these SVs exists even till now.

At the same time, we should note that nowadays no well-developed country exists which can manage without mobile communication, television, radio communication, techniques of observing dangerous regions of the ground surface, cosmic navigation, cosmic intelligence patrolling, and so forth.

Worldwide, only about six countries can construct spacecraft or rockets. From these well-developed countries, Great Britain, Germany, and France, cannot independently project, construct, or launch a large rocket. They have pooled their efforts by integrating into the European Space Agency created for this purpose. The same situation is with satellites—only about 10 countries worldwide can construct their own satellites.

In the previous chapters, we brought the reader's attention to the geophysical and radiophysical aspects of rocket launches (RLs) and flights, based on the physical explanation of the corresponding physical phenomenon or process accompanying such event and on the corresponding mathematical tools describing each phenomenon.

At the same time, we should also note that the research presented in these chapters has the so-called reverse side of the coin: RLs and distortion and falling of SV fragments that lead to serious ecological issues on the Earth and in the near-the-Earth space [1,2]. The main aim of this chapter is to show the "submerged part of the cosmic iceberg" and analyze all negative aspects of this activity, following the results obtained in Ref. [3]. As mentioned in the latter work, the problem of the ecology of the cosmic activity came up practically after the first launches of large rockets. But, several decades have passed before the complete ramifications of the problem were understood.

In the 1970s and later, one of the authors of this book took active part in the research on the physical processes of the near-the-Earth medium, accounting for the well-known effects of RLs. He found that RLs cause significant disturbances in the medium, which then travel to global distances from the rocket. During this research, it was suggested that RLs can cause running-down and leakages with energy much higher than that which was initially spent. This aspect was known for a much wider sphere of people regarding pollution of the medium in the proximity of cosmodromes, this aspect was known for much wider sphere of people.

14.2 Influence of RLs on the Planet's Surface

14.2.1 About Cosmodromes and Rockets

Cosmodromes exist in 12 countries worldwide [3]. More precisely, 16 cosmodromes are used for RLs, including four from Russia, and four from the United States. The area of the biggest cosmodrome (worldwide), constructed at Cape Canaveral, FL, is about 400 km², and the number of start or launch complexes is 48 (at Baikonur cosmodrome it is 15) [4]. The height of the largest Apollo rocket (which was used for flights to the Moon) exceeded 100 m and its mass was close to 3000 t. The Space Shuttle rocket photograph is seen in Figure 14.1.

In recent times, the mass of the largest rocket is 2000 t and height about 50 m. Such a rocket can burn about 10 t of fuel per second and emit harmful products of the burn into the atmosphere. The smallest SVs have a mass of about 100 t, and the fuel mass is about 90% of the rocket mass.

The most nonoffensive fuel is liquid hydrogen. Water is formed as a result of its burning. This fuel is used in rocket carriers such as the Space Shuttle (USA).

Figure 14.1   About 1000 t of products of rocket burn were ejected from the largest rocket, Space Shuttle, in the near-the-Earth atmosphere.

Further, solid fuel accelerators are also used in these rockets, which lead to high levels of pollution being emitted in the atmosphere. The Ukrainian rocket Zenit-2 and the Russian rocket Soyuz use gasoline as a comparably nonpolluting fuel. The Russian rocket Proton is very powerful and uses a highly toxic fuel—hydroxyl (or geptile). A photograph of this rocket is seen in Figure 14.2.

The ecological consequences depend on the mass of rocket starts, frequency of launches, that is, on cargo flows in the orbit. The latter comprises about 200, 700, and 600 t of fuel per year from Baikonur, Cape Canaveral, and Plesetsk cosmodromes, respectively.

14.2.2 Fall of the First Stages of Rockets

All rockets have different numbers of stages—from 2 to 6 [4]. The zero and the first stages of a rocket carrier work for about 1–2 min (see also Chapter 1). After fuel has been burnt, the stages shoot out and fall comparably close to the place of the rocket start (at distances of about 100 km). The second and the third stages fall far from the rocket start, at distances of about 800 and 2500 km, respectively. For RL, an area of the ground surface of 1500–2000 km² is isolated. Only in the Commonwealth of Independent States (CIS) countries such areas, selected for the fall of the rocket parts, have about 200,000 km², which corresponds to 20% of the total area of the state of Ukraine. The total mass of falling stages depends on the type of rocket and changes from several tons to several

Figure 14.2 The Russian rocket Proton with a most aggressive and toxic fuel—hydroxil.

hundred tons [4,5]. Thus, not only can the stages be dangerous by themselves (see examples shown in Figure 14.3a and b), but the other parts also are dangerous, especially the remaining fuel that is often toxic.

Thus, the zero stage of the Space Shuttle has kinetic energy equivalent to that released for an explosion of several tons of explosives. Only in the CIS countries have the remainder fuel from Proton, Cyclone, and Cosmos rockets polluted about 10,000 km² of the ground surface. An example of such pollution is shown in Figure 14.4.

14.2.3 Fall of SV Fragments

Enormous danger is introduced from fall of rocket fragments that have finished their utility in orbit and are out of control. Particularly, those satellites are dangerous which contain nuclear reactors or similar equipment. Let us give some examples.

(a)   (b)

Figure 14.3   (a) Fragments of the stages of a rocket carrier after its fall on the
ground surface. (Photo by Jonas Bendiksen.) (b) Fragments of the first stage of
Soyuz after its fall on the ground surface. (Photo by Jonas Bendiksen.)

Figure 14.4   Cosmic pollutions on the ground surface.

Owing to the flooding of the Russian cargo SV Progress, it was selected to
be submerged in an area in the southern part of the Pacific Ocean eastward of
New Zealand, in an area of several million square kilometers. The orbital station (OS) after finishing its working time, with a mass of about 100 t, can fully
destroy residential areas and natural objects.

A dangerous situation came up on July 11, 1979, during the fall of fragments
of the Skylab OS (USA). The fragments got scattered in an area several thousand
square kilometers, covering the north of Australia and the southern part of the
Indian Ocean.

The fall of Salyut-7 OS (former USSR) also was an uncontrolled event that occurred on February 7, 1991, near the eastern part of South America. Its mass was about 40 t.

The controlled splashdown of Mir OS (former USSR) with a mass of 120 t happened on March 23, 2001, about 3000 km far to the east of New Zealand. We should note that the cost to Russia for such controlled splashdowns were to the tune of around 400 billion rubles (at the exchange rate in 1997).

The currently operating International OS has a mass of about 400 t. It will currently, and in future, will bring new problems for the environment.

The effect of fall of large space objects was modeled (see, e.g., Figure 14.5) and shows that more dangerous threats can occur by the creation of the global network of space stations for new industrialization projects in the near-the-Earth space.

But, even the controlled fall of used-up OSs or other space objects (SVs, satellites, and spacecraft) can constitute a great danger by themselves. Thus, the polluted atmosphere can spread inside the OS, in which never-before studied micro-organisms can develop. During the past 11 years of functioning of the OS Mir, more than 140 types of new microorganisms have developed inside, some of which have changed by about 190,000 generations. According to some specialists in biology, these organisms and their mutants are a sample of biological weapons. They now affect cosmonauts who live in those extreme

Figure 14.5   A model of the effects of fall of large space objects (a model).

conditions, as well as people on the Earth. What will be the havoc wrought by these microorganisms when they fall along with the OS into the ocean? Nobody has an answer to this question till now.

At present, 58 objects with nuclear and radioisotope facilities and devices are in the near-the-Earth orbit [2]. Their fall (all SVs fall before or after finishing their working period) may lead to serious ecological problems.

In 1978, a Soviet SV Cosmos 954 with nuclear energy components was destroyed in the atmosphere above Canada, which led to radioactive pollution to about 100,000 km² of the ground surface.

14.3 Influence of Rocket Burn and Launch on Near-the-Earth Atmosphere

14.3.1 Effects in the Atmosphere

The path to space is via the atmosphere (located at altitudes from 10 to 20 km). The environment takes the main push of the rocket start. Here, maximal (in mass) outputs of the burning products occur. In addition, here maximally powerful acoustic and electromagnetic, as also optical, radiations are generated. Let us present some examples for this.

People who watched a Saturn-5 start told that the effect of the launch was similar to that caused by an earthquake. A crash level was compared with the noise accompanying Krakatau volcano eruption in Zondsky Gulf in 1883. The gaseous shock wave caused by the first stage registered 1770 km away from the RL [2]. Some people who watched the start of the same rocket stated the existence of a huge amount of wind and fog similar to that created by the energy from millions of horsepower.

The next example is regarding one of the most ecologically clean rockets—the Space Shuttle. As a result of the work of its engines, it emits about 1850 t of burn products into the atmosphere and about the same amount of this mass is ejected into the near-the-Earth atmosphere. Only about 160 t of acid salt is injected into the atmosphere, whereas 90 t is sent in the near-the-Earth atmosphere. After this, intense acid rains are recorded in large areas worldwide.

14.3.2 Influence on the Weather and Climate

Many elderly people, who are far removed from cosmic and space science, have gone on record to say that RLs can influence the climate. Nowadays, scientists who carry out special processing and computer modeling conclude that launches of just 60 rockets such as the Space Shuttle might change the meteorological conditions on both sides of the Atlantic Ocean. The influence of RLs from Baikonur cosmodrome is much more localized. These are usually accompanied by increase of precipitation. If this is true, we will investigate it in future. But if this influence exists, it is related first of all to the effects of the falls and to the process of self-correction of the atmosphere [6].

Except for the action considered here, SV launches are accompanied by thermal, gas-dynamic, electromagnetic action of a jet, the dynamic action of the body of the rocket, and other effects, as well as by generation of waves and fields of various physical nature.

14.3.3 Distortion of the Ozonosphere

This layer of the atmosphere is located at altitudes from 15 to 50 km with the maximum at altitudes of 20–30 km. The ozonosphere protects our planet from dangerous ultraviolet radiation (see, e.g., Refs. [4,6]). Thus, decrease of the content of ozone from 10% yields an increase in the number of skin cancer patients by 20%. Destruction of the ozone layer happens due to emission of chlorine and its compounds.

During the start of rockets, about 5000 t of chlorine and 100 t of nitrogen oxides are annually injected into the atmosphere. Investigations showed that solid-fuel rockets do greater harm to the ozonosphere than do liquid-fuel rockets. Fortunately, SV launches can destroy the ozonosphere only in the proximity of the place of rocket flight; the radius of the perturbed zone does not exceed several kilometers. The global influence of launches is sufficiently small to account for a small intensity of the RLs.

14.4 Influence of Cosmic Activity on Geospace

14.4.1 General Effects

14.4.1.1 Objective

Geospace is the near-the-Earth space that begins at altitudes of about 100 km, where the air pressure is a million times less than that near the ground surface. Satellite flights are usually undertaken at altitudes not less than 250–300 km, where the gas pressure is decreased by about 10 billion times. The upper boundary of geospace extends up to the altitude of geostationary artificial Earth satellites (AESs), that is, up to 36,000 km (see, e.g., Ref. [4]). Due to strong rarefication, geospace is more vulnerable with respect to the near-the-Earth atmosphere.

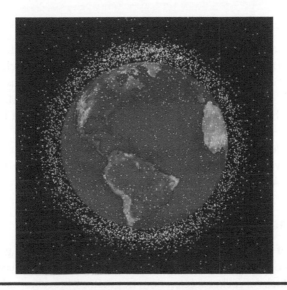

Figure 14.6   A model of the space debris distributed around the Earth.

Space activity influences the ecology of geospace through several channels, such as emission of large volumes of chemicals, which often are absent in natural conditions, injection of acoustic, electromagnetic, and thermal energy, littering of the near-the-Earth medium by fragments of rockets and SVs that are called space rubbish or debris. Let us consider this matter more in detail (Figure 14.6).

14.4.1.2 Gas Injection

In a Space Shuttle launch, more than 120 t of water, 4 t of hydrogen, 1 t of nitrogen, and 0.08 t of nitrous oxide are injected into geospace. We note that geospace contains only about 100 t of hydrogen. Thus, from about 25 launches of this SV the mass of hydrogen doubled. It is important also to note that water molecules quickly break into hydrogen and oxygen, sharply increasing the mass of anthropogenic hydrogen in geospace.

As a result of one launch of Proton more than 32, 37, 20, 1, 0.5, and 40 t of carbide-acid gas, water, fog gas, nitrous oxide, hydrogen, and nitrogen are emitted in space, respectively.

The anthropogenic hydrogen travels 10,000 km in geospace creating a characteristic mushroom cloud. The concentration of the atomic hydrogen in it from 1% to 10% exceeds the background magnitude for several weeks after the RL. The molecules of carbide-acid gas travel 1000 km for a week.

14.4.1.3 Energy Injection

The power of acoustic and electromagnetic radiation from the large rockets in geospace attains several gigawatts (the power of one nuclear power station is 1 GW). For hundreds of seconds, the energy of this radiation can attain 100 GJ, which is equivalent to 25 t of explosive material. It is clear that ejection of such an amount of energy into the geospace cannot be without bad consequences.

The SV that moves in geospace acts dynamically on the medium, and as the result, waves of various nature rise. They can propagate to distances of 1,000–10,000 km from the rocket trajectory.

14.4.2 Space Debris

The problem of space debris is illuminated in many works (see, e.g., Refs. [7–15]). These debris contain AESs, after their work is completed, the last stages of the rockets, the blocks of rocket acceleration, fragments of rockets and satellites, arising as a result of intentional and accidental explosions (Figure 14.7). We should note that 4%–10% of the RLs are accidental.

In about 60 years of cosmic era, about 24,000 artificial SVs, that is, 500 per year were launched in geospace. Nearly, 8000 of them orbit in the near-the-Earth space. It is a known fact that the higher the altitude of the orbit, the longer the "lifetime" of the SV. More than 16,000 of the mentioned objects have fallen to the ground surface. This means that 400 objects fell per year or about one

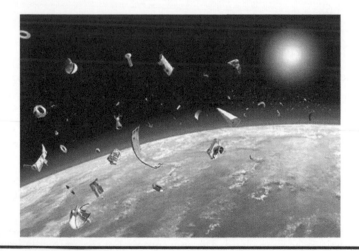

Figure 14.7 A model of about 85% of the space garbage is large parts of rockets and acceleration blocks, as well as satellites completing their work resource.

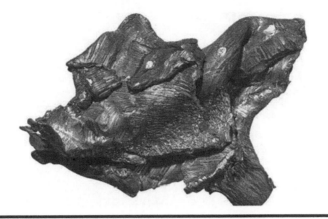

Figure 14.8   A view of typical fragmentation of a satellite or SV.

per day. Objects in an area of 1–10 m² fall with a frequency of once a week, and out of them one-tenth are not fully destroyed. Luckily, people were not practically hurt. Nor were there cases of animal extinction (namely, during Skylab fall). On average, there are about 600 working SVs in orbit around the Earth.

About 12% of space debris present as elements of constructions leaving the satellite in the process of its launch and working. About 3% of this debris comprise small fractions and fragments (Figure 14.8) occurring as a result of their interactions.

Such fragments have relative velocity of about 10 km s⁻¹. Therefore, even a "spatial microfragment" of 1 g mass has a kinetic energy of about 5000 J. The same energy is ejected during the explosion of 120 g of explosive material, that is sufficient, for example, to destroy a car. One of the astronauts mentioned that during a Space Shuttle flight it was hit by a microelement on its illuminator (as would happen if it was a piece of paint) of 0.2 mm diameter with a relative velocity of 6 km s⁻¹. It created a crater 2.4 mm in diameter and 0.63 mm in depth. A window was detected in the circumference of a circle 4 mm in diameter (Figure 14.9). We add here that the median frequency of SV interactions at an altitude of 400 km with a body of about 0.1 mm dimension is about one event per day.

The dimensions of space debris fragments change from dozens of millimeters to 5–6 m. Only fragments with dimension more than 10 cm are about 8000 in geospace. The constant observations of such fragments are made, and all data about them are written in a specific catalogue. The mass of these fragments exceeds 3000 t.

The number of fragments with dimensions from 1 to 10 cm exceeds 300,000. The number of particles with dimensions less than 1 cm is about several hundred million. Their erosion occurs following the interaction between

Figure 14.9 A crater at the glass of the illuminator of SV "Endeavour" after hitting a microparticle of space debris.

the two fragments. After some time this process gets extended. Consequent to that SV flights into space become impossible. Simultaneously, changes of the physical properties of the geospace will not revert. As a result, the gas content in geospace will be retransformed into the foreign elements that occur there. The possible consequences of these changes are difficult to predict till now.

We note that already now the mass of space debris is comparable with the total mass of gas in geospace at altitudes higher than about 450–500 km. Erosion of debris is occurring still now as a result of explosions (on average five explosions a year). The orbit that produce the most rubbish are of AESs that rise at altitudes of 500–600, 800–900, 1400–1500, and 33,600–40,000 km. The maximum density of debris occurs at altitudes of 800–900 km.

14.5 Minimization of Ecological Damage

14.5.1 Overview

In the long history of civilization development on the Earth, man has been forced to surrender all areas of the Earth's spheres: the planet surface (lithosphere), the hydrosphere, and the atmosphere. More than 50 years ago he started to surrender the last sphere—geospace sphere. Together with many advantages, this activity yields some drawbacks related to the "rubbish" in the near-the-Earth environment. Various serious ecological problems occurred during this half century.

Still, in the first half of the twentieth century, academician V. I. Vernadsky, from the former USSR, had dreamed that man will create a surrounding of a new sphere around all natural spheres—a "sphere of brains" or as he called a "noosphere." According to his idea, with its rise, all problems will disappear, including ecological problems. Unfortunately, Vernadsky's dream, finally, remained a dream. Instead of a noosphere, a "technosphere," appeared as a foreign structure that goes against nature, biosphere, and man.

The experience of human activity had shown that doubtlessly along with advantages new problems have arisen, including those having an ecological character. As of now, the ecological problem in geospace has become one of the most important for humanity. Whether this problem will be overcome or not over time—other researchers will write on not only the ecology of the near-the-Earth space but also on the ecology of the solar system, or even on the ecology of our galaxy—is the moot question.

This is an unavoidable result of the technological activity of man in space. Because scientific and technological progress cannot be stopped, only one way remains, that is to minimize its harmful ecological consequences.

14.5.2 Problems in Minimization of Ecological Damage

The problem for minimization of ecological damage should be resolved by a complex approach. First of all, it is necessary to try to decrease "pressure" of all types on the geosphere for new SV launches. Second, the geospace needs "help" for cleaning its space debris.

In our opinion, to decrease "pressure" on the geosphere it is necessary to realize the following steps:

1. It is necessary to move to smaller (in dimensions, volume, and mass) AESs and SV launches as is presented in Table 14.1. Examples of modern and prospective mini- and micro-AES are shown in Figure 14.10 and those based on nanotechnology are given in Figure 14.11.

    For launching such satellites, smaller rockets can be used that emit smaller amounts of harmful burn products. Miniaturization of AESs, their blocks and knots, should move in the future from micro- to nanotechnologies which will resolve all these problems that beset modern satellites, as well as open up new avenues of nonpollution in the latter.
2. It is necessary to totally ban the use of dangerous and harmful rocket fuels.
3. It is necessary to immediately ban explosive technologies in geospace.
4. It is necessary to keep away from industrialization of the near-the-Earth space that requires launchings of tens of thousands of large rockets with masses of 1,000–10,000 t each.

Table 14.1 Classes of the Prospective AESs and the Rocket-Carriers, and Their Main Parameters

Class of AES	Mass of AES	Dimension of AES	Power of Energetic System of AES	Mass of the Rocket Carrier
Large	20–100 t	3–5 m	2–10 kW	700–3000 t
Median	5–20 t	2–3 m	0.5–2 kW	200–700 t
Small	0.5–5 t	1–2 m	50–500 W	20–200 t
Minisatellites	0.1–0.5 t	0.5–1 m	10–50 W	4–20 t
Microsatellites	10–100 kg	0.2–0.5 cm	1–10 W	0.4–4 t
Nanosatellites	1–10 kg	0.1–0.2 cm	0.1–1 W	40–400 kg
Picosatellites	<1 kg	<0.1 cm	<0.1 W	<40 kg

Figure 14.10 Modern and prospective satellites: *left panel*—minisatellite **TOPSAT (mass 120 kg);** *right panel*—microsatellites (mass 25 kg each, dimensions 53 × 48 cm).

5. Via international agreements, prohibit militarization of cosmic space fully, testing of weapons against satellites in it, which are inevitably meant to destroy SVs and AESs.
6. Wait till the near-the-Earth space (altitudes not exceeding of 300–800 km) cleans debris by itself.

For acceleration of the process of cleaning up of geospace, we can recommend the following methods, techniques, and technologies:

XI-V   UWE-1

Ncube-2

Figure 14.11 Nanosatellites XI-V and UWE-1 (top panel) and nanosatellite Ncube-2 (bottom panel); all with volume 10 cm².

- After finishing its working resource, AESs are transferred to lower orbits and are forcefully splashed down into the world's oceans, as was done with the OS Mir.
- Merging of the remainder fuel from the stages of the rocket carriers that have finished their resource, as also from the engines of SVs. This will allow avoiding possible explosions in the remaining fuel and the subsequent formation of space debris clouds.
- Transportation of space debris to lower orbits with the help of cosmic robots, using ion engines
- Using electromagnetic techniques of inhibition of space debris fragments
- Using prospective ultrapower lasers for the inhibition and evaporation of space debris

All the above-mentioned suggestions should be carried out starting from education of ecology, spreading of ecological knowledge, and introduction of new courses and specialties in high schools and universities, as well as for the clear understanding of each government and community of various countries worldwide that the problems of geospace ecology—are problems existing for a long time, which require resolution soon.

14.6 Conclusions and Suggestions

1. Ecological problems from the space activity of humans have appeared still in the period of 50–60 years of the past century. Nowadays, they have become more critical for the citizens of the Earth.
2. The ecological consequences of space activity show on the planet's surface, in the near-the-Earth atmosphere, in the ozonosphere, in the upper atmosphere (ionosphere), and in geospace. The level of damage from these consequences depends on the mass of rocket starts, frequency of launchings, types of rocket fuel, use of technologies, and so forth.
3. The threat for citizens of the planet is from the fall of rocket and SV first stages after finishing their working resource.
4. Space debris has the property of duplication by itself, as the result of eroding of two fragments of debris during their interaction. Till now space debris posed a serious danger to piloting flights and normal functioning of SVs.
5. The problem of minimization of ecological damage from space activity of humans should be resolved completely. The main problem is of putting together ecological mediation.
6. In Ref. [3], one of the authors of this book had proposed of ways to decrease the "pressure" on Earth's EAIM spheres during the launch of new SVs.
7. The authors recommend using advanced techniques, methods, and technologies for accelerating the process of cleaning up space garbage from geospace.

References

1. *Strategy of Survival: Cosmism and Ecology*, Moscow: Nauka, USSR, 1007, 202 (in Russian).
2. Vlasov, M. N. and Krichevsky, S. V., *Ecological Danger of Cosmic Activity*, Moscow: Nauka, 1999, 240 (in Russian).
3. Chernogor, L. F., Ecology of space. On space activity without fanfares and salutes, *Universitates*, 3, 16–25, 2007 (in Russian).

4. Adushkin, V. V., Kozlov, S. I., and Petrova, A. V., Eds. *Ecological Problems and Risks from Rocket-Space Technique Affecting on the Surrounded Natural Medium: Handbook*, Moscow: Ankil, 2000, 640 (in Russian).

5. Murtazov, A. K., *Ecology of the Near-the-Earth Cosmic Space*, Moscow: Fizmatlit, 2004, 340 (in Russian).

6. Chernogor, L. F., *On the Nonlinearity in Nature and Science*, Kharkov: KhNU Publisher, 2008, 528 (in Russian).

7. Masevich, A. G., Ed., *The Problem of Polluting of Space (Space Debris)*, Issue of Scientific Works, Moscow: Kosmosinform, 1993 (in Russian).

8. Masevich, A. G., Ed., *Interactions in Near-the-Earth Cosmic Space (Space Debris)*, Issue of Scientific Works, Moscow: Kosmosinform, 1995 (in Russian).

9. Masevich, A. G., Ed., *Near-the-Earth Astronomy (Space Debris)*, Issue of Scientific Works, Moscow: Kosmosinform, 1998 (in Russian).

10. Boyarchuk, A. A., Ed., *Danger from Sky: Rock or Chance*, Issue of Scientific Works, Moscow: Kosmosinform, 1999 (in Russian).

11. *Near-the-Earth Astronomy of XXI Century*, Moscow: GEOS, 2002 (in Russian).

12. Kutirev, V. A., Cosmotization of the Earth as a danger for humanity, *Communitarian Sciences and Contemporariness*, 2, 127–135, 1994 (in Russian).

13. Shkolenko, Yu. A., Co-operation of the Earth and sky: Postindustrialism, biosphere, space, *Communitarian Sciences and Contemporariness*, 2, 141–147, 1994 (in Russian).

14. Morozov, I., Future revenge of cosmic space, *Literature Newspaper*, July 30, 1997 (in Russian).

15. Gavrilov, V., Space debris, *Popular Mechanics*, 45, 42–46, 2006 (in Russian).

Appendix

Appendix 1: Rocket Information

Rocket (Country)	Cosmodrome	Total Mass (kg)	Initial Force (MN)	Stages: Zero, First, Second, Third, Fourth		Low Orbit Strength (kg)/ Orbit Height (km)	Useful Geostationary Strength (kg)
				Working Time (s)	Engine Force (Vacuum) (MN)		
Energia (USSR)	Baikonur	2,524,000	35.11	145 (480)	4×7.9 (7.84)	88,000/200	22,000
				160 (−)	1.96 (−)		
Space Shuttle (USA)	Cape Canaveral	2,029,633	25.73	124 (480)	2×11.51^a (0)[b]	24,400/204	5900
				480 (−)	6.83 Shuttle (−)		
					− (10.46)		
Proton (Russia)	Baikonur	711,000	8.84	206 (238)	2.40 (0.63)	88,000/200	2400
				600	0.03		
Ariane 5 (France)	Kourou	710,000		123 (650)	6.47 (1.34)	21,000/200	
				1100 (−)	0.065 (−)		
Zenit (USSR)	Sea Platform	478,390	7.55	− (150)	− (8.18)	13,740/200	5180
				315 (−)	0.91 (−)		
Ariane 4 (France)	Kourou	470,000	5.39	142 (205)	4×0.75 (3.03)	7700/185	4520
				125 (759)	0.80 (0.63)		
Soyuz (Russia)	Baikonur/ Plesetsk	297,400	4.02	118 (286)	4×0.99 (0.98)	6855/200	
				250 (−)	0.30 (−)	Inclination 51.6	
Molnia (Russia)	Plesetsk						
Atlas (USA)	Cape Canaveral/ Vandenberg	234,000	3.54	56 (172)	4×0.48 (2.09)	8610/185	3630
				283 (392)	0.39 (0.18)		
Delta (USA)	Cape Canaveral/ Vandenberg	230,000	3.52	64 (265)	4×0.49 (1.05)	5089/185	1818
				444 (88)	0.42 (0.67)		
Titan (USA)	Cape Canaveral/ Vandenberg	150,530	1.90	− (139)	− (2.17)	3100/185	
				180 (−)	0.44 (−)		

(Continued)

Rocket (Country)	Cosmodrome	Total Mass (kg)	Initial Force (MN)	Stages: Zero, First, Second, Third, Fourth		Low Orbit Strength (kg)/ Orbit Height (km)	Useful Geostationary Strength (kg)
				Working Time (s)	Engine Force (Vacuum) (MN)		
Cosmos (Ukraine)	Kapustin Yar/ Plesetsk	107,500	1.48	− (130)	− (1.74)	700/1600	
				375 (−)	0.16 (−)	1400/400	
Rokot (Russia)	Plesetsk/ Kapustin Yar	97,170	1.55	− (121)	− (1.78)	1850/300	
				155 (−)	0.21 (−)	Inclination 74.0	
Pegasus (USA)	Launch from the air	24,000	0.49	4590 (73)	0.56 (0.59)	460/200	
				73 (65)	0.15 (0.35)		
2F (China)		480,000	5.923			84,000	
3B (China)		425,500	5.924			12,000	
4B (China)		254,000	2.971			4200	
3A (China)		241,000	2.962			8500	
2D (China)		232,000	2.962			3100	
2C (China)		192,000	2.786			2400	

[a] The solid-fuel accelerator.
[b] The tank of SV.

Appendix 2: Cosmodrome Information

Place of Launch	Geographic Latitude (°)	Geographic Longitude (°)	Magnetic Latitude (°)	Magnetic Longitude (°)	L-Overlay	Range to Observatory (km)	Minimum Inclination (°)	Maximum Inclination (°)
Kapustin Yar (Russia)	48.51 NL	45.80 EL	42.77 NL	126.9 EL	1.86	700	48.0	51.0
Plesetsk (Russia)	62.70 NL	40.35 EL	57.25 NL	128.3 EL	3.42	1470	59.0	83.0
Baikonur (Russia)	45.63 NL	63.26 EL	37.63 NL	141.5 EL	1.60	2050	49.0	99.0
Jiuquan (China)	41.10 NL	100.30 EL	30.50 NL	173.2 EL	1.35	4930	56.0	40.0
Chan'chenze (China)	28.10 NL	102.30 EL	17.47 NL	174.6 EL	1.10	5980	28.0	36.0
Sriharikota (India)	13.80 NL	80.30 EL	4.31 NL	153.0 EL	1.00	5630	44.0	47.0
Kagoshima (Japan)	31.25 NL	131.10 EL	21.34 NL	201.0 EL	1.15	7730	29.0	75.0
Tanegashima (Japan)	30.40 NL	131.00 EL	20.48 NL	201.0 EL	1.14	7800	99.0	99.0
Wallops (USA)	37.83 NL	75.48 WL	48.49 NL	4.30 WL	2.28	8220	37.0	70.0
Cape Canaveral (USA)	28.45 NL	80.53 WL	39.00 NL	9.80 WL	1.66	9330	28.0	57.0
Kourou (France)	5.20 NL	52.73 WL	15.28 NL	19.80 WL	1.07	9500	5.0	100.0

(Continued)

Place of Launch	Geographic Latitude (°)	Geographic Longitude (°)	Magnetic Latitude (°)	Magnetic Longitude (°)	L-Overlay	Range to Observatory (km)	Minimum Inclination (°)	Maximum Inclination (°)
Edwards (USA)	34.50 NL	117.50 WL	41.56 NL	51.90 WL	1.79	10,310	51.0	145.0
Vandenberg (USA)	34.63 NL	120.50 WL	41.24 NL	55.20 WL	1.77	10,370	51.0	145.0

Note: EL, eastern longitude; NL, northern latitude; WL, western longitude.
Kapustin Yar is the first cosmodrome in former USSR for launch of small-power rockets. Baikonur is the main cosmodrome in Russia (USSR) located in Kazakhstan for all types of rocket launches. Jiuquan is for high inclination satellite launches. Chan'chendze is located at the height of 1800 m for launch of geostationary satellites. Kagoshima is a sea center for launch of solid-fuel rockets. Tanegashima is the main cosmodrome in Japan. Wallops is a small cosmodrome. Cape Canaveral is the main cosmodrome in the United States. Kourou is located in French Guiana for launch of European rockets. Edwards is for launch from the air; the aircraft B-52 or L-1011 are used as the zero stage. Vandenberg is the testing polygon.

Appendix 3: Techniques and Methods of Observation

A3.1 Complex Technique of Partial Reflections

For investigation of the lower ionosphere (50–100 km altitudes), a radio complex based on the method of partial reflections (PRs) was used. It was arranged at the Radiophysical Observatory of Kharkov University (Ukraine) with coordinates 49°38′ NL and 36°20′ EL, the geomagnetic overlay is $L \approx 2$. Its main parameters are the diapason of frequencies $f = 1.5$–15 MHz, the transmitter power $P \approx 150$ kW in the pulse, and the pulse duration $\tau = 25$–100 μs. Usually, $\tau = 25$ μs is used which corresponds to altitudinal resolution of 3–4 km, the frequency of the pulses that follow is $F_r = 1$–10 Hz.

The transmitting antenna consists of two-phase grids of 300×300 m² area and operated at frequency band of $f = 1.5$–4.5 MHz. The gain of antenna is $G \sim 10$–100. Polarization of both antennas is linear. Antennas contain eight and four crossing elements, respectively.

The receiving antenna is the crossing grid, whose each shoulder is created by two elements. It has a circular polarization with a gain of $G \approx 1$–10. The elements of all antennas are made in double vertical diamond form. The sensitivity of the receiver is ~1 μV, and the bandwidth at the carrier frequency is 60 kHz. For detection of the signal from the noisy background, before radiation of each sounding pulse, two to six selections of noise take place.

The radio translator is constricted at the base of a two-channel wideband power amplifier with a common modulator for both channels. As for the carrier generator, standard Ch-31 of synchronization is used. It is possible to regulate the distance of the magnitudes f, τ, and F_r. Management and control of the radio complex is done with the help of a computer. During the process of measurement, the obtained amplitudes of the ordinary and extraordinary signal components are written on the hard disk for the altitudinal interval of 45–105 km with a step height of 3 km. The dynamic diapason of the signal is 40 dB. Existence of the attenuator allows the signal level to additionally vary in the limit range of 30 dB.

A3.2 Doppler Sounding Radar

The Doppler sounding radar is located at the same location of Radiophysical Observatory of Kharkov University, Ukraine. The main parameters of the radar are as follows: the diapason of frequencies is $f = 1$–24 MHz, the pulse power of the radio transmitter is 1 kW, the pulse duration is $\tau \approx 500$ μs, the frequency of pulse repetition is 100 Hz, and the filter bandwidth in the radio receiver is 10 Hz. The antenna system consists of a vertical thrombus with a gain of $G \approx 1$–10 (depending on the radiated frequency). The instrumental error of measurements of the Doppler frequency shift (DFS) is approximately 0.01 Hz. The signal-to-noise ratio (SNR) can achieve up to 10^5–10^6.

The radar is connected with a personal computer, creating a programming-apparatus setup that carries out measurements and a preliminary signal preprocessing in real-time scale.

The altitudinal range of the reflected signal essentially exceeds $c\tau/2 \approx 75$ km. Therefore, splitting along altitude is used with the discrete step of $\Delta z = 75$ km in the diapason of working altitudes of $z = 75$–450 km.

The signals of the oscillation beats of the basic generator and the reflected signal in the form of digits (frequency of polling is 10 Hz) are written on the hard disk. Then, with the help of the fast Fourier transform (FFT) Doppler spectra (DSs) are computed at the diapason of the possible Doppler shifting from -2.5 to $+2.5$ Hz at a time interval of 60 s (with the resolution on frequency ~0.02 Hz)

A3.3 Incoherent Scattering Radar

A single radar of incoherent scattering (IS) in Europe is located in the Ionospheric Observatory of the Academy of Science (AS) of Ukraine (49°36′ NL and 36°18′ EL). The radar is intended for investigating the ionosphere at the altitudinal range of 150–1500 km. The main parameters of the radar are the following: the working frequency is 158 MHz, the diameter of the parabolic antenna is 100 m, the effective area is about 3700 m^2, the antenna gain is about 10^4, the width of the main lobe of the vertically oriented diagram of directivity estimated at the level of half power is about 1, the impulse power of the radio transmitter setup is $P_t = 3.6$ MW, the average power is 100 kW, the duration of the pulse is $\tau = 65$–800 μs, and the frequency of the pulse repetition is 24.4 Hz. The bandwidth of the filter of the radio-receiving setup is $\Delta f = 5.5$–9.5 kHz. The effective noise temperature of the system depends on the day, time, season, and the level of the industrial disturbances, and changes in limits of 1300–1800 K.

The IS radar is connected to the local computer network, which is a measured computing complex that realizes preprocessing of the IS signal in real-time scale.

A3.4 Ionosonde

A standard performed ionosonde by the Institute of the Academy of Sciences, Ukraine, is intended for the common control of the ionospheric condition and calibration of the IS signal power. The accuracy of measurements of the critical frequency of the *F*-region is about 0.05–0.1 MHz, the power of the radio transmitting setup is 10 kW, the frequency range is 0.3–20 MHz, and the period to obtain one ionogram is 1–60 min. The ionosonde is located near the radar of the IS at the Ionospheric Observatory, Kharkov city.

A3.5 Magnetometer—Fluxmeter

A nonserial, high-sensitivity magnetometer, included inside the programming apparatus complex is located in the Observatory of Kharkov University, having the geographic coordinates 49°40' NL and 36°50' EL, and the geomagnetic coordinates 45°20' of geomagnetic latitude and 119°20' of the geomagnetic longitude, respectively.

The complex includes the induction magnetometer fluxmeter IM-II, a setup for registration at the base of a personal computer, having a map of the broadening of the multichannel analog–digital transformer (ADT) EC-1839.3004 and a block of the noninterrupted support with the accumulator capacity of 65 Ah. The latter supplies the autonomic functioning of the complex during up to 18 h at times of interruptions of electric supply.

An induction magnetometer constructively consists of taking out active induction sensors (up to three sensors in the complex), and a monoblock, having in its composition the following: a three-channel block of control and amplification (BCA); a block of the layered filters for the aim of traditional observing registration in narrow frequency bands. Additionally, a power supply, a calibrated low-frequency generator, as well as the registration device (recorder–writer and a slow analog magnetophone) can be included inside.

The main technical characteristics of the active induction sensor are the following:

■ A working frequency band from 0.001 to 15 Hz (the differential amplitude frequency characteristics (AFCs) is found at a frequency band of 0.001–2 Hz).
■ The inner noise at the frequencies 1, 0.1, and 0.01 Hz is not more than 0.5, 5, and 50 nT, respectively.
■ The coefficient of transformation at the output of the active sensor (differential AFC) is of 1 ± 0.2 mV/(Hz nT).

A BCA has the following parameters:

- A working frequency band (with deviation of -3 ± 1 dB from the pure differential AFC) is 0.001–2 Hz.
- A coefficient of transformation (with switching on of the active sensor) is 2–32 mV/(Hz nT) for frequencies is 0.001–2 Hz, respectively.
- A step of the continuous regulation of amplification is 6 ± 9.6 dB.
- The depth of regulation of amplification is 24 dB.
- A steep sharpness of AFC at the upper boundary of the bandwidth is not less than 24 dB/octave; the output voltage is ± 7V.

For digitization, a signal is taken exactly from the output of the BCA. Moreover, only sensors for measurements of the horizontal components of the geomagnetic field were used.

A special software tool based on Turbo Pascal is utilized for signal preprocessing with sufficient accuracy and low bit error rate or bit error ratio (with resolution of the digital data error of 10^{-4}). A special generator is used for synchronization of incoming data instead of a quartz clock, as well as additional electronic blocks for data equalization and error decision. All details of this programming-electronic complex are out of the scope of this book.

Index

Note: Locators "*f*" and "*t*" denote figures and tables in the text